AF876908

Region III Library
Environmental Protection Agency

RESOURCE RECOVERY AND RECYCLING HANDBOOK OF INDUSTRIAL WASTES

TD
794.5
.S57

RESOURCE RECOVERY AND RECYCLING HANDBOOK OF INDUSTRIAL WASTES

Marshall Sittig

NOYES DATA CORPORATION
Park Ridge, New Jersey London, England
1975

Copyright © 1975 by Marshall Sittig
No part of this book may be reproduced in any form
without permission in writing from the Publisher.
Library of Congress Catalog Card Number: 75-22685
ISBN: 0-8155-0592-2
Printed in the United States

Published in the United States of America by
Noyes Data Corporation
Noyes Building, Park Ridge, New Jersey 07656

2869

FOREWORD

This Environmental Technology Handbook intends to show the conversion of industrial wastes into resources. Increasing populations and ever expanding industrialization have increased waste output to a point where natural reclamation can no longer keep up. The synthesis pathways of nature are overloaded.

Wastes in the ecological sense are manufactured goods no longer useful to man, as are by-products and residua (dross, slag, offal) as well as liquid and gaseous effluents, including raw sewage.

This volume concentrates on the process technology available for resource recovery and recycling of industrial wastes. Financial and economic considerations and allied decisions must be left to the potential users of such processes, since their practicality varies widely with local conditions and the shortage or overabundance of raw materials.

As in our other handbooks on pollution control, vital data are condensed from U.S. Government-sponsored projects and recent U.S. patents. As can be seen from the large and impressive table of contents, entries have been arranged in an alphabetical and encyclopedic manner wherever possible.

CONTENTS AND SUBJECT INDEX

INTRODUCTION

The 1970s were introduced as a decade of conservation and environmental reclamation. It is fervently hoped that this will prove to be the case. The life-supporting resources of the earth and its atmosphere are being depleted and polluted at an alarming rate while the population increases ever more quickly. The use of the earth's ecosystem by man has resulted in the absolutely predictable problem of too much waste and too few resources. The historical role of industry has been one of processing raw materials into usable consumer goods. The inevitable fate of these goods has been to become wastes. Nature is then left with the task of reconverting these wastes into resources.

Increasing population and industrialization have increased waste output to a point where natural reclamation pathways cannot keep up. The synthesis pathways of nature are overloaded. It is clear that new methods of waste reclamation and reuse must be developed. The usable lives of raw materials must be lengthened, and more efficient ways of returning wastes to resources must be found, as pointed out by C.D. Callihan and C.E. Dunlap of the Department of Chemical Engineering at Louisiana State University in Report SW-24C, Washington, D.C., U.S. Environmental Protection Agency (1971).

Environmental concerns regarding air, water and solid waste pollution became major concerns in the late 1960s. Then the Arab oil boycott and related problems of resource control and availability took the spotlight in the 1970s.

Since pollutant removal has always been both a question of reducing environmental impact and of the economics of recovery of the pollutant, the new economic pressures introduced by limited availability of materials have made recycling a more desirable method of pollution control.

Recycling in the popular mind has been primarily associated with waste newspaper collection and segregation of clear and colored glass or aluminum from ferrous waste. These are all "postconsumer wastes". This volume will include some treatment of such problems but will also concentrate heavily on industrial wastes—both in-plant recycling and also particularly on recovery of metallic materials from drosses, slags and other liquid and gaseous industrial effluents.

Thus, as pointed out in the *First Report to Congress—Resource Recovery and Source Reduction*, Washington, D.C., U.S. Environmental Protection Agency (1974), "recycling should become more economical relative to other solid waste disposal options during the next several years. Energy costs are rising, making energy recovery more attractive and more economical. As pressures to bring about environmentally sound waste disposal increase, disposal costs will rise, and recovery will become a more attractive alternative. Finally, to the extent that air and water pollution control regulations are strengthened, the industrial incentives for using secondary materials will improve."

Thus the question: "to recycle or not to recycle" has been largely an economic question, varying with the economic times and varying with particular geographic location. This volume will concentrate on the technology of processes available for resource recovery and recycling, however, and will leave most economic determinations to be made by the potential user of such processes since the conditions which determine such applicability are so subject to fluctuation with time and so dependent on local conditions.

Up to the present, the portion of total consumption of various materials derived from waste sources has been declining as pointed out in the *First Report to Congress—Resource Recovery and Source Reduction,* Washington, D.C., U.S. Environmental Protection Agency (1974). The consumption of waste materials has not kept pace with total consumption in the case of paper, iron and steel scrap and rubber.

The percentage of reuse of major nonferrous metals (aluminum, copper and lead) has remained fairly constant as a percentage of total consumption. However, declining recovery has often been the rule as indicated in statistics presented at the 62nd annual convention of the National Association of Recycling Industries as reported in the *New York Times* for April 11, 1975, (Table 1).

TABLE 1: SELECTED USE OF RECYCLED MATERIALS IN 1974

Product	Amount Recycled	Change from 1973	Recycled Material, % of Total Use
Aluminum	1,250,000 short tons	+1.0%	20%
Copper	1,700,000 short tons	-0.06%	44%
Lead	815,000 short tons	-1.85%	52%
Silver	55,200,000 ounces	+63.0%	24%
Zinc	224,000 short tons	-1%	15%
Paper	14,000,000 short tons	-3%	22%

Source: *New York Times,* April 11, 1975

It is reasonable to assume that a secondary material, one that has already been processed, should be a more attractive raw material to industry than a virgin material that must be extracted or harvested and processed. The secondary material is already purified and concentrated. Scrap steel, for instance, is nearly 100% steel while the iron ore from which it is made contains large proportions of silicate materials, which must be removed. The low recycling rate is the result of a number of factors, among them the following:

(1) The delivery price of virgin raw materials to the manufacturer is almost as low in many cases as the cost of secondary materials, and virgin materials are usually qualitatively superior to salvaged materials. Consequently, demand for secondary materials is limited.

(2) Natural resources are abundant, and manufacturing industries have directed their operations toward exploiting them. Plants are generally built near the source of virgin materials (e.g., paper plants near pulpwood supplies). Technology for utilizing virgin materials has been perfected; because of adverse economics similar technology to exploit wastes has not been developed.

(3) Natural resources occur in concentrated form, whereas wastes occur in a dispersed manner. Consequently, acquisition of wastes for recycling is costly and is particularly sensitive to high transportation costs.

(4) Virgin materials, even in unprocessed form, tend to be more homogeneous in composition than waste materials, and sorting and upgrading mixed wastes is costly.

(5) The advent of synthetic materials made from hydrocarbons and their combination with natural materials cause contamination of the latter, thereby limiting their recovery. The synthetics themselves are virtually impossible to sort and recover economically from mixed waste.

(6) There are artificial economic barriers that favor the use of virgin materials over secondary materials. For example, depletion allowances, favorable capital gains treatments, and apparently favorable freight rates are available to processors of virgin materials but not to secondary material processors. Also, producers presently do not have to internalize all the costs of environmental pollution.

In most cases studied, the levels of atmospheric effluents, waterborne wastes, solid wastes, and energy and water consumption are substantially lower for resource recovery, as compared to virgin material utilization. However, the full environmental impact of these results is difficult to assess completely. Residuals and wastes produce different degrees of environmental damage, depending both upon their composition and where they are released. Emissions in areas with high populations could affect public health and welfare, whereas in rural areas, plant and wildlife ecology could be altered. Further research and analysis are needed to evaluate the overall environmental impact of the different mixes and locations of emissions due to increased levels of recycling.

A part of the cost differential between secondary and virgin raw materials is, in fact, artificially created by public policy actions. Virgin materials enjoy depletion allowances and other subsidies such as favorable capital gains treatments. For example, because of the 15% depletion allowance on iron ore, the ore producer could lower his selling price by 13.5% without reducing his profit margin. Publicly controlled freight rates appear to discriminate against the movement of scrap materials. To a large extent, virgin material prices do not reflect the full costs of environmental degradation that the materials create. Furthermore, the fuels required for the energy to extract and process the virgin materials—which are high energy consumers—are also subsidized by depletion allowances.

Environmental regulations will tend to internalize pollution costs and may partially close the relative cost gap between the use of virgin and recycled materials. However, the overall timing and impact of these measures is difficult to predict. Under present market conditions, indeed, some pollution control regulations may work to the detriment of recycling. For example, in the paper industry, many combination board mills (the major users of secondary paper) are already economically marginal operations and will find it difficult to absorb additional pollution control costs.

There are a number of waste materials of metal processing systems sometimes referred to as by-products. It is desirable to distinguish between them and those residues that are usually treated to recover contained metals. Actually, the classification is dynamic. The material classified as waste today may tomorrow be classified as a source of by-product metals because of changes in price or new technology. Nickel, bismuth, and lead contained in residues generated in cleaning up copper refinery electrolyte may be lost today because the metal content is too low to justify further processing.

This would be an example of a borderline case. Some plants recover the contained metals while others do not. A copper flotation mill recovers by-product molybdenite but sends a waste by-product material to tails that contain oxidized copper, 12% alumina (Al_2O_3), and 3½% iron. This material is regarded strictly as waste under current economic conditions and technology. Rhenium is lost up the furnace stack when rhenium-bearing molybdenum disulfide is converted to molybdic oxide without the availability of facilities to recover contained rhenium.

The alumina production system yields the by-product gallium, but it also produces a metal-bearing magnetic sand as a residue of the bauxite digestion process. This sand contains 40% iron oxide, 4% titanium oxide, minor amounts of manganese, magnesium, niobium, gallium, chromium, molybdenum, nickel, lead, and copper, and trace amounts of strontium, zinc, cobalt, platinum, and silver. These wastes of the mineral industry are in the form of mine waste, mill tailings, washing plant rejects, processing plant wastes, smelter slags, flue dusts, and overburden from surface mine operations.

Mine waste is the unsorted material that is removed with the ore. Mill tailings are the wastes from the upgrading or concentrating of the ore. Washing plant rejects and processing plant wastes are self-explanatory terms. Smelter slags are wastes resulting from smelting and refining processes. Overburden is the material stripped from underlying ore deposits to permit extraction of the mineralized material.

These wastes can be classified into groups based upon their economic value and the social cost inherent in their generation and existence. Examples of one group are coal-washing plant rejects with a high carbon content. Other wastepiles are important for use as building materials, for example, slag dumps and slack piles suitable for aggregates, road metal, or lightweight aggregate.

Another category of wastes not only has no economic value, but these wastes have definite social costs attached to their existence. Examples are burning culm banks in the anthracite areas and tailings ponds, which generate dust near populated areas, pollute water, offer a potential hazard with respect to their containment, or retard development; that is, they prevent land use for other purposes. Still others pose what is perhaps the most difficult problem to measure with respect to social costs, namely, those wastepiles which, because of substantial volume or lateral extent and their existence near urban centers or primary road systems, are an aesthetic problem.

In many cases in this volume, special attention has been given to the pollution-producing potentials of various resource recovery processes. Since resource recovery should have the dual potential of providing needed materials and reducing pollution, it would indeed be improper to consider resource recovery without considering the accompanying environmental effects of such resource recovery processes.

COMPLEX WASTE PRODUCTS

MUNICIPAL WASTE

Urban refuse (garbage or municipal waste as it is also known) is nearing a million tons per day in the U.S. alone and is a problem of the most major proportions the world over. Some major approaches to integrated handling of these wastes have been undertaken as follows:

(1) In Baltimore, Maryland, the first large scale application of Monsanto's Landgard system of solid waste pyrolysis has been made. The plant has a capacity of 1,000 tons per day of garbage feed.

(2) In Chicago, Illinois, a plant has just been completed by Commonwealth Edison Co. to burn 1,000 tons per day of refuse as supplemental fuel.

(3) In St. Louis, Missouri, Union Electric Co. is following an earlier prototype plant with a huge project which will handle 8,000 tons per day of wastes gathered from a 4,500 square mile area.

(4) A 200 ton per day demonstration plant is being constructed near San Diego using the Garrett Research and Development Co. process.

(5) A 200 ton per day demonstration plant of the Union Carbide Purox process is being operated at South Charleston, West Virginia.

(6) A 75 ton per day demonstration plant of the Carborundum Environmental Systems Torrax process has been operated in Erie County, New York since 1973.

(7) A pilot plant is being operated by Coors Brewery in Golden, Colorado to study the potential of a larger plant which could handle Denver's garbage and produce pipeline gas or a product.

(8) A 150-ton per day plant costing $5.3 million was dedicated on July 16, 1975 at Ames, Iowa. Serving a surrounding population of 65,000 people in a dozen communities, the plant is expected to produce annually: 2,750 tons of ferrous metals, 138 tons of aluminum and 20,000 tons of coal equivalent in heat from combustibles in the waste. As discussed in the *New York Times* for July 20, 1975, the operating costs and fixed charges of the plant are expected to run to $15.34 per ton of refuse. After deducting a fuel credit of $10.00 per ton and a materials credit of $3.45 per ton, the net disposal cost is expected to be $1.89 which is competitive with landfill costs without imposing the burden on land use.

As with many areas of new technology, however, all is not a success story. Thus, in Nashville, Tennessee, as reported in the *New York Times* for January 13, 1975, a plant was built to handle 350 tons per day of garbage (half the city's output). The purpose of the plant of the Nashville Thermal Transfer Corp. was to burn garbage and produce energy for heating and cooling the downtown area. Serious air pollution problems have plagued the initial operations of the plant and an evaluation of the disposal process awaits installation and testing of improved air pollution control equipment.

The reader is referred to *Recycling and Reclaiming of Municipal Solid Wastes* by F.R. Jackson, Park Ridge, N.J., Noyes Data Corp. (1975) for additional details of the proprietary processes referred to above. An earlier treatment of the recovery and utilization of municipal solid waste is that by N.L. Drobny, H.E. Hull and R.F. Testin of Battelle Memorial Institute in Publication SW-10C, Washington, D.C., U.S. Environmental Protection Agency (1971).

The economics of recycling metals and minerals from urban refuse have been discussed by P.M. Sullivan and M.H. Stanczyk in *Technical Progress Report No. 33,* Washington, D.C., U.S. Bureau of Mines (April 1971). The economics of resource recovery from municipal solid waste have been discussed by J.G. Abert, H. Alter and J.F. Bernheisel of the National Center for Resource Recovery in Washington, D.C., in *Science* for March 15, 1974, pages 1052-1058.

General Characteristics and Quantities

Many of the issues relating to resource recovery and source reduction could be brought into much sharper focus with the use of detailed and consistent base-line data on municipal waste quantities and compositions. Although there is no entirely satisfactory data base at present, Table 2 contains EPA's 1971 estimate (published in 1974) of the total quantity and material composition of postconsumer solid waste. This table also provides a cross-classification of the waste stream by product category.

With the exception of the food, yard, and miscellaneous inorganic fractions, all the product and material figures are derived from analyses of aggregate U.S. industrial material production and product marketing data, together with estimates of average product lifetime and material recovery rates. These estimates have thus been developed entirely independently of refuse collection data and sample refuse composition measurements. The food, yard and miscellaneous inorganics categories, on the other hand, are based on average proportions of these materials reported in published refuse collection studies. The data thus represent an estimate of aggregate U.S. postconsumer waste generation, not of current material recovery.

The "as generated" waste tonnages assume typical moisture content of the material prior to discard or collection. Thus, paper, textile and wood materials have an air-dry moisture of approximately 7 to 15%, food and yard wastes contain 50 to 70% moisture, and all other materials are assumed to contain zero moisture. Because most published refuse composition studies reflect the moisture content of mixed wastes delivered to incinerator or landfill sites, the "as disposed" tonnages have been adjusted to reflect moisture transfer en route to the disposal site. In making interpretations and comparisons, the as generated weight composition is thus more relevant from the standpoint of evaluating potential material recovery from solid waste, whereas the as disposed weight composition should be more comparable to other published composition studies based on measured samples.

The data presented will differ from measured municipal refuse collection samples in a number of respects. For example, most sampling has typically excluded the so-called bulky refuse—major home appliances, furniture, and automobile tires—which according to EPA estimates accounts for about 7 million tons. In addition, the given estimates attempt to account for most but not all container and packaging waste, including a portion of that which originates at industrial plant sites and is often disposed of privately. Furthermore, virtually all composition studies have been highly localized, and most have not covered the

TABLE 2: MUNICIPAL SOLID WASTE GENERATION BY MATERIAL AND SOURCE, 1971

Type of material	10^6 tons of waste by product source category							Total			
	Newspapers, books, and magazines	Containers and packaging	Major household appliances	Furniture and furnishings	Clothing and footwear	Food products	Other	As generated		As disposed	
								10^6 tons	Percent	10^6 tons	Percent
Paper	10.3	20.4	–	Trace	Trace	–	8.4	39.1	31.3	47.3	37.8
Glass	–	11.1	Trace	Trace	–	–	1.0	12.1	9.7	12.5	10.0
Metal:	–	6.1	1.9	.1	Trace	–	3.8	11.9	9.5	12.6	10.1
Ferrous	–	5.4	1.7	Trace	–	–	3.5	10.6	8.5	–	–
Aluminum	–	.6	.1	Trace	–	–	.1	.8	.6	–	–
Other nonferrous	–	.1	.1	Trace	–	–	.2	.4	.3	–	–
Plastic	Trace	2.5	.1	.1	.2	–	1.3	4.2	3.4	4.7	3.8
Rubber and leather	–	Trace	.1	Trace	.5	–	2.7	3.3	2.6	3.4	2.7
Textiles	Trace	Trace	–	.6	.5	–	.7	1.8	1.4	2.0	1.6
Wood	–	1.8	–	2.3	Trace	–	.5	4.6	3.7	4.6	3.7
Food	–	–	–	–	–	22.0	–	22.0	17.6	17.7	14.2
Subtotal	10.3	41.9	2.1	3.2	1.2	22.0	18.4	99.1	79.3	104.9	83.9
Yard waste								24.1	19.3	18.2	14.6
Miscellaneous inorganics								1.8	1.4	1.9	1.5
Total								125.0	100.0	125.0	100.0
Percent product source composition	8.2	33.5	1.7	2.6	1.0	17.6	14.7	–	79.3	–	83.9

Source: EPA Report SW-122

complete spectrum of residential, commercial, institutional and other postconsumer waste sources. Table 2 explicitly excludes sewage sludge, demolition and contract-construction-type wastes, as well as obsolete automobiles.

The total waste generation estimate of 125 million tons per year (3.32 lb per capita per day) for 1971 is significantly lower than the widely quoted 190 million tons per year (5.3 lb per capita per day) result estimated from the 1968 *National Survey of Community Solid Waste Practices.* A portion of this difference exists because the *National Survey* results included some industrial demolition and construction wastes not included in the present estimates. An additional small part of the difference could probably also be accounted for by the fact that some net moisture is added to the solid waste stream, both in the household from kitchen activities and in the storage and collection system en route to the disposal site.

Nevertheless, it appears that the National Survey tended to somewhat overestimate the national solid waste stream. In this regard it should be noted that the National Survey results were themselves principally based on collected tonnage estimates (as opposed to systematic measurements) prepared by the reporting local solid waste agencies. The present estimates for total waste are judged to be accurate to within 20%, given the waste categories covered.

The figures for product categories represent the first attempt to estimate physical weight quantities for the product sources of the solid waste stream. The first two categories (newspapers, books and magazines and containers and packaging) are judged to be accurate to within 10%. Other categories are judged to be accurate to within 25%.

The estimated values in Table 2 can serve as a satisfactory working basis for general analysis and policy formulation purposes at the national level. However, these composition estimates should not be utilized for detailed engineering system design purposes at the local level. The remainder of this subsection describes salient characteristics of postconsumer solid waste by examination of the generation weight data estimates.

Material Composition: 80% of the waste stream is organic (including synthetics) and 20% inorganic (9.7% glass, 9.5% metals, and 1.4% miscellaneous inorganics). Of the materials recoverable as recyclable materials, only the paper, glass and ferrous fractions each comprise more than 8% of the total waste stream. Other individual recyclable materials each comprise less than 3 to 4% of the total. 20 to 25% of the total material weight is contributed by moisture originating principally in the food and yard waste fractions.

Product Composition: About 80% of solid waste (generation weight data) is derived from market product sources (as opposed to yard and garden-type wastes). Excluding food wastes, market product sources account for about 60% of the waste flow. It is this 75 to 80 million ton fraction at which product source reduction and material recycling programs are principally directed. If one defines and measures the waste stream in terms of dry weight, the nonfood product sources account for 78%. Container and packaging materials currently contribute about one-third of total postconsumer waste, 42% of product waste, and 54% of nonfood product waste.

The container and packaging fraction currently accounts for about 72% of the total mineral (combined glass and metals) fraction. In terms of individual materials, this source category contributes well over 90% of the glass, 75% of the aluminum, and at least 45 to 55% each of the ferrous metal, paper and plastic fractions of the waste stream. Consumer durable goods, including household appliances, furniture, recreational equipment, and the like, account for 10 to 12% of total wastes. Newspapers, books, and magazines account for 8%.

Combustion and Heat Characteristics: Roughly 80% of the weight of typical raw municipal refuse is composed of combustible materials. The ash content of raw refuse, including the noncombustible metal and glass fractions, is about 20% by weight; excluding metal and glass, the ash content of combustibles is roughly 5%. Total weight reduction by burning is thus on the order of 80%. Volume reduction by burning under optimum conditions is close to 90% of gross refuse and over 95% of the combustible fraction. Approximate heat values are 4,600 British thermal units per pound for total raw refuse and 5,500 British thermal units per pound for refuse excluding metal and glass.

Waste Generation Projections: Very little is known regarding the factors that determine total and per capita solid waste generation rates, and there is a substantial lack of historical data in this area. Components of the municipal waste stream have certainly changed over the past several decades in response to shifts in consumer technology and expenditure patterns. For example, in the past, coal and wood ash were major components of municipal waste. Similarly, food wastes on a per capita basis have declined with increased consumption of industrially prepared foods, and the introduction of kitchen waste disposals has probably caused a reduction in food waste collected by diverting it to the sewage disposal system. On the other hand, packaging waste has increased very substantially in per capita terms, and its composition has also shifted considerably, particularly in the beverage container area.

Recognition of major historical shifts such as these, together with the general lack of detailed historical data and correlation analysis, suggests caution in the construction and interpretation of gross projections of future solid waste trends or the manner in which such projections might be modified by various public policy measures.

There is currently underway at EPA a research effort designed to project the major components of postconsumer waste over the next 15 to 20 years as a base line for policy analysis. Until the results of that project are available, the following discussion of future waste trends must be regarded as very crude and tentative.

If it is assumed as a first approximation that the total real consumer expenditure for durable and nondurable goods is the major variable determining postconsumer solid waste generation, then based on the experience of the past decade, solid waste should be expected to grow at a rate of about 4.5% per year. However, there are a number of reasons why this growth rate should probably be considered an upper limit. First, many of the major product source components of the waste stream (most notably foods) have been growing at a much slower rate than the average. The faster growing segments, such as containers and packaging, would probably by themselves not push the overall waste growth to a higher level. The fastest growing component of the consumer goods expenditure has been the durable goods category, but items in this category comprise only a relatively small fraction of the present waste stream (10 to 12%); and there is reason to suppose that the weight of durable goods does not necessarily increase in proportion to expenditure value.

Finally, fully 20% of the waste stream (Table 2) is composed of yard wastes, and these are not likely to grow very rapidly. Therefore, a 3.0 to 3.5% annual growth rate, in the absence of major material recovery or source reduction programs, would seem a more reasonable basis for projecting growth in the total solid waste stream. To project boundaries of future waste generation rates, Table 3 summarizes projections for total waste, assuming low, medium, and high growth rates.

TABLE 3: PROJECTED TOTAL SOLID WASTE QUANTITIES*

Assumed Annual Compound Growth (percent)	Waste (10^6 tons)		
	1980	1985	1990
2.5 (low)	155	175	200
3.5 (medium)	170	200	230
4.5 (high)	185	230	290

*1971 base = 125 million tons.

Source: EPA Report SW-122

Production of Useful Products from Mixed Solid Waste

Separation of Individual Waste Components: A number of integrated processes have been developed for the separation and utilization of complex municipal solid waste mixtures. As pointed out by *M.H. Stanczyk, P.M. Sullivan and M.J. Spendlove; U.S. Patent 3,848,813; November 19, 1974; assigned to the Secretary of the Interior* urban refuse can be considered as a mineral deposit or "urban ore" as the Bureau of Mines has characterized it since the waste generated annually contains over 10 million tons of ferrous metals, over one million tons of nonferrous metals and large quantities of glass, paper and plastics.

The economic solution to the recovery of these many diverse materials from the mountains of waste produced undoubtedly lies in large scale integrated recovery and disposal processes such as those described in the following pages.

One such process is that developed by *L.R.E. Dalberg and H. Dalberg; U.S. Patent 3,477,649; November 11, 1969.* It comprises a feed hopper which supplies the waste to a series of treating devices. The waste is first sized and then beaten to separate unlike materials; the mass of waste is then passed through apparatus for separating and removing in order magnetic material, film plastic, paper, rubber and rags. The remaining waste material is then supplied to a ball mill or the like for reduction to fine size to provide a fine ground material which can readily be treated for the separating and collecting of its individual components.

A process developed by *I. Anderson and V. Brown; U.S. Patent 3,524,594; August 18, 1970; assigned to Metropolitan Waste Conversion Corporation* is a process for mechanically separating municipal refuse into useful portions without manual sorting, comprising prebreaking of the largest objects in the refuse, screening and bypassing the fines in the refuse and grinding of the refuse to a suitable size. It then involves removing plastic objects from the refuse by a first air classifier, removing paper objects from the refuse by a second air classifier, separating ferrous metal and glass objects by a magnetic separator and an air balance separator. The remaining refuse containing the organic waste may be further processed, such as by composting. The removed plastic objects may be burned to produce heat for drying the removed paper objects so that the latter may be packaged and sold.

Figure 1 is an overall flow diagram of such a process. The incoming refuse, generally indicated at **10** is passed through a prebreaker **11**. The prebreaker may have a large clearance therein, for example, three feet, to permit large objects to pass therethrough. The purpose of the prebreaker is only for cutting down the size of extremely large objects. The prebroken refuse **12** is then fed onto a screen **13** to remove fines **14** which is bypassed around the primary grinder and further processed as described below. Screen **13** may conveniently have 1" mesh openings and fines **14** passing therethrough is made up of organic materials, such as leaves and food particles, dust and other broken and small objects.

The screened refuse **15** is then conveyed to a primary grinder **16** for shredding. Primary grinder **16** reduces the size of the refuse particles to no larger than about 6" square. The shredded refuse **17** is then passed through a screen **18** to remove fines **19**. Screen **18** is similar to screen **13** and also has 1" mesh openings. The screened and shredded refuse **20** is conveyed to a first air classifier **21** in which air under pressure lifts the lighter plastic objects in the refuse, such as plastic films and cellophane, and separates the same from the main body of the refuse.

The velocity of the air must be at least about 350 ft/min to entrain the plastic film. It should not exceed about 400 ft/min since higher velocities will entrain other unwanted materials. Separated plastic objects **22** are conveyed to a burner **23** to be more fully described below. The refuse from the first air classifier **21**, indicated at **24**, is then conveyed to a second air classifier **25**. In air classifier **25**, the paper objects **26** in the refuse are lifted and separated from the main body of the refuse by air and conveyed to a dryer **27** for removal of the moisture which is normally present in paper found in municipal refuse. The velocity of the air must be at least about 450 to 500 ft/min to entrain the paper.

FIGURE 1: SYSTEM FOR MECHANICAL SEPARATION OF REFUSE

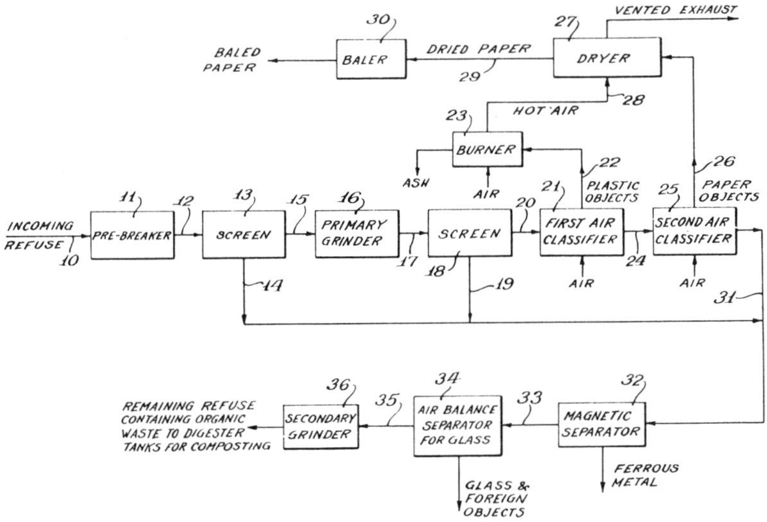

Source: I. Anderson and V. Brown; U.S. Patent 3,524,594; August 18, 1970

Screen **18** and air classifiers **21** and **25** may conveniently be constructed as compartments of one machine, for example, as a large screen 6' by 25' and having a bed depth of about 3', with screen **18** being the first compartment at a size 6' by 12' and air classifiers **21** and **25** dividing the remaining space. It should be noted that all sizes are given by way of an example only, since actually the size of any particular apparatus would depend upon such factors as the volume of refuse to be handled and the nature of the refuse.

Dryer **27** is operated by a stream of hot air **28** from the burner **23**. Thus, the plastic objects in the refuse are burned to provide heat energy for drying the paper which is separately removed from the refuse. The heat from burning the light plastic objects may be insufficient to dry the paper objects, particularly when the paper is very wet as happens on a raining day, and auxiliary heaters may be required to supply additional heat. Such auxiliary heaters may be conveniently of known design and burn any available fuel such as coal, natural gas, oil, etc. Alternatively, a portion of the paper may be burned to supply the additional heat required. The dried paper **29** is then conveyed to a baling apparatus **30** for further disposition.

The remaining refuse **31** emerging from the second air classifier **25** is then combined with fines **14** and **19** and the mixture conveyed to a magnetic separator **32** for the separation of ferrous metal objects from the refuse. The nonferrous refuse **33** issuing from the magnetic separator is then conveyed to an air balance separator **34** for removal of glass objects. Some foreign materials, such as ceramic objects, are also removed by the air balance separator. The construction of the magnetic separator and the air balance separator is known and they are available commercially. The remaining refuse **35** is then conveyed to a secondary grinder **36** for preparing the refuse particles to the proper size suitable for further processing such as composting. This remaining refuse contains the organic waste.

As indicated above, the purpose of the prebreaker is to reduce the size of the largest objects in the incoming refuse so that they can be handled by the primary grinder. Depending on the particular primary grinder employed, the prebreaker may be employed to reduce the size of the refuse particles to not more than, for example, 1 to 3 feet in diameter. The screening of the refuse is for the purpose of removing fines which may be ground glass, dust, small particles of paper, metal and plastics, etc. If these fines are not separated from the refuse, they may be classified and removed by the air classifiers. The screening operation also removes leaves and grass which are in the refuse.

The screened refuse is then shredded in a primary grinder which reduces particle size to a level suitable for subsequent processing. The shredded refuse typically may have a size about 6" square and smaller. The shredded refuse is then again screened before conveying to the first air classifier where the lighter objects, primarily plastic films and cellophane, but which may include small pieces of paper and other lighter materials, are lifted by air under pressure and separated from the main body of refuse. The remaining refuse is then passed into a second air classifier where air under stronger pressure is used to remove all extractible paper objects.

The lighter plastic materials removed in the first air classifier is burned to provide heat for drying the removed paper in a dryer, which may be of conventional construction, for example, the tumbler type. Heat in addition to that obtainable from the burning of the plastic objects may be required to dry the removed paper, particularly when the incoming refuse is wet and such additional heat may be provided by auxiliary burners or heaters (not shown). The remaining refuse is then mixed up with the fines previously screened out and conveyed to the magnetic separator for ferrous metal removal and air balance separation for glass and foreign object removal.

The first air classifier may be operated at an air pressure of from 6 to 8 psi and the second air classifier may also be operated at 6 to 8 psi. These air pressures may have to be varied and adjusted to suit the nature of the refuse collected. For example, on a raining day, the refuse may be wet and higher air pressure is needed to classify the refuse. It can be seen from the above that the process provides a mechanical method for handling municipal refuse substantially automatically and requires a minimum of manual attention. The elimination of manual labor results in a substantial saving thus making the process more economically attractive.

One such process is that developed by *W.H. Gardner and D.H. Graham; U.S. Patent 3,736,111; May 29, 1973; assigned to Hercules Incorporated*. In this process, domestic refuse is subjected to aerobic digestion. A crude humus product containing nondigestible materials is recovered. The crude humus is separated into a humus product and nondigestible materials. The nondigestible materials are pyrolyzed to produce valuable oils and gases. Industrial refuse can be pyrolyzed with the nondigestible material derived from domestic refuse.

Another such process is that developed by *A. Tempe; U.S. Patent 3,736,120; May 29, 1973* which involves disposing of solid refuse and converting it into usable end products. Two basic groups of method steps are involved. One is the sorting of the refuse to segregate metals, cardboard, cartons and papers; bottles and glass; cloth and rags; and plastics together with attending steps of treating the sorted components to reclaim them for usage. The second group of steps is the processing of the organic materials which remain after the above solids have been removed to convert it into usable organic fertilizer.

Still another process is that developed by *H.D. Brown; U.S. Patent 3,770,419; November 6, 1973.* It involves a pyrolysis process system for recycling of refuse where refuse is fed into a closed retort. The closed retort includes a moving molten lead bath for accomplishing the pyrolysis of the refuse. The pyrolysis process converts the organic portions of the refuse to a vapor state to facilitate ultimate recovery of tars, oils and other petroleum-related products, a part of which includes fuel that may be utilized to operate the system. The nonorganic portions of the refuse are ultimately separated into high grade char, ferrous metals, nonferrous metals and precious metals. The pyrolysis process exhausts no gases into the atmosphere and allows virtually all components of the refuse to be recycled into usable products.

The general process outline of this system is shown in Figure 2.

FIGURE 2: LEAD BATH PYROLYSIS SYSTEM FOR REFUSE TREATMENT

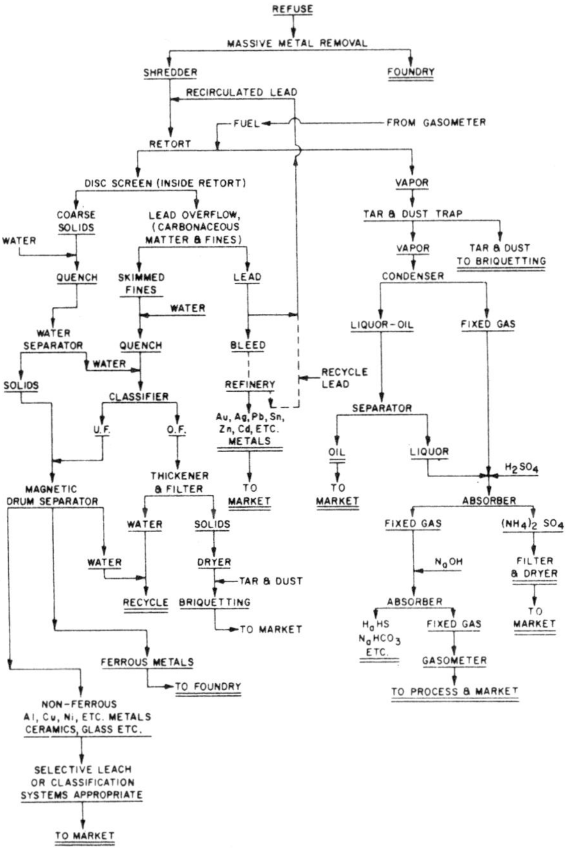

Source: H.D. Brown; U.S. Patent 3,770,419; November 6, 1973

A technique developed by *E. Law and J.H. Moore; U.S. Patent 3,790,091; February 5, 1974; assigned to National Recycling Inc.* involves separating solid waste material into identifiable classifications by recycling the solid waste material obtained from industrial or domestic source through a number of devices which reduces the solid waste material to a uniform workable particle size; then separating the presized particles by weight, size and magnetic properties. The classified material is then transferred to a number of disposal areas where the materials such as heavy paper material may be baled, the light paper and plastic may be pelletized for use as fuel, the metal and glass shipped to reprocessing plants, and the remainder of the solid waste material may be disposed of by landfill or incineration.

Figure 3 is an overall flow diagram of the process. The endless conveyor **10** transfers the solid waste material from a pit, or solid concrete slab (not shown), to a presizer **12** which functions to reduce the solid waste material to a uniform size so that it can be further classified in the later stages of the recycling process.

FIGURE 3: SYSTEM FOR PRESIZING AND SUBSEQUENT SEPARATION OF MUNICIPAL WASTE

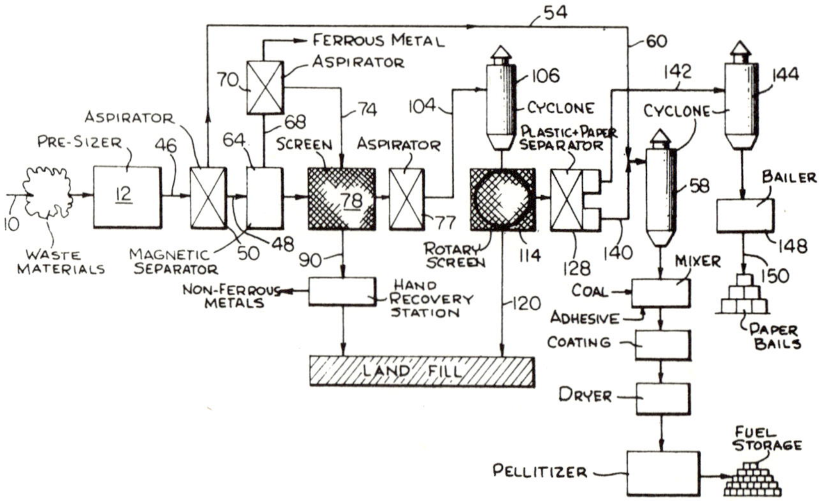

Source: E. Law and J.H. Moore; U.S. Patent 3,790,091; February 5, 1974

After passing through the presizer **12**, the solid waste material falls onto a vibrating conveyor **46** which transfers the solid waste material to an endless conveyor **48**. An aspirator hood **50** is disposed above the vibrating conveyor **46** to remove light material, such as thin paper, Kleenex and light plastic, for transfer through ducts **54** and **60** directly to the cyclone **58**.

The endless conveyor **48** passes under a crossbelt or in-line electromagnetic separator **64** having enough gauss force to remove the magnetic material from the other solid waste material. The magnetic material is released at a predetermined point and is transferred to an endless conveyor **68** which passes beneath an aspirator **70** which removes any paper or other light material which may have attached itself to the magnetic material, and transfers such material through a duct **74** to a vibrating screener **78**. The solid material remaining on the conveyor **48** is transferred to the vibrating screener **78**, which separates the solid waste material by particle size and transfers the solid particles to a conveyor **90** which conveyor may be monitored by a person to remove nonferrous metals such as copper, brass and aluminum. The remaining material on the conveyor **90** is then transferred to a landfill, or to an incinerator.

The screener **78** is provided with an aspirator **77** for removing plastic and paper material, and transfers the material through a duct **104** to a cyclone **106**. From the cyclone the material is transferred to a rotary screener **114** having a rotating tumbler which separates the dirt and glass, and other fine solid materials which are discharged by a rotary screw conveyor **120** to landfill. The paper and plastic are discharged from the rotary screener to a plastic separator **128** which separates the plastic and light paper from the heavy paper. The light paper and plastic are transferred through duct **140** to the cyclone **58**, and the heavy paper is transferred through duct **142** to the cyclone **144**.

From the cyclone **58** the light paper and plastic is transferred to a mixer for the addition of fine coal thereto. If desired, an adhesive may be added to the mixer for coating the material and thereafter passes through a drier to dry the coating and set the adhesive with respect to the material. Thereafter the material is passed through a pelletizer for compacting the material to increase the density of the material so that it may be used as fuel. The pellets may be stored, or fed directly into a power generator plant. In the event the pelletized material is fed directly into a power generating plant it is not necessary to add the adhesive coating, which is added primarily to prevent the deterioration of the pellets. Such a coating will also protect the pellets from insect attraction since the adhesive material is pathogenic free.

When the adhesive is not applied, the drier can be eliminated from the recycling process. From the cyclone **144** the heavier paper is transferred to a baler **148** which compacts the material into bales which are transferred by the conveyor **150** to a storage area. The bales of paper may be classified in different categories and may be used for the production of plasterboard, roofing, and other various products in accordance with its particular classification.

Another such process has been developed by *J.A. Boyd; U.S. Patent 3,802,631; April 9, 1974.* It provides a method and apparatus for separating waste material into its various component parts for the recycling and reuse of each of those component parts. The various component parts of waste material which can be separated by this method and apparatus are soluble wastes, plastic material, glass, ferrous metal and nonferrous metal.

A technique developed by *M.J. Gilberto; U.S. Patent 3,817,458; June 18, 1974; assigned to Chicago Hydraulics Inc.* provides a recycling system for solid municipal waste, in which the garbage is milled to given maximum size, then fed to a three-level flotation separation vessel; in the flotation separation vessel, paper is raked from the surface, metal, glass and like heavy solids are removed from the vessel bottom by a deep drum separator, and light organic materials are extracted from an intermediate level by a dewatering screw. The paper materials are dried and compressed; the organic materials are treated separately but similarly. The heavy solids are passed through a magnetic separator to recover ferrous metals, which are washed in a vibratory cleaner. The residual heavy solids are fed to a two-level flotation separation vessel to segregate metal from glass on a surface-differential basis; the two outputs of this second vessel are washed in vibratory cleaners.

Figure 4 is an overall flow diagram of this system. The equipment includes a hammermill **11** preferably equipped with an input conveyor for charging the hammermill from dump trucks or other vehicles utilized to collect solid municipal waste. Hammermill **11** discharges to a conveyor **12** that introduces the output of the hammermill to a first flotation separation vessel **13.** Vessel **13** is filled with a liquid flotation medium; the practical and economical flotation medium is water.

There are three output devices associated with the first liquid flotation separation vessel **13.** The first of these output devices is a rotary paper rake **14** that discharges to a conveyor **15.** Material from conveyor **15** is supplied to a hydraulic dewatering densifying device **16** which is one of the output devices of system **10.** The output of device **16,** a principal output of system **10,** is a relatively dense and dry product consisting primarily of paper.

The second output device for the first flotation vessel **13** is a dense material separator **17** which, as described hereinafter, preferably comprises a large drum extending to the bottom of vessel **13** and projecting above the liquid level in the vessel. The dense material dredged from vessel **13** by separator **17** is supplied to a conveyor **18** that feeds a magnetic separator **19.** Magnetic separator **19** has two outputs, a conveyor **21** and a conveyor **22.** Magnetic materials are deposited on conveyor **21** and fed by that conveyor to a vibratory cleaner **23.** The cleaner is one of the end output devices of system **10,** producing a relatively clean supply of iron, steel and other ferrous metals suitable for recycling. Conveyor **22** deposits the nonmagnetic output of separator **19** into a second liquid medium flotation separation vessel **24.** Vessel **24** comprises a glass-metal separator that separates the remainder of the dense material into primarily glass material and primarily light metal material.

FIGURE 4: MUNICIPAL WASTE SEPARATION SCHEME FEATURING FLOTATION SEPARATION

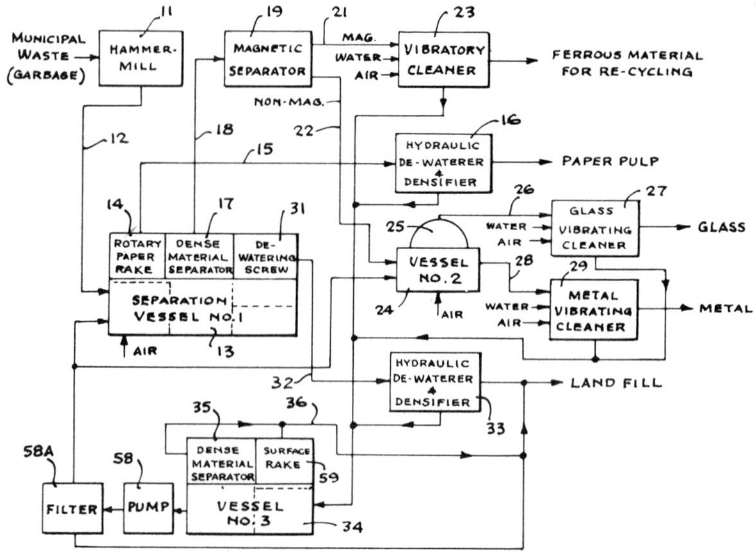

Source: M.J. Gilberto; U.S. Patent 3,817,458; June 18, 1974

The glass is removed from vessel **24** by a dense material separator **25** and supplied to a conveyor **26** that deposits the material in a vibratory cleaner **27**. The output of cleaner **27** is a relatively clean material consisting primarily of glass suitable for reuse. The light metal materials from vessel **24** are deposited on a conveyor **28** that feeds another vibratory cleaner **29**. The output of cleaner **29** consists primarily of aluminum and other light metals, cleaned and essentially ready for reuse.

The third output from the first fluid flotation separation vessel **13** is a dewatering elevator **31** that projects a short distance below the surface level of the liquid in vessel **13**, and removes from the vessel solid organic materials that are of lower specific gravity than the liquid in the vessel. The output of elevator **31**, which is almost entirely organic material, is discharged to a conveyor **32** which supplies this material to a second hydraulic dewatering and densifying apparatus **33** essentially similar to the device **16**. The output of densifier **33** is a milled mixture of organic materials, relatively dense and dry, suitable for use as a raw material for landfill purposes.

The vibratory cleaners **23, 27** and **29** each have an excess water drain; these drains are all returned to a third flotation separation vessel **34**. Similarly, the dewatering and densifying devices **16** and **23** have drains that are connected back to vessel **34**. Heavy solids are removed from the bottom of vessel **34** by a separator drum **35** to a conveyor **36** that conveys these materials to the landfill outlet of system **10**. A surface rake **59** removes floating solids to the same conveyor **36** for disposition as landfill material. The liquid in vessel **34** is removed by a pump **58** that supplies process fluid to vessels **13** and **24**, through a filter **58A**. Solids from filter **58** may be combined with the output of densifier **33**.

A system developed by *M.H. Stanczyk, P.M. Sullivan and M.J. Spendlove; U.S. Patent 3,848,813; November 19, 1974; assigned to the U.S. Secretary of the Interior* is one in which urban waste, such as that collected from households in urban areas, is processed in the dry as collected state to recover ferrous metals, nonferrous metals, glass and paper fractions of suitable quality for sale and recycling. Waste is first shredded without balling or crushing tin cans and is thereafter subjected to a series of separations based upon differences in the physical properties of the waste materials.

Figure 5 is an overall diagram of the process. Unburned refuse **10**, such as that collected from households and businesses in an urban area, is fed into a primary shredder **11**. The purpose of this initial shredding step is to physically liberate components of the refuse contained in bags and boxes while at the same time performing a tearing action to produce a relatively coarse shredded product. Proper control of the shredding step is mandatory in order to obtain efficient separation of refuse components in later processing steps. Glass containers and bottles must be shattered but an excessive amount of fines production is undesirable.

FIGURE 5: OVERALL SCHEME FOR MUNICIPAL WASTE SEPARATION INTO INDIVIDUAL COMPONENTS

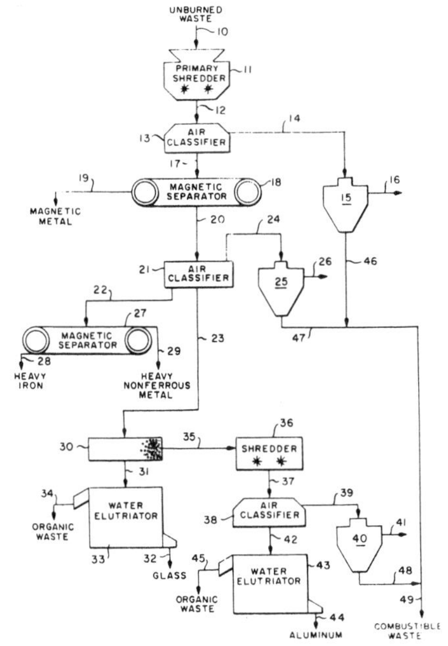

Source: M.H. Stanczyk, P.M. Sullivan and M.J. Spendlove; U.S. Patent 3,848,813; Nov. 19, 1974

Metal cans are damaged as little as possible. If they are balled up or folded, the cans trap other components of the refuse and are difficult to separate and further treat. Solution detinning, for example, is extremely difficult to accomplish if the tin-coated cans are sharply folded or balled up. Folded paper objects such as newspapers tend to be separated into shredded sheets but tightly bound magazines and the like may survive this shredding step in a somewhat more complete form, being ripped into a number of sections. In many cases, such partial comminution of magazine stock is desirable since it allows for a later separation of newsprint from magazine stock. In all cases, excessive production of very fine fragments is undesirable. All urban wastes contain occasional massive metal objects such as cooking utensils, discarded appliances, auto parts and the like. It is necessary that objects such as these can pass through shredder **11** without causing damage to the machine.

All of these criteria are met by a shredding device known as a double-opposed flail mill or chain mill. Such mills comprise a series of pivoted flails or chains attached to rotating cylinders. Two sets of counterrotating flails or chains produce a shredding or shattering action as refuse is passed through the mill between the flails. Conventional types of impact mills, such as hammermills, are unsuitable since cans, sheet metal, foil and the like are balled up trapping other components of the refuse.

Shredded waste **12** is then passed to air classification or pneumatic separation means **13** where a light fraction comprising mostly shredded paper and plastic film is removed from the shredded waste. Separation means **13** may conveniently comprise a belt conveyor positioned beneath the discharge of shredder **11** and having a covering hood. The hood has the secondary purpose of controlling dust produced by operation of shredder **11**. Suction is created by a fan pulling through a cyclone **15** where the entrained light refuse is collected. Conduit means **14** communicates between cyclone **15** and the hood portion of separation means **13**. The air stream **16** exhausted from cyclone **15** is preferably further treated to remove fine particulates, as in a bag house (not shown) before release to the atmosphere.

After passing through separation means **13**, the remaining shredded waste **17** is subjected to a magnetic separation **18** where most of the ferrous metals are removed. In practice, a suspended belt magnetic separator has been found appropriate. Essentially all of the light ferrous metals comprising mostly cans and similar items are removed at this point and comprise a product fraction **19** which can be further shredded into potato chip shaped pieces and washed free of labels and residual food resulting in a product that is ideal for detinning or other processing.

The nonmagnetic fraction **20** is passed to a primary air classifier **21** where it is separated into three fractions: a heavy fraction **22** comprising mostly massive metals, an intermediate fraction **23** comprising glass, aluminum, food wastes and other heavy organic materials and a light fraction **24** comprising mostly shredded paper and plastics not removed at classifier **13**. Fraction **24** is collected in cyclone **25** and air stream **26** exhausted from the cyclone may be further treated to collect fine particulates before being released to the atmosphere. Primary classifier **21** may conveniently comprise a horizontal air elutriating system in which refuse stream **20** is dropped vertically into a horizontally moving air stream and is thereby assorted into various fractions. A description of such a horizontal air classifier appropriate for use with solid wastes is found in the Bureau of Mines Technical Progress Report No. 34, June, 1971.

The heavy fraction **22**, comprising mostly massive metals, is preferably treated in magnetic separator **27** to obtain an iron or ferrous metal fraction **28** and a heavy, nonferrous metal fraction **29**. Fractions **28** and **29** represent concentrated scrap metal fractions and, when suitably cleaned, can be sold or further refined. Intermediate fraction **23** may be subjected to a further magnetic separation step to remove residual ferrous metal contaminants before it is passed to a physical sizing means such as trommel **30**.

While other sizing means may be used, it has been found that a trommel works well. The trommel comprises a rotating cylindrical screen open at the ends. Fraction **23** is delivered inside the trommel at one end. Fine material drops through the holes; coarse material is delivered out the other end. A hole diameter of about 2 to 2½ inches has been found appropriate. Trommel undersize stream **31** consists mostly of glass, food wastes, dirt and small amounts of other materials. Stream **31** may be further processed to yield a clean glass fraction **32** by treatment in water elutriator **33**. An organic waste stream **34** is also produced in the water elutriation treatment. In some cases it is advantageous to subject trommel undersize stream **31** to a second screening through ¾" on 1" mesh. A second screening further concentrates the glass in the fine fraction while scalping off most of the organic waste that would otherwise go to the water elutriator. It is also possible to clean and concentrate the glass contained in stream **31** by other techniques. For example, the stream may be treated using standard mineral beneficiation methods and devices such as jigs and log washers. Stream **31** may also be subjected to a drying step and the glass fraction concentrated by air tabling.

Trommel oversize stream **35** is composed of a variety of materials including bones, leather, heavy plastics and paperboard, wood, rubber, unshredded magazine fragments and the like. This stream is passed to a second shredder **36** for additional comminution. Shredder **36** may be a cutter-type machine which chops the refuse into pieces having a size preferably in the range of 1 to 3 inches. Shredded product **37** is discharged into an air classifier or pneumatic separator **38**. Light material, capable of being suspended and transported in an air stream, is removed from classifier **38** by way of conduit **39** and is collected in cyclone **40**. Air stream **41** from the cyclone is preferably passed through a dust collector before being released to the atmosphere. Classifier **38** may be similar to classifier **13** or it may be of any other common type. A commercially available 3-stage aspirator has been found to work well in this application.

Heavy material stream **42** contains most of the aluminum metal content of the original urban waste in admixture with wood fragments, heavy plastics, bone fragments, rubber, leather and the like materials. Treatment of stream **42** in water elutriator **43** yields an aluminum concentrate **44** and an organic waste fraction **45**.

Materials collected from the three cyclones are in dry form, are essentially completely combustible, and represent most of the paper, light plastics and fabric contained in the original waste. Combustible fraction **46**, collected by cyclone **15**, may be merged with fraction **47** collected by cyclone **25**. These two fractions in turn may be merged with fraction **48** collected by cyclone **40**. The three fractions comprise combustible waste stream **49** which may be burned for the recovery of heat energy. Alternatively, combustible waste **49** may be utilized for other purposes or converted to other forms. This waste stream, for example, may be gasified by known techniques to provide a fuel gas and other organic fractions. Organic waste streams **45** and **34**, which otherwise present a disposal problem, can also be added to combustible waste **49** raising the total energy value of the mix with a small sacrifice in heating value per unit of mix.

The following example further illustrates this process and represents data obtained from a specific test on refuse collected along a typical route in Marlow Heights, Maryland. The raw refuse sample weighed approximately 1,730 lb and was processed generally in accordance with the flowsheet. Data obtained are set out in the following table.

Product	Percent of Original Sample (weight)	Location*
Ferrous metals	7.9	19
Air classifier heavies	2.2	22
Trommel undersize	22.3	31
First cyclone combustibles	14.8	46
Second cyclone combustibles	36.0	47
Third cyclone combustibles	15.1	48
Air classifier heavies	1.7	42

*Denotes location by element number on Figure 5.

The trommel undersize stream **31** was further analyzed to determine the distribution of materials contained. Distribution of materials in this stream in weight percent was found to be: glass, 33.5.; putrescible, 44.3; metals, 10.1; paper products, 10.1; plastics, 1.7 and miscellaneous, 0.3. Distribution of materials in heavies stream **42** separated in classifier **38** was: corrugated board and paper, 25.2; leather and rubber, 19.1; metals, 24.4; heavy plastics, 13.0; wood, 7.0, and fabric and putrescibles, 11.3.

Conversion to Agricultural Mulch: A process developed by *K. Wandel; U.S. Patent 2,699,629; January 18, 1955; assigned to Wilbro Corporation* involves utilizing for mulch, organic refuse material such as garbage, stable manure, dried leaves, grass clippings and all other organic material waste available in the maintenance of a farm or country residence, and in distintegrating such refuse in water to thereby produce a water suspension of the fibrous material and other insoluble components with the water-soluble ingredients in solution.

When this mulch is spread, the water which will soak into the ground and the soluble matter in the organic material from which the mulch is made will be immediately utilized by the plants, leaving on the surface a mulch of fibrous material similar in its action to the mulch of paper pulp, but differing therefrom in that, in addition to the cellulose fibers which are resistant to decay, the mulch contains a high percentage of finely comminuted organic material which gradually decays and continuously adds nutrients to the soil.

Conversion to Construction Products: A process developed by *A.A. Aintablian; U.S. Patent 3,734,988; May 22, 1973; assigned to International Systems Management Corporation* is a process for converting municipal refuse into useable products such as building blocks, wall board, and building bricks which consists of the steps of reducing the size of the refuse to a chip size by shredding; decomposing the refuse in the presence of nitrogen, water and air; drying the chemically processed material in a gas fired dryer to render it biologically inert; grinding the dried material to a fine powder; mixing the powder with fillers and binders such as hydraulic setting cements and glues; adding water and other chemicals as required; forming the blocks or other building materials by pressing; and curing the product until a specified strength has been achieved.

A process developed by *R.A. Penque; U.S. Patent 3,850,771; November 26, 1974; assigned to Biocel Corporation* is one in which useful structural products such as blocks, boards, strips and pellets are made from garbage by utilizing the cellulose content of the garbage as a binder for the garbage proper. A small portion of the garbage to be made into a useful product is removed from the mass of the garbage and is subjected to a xanthation process, converting the cellulose to cellulose xanthate. The thus converted cellulose xanthate is then mixed with the main mass of garbage, and that mixture is then subjected to conditions suitable to convert the cellulose xanthate to cellulose. The thus-formed cellulose acts as a binder for the mass of garbage, ensuring that it will retain its shape after formation by molding, extrusion or the like.

Conversion to Electric Power: In this connection, the reader is referred to more extensive treatment of this topic than space permits here in *Energy from Solid Waste* by F.R. Jackson, Park Ridge, N.J., Noyes Data Corp. (1974). A few specific examples of processes will be cited here.

A process developed by *H.W. Olsen; U.S. Patent 3,733,271; May 15, 1973* is one in which solid waste in garbage, trash and sewage is subjected to progressive incineration and reduced to inert residue capable of reclamation. Combustion gases are utilized for generation of steam and electricity, and are freed of sulfur and other atmospheric pollutants. Liquid sewage effluent is made potable by chemical treatment.

A process developed by *R.J. White; U.S. Patent 3,759,036; September 18, 1973; assigned to Chevron Research Company* is a process for gasifying waste material and producing energy which comprises contacting a solid organic waste material containing at least 10 weight percent oxygen with steam in the presence of an alkali metal carbonate catalyst at an elevated temperature to obtain a gas, combusting at least a portion of the gas to obtain combusted gases, and expanding the combusted gases through a turbine to rotate the turbine. Preferably, the energy of the rotating turbine is used to drive an electricity generator. It is strongly preferred to use a potassium carbonate catalyst in the gasification of the oxygen-containing waste material. Preferably feedstocks include ordinary municipal refuse or garbage having the requisite oxygen content and usually having a more preferred oxygen content above at least 20 weight percent oxygen.

A process described in U.S. Patent 3,850,755 will be utilized in a plant being constructed at Riverside, California as described in *Chemical Week* magazine for July 23, 1975, p. 31. A pilot unit is expected to convert 50 tons per day of solid waste into 790 kw of electrical energy and a full scale unit will eventually convert 300 tons of refuse per day into 4 megawatts of electrical energy.

Shredded solid waste is continuously fed to a 75-foot long, ceramic-lined pyrolysis reactor operating at 1560°F. Gaseous fuel with a heat content of 486 Btu/ft^3 is generated from the unit and fed to an internal-combustion engine to generate electricity. Some of the gas is used to fuel the pyrolysis unit. Solids, reduced from 160 to 10 ft^3, are air classified, cooled and separated.

Conversion to Heavy Oil or Bitumen: A process developed by *H.R. Appell and J.S. Tosh; U.S. Patent 3,733,255; May 15, 1973; assigned to the U.S. Secretary of Interior* is one in which municipal sewage sludge, refuse and other wastes are converted to heavy oil or bitumen by heating under pressure, in the absence of air, in the presence of carbon monoxide and water. The reader is also referred to Report of Investigation No. 7428 by W.S. Sanner, C. Ortuglio, J.G. Walters and D.E. Wolfson, Washington, D.C., U.S. Bureau of Mines (Aug. 1970) for details of a refuse pyrolysis process yielding oil, gas and tar.

Conversion to Pipeline Gas: A process developed by *H.F. Feldmann; U.S. Patent 3,733,187; May 15, 1973; assigned to the U.S. Secretary of the Interior* is one in which solid wastes are heated to 500° to 675°C at 1,000 to 3,000 psig in presence of gas containing at least 90 mol percent hydrogen to convert the wastes to a methane- and ethane-containing gas having no condensable oils and tars. The carbonaceous residue is also free of tars and oils.

A process developed by *J.C. Janka; U.S. Patent 3,820,964; June 28, 1974; assigned to Consolidated Natural Gas Service Company* is a refuse gasification process for producing high-Btu pipeline gas which is interchangeable with natural gas. The gasification process is a relatively low pressure process, and utilizes a gasifier with a slagging bed rather than a fluidized bed as in many existing processes. The raw refuse is gasified with steam and oxygen to produce a raw synthesis gas, and the H_2/CO ratio is controlled, by proper adjustment of the steam supply to the gasifier, in such a way as to yield a raw synthesis gas from the gasifier having sufficient H_2 for methanation. In this fashion, recourse to a water-gas shift reaction is completely eliminated.

Conversion to Synthesis Gas: A process developed by *C.F. Teichmann and A. Brent; U.S. Patent 3,671,209; June 20, 1972; assigned to Texaco Development Corporation* is one in which garbage and other inorganic and organic solid waste matter is reduced to small size and mixed with a fluid selected from the group consisting of H_2O, liquid hydrocarbon fuel, liquid hydrocarbon fuel-particulate carbon slurry or particulate carbon-water slurry produced subsequently in the process, and mixtures thereof to produce a feed mixture stream. The feed mixture is preferably preheated to produce a feed dispersion stream comprising bits of garbage and solid waste matter, particulate carbon, liquid and vaporized hydrocarbon fuel, and steam. By the partial oxidation of the feed mixture or dispersion stream with a stream of oxygen-rich gas in the reaction zone of a free-flow unpacked noncatalytic synthesis gas generator at an autogenous temperature in the range of 1500° to 3000°F and at a pressure in the range of 1 to 250 atmospheres, the garbage and solid waste matter are converted into a stream of synthesis gas or fuel gas, and other valuable by-products.

Conversion to Soil Conditioners: A process developed by *J. Bassick; U.S. Patent 3,776,474; December 4, 1973* comprises the steps of sorting processable and nonprocessable materials from waste materials, sorting ferrous materials from processable materials, shredding processable materials, evaporating and dehydrating processable materials at substantially high temperatures while avoiding burning or incineration of the materials, and pulverizing or grinding the materials into a spreadable grain-size fertilizer and soil conditioner product.

Production of Fertilizer by Chemical Addition: A process developed by *J.N. Karnemaat; U.S. Patent 3,713,800; January 30, 1973* is one in which a fertilizer is made by adding dry fertilizer chemicals, including urea, to dry composted organic waste materials, following which a mixture of prepolymer of urea and formaldehyde suspended in an excess of formaldehyde is added and then phosphoric acid or other mineral acid is added in order to produce the fertilizer product.

A process developed by *H.P. Scheel; U.S. Patent 3,758,287; September 11, 1973; assigned to Western Minerals, Inc.* is one in which conversion of organic waste into organic plant food is carried out by subjecting the waste to heat sufficient to sterilize, deodorize, and dewater the waste, the heat being generated by the chemical reaction of a mixture principally of magnesium oxide and sulfuric acid. Although commercial magnesium oxide may be used, a preferable source is finely ground olivine, a less highly refined and less expensive material containing a large percentage of magnesium oxide. Olivine is a mineral composed essentially of the oxides of magnesium and iron but also containing aluminum, nickel and other trace elements. When the olivine is mixed with concentrated sulfuric acid, a paste is formed which is then mixed with the waste. The exothermic reaction which takes place produces a compost-like material which is valuable as a plant food. Figure 6 is a block flow diagram of this process.

FIGURE 6: SCHEME FOR CONVERTING WASTE INTO ORGANIC PLANT FOOD

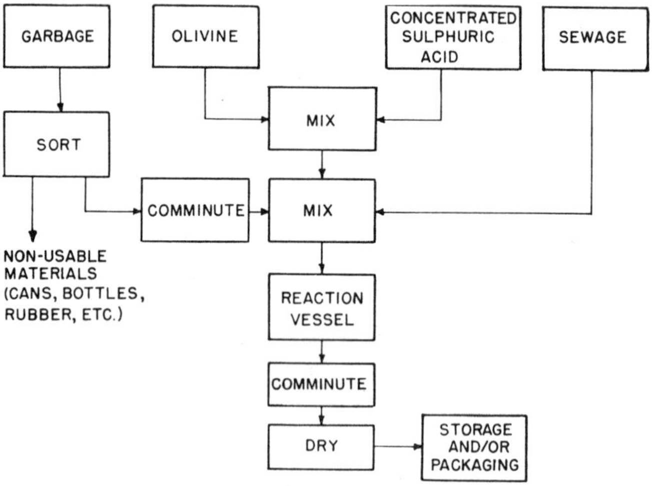

Source: H.P. Scheel; U.S. Patent 3,758,287; September 11, 1973

Recovery of Organic Materials by Pyrolysis or Hydrogenation: A process developed by *P.G. Marsh; U.S. Patent 3,714,038; January 30, 1973; assigned to The Black Clawson Co.* is a process for chemically changing solid waste materials to obtain valuable organic products therefrom which consists of pulping a mixture of organic and inorganic wastes in a liquid such as water to form a slurry, removing inorganics from the slurry, dewatering the resulting slurry of organics and either pyrolyzing or hydrogenating the dewatered slurry. Additionally, if it is desired to sort the organics by types or by specific materials for separate pyrolysis or hydrogenation, this may be done by one or more screening, classifying or separating steps intermediate the pulping and pyrolysis or hydrogenation steps.

Recovery of Proteins by Biodegradation: A process developed by *J.B. Metzger; U.S. Patent 3,711,392; January 16, 1973* is one in which biodegradable waste materials are utilized by forming a suspension in a fermentation tank of comminuted waste materials mixed with a microbiological system of a plurality of selected types of microorganisms under conditions to decompose the waste materials. After sufficient microbial metabolic action has taken place, gaseous products and microbial-resistant portions of the suspension are removed. The solid proteinous materials are then separated from the liquor formed in the suspension and the components of the liquor are separated by electrodialysis.

PACKAGING MATERIALS

Packaging is the largest and one of the fastest growing product classes in municipal solid waste. Because of its predominance in the waste stream, it has become the focus of a great deal of public attention in recent years.

Packaging activity in the United States has been growing at a rapid rate over the past decade. Shipments of containers and packaging were valued at $19.7 billion in 1971, an increase of 5% since 1970, and an increase of 82% since 1960. Table 4 shows that in 1958 packaging material consumption equaled 412 lb per capita. By 1971 per capita consumption had risen to 591 lb, a growth rate of 43% per capita.

The growth of packaging consumption has led to increased consumption of raw materials and energy (with attendant adverse environmental effects) and an increased rate of generation of solid waste. Table 5 shows that packaging accounts for approximately 47% of all paper production, 14% of aluminum production, 75% of glass production, more than 8% of steel production, and approximately 29% of plastic production.

Postconsumer solid waste resulting from the discard of packaging material was estimated at between 40 and 50 million tons in 1971. Packaging was thus estimated to be between 30 and 40% of municipal solid waste, based on the EPA estimate of 125 million tons of municipal solid waste in 1971.

TABLE 4: CONSUMPTION OF PACKAGING MATERIAL*

Type of material	Total consumption: Weight (10^3 tons) 1958†	Total consumption: Weight (10^3 tons) 1971	Total consumption: Change, 1958-71 (percent)	Per capita consumption: Weight (lb) 1958†	Per capita consumption: Weight (lb) 1971	Per capita consumption: Change, 1958-71 (percent)
Paper	16,552	‡27,700	67.3	193.0	‡271.3	40.6
Glass	5,933	§11,100	87.1	69.2	§108.7	57.1
Steel	6,198	¶7,255	17.1	72.3	¶71.1	-1.7
Plastic	368	§2,900	688.0	4.3	§28.4	793.0
Aluminum	97	§757	680.4	1.1	§7.4	572.7
Wood and miscellaneous	6,212	§10,613	84.2	72.4	§103.9	43.5
Total	35,360	60,325	70.6	412.4	590.8	43.3

*Paper figures represent paper packaging produced rather than paper packaging consumed. Other material categories reflect consumption of packaging material.

†Darnay, A., and W. E. Franklin. The role of packaging in solid waste management, 1966 to 1976. Public Health Service Publication No. 1855. Washington, U.S. Government Printing Office, 1969. 205 p.

‡The statistics of paper. Washington, American Paper Institute, 1972.

§U.S. Department of Commerce. *Containers and Packaging*. v.24-25. Washington, U.S. Government Printing Office, [1971-1972]. (Published quarterly.)

¶Shipments of steel products. Washington, American Iron and Steel Institute, 1972.

Source: EPA Report SW-122

TABLE 5: PACKAGING MATERIAL CONSUMPTION IN RELATION TO TOTAL MATERIAL CONSUMPTION, 1971

Type of material	Consumption: Packaging (10^3 tons)	Consumption: Total (10^3 tons)	Packaging as a percent of total consumption
Paper*	27,700	58,652	47.2
Glass	11,100	14,900	74.5
Steel†	7,255	87,038	8.3
Plastic‡	2,900	10,000	29.0
Aluminum§,¶	757	5,074	14.1

*The statistics of paper. Washington, American Paper Institute, 1972.

†Shipments of steel products. Washington, American Iron and Steel Institute, 1972.

‡[Arthur D. Little, Inc.] Incentives for recycling and reuse of plastics; a summary report. [Cincinnati], U.S. Environmental Protection Agency, 1973. 18 p.

§U.S. Bureau of Mines. Minerals yearbook, aluminum chapter reprint. Washington, U.S. Department of the Interior, 1973.

¶Aluminum statistical review, 1971. New York, Aluminum Association, 1972.

Source: EPA Report SW-122

There are three technical approaches that have been considered to reduce material and resource utilization and reduce the waste generation resulting from packaging consumption: using larger containers, eliminating excess packaging of particular products, and using reusable containers. The remainder of this section provides a discussion of these approaches.

Increasing Average Package Size: The trend toward increased use of convenience-sized containers has been one of the contributing factors to increased consumption of packaging materials. Utilization of a greater quantity of smaller containers to fulfill consumption needs results in increased resource use and waste generation. It has been estimated, for example, that elimination of all tomato juice cans smaller than 32 ounces in 1971 would have resulted in a reduction in steel use of 19.6% for this product. This example illustrates how use of larger sizes would have significant resource consumption and solid waste generation implications. Reduction of the convenience-sized container could also, however, be attended by impacts on the consumer, the package manufacturer, and the product manufacturer.

With respect to the consumer, it is anticipated that whereas the cost per unit of product will decline under a system of larger sized containers, there would be a significant reduction in consumer choice and convenience (the consumer will not have the choice of different-sized packages to meet his individual needs).

Industry impacts could be significant. Material suppliers would be negatively affected as the demand for packaging material would decrease. Package fabricators would also be adversely affected as the number of containers produced would decline. If product consumption decreased as a result of a shift in container size, product manufacturers would be adversely affected. The warehousing and trucking sectors of the product industries would also experience some disruption if changes in sizes and product mixes necessitated changes in storage and delivery systems.

Eliminating Overpackaging: As packages have come to fulfill more and more functions beyond product containment and protection, their complexity has increased. This complexity has surfaced mainly in multilayer and multimaterial packaging. Thus many single products are now sold in two packages, one which may be necessary for containing the product and one which is utilized to distinguish and advertise the product. Many premium wines, for example, are now sold in bottles that have been placed in sculptured cartons for shelf appeal. Many toiletry containers have also been packaged in highly elaborate cartons for marketing purposes.

Increases in packaging layers have been accompanied by increases in the use of multimaterial packaging. While packaging manufacturers have always combined dissimilar materials, the number of materials suitable for combination has increased dramatically with the advent of plastic packaging. Thus steel and glass, which had traditionally been combined with paper, have now also been combined with plastic. Although the particular dimensions of the overpackaging issue are difficult to quantify in particular solid waste generation terms, it appears that overpackaging will be increasing at an extremely rapid rate as marketing takes a firm place beside protection and containment as key motives for packaging consumer products.

Reusing Packaging: Approximately 90% by weight of all packaging is discarded by the consumer within 1 year of purchase. Most packages are designed for short lifetimes, with little attention given to the possibilities for reusing the package. There are, however, positive environmental effects resulting from the reuse of various container types. These effects include a decrease in the environmental discharges associated with production of the packaging (e.g., air pollution and water pollution), a reduction in material and energy consumption, and a decrease in the quantities of solid waste generated.

Two examples may be cited here to illustrate the potential environmental effects. In the first, the use of 1,000 tons of single-use corrugated containers is compared to the use of reusable corrugated containers designed to ship an equivalent volume of product. Each of the reusable containers is assumed to be used five times. Table 6 provides the data for the detailed comparison. These data reveal that the reusable container system utilizes approximately 80% less energy than is used by the single-use container system, that air pollution decreases by 57% under the reusable system, that water pollution decreases by approximately 98%, and that solid waste savings of almost 77% accrue in the use of reusable corrugated containers.

In the second example, a refillable bottle system is compared with four different single-use beverage container systems. Table 7 provides the data for the detailed comparison. The refillable system, for example, provides resource consumption savings, energy consumption savings, and air and water pollution reduction. Although the environmental effects that can be derived from reusable packaging systems are positive, there are technological and economic issues that are likely to affect the establishment and implementation of these systems. A brief discussion of the major barriers follows.

TABLE 6: COMPARISON OF THE USE OF 1,000 TONS OF SINGLE-USE CORRUGATED CONTAINERS WITH THE USE OF REUSABLE CORRUGATED CONTAINERS PROVIDING THE SAME DEGREE OF SERVICE*

Environmental impact	Single-use container system	Reusable container system†	Difference (percent)
Production energy consumption (10^ Btu)	25,554.0	5,017.0	-80.3
Production air pollution generation (lb)	38,183.7	11,403.9	-57.0
Production water pollution generation (lb)	849,976.0	18,998.0	-97.8
Post-consumer solid waste generation (tons)	1,000.0	‡231.0	-76.9

*Source: Gordian Associates. Energy consumption for six basic materials industries. U.S. Environmental Protection Agency Contract No. 68-01-1105, Task No. 68-01-1111, 1973. (Unpublished data.)

†Assumes that each reusable container is utilized five times prior to discard.

‡To allow for reuse five times, each container has been designed to utilize 25 percent more linerboard than the single-use container.

Source: EPA Report SW-122

TABLE 7: COMPARISON OF FIVE DIFFERENT CONTAINERS* FOR DELIVERING 1,000 GALLONS OF BEVERAGE†

Environmental impact	10-trip returnable glass‡	All steel§	Bimetallic§	One-way glass‡	Aluminum§
Energy (10^6 Btu)	24	41	57	72	91
Virgin raw materials (lb)	1,538	2,029	1,677	7,515	578
Water volume (10^3 gal)	11	38	34	37	16
Waterborne waste (lb)	45	349	335	68	249
Atmospheric emissions (lb)	111	157	234	328	381
Post-consumer solid waste (ft^3)	12	4	3	41	3
Industrial solid waste (lb)	8	71	61	32	29

*All containers are 12-ounce beer.

†Source: Preliminary data prepared by the Midwest Research Institute for U.S. Environmental Protection Agency Contract No. 68-01-1848.

‡Capped with steel closures; solid bleached sulfate paperboard carriers are included.

§Plastic ring-type carriers are included.

Source: EPA Report SW-122

At this time, there are relatively few package types designed for reuse. In the consumer product sector, the refillable bottle and the reusable carton are the only systems in use. In the industrial sector some drums, pallets, and boxes are designed for reuse. The institution of a large number of reusable packaging systems would therefore require some product design and development activity.

If a package is to be reusable, it must be obtained from the ultimate user and returned to the point at which the container can be refilled. For consumer packaging, this might involve establishing either a deposit system that would provide an incentive for the consumer to return the container to a particular location where it could be obtained by the filler (as with the refillable soft drink bottle) or a system in which containers would be separated by the household and collected for return to the filler.

With respect to commercial or shipping packaging, a system similar to those currently in operation might be employed. At this time, for example, the distributors of commercial packaging to large outlets often return to collect the used packages for reuse. This type of system could be employed on a large scale if reusable packaging became more widespread. For either commercial or consumer packaging users, then, it is clear that systems for obtaining and returning the used containers would have to be developed and instituted prior to the widespread acceptance and use of refillable containers.

Substantial economic effects are likely to result from the institution of reusable packaging systems. These impacts will affect all major industries from material suppliers to retailers and might include production and sales volume changes, employment dislocations, capital investment requirements, and obsolescence of existing equipment. The extent of economic dislocation is expected to be dependent upon the mechanism used to either create incentives or require the utilization of reusable containers. The section on beverage containers under the "Food & Beverage Industry" heading in this volume provides more information on the economic impacts that might occur from a shift to a refillable beverage container system.

Both fiscal and regulatory measures could be used to implement the approaches described to reduce material and resource utilization and reduce the waste generation resulting from packaging consumption. Fiscal measures could be used to provide incentives for decreasing packaging material consumption, encouraging the reuse of packaging, or encouraging the use of recycled material in packaging. Regulatory mechanism could set standards that require consumption decreases, reuse, or the use of recycled materials in packaging.

An EPA study is currently underway to assess the effectiveness and impact of four mechanisms designed to decrease the generation and increase the recycling of packaging waste: a tax on package weight, a packaging weight tax with a rebate for recycled material use, a packaging unit tax for rigid containers, and required utilization of recycled material in packaging.

Although this study is not yet complete, some tentative findings can be presented at this time. A tax on the weight of packaging is likely to have some source reduction effects because of absolute decreases in material consumption at the manufacturer level and shifts in material use for certain products. The total waste weight reduction for a tax of $20 per ton of packaging used is not expected to exceed 4 million tons. Energy use reductions could be expected on the order of 1 to 2% of current energy use in packaging production. A unit tax on rigid containers is likely to have a slightly greater source reduction effect. Reductions of packaging waste of 4 to 5 million tons are anticipated from a $0.01 per unit tax.

Both of these broad-based fiscal measures are likely to discriminate against certain packages and may even result in shifts to materials and packages that are less desirable from an environmental point of view. For example, a tax on packaging weight would discriminate against refillable glass bottles and favor aluminum beverage cans. With a unit tax, all containers would be taxed equally regardless of their material content or weight.

An alternative to fiscal incentives for packaging reduction is a comprehensive regulatory program for all products. Such an approach would be extremely difficult and cumbersome to administer given the wide variety of packaging end uses and configurations. Studies of the tradeoffs between fiscal and regulatory approaches to packaging control have not been completed. Furthermore, it is necessary to evaluate the environmental benefits and the costs of these measures before one can confidently provide specific views on packaging waste reduction.

The reader is referred to the individual sections of this volume dealing with aluminum, glass, paper, plastics and steel for further details on resource recovery and recycling of those materials when they are used as containers.

SCRAP AUTOMOBILES

As described by the Office of Solid Waste Management Programs in *Second Report to the Congress—Resource Recovery & Source Reduction,* Washington, D.C., U.S. Environmental Protection Agency (1974), disposition of obsolete automobiles can take several forms as indicated in Figure 7. The automobile wrecking industry acquires discarded vehicles for spare part value and sells the stripped hulks to scrap processors. Most retired vehicles that enter the scrap cycle follow this route. A small percentage of vehicles that are very old, or have no part value, are conveyed directly to scrap processors by their final owners. Some cities designate disposal sites where obsolete vehicles are accumulated prior to shipment to wrecker, or processor, or where vehicles are disposed of by landfill.

Obsolete vehicles are abandoned when the owner does not know another means of disposal or if he is unwilling to incur the cost of delivering the vehicle to a disposal site, automobile wrecker, or scrap processor. Automobiles that are abandoned on public roadways are generally collected because they present traffic problems or hazards. Vehicles that are abandoned in remote locations or in locations with difficult access are generally not collected. There are three basic environmental and social problems that result from improper disposition of obsolete automobiles:

(1) The degradation of the aesthetic quality of the physical environment caused by abandoned vehicles. (This is essentially a litter problem causing considerable public annoyance.)

(2) The financial burden imposed on the general public for collection of

abandonments. (This is an inequitable situation because the collection costs are not borne by the abandoner.)

(3) The aesthetic, resource consumption, and environmental consequences of the failure to recycle obsolete automobiles. (This is an aesthetic problem in that vehicles that are not recycled accumulate in automobile wrecker yards or storage sites that are popularly referred to as auto graveyards. In addition, such vehicles also represent an untapped resource that if recycled would reduce consumption of virgin resources and reduce the environmental damages caused by virgin material mining and processing.)

FIGURE 7: AN OBSOLETE AUTOMOBILE MAY FOLLOW ONE OF SEVERAL PATHS ON THE WAY TO BECOMING PROCESSED SCRAP

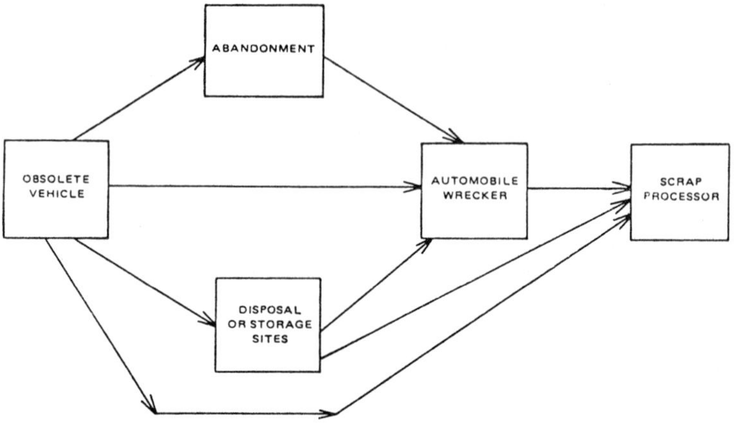

Source: EPA Report SW-122

Some of the problems presented by junk automobiles and some alternatives to those problems are shown in Table 8 and are graphically presented in Figure 8 as taken from *Automobile Scrapping Processes & Needs for Maryland,* by Management Technology, Inc., Report SW-10d, Washington, D.C., Bureau of Solid Waste Management, U.S. Public Health Service (1970).

TABLE 8: THE PROBLEMS AND APPROACHES

	The Problems	The System Component	The Alternatives
1)	Abandoned vehicles continue to cover the countryside in	Abandoned Auto	The abandonment of vehicles can be stopped, or minimally reduced, by: a) Instituting heavy fines for those who abandon a vehicle b) Placing an artificial value on a junk vehicle by paying a fee to all who turn in their obsolete auto c) Subsidizing those who retrieve abandoned autos

(continued)

TABLE 8: (continued)

The Problems	The System Component	The Alternatives
ever-increasing numbers. This spreading pall must be stopped 2) The vehicles already abandoned require an integrated system of retrieval	Abandoned Auto	d) Penalizing those who create scrap by demanding cars which become junk—taxing new car purchasers e) Charging manufacturers who produce items which become obsolete for the "social" costs involved in the retrieval of these items f) Informing the public of how and where to dispose of vehicles properly g) Licensing all junk dealers and defining them so that anyone who leaves an "abandoned" vehicle on his property is subject to licensing requirements
3) Current statutes often restrict both government and private action in obtaining, storing, and disposing of junked autos	Impounded Auto	Institute state and local legislation to: a) Permit freer access to abandoned vehicles for the purpose of retrieving them b) Reduce title clearance requirements c) Stipulate maximum time allowance for owner to retrieve vehicle d) Increase charges to those who do not retrieve vehicles immediately e) Inform public of locations where vehicles can be recovered
4) The auto dismantler's facility: a) Is frequently offensive in appearance, particularly in certain areas, and b) Presents a pollution problem due primarily to the requirement for burning vehicles when they are to be processed as scrap by most current methods 5) The current inventory of hulks, that is vehicles stripped of saleable used parts, in auto dismantlers' yards, is extensive. The demand by processors for these hulks must be increased.	Auto Dismantler	Require that: a) Extant auto dismantlers facilities be shielded b) Future facilities be located away from areas of public congestion and/or shielded c) Effective pollution control be implemented a) Artificially support the scrap industry to permit them to expand their sales to the scrap users and to purchase more vehicle hulks from the auto dismantler
6) The scrap processor's facility is frequently not acceptable either in appearance or as a safe operation 7) The quality of scrap produced by the processors from auto hulks is frequently such that it is unacceptable to user except at extremely low prices	Scrap Processor	a) Require that scrap processors shield their facilities in accordance with standards for the zone in which they are located a) Artificially support the scrap processing industry so that the price of technically low-quality scrap will be acceptable to scrap users b) Improve the quality of scrap by utilizing more advanced technology in its production
8) The current methods of production used by scrap users do not, at most times, allow them to utilize sufficient scrap to clear the junk auto hulks from the system	Scrap Users (Mills and Foundries)	a) Increase the amount of scrap consumed by expanding the use of steel and casting production methods which can utilize greater amounts of processed scrap b) Decreasing the price of scrap by artificially supporting the scrap processing industry

Source: EPA Report SW-10d

FIGURE 8: DISPOSAL/RECLAIMING CYCLE FOR JUNKED CARS

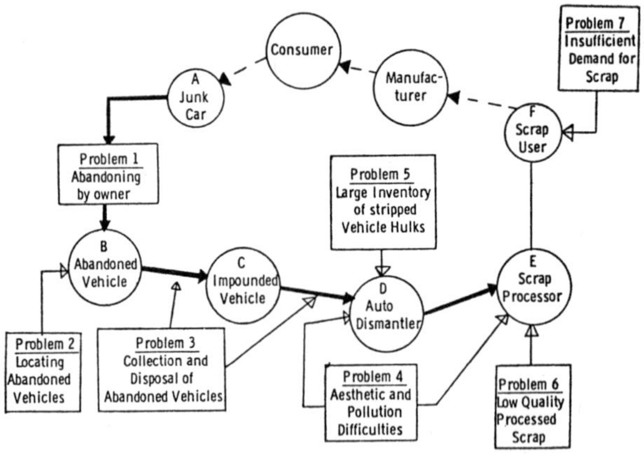

Source: EPA Report SW-10d

FIGURE 9: THE STRUCTURE OF THE JUNK AUTOMOBILE RECYCLING SYSTEM

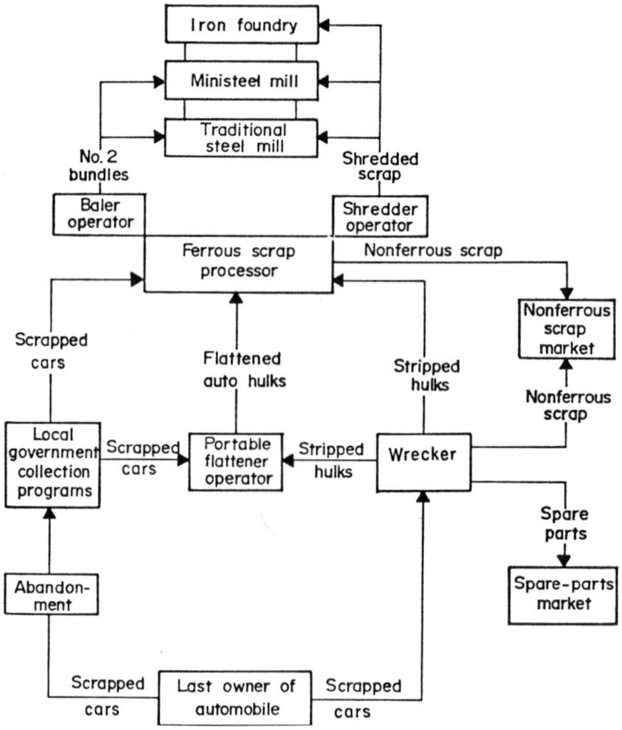

Source: Bureau of Mines I.C. 8596

According to R.L. Adams in Information Circular 8596, Washington, D.C., U.S. Bureau of Mines (1973), the junk automobile recycling system began changing in the mid-1950s. First, traditional steel production technology began to switch from the relatively high scrap-consuming open hearth furnace to the low scrap-consuming basic oxygen furnace. A second change was the growth of high scrap-consuming electric furnace steel production. Much of the increased electric furnace capacity was in increasing numbers of "ministeel" mills, which use the furnaces to melt scrap and produce steel for local consumption.

Scrap processing technology also changed. The scrap shredder was introduced into the junk automobile recycling system in the early 1960s. Instead of compressing the junk car into a bale, the shredder tears the hulk into small uniform pieces of metal that seldom reach a length of more than 2 feet. Like the No. 2 bundle, shredded scrap can be consumed by traditional and ministeel mills. In addition, however, shredded scrap's quality and size has resulted in its being consumed in the cupola furnaces of the iron foundry. No. 2 bundles were generally not consumed by foundries.

Finally, in the late 1960s, the portable flattener was introduced into the junk automobile recycling system. The portable flattener is a machine that is moved from location to location, by truck or train and used to flatten hulks before they are hauled from wreckers and/or municipal collection points to scrap processors. The portable flattener industry is still in its early stages of development at this time. The structure of today's automobile recycling system is shown in Figure 9.

FIGURE 10: TYPICAL SHREDDER SYSTEM SIMPLIFIED FOR 100 TONS FERROUS SCRAP OUTPUT

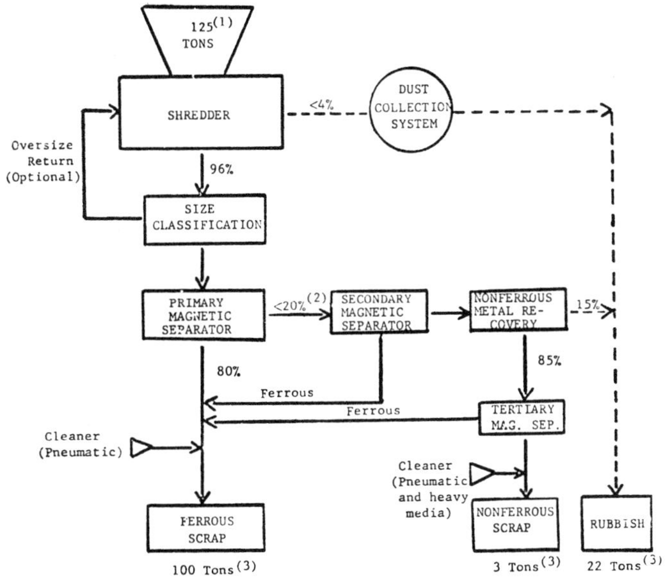

(1) Shredder Input - 125 Tons
85% automotive (exclusive of tires, wheels, power trains, gas tanks, valuable nonferrous)
15% consumer durables (appliances, etc.)

(2) Main Nonferrous Stream
ferrous - 5%
nonferrous - 15%
rubbish - 80%

(continued)

FIGURE 10: (continued)

(3) System Output - Tons

100 - ferrous scrap

3 - nonferrous

	Percent
zinc die castings	65
aluminum (cast & wrought)	15
red metals	13
nonmetallics - red metals	5
nonmagnetic stainless	<2
entrapped ferrous	<1

22 - rubbish (rubber, glass, plastics upholstery "fuzz")

(4) Ferrous Scrap Chemistry

Cu 0.22%
Cr 0.16%
Ni 0.10%
Sn 0.025%
Mo 0.015%

Source: Report PB 213 577

The shredding process uses huge hammermills which are capable of tearing an automobile, less the engine and gas tank, into fist-size chunks in about 1 minute. The shredded metal is processed magnetically to separate the steel from the nonferrous metals and nonmetallic materials. Shredding features rapid processing of unincinerated cars and production of scrap admirably suited for use in electric furnaces. A shredding plant represents a large capital investment, and low-cost operation is possible only with a constant large supply of automobiles. For this reason, economic operation of shredder plants is usually limited to the Nation's major urban areas.

Figure 10 shows the flow in a typical shredder system as described by W.J. Regan, R.W. James and T.J. McLeer of the Institute of Scrap Iron & Steel, Inc. in Report PB 213 577, Springfield, Va., National Technical Information Service (1972).

The baling method comprises removing tires, battery, radiator and engine; incineration; stripping to remove nonferrous metals; and finally baling to prepare a clean and marketable steel. The bales can also be sheared after baling if smaller size scrap is desired. Compared to a shredder plant, the capital investment and capacity of a baler-type operation are modest. A supply of about 250 cars per week is adequate for a baler plant. Unit operating costs are higher than for a shredder-type plant; and incineration is mandatory. As open-air burning with resulting air pollution is rapidly being forbidden in most states, baler-type operations are faced with installing smokeless incinerators.

A study by K.C. Dean and J.W. Sterner in Report of Investigation 7350, Washington, D.C., U.S. Bureau of Mines (1969) indicates that removal of combustibles in a smokeless incinerator followed by dismantling is an economically attractive method for producing quality scrap which can be either baled, baled and sheared, or shredded. As incineration is necessary prior to baling and is desirable prior to shredding, a study was initiated of the problem of designing, building, and operating a low-cost, smokeless incinerator of moderate capacity. An incinerator report was designed, built, and operated by the Bureau of Mines in cooperation with the Wasatch Metal & Salvage Co. of Salt Lake City, Utah and the results of its testing are reported by C.J. Chindgren, K.C. Dean and J.W. Sterner in Technical Progress Report No. 21, Washington, D.C., U.S. Bureau of Mines (Feb. 1970).

The selective oxidation of automobile scrap to produce iron oxide and a dense residual metal of high purity has been studied by R.E. Peterson and C. Prasky in Report of Investigations No. 7614, Washington, D.C., U.S. Bureau of Mines (1972).

The possibility of upgrading junk auto shredder rejects by water classification and heavy media separation after an initial air elutriation step has been described by L.J. Froisland, K.C. Dean and C.J. Chindgren in Technical Progress Report No. 53, Washington, D.C., U.S. Bureau of Mines (March 1973). Figure 11 is a flow diagram for such a system as described by R.A. Boettcher of Stanford Research Institute in Report SW-30c, Washington, D.C., U.S. Environmental Protection Agency (1972).

FIGURE 11: PROCESS FLOW BLOCK DIAGRAM: MODIFIED SYSTEM FOR AUTOMOBILE BODY SHREDDING EMPLOYING AIR CLASSIFICATION AND INCLUDING NONFERROUS METAL RECOVERY

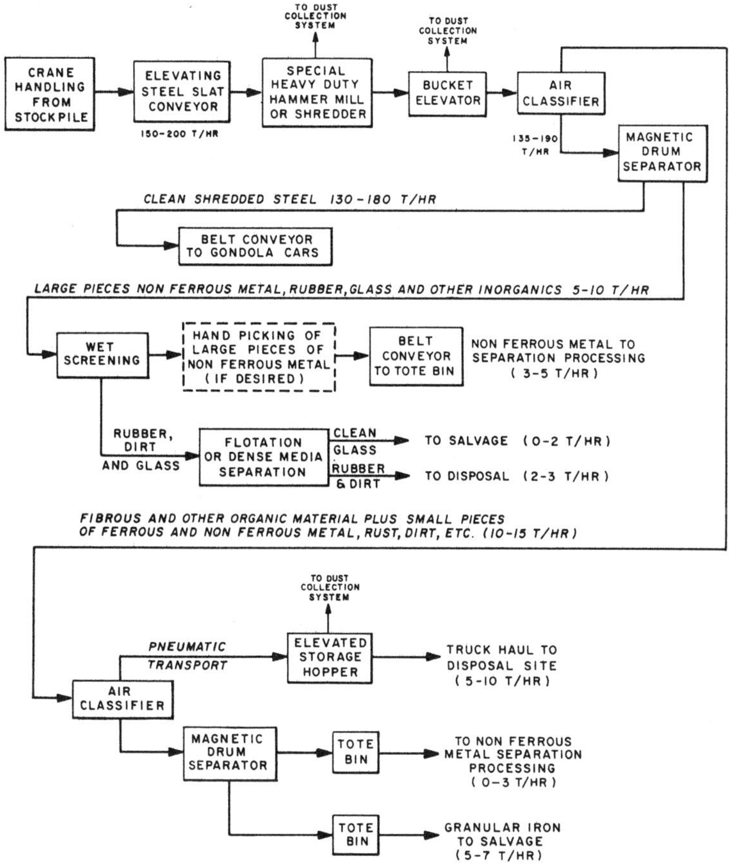

Source: EPA Report SW-30c

Table 9 presents the metals recoverable from obsolete automobiles. Steel is the primary component. Steel scrap can be processed in three different ways prior to recycling; it can be (1) baled into No. 2 bundles (removal of the engine, seats, and gas tank and compression of the hulk into a block), (2) slabbed (slicing flattened automobile bundles into slabs), (3) shredded (shredding of automobile hulks and magnetic separation of ferrous fraction).

The first two processes produce a low-value contaminated steel scrap, while the third process generally results in a high-value material for which there is considerable demand. It is estimated that in 1972 approximately 4.4 million tons of bundled automobile scrap and 3.3 million tons of shredded automobile scrap were processed in the United States.

TABLE 9: VALUE OF RECOVERABLE METALS IN COMPOSITE AUTOMOBILE, 1972

Metal	Weight* (lb)	Price† (dollars)	Value (dollars)
Steel	2,614.0	25.00-38.00‡ (per ton)	32.68-49.67
Cast iron	429.3	55.00 (per ton)	11.81
Copper:			
Radiator stock	15.4	0.25 (per pound)	3.85
No. 2 heavy and wire	13.8	0.32 (per pound)	4.42
Yellow brass solids	2.7	0.23 (per pound)	.62
Zinc, die castings	54.2	0.04 (per pound)	2.17
Aluminum, cast, etc.	50.6	0.12 (per pound)	6.07
Lead:			
Battery	20.0	1.40 (per battery)	1.40
Battery cable dumps	.4	0.11 (per pound)	.04
Total			63.06-80.05

*Dean, K. C., and J. W. Steiner. Dismantling a typical junk automobile to produce quality scrap. U.S. Bureau of Mines Report of Investigations 7350. [Washington], U.S. Department of the Interior, Dec. 1969. p.17.

†Typical secondary material prices from *Secondary Raw Materials*, 10(10):123-125, Oct. 1972.

‡$25.00 per ton is the price as a No. 2 bundle (crushed flattened hulk); $38.00 per ton is an estimate of the price of shredded and magnetically separated automobile steel—comparable to No. 1 heavy melting scrap.

Source: EPA Report SW-122

The possibility of dismantling junk automobiles to produce quality scrap consisting of nonferrous metal, rubber, plastic and glass fractions as well as steel has been studied in some detail by K.C. Dean and J.W. Sterner in Report of Investigation No. 7350, Washington, D.C., U.S. Bureau of Mines (December 1969).

In an economic analysis of the junk automobile problem, R.L. Adams in Information Circular 8596, Washington, D.C., U.S. Bureau of Mines (1973) has pointed out that the new shredder processing technology and the new portable flattener technology have altered the structure of the junk automobile recycling system. The shredder technology has increased the demand for processed automobile scrap. This increased demand is reflected in higher prices paid for hulks by scrap processors. The portable flattener has reduced the transportation cost of shipping hulks to processors. The two new technologies have eliminated, at least potentially, the net cost of shipping stripped hulks so frequently incurred by wreckers in the past. The elimination of this cost eliminates the economic incentive for excessive wrecker inventories.

Unfortunately, the incentive for the last owners to abandon their cars does not appear to have been eliminated by the new technologies. The last owner of an automobile with no spare parts value is still likely to incur a cost by delivering his automobile to the recycling system. Although a subsidy payment is probably unnecessary to reduce excessive wrecker inventories, such a payment might be one solution to the problem of automobile abandonment. However, there are a wide variety of subsidy programs that could be introduced. Before any Federal subsidy program is introduced to reduce automobile abandonment, it is recommended that the program be tested on a limited basis in many different parts of the country.

J.W. Sawyer, Jr. of Resources for the Future, Inc. in *Automotive Scrap Recycling—Processes, Prices & Prospects,* Baltimore, Maryland, John Hopkins University Press (1974) has presented an in-depth analysis of the processes employed in that part of the scrap industry that uses obsolete automobiles as its raw material. He examines the rate of formation of obsolete autos, the role of impurities in scrap, the supply of scrap and its elasticity, and the costs of processing methods.

He concludes that, at present, steel scrap derived from obsolete steel forms a very small portion of the total raw material input to the ferrous materials industry—about 3 to 4% by weight of the total metallic feed. In addition to showing that the elasticity of supply of automobile-derived scrap is high, he suggests ways to reduce residuals production by using simple and relatively inexpensive control equipment. This research report combines formal mathematical models with primary data in a way that not only sheds new light on elasticity of ferrous scrap supply, but also provides a basis for evaluating the effects of proposed governmental policies.

A somewhat similar study entitled *The Automobile Cycle—An Environmental & Resource Reclamation Problem* has been prepared by the Federal solid waste management program, Washington, D.C., U.S. Environmental Protection Agency (1972). The problem of *Solving the Abandoned Car Problem in Small Communities* has been discussed by W.T. Dehn in Report SW-70ts.1, Washington, D.C., U.S. Environmental Protection Agency (1974).

The production of secondary pig iron from shredded auto scrap has been discussed by J.E. Tress and W.L. Hunter in Report R.I. 7808, Washington, D.C., U.S. Bureau of Mines (1973). The reader is also referred to the section of this volume on Iron, Recovery from Solid Wastes and Recovery from Scrap Automobiles for additional information on this topic.

The reduction roasting of nonmagnetic taconites using automobile scrap as the reductant has been described by C. Prasky, R.E. Peterson and D.L. Siebert in Report of Investigation No. 7389, Washington, D.C., U.S. Bureau of Mines (May 1970).

Quality magnetic concentrates containing about 66% iron and 6% silica with iron recoveries of over 90% were obtained after roasting fine-grained oxidized or nonmagnetic taconite with unburned, sheared automobile hulks in a rotary kiln. The relatively complete oxidation of the lighter gage metal in the automobile during roasting resulted in a metallic byproduct consisting of the unconsumed heavier gage metal scrap. This residual or tailored scrap is potentially useful as heavy melting stock and has more value than the original automobile hulk.

SCRAP RAILROAD FREIGHT CARS

Railroad car dismantling is a unique problem in solid waste disposal. Approximately 70,000 rail freight cars are dismantled per year for the purpose of salvaging reusable parts, scrap iron, and steel. About 50% of these cars contain 3 to 7 tons of wood each, which must be removed before the scrap can be recycled. However, the conventional means of removing the wood, i.e., open burning, produces the very dense smoke emissions prohibited by the Clean Air Act. In an effort to control or eliminate this environmental pollution, the Bureau of Solid Waste Management contracted with Booz, Allen Applied Research Inc. to investigate alternative methods for railroad car dismantling and salvage.

Information on the subject was compiled by means of mail solicitation, site visits, and observation/discussion. Possible solutions were carefully screened, evaluated, and subjected to a cost effectiveness analysis. The mathematical model used was developed by the Bureau of Solid Waste Management with the contractor and produces ordinarily weighted scores. In itself it is an excellent decision-making tool as reported by D.M. Butler and W.M. Graham in Public Health Service Publication No. 1850, Washington, D.C., U.S. Government Printing Office (1969).

SCRAP SHIP HULLS

A process developed by *A.B. Butterworth; U.S. Patent 3,393,520; July 23, 1968* involves building a breakwater by utilizing a section of a ship by removing the bow and stern ends of a ship and also all superstructure and then cutting the remaining hull into sections as containers followed by welding plates to the cut sections to provide approximately rectangular sections to be placed end to end to form the breakwater and pinning the sections to the bottom of the sea.

SEWAGE

Production of Useful Products from Sewage

Production of Active Carbon from Sewage Sludge: A process developed by *R.D. Nickerson and H.C. Messman; U.S. Patent 3,887,461; June 3, 1975; assigned to Combustion Engineering, Inc.* involves the steps of subjecting the sewage to primary treatment which includes allowing solids to settle out in a settling tank, conveying the solids to a furnace, subjecting the solids to pyrolitic treatment in the furnace at temperatures within the range of 500° to 1000°C by passing a stream of hot gases therethrough, to form active carbon and char, separating the active carbon from the char by entraining it in the gases leaving the furnace, separating the active carbon from the gases, and introducing the active carbon into the settling tank to act as an adsorbent.

Production of Portland Cement Using Sewage: A process developed by *N.S. Davis; U.S. Patent 3,442,498; May 6, 1969; assigned to North American Rockwell Corp.* is one in which an activated sludge sewage treatment process is combined with a cement-making process for making cement economically from normally waste materials. Activated sludge is flocculated with sequential additions of carbon dioxide and calcium oxide, filtered, and fed to a cement kiln. The organics in the sludge reduce fuel requirements in the kiln and the resultant clinker is in the Portland cement composition range.

Production of Synthesis Gas from Sewage: A process developed by *A. Brent and C.F. Teichmann; U.S. Patent 3,687,646; August 29, 1972; assigned to Texaco Development Corporation* is one in which sewage is converted into a useful nonpolluting product gas, e.g., synthesis gas or fuel gas by dewatering municipal sanitary sewage, producing a thickened sludge. The thickened sludge is mixed with a fluid selected from the group consisting of H_2O, liquid hydrocarbon fuel, particulate carbon slurries in water or in a liquid hydrocarbon fuel as produced subsequently in the process, and mixtures thereof to produce a feed mixture stream. Preferably the feed mixture is heated to a temperature in the range of 212° to 600°F to vaporize volatile constituents in the feed mixture and to produce a feed dispersion stream.

The feed dispersion is then reacted by partial oxidation with an oxygen-rich gas in an unpacked free-flow noncatalytic synthesis gas generator, at a temperature in the range of 1500° to 3000°F and at a pressure in the range of 1 to 250 atmospheres to produce synthesis gas. Water separated from the sewage is purified and used for industrial applications or converted into by-product steam by indirect heat exchange with the product gas. In a second embodiment of the process, garbage is disposed of by mixing bits of garbage with thickened sewage sludge and the aforesaid fluid thereby forming a feed mixture which is converted into synthesis gas by the partial oxidation reaction with an oxygen-rich gas in the flow type gas generator.

Production of Yeast from Treated Sewage Effluent: A process developed by *K.B. Farris; U.S. Patent 3,751,338; August 7, 1973* is a process for the production of yeast utilizing as a growth medium for yeast organisms, a nonchlorinated, sterilized and filtered sewage effluent, in which the pH is in the range of 5.0 to 8.0 and the osmotic pressure is in the range of 0.05 M to 1.5 M NaCl. Fermentation of yeast species *Saccharomyces cerevisiae* and *Candida utilis* individually or in combination, and of yeast species *Endomycopsis fibuliger* in combination with either one or both of the aforesaid species is conducted under aerobic conditions at a temperature in the range of 20° to 35°C for a period of from 12 to 30 hours. Centrifugation of the fermented liquid separates out the solid yeast particles from liquid waste. Repetitive washing and centrifugation provides for purification of the final end product.

Recovery of Fertilizer from Activated Sludge Process: A process developed by *C.H. Carothers; U.S. Patent 3,728,254; April 17, 1973* is one employing on site utilization of the final effluent discharge from an activated sludge treatment process through a separation of the sludge component and the nutrient rich liquid component for respectively forming dried commercial fertilizer and/or commercially saleable compost and reducing the nutrient rich liquor to nutrient free water by conveying it as a feeder liquor through a suitably formed hydroponic garden containing commercially saleable plants arranged to extract the nutrients.

A process developed by *P. Fisk; U.S. Patent 3,847,803; November 12, 1974* involves recycling raw sewage and unsorted solid waste by adding the solids removed from sewage to ground unsorted waste forming a sludge, treating the sludge first anaerobically, sterilizing the anaerobically treated sludge, forming the sludge into a slurry. The process next involves separating the biodegradable from the nonbiodegradable material in the slurry by aerobic fermentation, composting the biodegradable material in batches, curing the compost under control conditions to stabilize its fertilizing properties and classifying the biodegradable and nonbiodegradable material to provide useable products. The colloidal raw sewage liquor remaining after separation of the sewage solids is converted into pure water.

FERROUS METALS

IRON

A comprehensive report on the *Identification of Opportunities for Increased Recycling of Ferrous Solid Waste* has been prepared by the Institute of Scrap Iron and Steel, Inc. under contract with EPA and issued as PB 213 577, Springfield, Va., National Technical Information Service (1972). This study is an in-depth look at the recycling of ferrous solid waste. The study examines: the sources, types, and quantities of iron and steel scrap; the markets for iron and steel scrap; the functions of the ferrous scrap processing industry; and the technology of iron and steel scrap utilization. Problems inhibiting greater use of iron and steel scrap are identified and discussed, and recommendations are made for solutions to these problems. A survey of the ferrous scrap industry was the basis for many of the identified problems and proposed solutions, and also provided numerous statistics on the ferrous scrap industry.

Recovery from Ferrous Alloy Scrap by Zinc Treatment

A process developed by *E.T. Martyniuk; U.S. Patent 3,476,547; November 4, 1969; assigned to General Motors Corporation* provides a method of reducing certain soft scrap metals such as iron and copper to a useful powder form. The metallic scrap is dissolved in molten zinc and solidified to form a friable ingot which is comminuted to a fine powder. The zinc is separated from the powder by sublimation to leave a residual metal sponge which is readily comminuted to a powder of suitable size.

Recovery from Particulate Forms of Scrap Iron

A process developed by *J.E. Allen; U.S. Patent 3,870,509; March 11, 1975; assigned to Ferro-Carb Agglomeration Ltd., Inc.* is concerned with the recycling of particulate scrap iron products such as foundry chips, and machine shop turnings or borings. Scrap iron is one of the pervasive sources of pollution throughout the world. It is generally collected and remelted in various iron- and steel-making processes. Small iron pieces, such as foundry chips and machine shop turnings or borings, have been particularly troublesome, since they are difficult to collect, tend to float in the metal-making processes and often have contaminants on them which make their use in steel-making difficult.

In cupola operations, for example, the use of such particles results in irregular operations. It has often been necessary to find rather costly ways of disposing of such scrap, in order to prevent serious pollution problems, despite the fact that such scrap often contains hydrocarbons which could be a useful source of fuel values in the metal-making process.

In this process, particulate scrap iron is disposed of by blending it with a bituminous, hydrocarbonaceous binder, forming briquettes from the mixture, and treating the briquettes with an oxygen-containing gas at 350° to 600°F to selectively react with the hydrogen and unsaturated hydrocarbon in the binder, to obtain strong briquettes which can be used as a source of iron in steel-making processes.

Recovery from Red Mud from Bayer Process

A process developed by *T.E. Ban and H.E. Goetzman; U.S. Patent 3,495,973; February 17, 1970; assigned to McDowell-Wellman Engineering Company* is useful in recovery of iron values from alumina-depleted bauxite and will be illustrated in reference thereto. Bauxite is a principal ore of aluminum. Bauxite contains trihydrates, monohydrates and typically about 60 weight percent alumina. Methods for refining bauxite to obtain the alumina generally take advantage of the solubility equilibria of the alumina hydrates in a caustic soda solution.

One process, universally used today, for refining bauxite is the Bayer process. By this process bauxite, as mined, is finely ground and charged to digesters with a hot caustic soda solution to put the aluminum in solution as sodium aluminate. After digestion, the caustic and soluble sodium aluminate are separated by settling and filtration from insoluble residue. The residue is red mud about 3 tons of which are formed for each ton of aluminum product. The mud appears as fine particles (less than a micron size).

The red mud, depleted of alumina, ordinarily imposes severe disposal problems inasmuch as the caustic content is detrimental to plant and animal life. Also large unsightly ponds are needed as settling ponds for storage of the waste material. Red mud generally comprises oxides of aluminum, iron, silicon, titanium, calcium and sodium. Attempts have been made to utilize the red mud, generally without much success.

It has been found in the development work leading to this process that iron values can be extracted from red mud by superimposing upon a first layer of pellets formed from the red mud and a solid carbonaceous material, a layer of gas-releasing material, such as for example, limestone or oyster shells. When this is done, and gases at elevated temperatures passed through the gas releasing material, prior to contacting the layer of pellets derived from the red mud, the gaseous medium is conditioned both with respect to its temperatures, and its chemical composition, so as to be suitable for favorable reaction with the red mud derived pellets to form a carbonized product suitable for use in the recovery of iron values therefrom, as well as providing a fluxing material capable of carrying into slag the impurities which are undesirable and also contained within the red mud.

Red mud of a type contemplated for use in this process has approximately the following composition:

	Percent (dry basis)
Al_2O_3	22.0
Fe_2O_3	46.8
SiO_2	5.0
TiO_2	6.1
CaO	4.9
Na_2O	2.8
LOI	11.0

Red mud of such a composition as above described and containing about 50% moisture is retained in reservoir **10** before being combined with approximately equal amounts of dried red mud of about 98 wt % solids from reservoir **11**. The ingredients are thoroughly blended and are sent to a nodulizing or pelletizing apparatus **12** of the rotating inclined pan or drum type. The particular material from **11** and the slurry from **10** after nodulizing or balling takes on the form of nodules of various small sizes. The nodules generally range in size from about ⅜" to 1⁄16".

FIGURE 12: PROCESS FOR RECOVERY OF IRON FROM RED MUDS

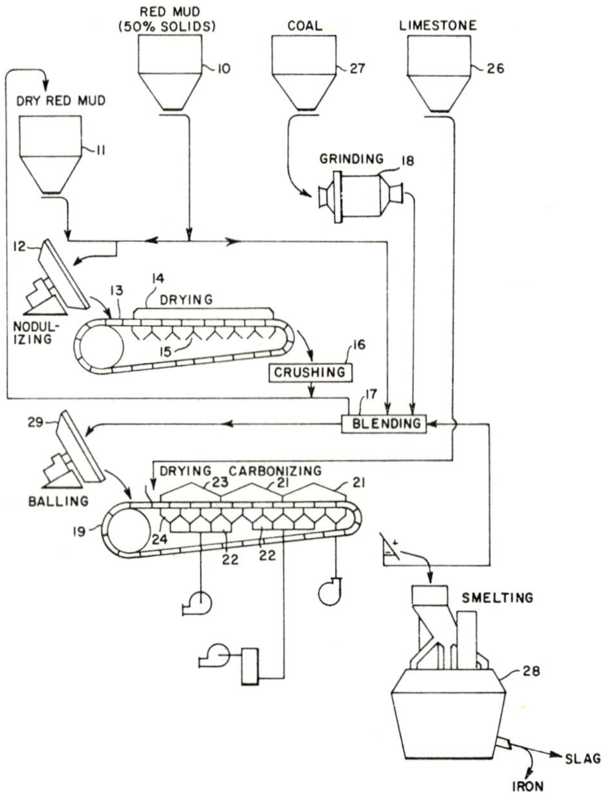

Source: T.E. Ban and H.E. Goetzman; U.S. Patent 3,495,973; February 17, 1970

The green nodules (nodules which have not been dried) are deposited about 5 to 10 inches deep on a traveling grate apparatus **13** for drying. As is well known to those skilled in the art, the traveling grate is an elongated trackway having supported thereon a plurality of individual traveling pallets each having a bottom formed of a plurality of grate bars and upstanding side walls. When the traveling grate members are progressing along a level portion of the trackway, they form a continuous channel.

At the discharge extremity of the grate, the individual pallets are carried by a terminal guide member in an inverted position and returned in an inverted position to the opposite or charging extremity of the trackway where they are again inverted to an upright position to repeat the cycle. The trackway is provided with a hood portion **14** for the admission (not shown) or removal of gaseous medium. In coacting relation beneath the pallets and the burden are a plurality of contiguous windboxes **15** adapted for the admission of gaseous material to or withdrawal of gaseous medium through the material on the grate.

The gases traversing the material on the grate can be spent gases from another portion of the process and enter a drying chamber at a temperature in the range of about 400°F to

1000°F at a flow rate of about 250 scfm/ft^2 of grate area. Because spent gases are often at temperatures too high for proper drying, these gases can be diluted with ambient air so that the temperature of the drying gases contacting the nodule burden is within the desired range. The nodule burden on the grate is subjected to the drying phase for about 12 to 14 minutes and are reduced in weight by about 25 wt % due primarily to water loss.

Following the drying phase, the nodules enter conventional crushing apparatus **16** where the nodules are shredded and reduced in size. The shredding operation produces a pulverized product of the approximate original size of red mud. It is preferred to recycle a portion of the dry pulverized material to reservoir **11**, e.g., about half, for subsequent blending with red mud slurry for the nodulizing and drying operation.

After crushing, the product enters a blending zone **17** where it is mixed with the metallurgically required amounts of ground, solid carbonaceous material, such as for example, coal from reservoir **27** ground at **18**, and significant amounts of liquid red mud from the reservoir **10** and returns from the smelting operation **28**. The additional liquid red mud is added to provide the moisture for subsequent balling or pelletizing. Carbonaceous material is added to provide a reducing agent for the pellets. The blend of red mud and coal is conveyed to a balling apparatus **29** where it is pelletized. The green pellets, containing moisture to the extent of about 17 to 19% by weight and of a size of about ¼ to 1 inch in diameter, pass from the balling apparatus onto traveling grate **19** where they form a layer about 5 to 10 inches deep.

A layer of oyster shells (calcium carbonate) is superimposed from hopper **26** over the pellet burden. This burden is first submitted to a drying phase defined by hood **23** coacting with windboxes **24** by means of which gases heated to 400° to 600°F traverse the burden for about 12 to 14 minutes at a flow rate of 250 scfm/ft^2. These gases also can be spent gases from another portion of the process. The temperature of the drying gases is adjusted so that preignition of carbonaceous matter is avoided, and spalling of the balls is minimized. Spalling causes the formation of undesirable quantities of fines. The calcium carbonate layer is imposed on the pellet layer to a depth of about 1 to 4 inches.

The grate loaded with the composite burden then passes into a carbonizing and reducing zone defined by hoods **21** and windboxes **22** where the burden is contacted with a hot downwardly directed draft. The hot draft is directed downwardly through the burden at a rate of about 100 to 222 scfm/ft^2 at a temperature of about 2000° to 2400°F or at a temperature sufficient to cause rapid reaction between $CaCO_3$-containing material and the incoming hot gases. This hot draft causes calcination of the calcium carbonate-containing material. Because of the endothermic nature of the decomposition reaction, the gaseous medium is cooled as it passes through the burden, and upon reaching the pellets has a temperature of about 1500° to 2000°F, which is ideally sufficient to carbonize and prereduce iron in the pellets. The burden is subjected to carbonizing for a period of 10 to 12 minutes.

The hot gases are heated externally of the machine by burning combustible gases and also by igniting the gas after passing through the burden and recycling of such gases. The heat of the burden then results from the passage of the hot gases therethrough and from some oxidation of a portion of the carbonaceous material. The abundant quantity of gangue in the burden fuses at a relatively low temperature and causes slag or ceramic bonds to be formed. Thus, the pellets produced in accordance herewith are different from other iron ore pellets in that they are ceramically bonded and therefore, very hard and strong and able to withstand rough handling after the establishment of the ceramic bonds. In addition, a slight degree of carbon bonding takes place.

While it is not practical to use entirely inert gases in the gaseous medium traversing the burden on the grate, some inert gases could be used. However, from a practical standpoint, oxygen-depleted air wherein there are minor amounts of oxygen present, are used. If the passage of such inert gases is continued for sufficient periods of time, the temperature of the burden may be raised to the incoming gas temperature at which temperature the reduction of the iron oxide is carried on to the extent of from about 10 to 90% of the available iron and the pellets are both metal and ceramic bonded. By this method, substantial quantities of limestone can be introduced. Prior art has shown that introduction of large quantities of $CaCO_3$-containing material with the pellets interferes with chemical beneficiation. By this method the $CaCO_3$ is not incorporated into the burden. By providing layers, the limestone becomes calcined and the ore of the red mud reduced, discharging from the traveling grate as a preheated, highly beneficiated burden.

While the process may be successfully practiced with just the downdraft operation, it has been found that there is benefit in reversing the direction of the hot gases (not shown). By passing the gases through the burden of pellets in an opposite direction further down the path of travel, a uniform degree of reaction is insured for any particular level within the burden. The updraft reaction is conducted at about ambient temperature at a flow rate of 25 to 75 scfm/ft^2 and the burden is subjected to the updraft for a period of 1 to 2 minutes. Following the change in direction of flow of draft, further reduction may be done in a zone immediately following the reversed gas zone.

In the methods described, any type of carbonaceous matter may be incorporated as the reducing agent. To produce a char-bond product, the reducing component should also include a tarry or asphalic material as the binder forming agent. The pellets and limestone treated in accordance with this process may be charged directly to a smelting furnace **28** or cooled and transferred to a remote smelting installation. The product resulting by the method of this process is a carbonized and prereduced pellet and calcined limestone. The resulting pellets have an iron content of about 25 to 35%.

A process developed by *G. Horváth, Z. Felföldi, Z. Oswald, K. Solymár, A. Juhász, G. Dobos and G. Kaptay; U.S. Patent 3,876,749; April 8, 1975; assigned to Aluminiumipari Troszt, Hungary* is one in which the red mud by-product of the Bayer aluminum manufacturing process is utilized by mixing the red mud with a reducing agent, reducing the red mud, separating it into molten steel and molten slag, reacting the slag with CaO, leaching out the useful aluminates for recirculation to the Bayer process, and utilizing the remaining calcium silicates in cement manufacture. The process and equipment arrangement is shown schematically in Figure 13.

FIGURE 13: ALTERNATIVE PROCESS FOR RECOVERY OF IRON FROM RED MUD

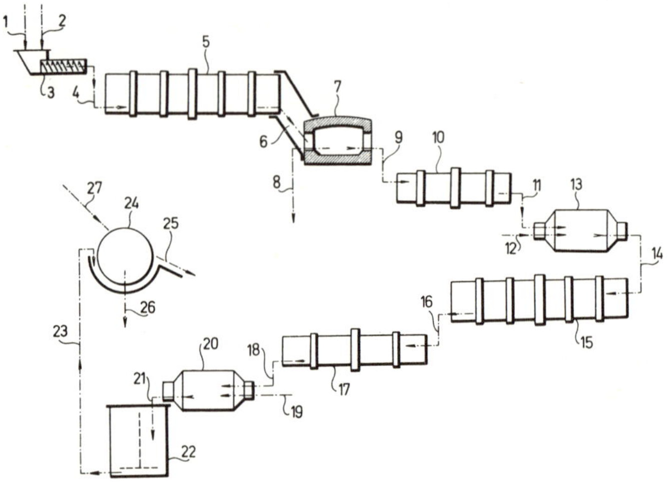

Source: G. Horváth, Z. Felföldi, Z. Oswald, K. Solymár, A. Juhász, G. Dobos and G. Kaptay; U.S. Patent 3,876,749; April 8, 1975

The red mud **1** which contains not more than 10% (wt) CaO, is mixed with the reducing agent **2** in a mixing feeder **3**. The reducing agent, such as anthracite, suitably has a particle size under 3 mm. The homogenized mixture **4** is fed from the mixing feeder **3**, into a rotary kiln **5** for reduction. Here the material is heated to about 1000°C and its iron oxide content becomes reduced to metallic iron with a conversion efficiency of about 80 to 90%. The reduced material **6** is transferred in a closed system into a melting furnace **7** which is suitably fired with oil or natural gas, the combustion gases of which are suitably utilized for the heating of the reduction kiln **5**. Thus the heat of combustion of the solid reducing agent **2**, together with the hot combustion gases from the melting furnace **7** may be sufficient to require the introduction of only a small amount of additional external heat.

The reduced material separates in the melting furnace **7** into molten steel **8** and molten slag **9**, which are both continuously or periodically tapped. The molten steel **8** can be further treated, such as by desulfurizing, before utilization. The molten slag **9** is cooled, suitably in a rotary cooler **10**, such as by sprinkling with water and, thereafter, the solidified slag **11** is mixed and ground with crushed burnt lime in a mill **13** to a maximum particle size suitably of 0.1 mm.

The burnt lime is mixed with the slag in molar proportions (with a tolerance of about ± 10%) corresponding to $CaO:SiO_2 = 2$; $CaO:TiO_2 = 1$ and $CaO:Fe_2O_3 = 2$. The slag-CaO mixture **14** is introduced into a sintering furnace **15**, in which the formation of sodium aluminate and of dicalcium silicate occurs between 1000° and 1400°C. The sintered material **16** is cooled in a cooling drum **17** and the cooled product **18** is ground with sodium aluminate **19** obtained from the Bayer process, in a mill **20**. The resulting slurry **21** is introduced into an agitator-equipped leaching vessel **22** to complete the dissolution of the sodium aluminate content of the slag.

The slurry **23** which is obtained in the leaching vessel **22**, is transferred into a vacuum filter **24** where the solid mud phase **25** is separated from the liquid phase **26** which contains the basic leaching solvent and the dissolved sodium aluminate. The mud **25** is washed in the filter **24** with a washing liquid such as water **27** and, if desired, the washwater is combined with the liquid phase **26**. The liquid phase can then be returned to the Bayer process, while the dicalcium silicate mud **25** can be utilized in the manufacture of cement.

Pilot plant tests on the utilization of red mud residues from the production of alumina from both Jamaican and Surinam bauxites have been reported by O.C. Fursman, J.E. Mauser, M.O. Butler and W.A. Stickney in *Report of Investigation No. 7454,* Washington, D.C., U.S. Bureau of Mines (November 1970).

Both the carbon-lime-soda sinter process and electric arc smelting with coke and limestone were employed to recover iron and alumina. Alumina recoveries of 85 to 90% were obtained by the sinter process; although iron recovery was about 80%, the magnetic product was finely divided and difficult to separate from the gangue. Smelting red muds in the electric-arc furnace produced a pig iron with 98% recovery of iron; up to 84% of the Al_2O_3 was recovered by leaching the calcium aluminate slag with Na_2CO_3 solution.

Recovery from Scrap Automobiles

A process developed by *S. Proler; U.S. Patent 2,943,930; July 5, 1960; assigned to Proler Steel Corporation* involves upgrading junkyard type raw ferrous scrap such as automobiles, having at least a large proportion of sheet material of at least 2S gauge and no more than a small proportion of plated material, to a fluent bulk furnace charging material. This process involves milling the raw scrap to a piece size small enough to pass a grate of no larger than one foot opening size, magnetically separating the more ferrous milled scrap from the less ferrous milled scrap, roasting the more ferrous milled scrap at a temperature in the range of 1300° to 1800°F and individually compacting and balling up the pieces of the roasted scrap while below the welding temperature for steel to increase the density of the bulk roasted scrap while maintaining the individuality of the separate pieces sufficient to keep the final product fluent.

A particular object of the process is to convert material heretofore suitable only for making what is known in the trade as a number 2 or number 3 scrap bale into a material equivalent or superior to a number 1 bale of scrap.

A process developed by *R.M. Eidal; U.S. Patent 3,356,016; December 5, 1967; assigned to Southwest Factories, Inc.* utilizes a metal working machine particularly adapted to flattening, shearing, chopping, cleaning and compacting automobile bodies into a highly acceptable and usable form of high quality scrap for steel furnaces, electric furnaces and the like.

This destructing apparatus progressively effects flattening and crushing of the body to manipulatable relatively thin slabs; laterally shearing chunks from the slabs, chopping laterally the chunks, and compacting the chunks to pulverize and clean the material and reduce it to a relatively fine aggregate of highly acceptable and usable scrap metal.

A process developed by *C.C. Schott; U.S. Patent 3,367,769; February 6, 1968* is one which provides for continuously melting scrap automobiles. Scrap vehicles are partially compressed in a squeezing apparatus and then lifted by a hoist and deposited on end, one on another, in the top opening of a furnace stack. An annular, water cooled ledge, located between the bottom of the stack and the top of the combustion chamber, has an opening smaller than the stack for engaging the lowermost scrap vehicle and supporting the entire column of vehicles in the stack. Progressive melting of the lowermost scrap vehicle results in the descent of the others in the column, the molten metal being collected in the hearth. The molten metal is withdrawn from the hearth into a trough, the slag separated from the molten metal and the molten metal deposited in a series of molds carried by the endless conveyor of a continuous pigging machine.

A device developed by *W.D. Ballard; U.S. Patent 3,413,914; December 3, 1968* includes a base with bed therearound, successive bodies, as automobile bodies, placed on the base being smashed by a hammer with forward end mounted on a transverse pivot across the base. A support, cantilevered forwardly, has forward sheave means, cable dead end, and clutch-connected, engine driven reel mounted thereon, with reel cable extending to a rear sheave means on a frame upstanding from hammer and thence back and forward sheave means. Engine drive is declutched for reel free-wheeling to run off cable as hammer descends, and is enclutched to elevate the hammer.

A process developed by *E.A. Uzdavines; U.S. Patent 3,484,231; December 16, 1969* is one in which scrap automobile bodies are stripped of contaminant chromium and copper, placed on a conveyor moving angularly upward through a thermally insulated housing and ignited so that all combustibles and low temperature melting metals are removed by the time that each car reaches the upper end of the housing which latter opens into a cupola. The cars are deposited successively into the cupola and melted down to high grade steel which is tapped off and pigged. Incomplete combustion products produced in the housing in the form of smoke are drawn off and forced through the cupola bed to complete the combustion and eliminate air pollution.

A device developed by *J.L. Chazen; U.S. Patent 3,518,078; June 30, 1970; assigned to SMC Ind.,* provides a method in which junked vehicles have portions sheared off by a shearing plunger which also causes intermittent advance of the sheared portions into and through an incinerator where the combustible contaminants are burned off.

A process developed by *J.C. Brewer; U.S. Patent 3,613,562; October 19, 1971; assigned to Garbalizer Corporation of America* is one in which a vehicle body shell processing plant is provided wherein vehicle shells are fired and quenched prior to baling in a manner such as to avoid air pollution difficulties. Exhaust gases from the furnace and quencher stages of the system are routed through an air-cleaning system such that only clean gases are exhausted from the system. Quenching is done in the system such that steam, entrained particulate matter, and residual exhaust gases are conveyed to a gas scrubbing system for cleansing purposes. Independent conveyors are utilized at feed-in, furnace, and quencher points. The quencher conveyor is designed to have a sloping construction such that there can be an automatic and continuous process to shell bodies for each intermittent group processed.

Another very similar process is described by *J.C. Brewer; U.S. Patent 3,706,289; December 19, 1972; assigned to Garbalizer Corporation of America.*

A process developed by *C. Prasky and W.M. Mahan; U.S. Patent 3,630,675; December 28, 1971; assigned to the U.S. Secretary of the Interior* involves utilizing ferrous scrap, particularly automobile scrap, comprising selectively oxidizing the scrap under conditions such that portions of the scrap having a thickness less than about one-eighth inch are substantially completely oxidized and portions having a thickness greater than about one-eighth inch are less than completely oxidized. The unoxidized portion of the scrap is then recovered as a dense residual scrap which may be used as a heavy melting stock for an electric furnace or foundry. The oxidized scrap, consisting of a high-grade iron oxide material, finds utility in processes such as production of powder iron by various sponge iron processes.

A process developed by *J.W. Snelling; U.S. Patent 3,646,897; March 7, 1972; assigned to Amex Corporation* relates to a method and apparatus of burning automobile vehicles. Before a body of a vehicle can be salvaged and the scrap metal properly utilized, it is necessary that the upholstery, seat cushions, plastic and other nonmetallic portions of a disposed automobile be removed. The more of these nonmetallic materials of the vehicle that are removed, the more valuable is the remaining scrap. This is because if the metal or body of the vehicle is shredded, otherwise broken up or pressed for scrap it is important that the above hydrocarbon materials be removed. In this manner the metal can go almost directly to a smelting process. If the nonmetallic materials are left, the scrap and shredded metal is not as readily usable.

There have been numerous attempts at providing means for disposing of the junked vehicles with and without burning so as to eliminate pollution from the air. Unfortunately, these approaches have not entirely been successful, and they are very expensive. In the past it has been common practice to burn the vehicles in the open. This procedure has the obvious disadvantage of contaminating the polluting the air to an extremely high degree in the areas surrounding such burning operations. It is not only objectionable from the standpoint of the appearance of smoke filled sky, but is also a definite health hazard. Obviously, the more grease and oil in the auto, the worse the smoke and pollution. The same is true of tires.

An object of the instant process is to provide a low-cost means for the reduction of contaminates released in the atmosphere by burning disposed automobile bodies. It is another object of this process to provide a relativley small, compact device which is in a price and size range wherein the large number of small, independent scrap metal yards and the like can utilize and afford it.

A process developed by *G.C. Patros; U.S. Patent 3,745,916; July 17, 1973; assigned to Kar-Denser, Inc.* utilizes a scrap processing machine for crushing scrap, such as stripped automobile bodies, into rectilinear laminated slabs which comprises a crushing press and a loading hopper. The press comprises a pair of spaced apart walls, a stationary inclined compression gate between the walls, and a pivotably movable hydraulic ram operated inclined compression gate between the walls. The gates and walls cooperate to define a compression chamber having an upper scrap receiving opening and a lower slab discharging opening.

The hopper is pivotably movable by a hydraulic ram from a lowered loading position to a raised unloading position adjacent the upper opening wherein it deposits scrap into the chamber while the compression gates are separated and then returns to a lowered position. The scrap is crushed into a slab by closing movement of the movable compression gate. Opening movement of the latter gate allows the slab to slide out of the press through the lower opening under the force of gravity. A motor driven slab ejector drum adjacent the lower opening assists in removal of the slab. The corners of the compression chamber formed by the side walls and the stationary gate are provided with fillets which prevent scrap from being jammed into those corners. The surface of the movable gate has spaced apart parallel members along its side edges which cooperate with the fillets to force the scrap inward and further prevent the slabs from lodging between the movable gate and side walls.

A process developed by *R.L. Flanagan; U.S. Patent 3,802,629; April 9, 1974; assigned to Mobil Auto Crushers Corporation of America* is one for processing scrapped internal combustion engines wherein the various component engine parts are separated for reclamation according to material. The method includes applying a shearing force to the cast iron engine block causing the block to be split open generally along a line parallel to the crankshaft so as to release and separate the internal steel components such as the pistons, rods, crankshaft, and camshaft from the block. The associated transmission may also be severed in an initial step and subsequently fragmented along with the separated internal engine components.

The apparatus includes an anvil or die block for supporting the engine in the proper alignment for processing. A blade powered by a hydraulic cylinder is positioned to engage the engine block along a line generally parallel to the engine crankshaft and hold the engine while a plate is pivoted downwardly by a hydraulic cylinder to sever the transmission. A compartment located below the anvil receives the severed transmission. The blade is then lowered farther to split the block and strip the crankshaft, connecting rods, pistons, etc. into the compartment. Another plate movable within the compartment then crushes the components over a conveyor which removes the separated and crushed materials.

A process developed by *S.M. Coulter and R.P. Lombardi; U.S. Patent 3,868,064; February 25, 1975; assigned to Dravo Corporation* involves reducing automobile bodies to scrap suitable for use in metallurgical furnaces. Automobile bodies, front end forward are placed on an inclined chute, one behind another, the chute accommodating two bodies in tandem. Power-driven feed rolls, the upper one of which is vertically movable, control the rate of feed of the lead body into a reducing mill having an upright rotor with a shell of downwardly decreasing diameter having a replaceable lining therein.

Spaced knocker arms with cutters at the free ends, forming the top of the rotor, progressively shear and rend the lead ends of the car body as it projects into the mill and the pieces are further broken up and rolled into balls and otherwise irregularly shredded and compacted as they then move down between the rotor and the shell lining, the rotor having ring hammers at intervals around its periphery from a level just below the arms to its bottom. The end product drops into an annular trough from which it is expelled by vanes at the bottom of the rotor.

Recovery from Scrap Iron in Basic Oxygen Furnace

A process developed by *J.J. Bowden; U.S. Patent 3,479,178; November 18, 1969;* involves preheating scrap before it is charged into a basic oxygen vessel by placing the scrap on cars which are driven through a tunnel-type furnace. A nonoxidizing atmosphere is maintained in the furnace and radiant heat is utilized to minimize the addition of objectionable oxides to the scrap. The scrap is heated to a temperature just short of the point where it loses its magnetism, and a magnet is used to lift the scrap from the cars and deliver it to the vessel.

Recovery from Slags

A process developed by *L.S. Moussoulos; U.S; Patent 3,002,832; October 3, 1961* involves treating molten iron bearing slag for the recovery of iron. In some metallurgical processes, such as in the production of FeNi alloy from nickel bearing iron ores, slag is produced at high temperatures (1600°C) and this slag also contains a considerable iron content (25 to 30%).

The proposed method refers to the reduction treatment of such molten slag, resulting in the recovery of the contained Fe. Further the heat content of the molten slag is also utilized, reducing the consumption of heat energy. Under these conditions, the proposed process appears as a supplementary step of a preceding metallurgical treatment. It is conducted as follows:

The molten iron bearing slag produced during the preceding metallurgical treatment, which for reasons of simplicity is named slag A, is introduced into an electric (or other kind) furnace. To the same furnace, a mixture of iron ore, flux and reduction fuel is simultaneously charged. This simultaneous charging is regulated in such a way, as to obtain, as far as possible, a uniform mixture of the solid materials with the molten slag A. Slag A provides heat and forms together with the solid materials a semimolten or plastic mass, in which the reduction fuel (coke breeze, etc) remains dispersed, so that reduction starts immediately.

The added ore and flux are regulated in quantity and quality so as to permit the resulting mass to have as high a temperature as possible (1200°C) and to provide with regard the new slag, about to be formed, which is named slag B, the required characteristics of fusibility. In order to cause a more drastic reduction, the temperature of above mass is raised to the highest possible level, which the fusibility characteristics of slag B permit, without bringing about the development of a liquid phase. Thus complete reduction is achieved in the considerably short time of about two hours.

After reduction is completed, the temperature is raised further, to the level where the whole mass becomes fluid and specifically to the temperature level where the reduced metal is molten (1300°C). At this stage the metal is separated by gravity from slag B and is collected in the form of cast iron or semisteel. The final highest temperature level required depends eventually on the fusibility of slag B and influences considerably the P, Cr, etc. content of the produced cast iron, in the event that such elements are present.

A process developed by *A.F. Johnson; U.S. Patent 3,015,553; January 2, 1962; assigned to Strategic Materials Corporation* relates to the treatment of slags resulting from the burning of coal, and has for its object the provision of an improved process for the reduction of the iron oxide contained in the slag and the conversion of the reduced slag into useful products. More particularly, a process is provided for heating the slag with a reducing agent at a temperature sufficient to reduce the iron oxide and melt the remaining slag, followed by rolling of the reduced liquid or plastic slag and its contained metallic iron into thin sheets. After suitably crushing or shattering the resulting sheets the iron is separated from the flat slag flakes thus produced to recover the iron as well as the slag component in the form of separate valuable commercial products.

Recovery from Slimes from Riverbeds

A process developed by *L.A. Duval; U.S. Patent 3,439,803; April 22, 1969* involves removing iron from slime. The slime enters the plant and is roughly screened to remove large noniron producing debris which are then discharged. The slime is then passed through magnetic separation stations which act in series to collect the course category iron particles and some of the intermediate category particles. The remaining tail is then passed into a gravity separation station operable to collect most of the intermediate particles. The tails from the gravity separation station may then be operated upon by froth-flotation and techniques to recover fine category particles remaining in the tails solution.

A valuable supply of iron exists in the riverbeds adjacent most basic industrial plants particularly iron and steel-making plants. The iron is discharged from the plants in waste material from the various manufacturing operations. The most prevalent forms of iron bearing waste material are scale discharged from such plants as the rolling mill, billet mill and blooming mill and dust emission from the blast furnace, sinter plant, open hearth and the oxygen steel-making process. The iron dust is discharged in the atmosphere and settles to the river bottom while the scale is generally discharged directly into the river and settles to the bottom and forms a layer of slime. The problem heretofore encountered in recovering this iron has been to devise an economical system for recovering same. This process offers a solution to this long-standing problem by use of a recovery plant which is located on a barge or the like so that the plant can be near the source of the iron-rich slime.

Recovery from Steelmaking Process Dusts

A process developed by *J.E. Cooper; U.S. Patent 2,810,633; October 22, 1957* involves the recovery of iron values from blast furnace dust. It has been the practice of steel producers in the past to dispose of waste products either to the atmosphere or to adjacent water courses. In recent years the necessity of finding more satisfactory methods of disposing of waste products has driven the steel industry to adopt rather expensive expedients. Two of the most troublesome wastes incidental to the operation of an integrated steel plant are the extremely fine dusts produced by the blast furnace and the pickle liquor produced by the sulfuric acid pickling of steel. The dust blown out of a blast furnace during operation has substantially the same chemical composition as the charge, and varies in size from sizable chunks down to particles of submicron size.

By weight, by far the greater portion of the blast furnace dust is recovered in the initial dry dust catcher. However, to clean the blast furnace gas sufficiently to permit its efficient utilization as a furnace fuel, resort is had to three further cleaning apparatuses. Arranged in series, these are first, an ordinary wet gas scrubber, second, a so-called disintegrator, and third, an electrostatic Cottrell precipitator. In each of these pieces of apparatus, the dust is finally removed as a suspension in water. In a typical blast furnace installation producing 3,000 tons per day of pig iron, the water flow through these three final gas cleaners amounts to about 8,250 gallons per minute and is burdened with approximately 229 grains per gallon of solid matter. Over a 24-hour period these wash waters remove from the three final cleaners about 194 tons per day of dry solids.

In conventional blast furnace practice these wash waters are combined and processed through a settling chamber and the effluent from the settling chamber is discharged into the sewer or other adjacent water course. The average dry solids content of this effluent which is discharged into the sewer is about eight grains per gallon or, in the course of 24 hours, about 7 tons of dry solids are so lost. The iron value of these dry solids is not particularly serious economically even though its recovery would of course be desirable.

The unfortunate aspect of this process is that the material escaping in the effluent from the settling chamber represents the very finest dust produced in the blast furnace and hence the dust which is the most objectionable from a standpoint of disposal in natural water course, since these dusts are the most objectionable from a chromogenic standpoint. These dusts impart a persistent iron red color to the water course which is highly objectionable to the general public, despite the fact that quantitatively their effect upon the water course may be insignificant. This process has been developed in an effort to solve this particular pollution problem. Figure 14 is a block flow diagram which shows the principal features of this process.

The combined flows of dust laden water from the blast furnace wet washer, disintegrator and Cottrell precipitator are emptied into mixer **10**. While being intensively agitated in mixer **10**, pickle liquor containing ferrous sulfate and sulfuric acid is slowly added. The agitating apparatus employed in mixer **10** is preferably one which is capable of inspirating air into the liquid being mixed. This type of agitator is well known to the art and any of several types may be employed.

It is essential for this mixing operation that the pickle liquor be added slowly so that at no time does the concentration of calcium sulfate produced by the interaction of the calcium carbonate in the blast furnace dust and the free acid in the pickle liquor exceed the solubility of calcium sulfate in the existing liquid. In this mixer the natural alkalinity in the incoming water as well as the alkalinity available from the iron, lime and magnesium compounds from the blast furnace serve to at least partially neutralize the acidic constituents of the pickle liquor.

FIGURE 14: PROCESS FOR IRON RECOVERY FROM BLAST FURNACE DUSTS

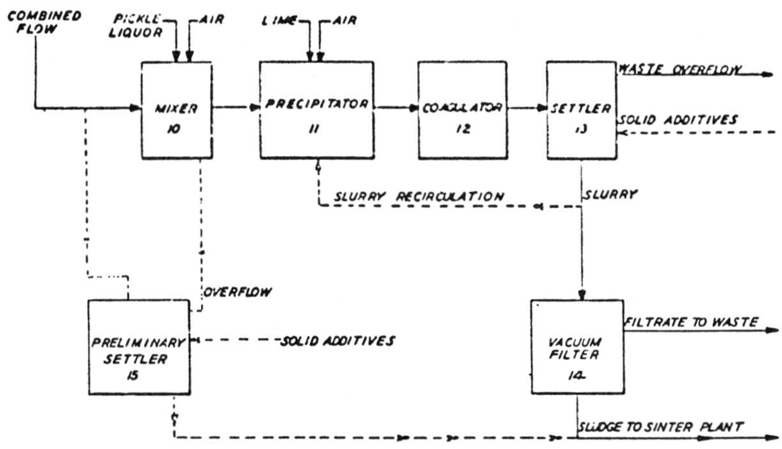

Source: J.E. Cooper; U.S. Patent 2,810,633; October 22, 1957

It is to be understood that the free acid in the pickle liquor will be first neutralized with the resultant production of carbon dioxide. The air added to this mixer **10** serves the purpose of scrubbing the carbon dioxide from the mix as well as partially oxidizing any iron hydrates which may be precipitated at this stage. The removal of carbon dioxide, of course, permits the neutralization of more of the acidic constituents in the pickle liquor by the calcium carbonate and bicarbonate available in mixer **10**.

It is difficult in mixer **10** to raise the pH of the liquid to a value between 7 and 8 which is necessary for the quantitative precipitation of the iron compounds. Accordingly, the liquor from mixer **10** is flowed to a precipitator **11** which may be structurally similar to mixer **10**. Here calcium hydroxide, preferably in the form of slurry is slowly added. The agitation in precipitator **11** is accompanied by intensive aeration again to remove carbon dioxide from the slurry and to oxidize precipitated ferrous hydrate to ferric hydrate. Contrary to most literature teachings, the oxidation of ferrous to ferric iron in this very dilute solution proceeds quite well and yields a precipitate much more amenable to the subsequent steps of coagulating, settling and filtering. A typical time in mixer **10** and precipitator **11** would be two to five minutes each.

The effluent from precipitator **11** is quite free of soluble iron and is passed to coagulator **12**. In this vessel the material is very gently agitated for a period of time of about twenty minutes. During the residence in the coagulator the precipitated iron hydrates and the iron oxides, lime compounds and coke carried over from the blast furnace coact to produce a mass which will settle in a practicable length of time and which will produce a sludge which can be handled on a vacuum filter.

The effluent from coagulator **12** is conducted to any conventional quiescent settling apparatus of which the Dorr Thickener is a typical example. The size of the thickener is selected so that the residence of the material therein will be sufficiently protracted to produce an effluent having a solids content below the desired minimum. In most cases it is possible to produce an effluent from settler **13** having a solids content below ten parts per million. An effluent of this caliber may be added directly to streams with no adverse effects.

The underflow or slurry produced by settler **13** is conducted to vacuum filter **14**. This vacuum filter may be any of the well-known continuous rotary vacuum filters such as the Oliver or Emico, or resort may be had to continuous centrifuges for this separation. A particularly satisfactory filter for this separation is known as the string or FLine filter. Any of these devices will produce a substantially clear filtrate and a sludge having a water content between fifty and eighty percent which is sufficiently dense to enable its transportation by conveyor to a sintering plant. Here it may be sintered along with the dry pulverulent solids recovered from the blast furnace gas by the dust catcher.

Modification of the above described flow sheet may be necessary to accommodate the conditions obtaining at individual plants. For example, a preliminary settler **15** may be interposed ahead of mixer **10** to give a preliminary separation of the more coarse solids and to relieve vacuum filter **14** of a portion of its load. The underflow from preliminary settler **15** may be added directly to the sludge from vacuum filter **14**. Similarly a portion of the slurry obtained from settler **13** may be recirculated back to precipitator **11**. This results in a further utilization of the alkalinity available in the slurry which would otherwise be wasted in the vacuum filter filtrate. This recirculation also provides solids for the nucleation of the ferrous and ferric hydrates.

To either settler **13** or preliminary settler **15** may be added any carbonaceous or ferruginous materials which are either unobjectionable or desirable in the blast furnace after passage through the sintering plant. Typical examples of material which may be added at these points and hence disposed of to advantage, are coke breeze or fines, mill scale, fine iron ore, fine coal, or wood flour. These materials will aid in the settling or filtration or add valuable iron to the blast furnace or provide fuel for the sintering process.

This process has been described particularly with reference to treatment of whole pickle liquor. However, it is equally applicable to the disposal of copperas solutions which are comparatively acid-free. The necessity of disposing of copperas has hindered the utilization of many of the processes which have been suggested for the recovery of sulfuric acid from spent pickle liquors. These processes would be more feasible economically if the copperas produced could be considered as an asset if handled by the above described process and not as an industrial liability.

In any event, it is necessary that the concentration of the reacting ingredients be kept sufficiently low and the agitation sufficiently intense that at no time is calcium sulfate precipitated. This is in accordance with well known chemical engineering techniques. In this way all of the sulfate ions originating in the pickle liquor are disposed of harmlessly as soluble sulfates and no sulfur is carried into the sintering plant or blast furnace.

A process developed by *H. Werner; U.S. Patent 3,169,054; February 9, 1965; assigned to Metallgesellschaft AG, Germany* is based on the discovery that it is possible to moisten the dust taken from the precipitator hopper and then introduce the moistened dust into the furnace. A greater portion of the dust is then used in the furnace, because the moisture prevents the otherwise very light, dry dust from being carried away by the hot gases. In accordance with this process, the flue dust contained in the waste gases of metallurgical furnaces, preferably converters for the manufacture of steel, is precipitated in an electric filter, and the precipitated dust, in accordance with its hydrated component parts, is pelletized and the dust in the form of fresh pellets immediately recycled into the furnace.

When the fresh pellets contain more than from about 8 to 12% of water, they then must be dried in order to increase their mechanical hardness or stability, but the extent of drying is limited to a removal of from about 1 to 2% of the moisture content of the fresh pellets. It has been found, however, that sometimes the pellets so dried would burst either during the drying period or while being dumped and stored so that they revert to dust. It has been further found that this lack of pellet durability is due to the calcium content (CaO) of the dust as a hydrated component part. The fresh green pellets as formed contain about 8 to 10% water. Pellets with a calcium content of 15% will burst during drying because of the high heat developed during the slaking reaction of the calcium within a period of two hours.

However, in steel mills, it is impossible to store the pellets for a longer period of time. The slaking reaction of calcium according to the formula $CaO + H_2O = Ca(OH)_2$ depends upon the temperature and quickens the higher the temperature. During drying and hardening of the fly dust pellets having a normal calcium content of, for example, 10% and a moisture content of from 8 to 12%, such high temperatures, by reason of the slaking reaction, occur in the hopper so that the speed of the slaking reaction is increased, which again leads to a further increase in temperature and the destruction of the pellets.

According to this process, the fresh pellets made from dust having a calcium content from about 6 to 20% are dried by gas having a temperature of from about 15° to 30°C under a forced draft and with a slightly above atmospheric pressure of about 200 mm for a period of time until they have lost from about 1 to 2% of their moisture content so that the heat of reaction is at a temperature of from about 40° to 55°C, at which temperature, the pellets are hardened without bursting.

This process has the advantage in that the drying gas such as air and/or waste gases, such as cooled converter gas, are not at such a high temperature as would lead to the damage and destruction of the pellets, but nevertheless the moisture content of the pellets is reduced within prescribed limits. The hardening of the pellets is not spontaneous and can be adjusted to any desired period of time, with the pellets receiving such a degree of hardness or durability so that they can be stored for a longer period of time without damage.

When the flue dust has a calcium content (CaO) between 1 to 6%, additional CaO is added to the dust before or during the pelletizing step to bring the pellets up to a 10% calcium content, so that good pellets are formed which only require a limited drying time. When the pellets have a calcium content of not more than 3%, warm air and/or waste gases at a temperature of from about 50° to 80°C can be drawn through a drying hopper for the limited drying and hardening of the pellets. It has been found that, with this temperature range for the drying gases, there is no spontaneous slaking reaction for the calcium, and accordingly, no pellet destructive high temperatures.

A process developed by *J.K. Stone; U.S. Patent 3,374,085; March 19, 1968; assigned to Kaiser Industries Corporation* is one in which a usable ferrous material, especially for feeding to an oxygen steelmaking furnace, is produced by agglomerating dust recovered from the gases issuing from a basic oxygen steelmaking furnace by admixing therewith from 10% to 70% of particles of not less than 20 microns diameter and comprising a material compatible with iron in steelmaking operations (iron ore, limestone, fluorspar or aluminum oxide) and with a cementitious agent, mixing thoroughly to form a uniform mass and then forming the mass into shapes such as pellets or others. The cementitious agent is advantageously a cokable carbonaceous material or a calcium aluminate cement.

A process developed by *P.G. Barnard, A.G. Starliper and J.B. Stephenson; U.S. Patent 3,676,107; July 11, 1972; assigned to the U.S. Secretary of Interior* is one in which iron-bearing wastes, such as steelmaking dusts, are treated with sulfuric acid to form sulfates of the iron, copper, lead, manganese and zinc in the dust. The sulfated material is roasted at a temperature of about 550° to 700°C to convert the iron sulfate to iron oxide and is then leached with water to remove the soluble copper, zinc and manganese sulfates. The lead is removed from the iron-bearing residue by roasting in a reducing atmosphere whereby the lead sulfate is decomposed, the decomposition products volatilized and the lead collected in the form of lead oxide. The iron-rich residue is then reduced to a metallized product suitable for recycle to a steelmaking plant.

A process developed by *J.F. Cachat; U.S. Patent 3,834,895; September 10, 1974; assigned to Park-Ohio Industries, Inc.* is one in which iron-bearing particulate waste material is introduced into a plasma-arc furnace and heated in the furnace to separate the iron from the particulate material and to melt the iron. The molten iron is collected in a trough in the bottom of the plasma-arc furnace and is discharged from the furnace directly into an ingot mold or steelmaking furnace.

A process developed by *J.E. Allen; U.S. Patent 3,870,507; March 11, 1975; assigned to Ferro-Carb Agglomeration Ltd., Inc.* is one in which solid particulate steel mill wastes, such as blast furnace flue dust, basic oxygen furnace dust, mill scale, coke braise, fluxes and the like, which contain iron, iron oxides, carbon and other constituents, are briquetted with 2 to 15% (preferably 3 to 10%) hydrocarbonaceous binder of the total dry weight of the waste fines, and the raw briquettes are treated with an oxygen-containing gas at 350° to 600°F to selectively react with the hydrogen in the binder, to obtain strong briquettes which can be fed back into iron- or steelmaking process as a source of iron and/or carbon.

As described by P.G. Barnard, A.G. Starliper, W.M. Dressel and M.M. Fine in *Technical Progress Report No. 52,* Washington, D.C., U.S. Bureau of Mines (February 1972), an iron product can be produced from steel furnace dust which would be equivalent to high-grade oxide or reduced iron pellets. However, the economics of recycling the iron depends primarily on the value attached to elimination of a waste disposal problem and in some cases on the value of recovered base metal compounds from the dust. Because relatively small quantities of steelmaking dust may be generated at any particular plant, the economics will require processes which can be used on a mixture of dusts from various sources.

The work reported here is slated to be merely a beginning. A total solution to the problem of recycling iron- and steelmaking dusts is not at hand, and the research is continuing. Nevertheless, for selected particulates, existing technology can be successfully applied to yield acceptable products. This thought is particularly appropriate to basic oxygen furnace dusts, which are usually high in iron (50 to 60%), where lead and zinc are the excessive contaminants, and where the copper content is low enough to be dealt with by dilution. BOF dusts can first be pelletized (with addition of some coarsening agent as mill scale if necessary). Both lead and zinc can then be readily evolved in a heated reducing atmosphere generated from either gaseous or solid reductants. High-iron products (with iron contents in the 80's) with less than 0.1% each of lead and zinc could be expected.

This investigation was concerned, in part, with elimination of copper contamination, principally from electric furnace dust. Although sulfatization (with H_2SO_4 or SO_2) and leaching can remove both Cu and Zn, at this stage the potential rewards do not warrant the effort. The extraction technology is complex, and the end product is enriched in sulfur, a most undesirable constituent. The sulfur is bound to calcium (a usual component of dusts) as CaS or $CaSO_4$. Although the sulfur in this form might not be detrimental in steelmaking, its presence would not enhance marketability.

It appears that electric furnace dusts (at least from integrated or other large plants) should be considered in a package with other metalliferous steelmaking wastes. Materials such as BF, OH, BOF, and EF dusts plus mill scale complement each other by supplying iron and carbon units and diluting the copper. As reported, blends of these materials processed by reduction-volatilization yield low-impurity, highly metallized pellets (85% Fe) which should be useful in any steel mill. Research is continuing to optimize operating conditions and to appraise the economics. Blending plus reduction-volatilization offers good possibilities for development into a technique for coping with a variety of steelmaking wastes, conserving a resource, and helping solve a pollution problem.

Recovery from Waste Phosphatizing Sludges

A process developed by *R.F. Waters, H.E. Powell, L.N. Ballard; U.S. Patent 3,653,875; April 4, 1972; assigned to the U.S. Secretary of the Interior* is one in which waste phosphate sludge containing metals including iron and zinc are mixed with an alkali solution, and the mixture is dried and sintered at a temperature of about 500° to 1300°C in the presence of a reductant, which may be either a reducing gas or a solid reductant. This results in conversion of the phosphate to a water-soluble salt, reduction of zinc to the metallic state and volatilization of the zinc from the reaction zone, and reduction of iron and other metallic elements to the metallic state. Leaching of the sinter cake with water recovers the soluble phosphate and frees the reduced metals for recovery in the form of fine-grained iron or iron alloy powder.

Removal of Copper in Ferrous Scrap Melting

A process developed by *K.H. Ivey and S.J. Chastain; U.S. Patent 3,441,404; April 29, 1969; assigned to the U.S. Secretary of the Interior* relates to the removal of copper in the reclamation of metal scrap. For some years now, the U.S. has imported large quantities of iron ore since foreign ore possesses higher quality than domestic ore and can be delivered at competitive costs. While there are large domestic supplies of scrap iron available in junk cars, it is considered unusable as clean basic cast iron because copper, in the form of small wires and tubing, is not removed by the junk dealer for economic reasons. When this scrap is melted, the ingot contains enough copper to cause the metal to be unusable as a source material for metallurgical industries.

It has now been discovered that copper can be removed from such metals by treating it in a molten state with sulfur and a halide-containing compound selected from the group consisting of alkali earth metal halides, alkali metal halides, alkali earth metal silocohalides, alkali metal silicohalides, $KMgF_3$, $KAlF_4$, K_2TiF_6, K_2ZrF_6 and K_2SnF_6. One theory advanced with regard to the process is that the sulfur forms copper sulfide with copper in the metal, and the halide vaporizes and strips off the sulfide.

In the practice of this process, the halide-containing compound is mixed with sulfur to form the treating agent, and the agent combined with the copper-containing molten metal by (1) placing the agent in the bottom of the furnace or crucible prior to melting the metal; or (2) adding the agent to the molten metal; or (3) pouring the molten metal into a ladle containing the agent. When the agent contacts the molten metal, heavy fumes containing the copper are given off. After the fuming ceases, the molten mass can be cast or poured. An easily removable slag, which contains elements from the agent, forms on top of the resultant ingot. After slag removal, the ingot, substantially free of copper, can be worked in the conventional manner.

To obtain substantial removal of the copper, sulfur is preferably added in amounts ranging from about 2 to 6 parts per part of copper; the halide-containing compound being added in amounts ranging from about 4 to about 8 parts per part of sulfur. If lesser amounts of treating agent are employed, smaller percentages of copper will be removed.

STEEL

Ferrous solid waste, primarily in the form of food and beverage containers and discarded consumer appliances, constitutes 7 to 8% of collected municipal solid waste and totaled roughly 14 million tons in 1970. However, a much more sizable amount of used and discarded ferrous products (an estimated 38 to 54 million tons) is generated annually and appears on our landscape in such visible forms as abandoned automobiles, discarded farm implements, out-of-service rail cars, construction and demolition waste, and other steel products.

In 1967, American industry consumed about 85.4 million tons of iron and steel scrap, and 7.6 million tons were exported. The domestic scrap consumption represented about 65% of the raw steel production. 50 million tons of this domestic scrap consumption was "home" scrap that was generated in the ironmaking and steelmaking process and was fed back into the furnaces. Excluding home scrap and exports, 35 million tons of scrap, or about 20% of the iron and steel consumption, was recycled in 1967.

For the past 25 years, scrap as a percent of total metallic input to steelmaking has remained essentially constant. However, the amount of this scrap that is purchased by the steel industry (originating from outside the steel plant) has been decreasing slightly, while that generated within the steel mills has increased. Purchased scrap as a percent of total scrap input to steelmaking has decreased from 44.9% for the period 1949 to 1953 to 40.0% for 1964 to 1968. In absolute terms, while total steel production increased 35% over the period 1950 to 1969 and total scrap consumption increased 30%, purchased scrap increased only 8%.

There are two basic types of iron and steel scrap: home and purchased scrap. Home scrap, the ferrous waste product generated during iron and steel production, includes ingot croppings, sheet trimmings, and foundry gates and risers. Because it is generated in the steel mill, the scrap is of known composition and purity, and the total amount generated is normally consumed. Home scrap represented 60% of the domestic scrap consumption in 1967.

Purchased scrap is further classified as "prompt" or "obsolete". Prompt industrial scrap is generated by metalworking firms in their fabrication of products. Standard procedures have been developed for the recycling of prompt scrap, and it never really enters the waste stream. At least 90% of the available prompt scrap is estimated to be recycled. The scrap is desirable because of its known composition, condition, and freedom from contaminants. In addition, it is considered a reliable material source because the quantities available are predictable and because recycling channels have been established. Prompt scrap represented about 16% of the domestic scrap consumption in 1967.

Obsolete scrap comes from discarded iron and steel products. Major sources are structural steel from building demolitions, ships, railroad equipment, and abandoned motor vehicles. Ferrous solid waste, of course, occurs in many other forms, such as food and beverage cans and home appliances, which are not generally recovered because of logistics, contamination, or other factors. Obsolete scrap represented about 25% of the domestic scrap consumption in 1967.

Not all of the steel consumed flows immediately into the waste stream and is available as scrap. Considerable portions go into semipermanent use (buildings, machinery, etc.) and enter the waste stream years later. It is estimated that the 21.4 million tons of obsolete scrap purchased or exported in 1967 was 43 to 56% of that available in the solid waste stream. Taking into account scrap located in remote locations and probably not recoverable as well as scrap disposed by individuals, it is estimated that roughly another 24 to 39 million tons of ferrous scrap could feasibly have been recovered in 1967.

The major markets for iron and steel scrap are the domestic steel industry, the domestic foundry industry, and exports. In 1969, the percent of total scrap consumption by each was 73.8, 17.5, and 8.7, respectively. However, in terms of purchased scrap, both prompt and obsolete, foundries and exports weigh more heavily. For the steel industry, about 35% of scrap consumed is purchased, whereas foundries purchase about 60% of their scrap consumption, and exports are, of course, purchased scrap.

The American steel industry is composed of approximately 110 companies; of these 21 are fully integrated (coke ovens, blast furnaces, and steelmaking furnaces); 9 operate mostly blast and 80 operate only steelmaking furnaces, with electric steelmaking predominating. These 80 companies currently produce less than 10% of the nation's steel output, but they are a significant outlet for ferrous solid waste.

The type of furnace used in steelmaking has a direct bearing on scrap usage. Three types of furnaces are used: open hearth, which uses an approximately 45% scrap charge; basic oxygen, 30% scrap charge; and electric, 100% scrap. (These charges are based on standard operating conditions, which take into account both technological and economic factors.) Basic trends have included (1) the decline of the open hearth from 87% of steel production in 1960 to 50% in 1968; (2) rapid rise of basic oxygen furnaces from 3.3% of production in 1960 to 37.1% in 1968; (3) moderate growth of electric furnace steel production, 8.4% in 1960 to 12.7% in 1968. To date, declines in scrap requirements from decreased open hearth steelmaking have been balanced by increased scrap needs from rising electric furnace production.

In the foundry industry, scrap already accounts for about 85% of the metallic input, and product specifications dictate that pig iron be a portion of the charge in some cases. The cupola furnace, which uses an 84% scrap charge, dominates; it comprises about 90% of the furnaces. Electric furnaces, which make up most of the remainder and which use scrap

100%, have been making inroads, however. The potential for increased scrap consumption by foundries is limited, but factors such as the increasing trend toward replacing cupola facilities with electric furnaces, geographic dispersion of foundries to put them closer to scrap sources, and a growth rate in excess of the domestic steel production indicates that use of scrap by foundries should at least hold its own and may even increase slightly. However, foundries do not have potential as major markets for increased scrap consumption.

Exports are a significant market for iron and steel scrap, constituting 24% of total purchased steel in 1970. Exports are particularly important for the movement of obsolete scrap, because a large portion of the exports are from obsolete sources. Japan is the largest consumer of export scrap, taking 48.8% of the market in 1970. Copper precipitation is the major market for steel can scrap at present, but it is quite limited. Only about 300,000 to 400,000 tons of old steel cans and canmaking wastes–a small percentage of the estimated five million tons of cans produced each year–are consumed annually by this market.

The replacement of open hearth furnaces by basic oxygen furnaces has tended to reduce scrap requirements. However, the increased usage of electric furnaces has kept total scrap consumption roughly constant overall. Future scrap consumption is tied closely to continued increases in electric furnace melting. Investment decisions depend on the comparative returns on investment from various types of furnaces. The return on investment from an electric furnace that uses 100% scrap obviously is strongly influenced by the price of scrap.

The technical feasibility of using increasing proportions of scrap in other steelmaking furnaces has been demonstrated. The basic oxygen furnace charge, for example, can be increased by preheating the scrap, but because this entails additional costs, it can only be justified if scrap cost decreases relative to ore cost.

The costs to an integrated steel producer of using scrap versus ore are difficult to determine. The steel industry does not maintain, or at least does not report, such figures. Estimates have been made, however, which indicate that the cost of using high-grade scrap is higher than the cost of using ore.

The point of equivalency of scrap and ore in the production process is the point where either hot molten pig iron or melted scrap is used to charge a basic oxygen furnace. The total cost of scrap at this point was estimated to $44.00 per ton, including $33.50 purchase price of the scrap, $6.00 melting cost, $3.50 for scrap handling, and $1.00 for increased refractory wear caused by scrap usage. The cost of molten pig iron (which competes directly with scrap) was estimated at $37.50 per ton, including $28.50 for the ore and associated raw materials and $9.00 for melting cost.

Thus, the cost of scrap ready for charging a basic oxygen furnace is about $6.50 per ton greater than that of the hot metal derived from ore at the same point. Thus, without a reduction in scrap cost of at least $6.00 to $7.00 per ton, it is unlikely that the utilization of scrap by existing steel mills in basic oxygen furnace steel production will substantially increase. Nonintegrated steel mills using electric furnaces (which operate on a virtually 100% scrap charge), of course, find the use of scrap economical. These scattered mills are usually located near metropolitan areas, and transport cost of scrap is not a major expense.

The above cases illustrate the fundamental economic barriers to the increased utilization of secondary materials. The economics of recovery today result in the recovery of all waste materials that are of high quality and that can be obtained from reasonably concentrated sources. Extraction of materials from solid waste is limited both by the relatively low quality of such wastes, due to contamination and admixture with foreign materials, and also limited by the relatively greater effort required to acquire such materials.

Rigid steel production specifications require that scrap be processed to remove contaminants and impurities. Home and prompt scrap are from known sources and are generally higher in quality than obsolete scrap, with the exception of certain obsolete scrap such as rail, ship, and structural scrap. Cans present a special problem because of their contamination with tramp elements—aluminum from tops, lead from the seams, and tin. For example, lead can be harmful to furnace refractories, and too much tin causes undesirable properties in finished steel. Thus, except in periods of peak demand or hot metal shortages, the availability and low cost of higher quality raw materials tend to reduce the steelmakers incentive to use the lower quality portion of obsolete scrap.

There are substantial differences in environmental impact when recycled steel is utilized, as opposed to steel produced from iron ore. A comparison of the impacts of producing 1,000 tons of steel reinforcing bars from virgin ore and from scrap indicates that 74% less energy and 40% less water are used in the recycling case. In addition, air pollution effluents are reduced by 86% and mining wastes, by 97% (Table 10).

TABLE 10: ENVIRONMENTAL IMPACT COMPARISON FOR 1,000 TONS OF STEEL PRODUCT*

Environmental effect	Virgin material use	100% waste use	Change from recycling (%)+
Virgin material use	2,278 tons	250 tons	−90
Water use	16.6 million gal	9.9 million gal	−40
Energy consumption	$23{,}347 \times 10^6$ Btu	$6{,}089 \times 10^6$ Btu	−74
Air pollution Effluents	121 tons	17 tons	−86
Water pollution	67.5 tons	16.5 tons	−76
Consumer wastes generated	967 tons	−60 tons	−105
Mining wastes	2,828 tons	63 tons	−97

*Midwest Research Institute. Economic studies in support of policy formation on resource recovery. Unpublished report to the Council on Environmental Quality, 1972.

+Negative number represents a decrease in that category resulting from recycling.

Source: EPA Report SW-118

Recovery from Cans

Ferrous materials constitute roughly 7% of municipal solid waste (excluding automobiles) and approximately 60 to 80% of this fraction is steel cans. It is estimated that in 1972 approximately 5 million tons of cans entered the solid waste stream. About 70%, or 3.5 million tons, were generated in Standard Metropolitan Statistical Area, where recovery should be possible. The current recovery of cans from municipal solid waste is small. In 1972 approximately 70,000 tons of cans were recycled; and of this, 50,000 tons were consumed in one market, copper precipitation, which is exclusive to the western United States.

Of major importance to can recycling is the fact that the steel can is in reality a composite can consisting of tin-plated steel (thus the term tin can) and possibly lead, organic coatings, and aluminum. More specifically, of the cans in solid waste, about 63% are tin-plated food and nonfood containers that have lead-soldered side seams. The other 37% of the cans are beer and beverage cans, three-quarters of which are also tin-plated and have soldered seams. (Roughly 22% of steel beverage containers are tin-free steel.) Most of the beer and beverage cans have aluminum tops to allow use of the easy-to-open pull-tab. Taken as a composite, this can fraction contains approximately 92% steel, 0.4% tin, 1.5% lead, 3.7% aluminum, and 1.8% organic coatings.

Nonferrous residuals as high as these present serious metallurgical problems for the steel industry and also for certain other markets. This is the major reason can scrap is often considered to be bad scrap. However, for some markets the residuals may be as valuable as the steel. In particular, the small quantity of tin in the 3.5 million tons of cans available in SMSA's would be worth almost $60 million, roughly one-half the total value of the steel in the cans.

There are three major potential markets for old cans: the steel industry, the detinning industry, and the copper precipitation industry. (The detinning industry is really an intermediate processor, extracting tin from the cans and selling the detinned scrap to the steel industry.) Several steel mills have made promising and energetic efforts toward can recycling over the past 2 years. However, in 1972 the steel industry only consumed an estimated 11,000 tons of old cans. This is a small amount relative to that industry's raw material inputs. Old cans have not been considered a desirable raw material by the steel industry in the past because of tin, aluminum, and lead residuals. These contaminants can cause either quality loss in the steel product and/or refractory damage to the melting furnace.

For steelmaking, tin is probably the most serious contaminant. Although specific estimates differ, it is generally accepted that extremely small quantities of tin on the order of 0.01%, 0.03%, and 0.05% should be the maximum quantity of tin allowable in medium grade steel, plate steel, and reinforcing bars, respectively. Therefore, the maximum consumption of old cans by the steel industry on a national basis would be between 1.5 and 7.2 million tons per year because the furnace charge could contain only a small quantity of old cans. However, suitable supply arrangements and logistics would be required to achieve even this level of use.

Widespread acceptance and consumption of old can scrap by the steel industry is possible but severely limited by a host of problems. Consumption is expected to grow, but at a very slow pace, over the next 10 years. This situation is completely changed if the cans have first been processed through a detinning plant to remove (and recover) the tin. Detinned can scrap is readily marketed to the steel industry at or near the price of No. 1 heavy melting scrap.

Detinners chemically process tin plate to remove and recover tin. Detinned ferrous scrap is readily marketed to the steel industry. The tin, valued at about $4,000 per ton, contributes about one-third of the detinners' revenue, and the steel scrap sold accounts for the other two-thirds. The detinning industry is not large. Two companies represent over 90% of the industry's sales. The industry presently processes a negligible quantity of post-consumer cans, obtaining most of its raw materials from can manufacturing scrap. In 1971, 752,000 gross tons of tin plate were processed, essentially the same as in 1950.

The reason that more old can scrap has not been processed by detinners is largely because of contaminants. Entrapped organics, labels, paper, and plastics are all troublesome, as is lead. Aluminum (from bimetal cans) is by far the most serious contaminant and is the major deterrent to use of old cans by detinners. Aluminum can be removed by an additional processing step in the detinning operation, but it has been estimated that this would increase processing costs roughly $10 to $15 per ton. Thus, coping with the aluminum in cans is ultimately more of an economic than a technological problem. Despite this added cost, it appears that detinning plants might still be able to expand capacity to handle post-consumer cans if supplies could be made available at $10 to $20 per ton, depending upon the market price for No. 1 bundles. To date, supplies of postconsumer cans have been difficult to obtain from municipalities, which do not generally attempt to recover and market these items. It is also significant that altering the design of cans to eliminate the aluminum top would essentially alleviate the problem of detinning postconsumer cans.

Copper precipitation accounted for 50,000 of the 70,000 tons of recycled cans in 1972. Old cans represented approximately 10% of the 500,000 tons of scrap consumed by the industry in 1972. This scrap was consumed in facilities in Montana, Nevada, Arizona, Utah, and New Mexico. These markets are obviously remote from major centers of waste generation in the midwestern and eastern United States. The market for "precipitation iron," as it is called by the industry, is expected to nearly double over the next 15 years; but the use of old cans in this context will be largely dependent upon getting supplies to these markets.

Although it is a desirable outlet for old cans, copper precipitation is a very different form of can recycling. The steel is not actually recovered but is used in a chemical displacement reaction to precipitate copper. The potential for increased old can recycling in the copper industry is considered good, but moving very slowly. Based on a recent study conducted for EPA, a subsidy of $6 per ton of postconsumer cans to the steel, detinning, and copper precipitation industries would result in a 9 million ton increase in the recycling of cans from 1976 to 1985. (Extending the subsidy to the steel and copper precipitation industries would be virtually ineffective compared to its application to detinning and would have no predictable cost effective benefits.) The cost of this subsidy to the federal government would average about $6.5 million per year. Windfall, based on present levels of postconsumer can recycling, would be about 30%.

Ferrous materials are relatively easy to extract from waste because they can be separated magnetically. However, there is presently very little postconsumer can scrap being extracted from solid waste. In 1972, the American Iron and Steel Institute published a list of 17 cities magnetically separating ferrous material. However, many of these installations have not been successful, and many others are operating far below capacity, largely because of design problems. Shredding is required prior to ferrous material extraction, but shredders were often not properly designed for maximum efficiency, or the ferrous material was shredded in such a configuration that it was not acceptable for the local market. In many cases the ferrous product was dirty or contaminated, and thus not marketable.

It appears that shredding of waste for ferrous material extraction alone is not economical. At $20 per ton for cans and $12 for other miscellaneous ferrous materials, revenue from ferrous material extraction would total less than $1.25 per ton of refuse processed, hardly enough to cover shredding costs in almost any size facility. However, refuse shredding is often justified in itself by virtue of densification for improved landfill efficiency or increased freight payloads where transfer stations are involved. In these instances, the incremental costs of ferrous material extraction should be easily covered by the revenues.

An excellent opportunity for ferrous material extraction is to be found in energy recovery or other types of comprehensive recovery facilities that are emerging. In all instances, shredding of waste is required; and in most of the energy recovery facilities, organics must be separated from inorganics to maximize burning efficiency. Ferrous material extraction is easily justified in these instances.

The fact that shredding is becoming more common as an element of solid waste disposal or recovery will lead to increasing potential opportunities for ferrous material extraction. If the planned systems are actually put into operation, there will be a marked increase in can supplies, which could exceed demand. This would occur if the cans were not properly shredded and prepared to meet the needs of particular markets, if markets for cans were not stimulated to some degree, and if buyers and sellers were not brought together. If demand did not rise to meet supply, the long-run impact would be a phasing-out of some can extraction operations.

The barriers to steel can recycling include both demand and supply constraints. The technological barriers to use of undetinned cans by the steel industry are obvious and will undoubtedly work to limit recycling through this channel even if price or cost relationships are altered or more abundant supplies are made available. Higher consumption in copper precipitation is promising, but limited by the location of the markets and the difficulty of transporting supplies to those market areas.

Detinning offers potential for steel can recycling. It upgrades the can scrap into a form of high-grade steel scrap and recovers a valuable resource, tin, which would otherwise be considered a contaminant. Also, the economies of scale of detinning plants are such that new small-scale plants could be built near cities or resource recovery plants where cans are generated. Aluminum contamination is the major significant economic barrier to detinning of old cans. However, the difficulty of obtaining reliable supplies is also an important consideration. Major expansion of the detinning industry will probably require that the economic burden of dealing with aluminum (or the aluminum top itself) be reduced and that new supplies of properly prepared scrap be made available.

Recommendations to stimulate steel can recovery, to the extent appropriate, must obviously respond to the barriers just discussed. One obvious need is to ensure adequate supplies of properly extracted and processed cans where markets exist. More and better information is needed by municipalities on the configuration and economics of optimum techniques for recovering cans from the waste stream, the necessary form and quality of the scrap to ensure its suitability for available markets, and appropriate means of establishing contacts with these markets. To a large degree, these needs could be met through proper communication between industry and municipalities. Bringing the supply and demand sectors together is believed to be an appropriate application of a federal technical assistance and information dissemination program, and EPA will continue efforts in this direction in the future.

The type of action needed to stimulate demand is somewhat less clear. Some increased recycling through the steel, detinning, and copper precipitation industries could be expected to result from availability of a broader supply base even with no demand stimulation. However, the real key to long-term increases in can recycling is expansion of the detinning industry. It is presently questionable whether this industry would significantly expand capacity in the face of the additional \$10 to \$15 processing cost required to remove aluminum contaminants. Possible federal actions for encouraging detinning expansion range from providing direct subsidies to offset the cost of removing the aluminum contaminant, to establishing product controls, to eliminating aluminum tops on beverage cans.

The ultimate decision of whether the federal government should consider such measures hinges primarily on two issues, both of which are difficult to resolve at present. One is the degree to which can recycling might increase with supply creation through a technical assistance effort. If a significant increase in recycling could be achieved by such a technical assistance effort, the added value of a subsidy would be small. The other major issue is the question of beverage container legislation. State programs such as Oregon's beverage container deposit law could have the ultimate effect of sharply reducing the quantity of cans and other nonreturnable beverage containers in waste. This would eliminate the aluminum content of steel cans (because food cans do not contain aluminum) and make detinning of old cans more attractive.

Recovery from Grinding Swarf

A process developed by *J.D. Dankoff and D.K. Snyder; U.S. Patent 3,865,629; February 11, 1975* is a process for the recovery of components of grinding swarf comprising screening the swarf when necessary, mixing and washing it with a hydrophilic solvent and the removal of the bulk of the resulting solution therefrom, followed by the mixing and washing one or more times with a second solvent for grease and oil, with the separation of the bulk of the resulting solution and the removal of the remaining trace of such solution by volatization at elevated temperature. Depending on the component to be recovered, magnetic separation of the steel from abrasive substances may be affected at some stage, and abrasive grains of different composition may be separated, and lubricant can be recovered from the second solvent as this solvent is prepared for reuse. Figure 15 is a block flow diagram of the process.

FIGURE 15: PROCESS FOR RECOVERY OF STEEL FROM GRINDING SWARF

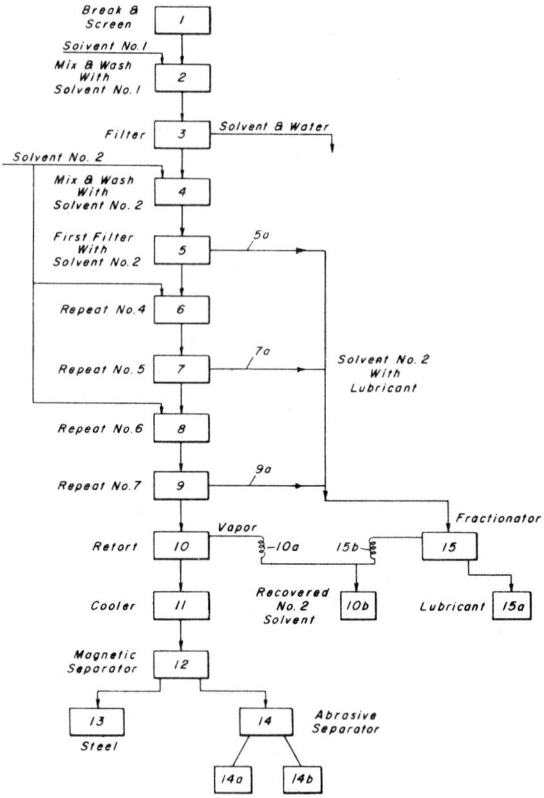

Source: J.D. Dankoff and D.K. Snyder; U.S. Patent 3,865,629; February 11, 1975

In the drawing, **1** designates a lump breaking and screening apparatus where agglomerated masses of swarf are broken down to a size where the material may be effectively washed and foreign bodies, such as stray pieces of iron or steel are removed. The screened material is discharged from screening unit **1** into the first mixer and washer **2** where it is mixed and agitated with first solvent, designated in the drawing as the No. 1 solvent. This solvent is a hydrophilic solvent, methanol being a desirable first solvent, but other solvents for water, but not for grease or oil, may be used, ethanol and isopropyl alcohols being other examples.

After being thoroughly washed with its first solvent, the mixture is transferred to a filter or separator **3**. The filter may be of a well-known type where air pressure or vacuum is used. The bulk of the solvent-water solution is removed from the swarf in this filter and the swarf as indicated at **4**, is then mixed for the first time with the grease and oil solvent, that is, the second solvent. This second solvent is kerosene or some other hydrocarbon solvent, preferably one which is not too explosive or dangerous, or perhaps a chlorinated hydrocarbon, such as trichlorethylene, trichloroethane and carbon tetrachloride. Kerosene being relatively inexpensive and quite satisfactory, is referred to the second solvent as kerosene, but without exclusion of other mineral oil solvents, unless otherwise indicated.

After this first mixing and washing with the second solvent, the mixture is processed to remove the bulk of the kerosene-lubricant solution, as for example, by filtering under pressure or vacuum as indicated at **5**. The filtrate is removed at **5a**. As previously indicated, two or more washings with kerosene are desirable, and **6** designates another mixing and washing stage. The second mixture is also processed at **7** to remove the bulk of the solvent with additional lubricant, again filtering under pressure or vacuum, and then the filtrate removed at **7a**. Generally a third mixing with the second solvent is desirable. This is indicated at **8** and the bulk of the solvent is again extracted, desirably by pressurized or vacuum filtering, indicated at **9**. The filtrate is removed at **9a**.

Since the final step of filtering cannot completely remove every trace of solvent and lubricant, the residue from the third filtering step, which may perhaps be described as kerosene moist solid residue, is sealed inside an air-tight retort and heated from an external heat source until this last trace of kerosene-lubricant mixture is substantially entirely, if not completely eliminated. This occurs when the mass reaches a range between 600° and 1000°F. In the flow diagram, **10** designates this step. Vapors from the retort may be conducted to a condenser **10a** and the condensate collected in a receiving vessel **10b**.

The dried and degreased solids are cooled at **11** and then at **12** processed through a magnetic separator with the steel particles being collected in receiver **13** and the abrasive particles into receiver and/or separator **14**. Normally these abrasive particles will be a mixture of silicon carbide and aluminum oxide and for reuse, they may be electrostatically separated by known methods, the separated particles being discharged at **14a** and **14b** respectively. The filtrate removed at **5a**, **7a** and **9a** may be carried through a common duct to a fractionating unit or other equipment known in the petroleum refining industry as indicated at **15** to separate the kerosene and the lubricant. The kerosene from which the lubricant has been stripped is discharged through **15b** to the solvent receiver **10b** and the lubricant collected in a receiver **15a**.

As indicated above, the order of the steps, especially insofar as magnetic separation is involved may be varied, particularly if the recovery of steel particles only should be important. For example, the magnetic separation may occur following the removal of water from the swarf, such that the subsequent steps then apply only to the steel particles. In many instances, the grinding swarf results from the grinding of specialty steels or steel alloys, tool steel, for example, being one of these, and the recovered steel particles may be highly useful in powdered metallurgy or some other sintering or forming operation and often it is briquetted for further use. In a typical grinding swarf resulting from high speed tool steel, the content averaged about 75% steel particles, 20% to 23% abrasive particles and the remainder was the lubricant-coolant mixture.

Recovery from Scrap

A process developed by *S.E. Khalafalla and S.L. Payne; U.S. Patent 3,620,715; November 16, 1971; assigned to the U.S. Secretary of the Interior* is one in which the surface of scrap steel from discarded auto bodies, etc. is impregnated with minute amounts of an oxide of an oxidation acceleration agent containing the agent in its highest valence state by immersion in a solution of the agent or by spraying. After such treatment the steel will be rapidly oxidized during a reduction-roasting operation.

The pretreatment of steel scrap to speed oxidation has also been described by S.E. Khalafalla, S.L. Payne and G.W. Reimers in *Report of Investigations No. 7332,* Washington, D.C., U.S. Bureau of Mines (January 1970).

Recovery from Spent Pesticide Containers

A process developed by *J.H. McKean; U.S. Patent 3,701,355; October 31, 1972; assigned to Baychem Corporation* involves recycling, as high grade steel scrap, spent metal pesticide containers by shredding the containers into reduced nugget-sized pieces followed by incinerating at a temperature sufficient to burn off all organic residual matter in the can using as combustion air for incineration the pesticide contaminated exhaust air generated by the shredder. A further step includes the removal by afterburning and scrubbing of any residual pesticide entrained in the incinerator off-gases. A suitable form of apparatus for use in the conduct of this process is shown in Figure 16.

FIGURE 16: APPARATUS FOR RECOVERY OF STEEL FROM PESTICIDE CONTAINERS

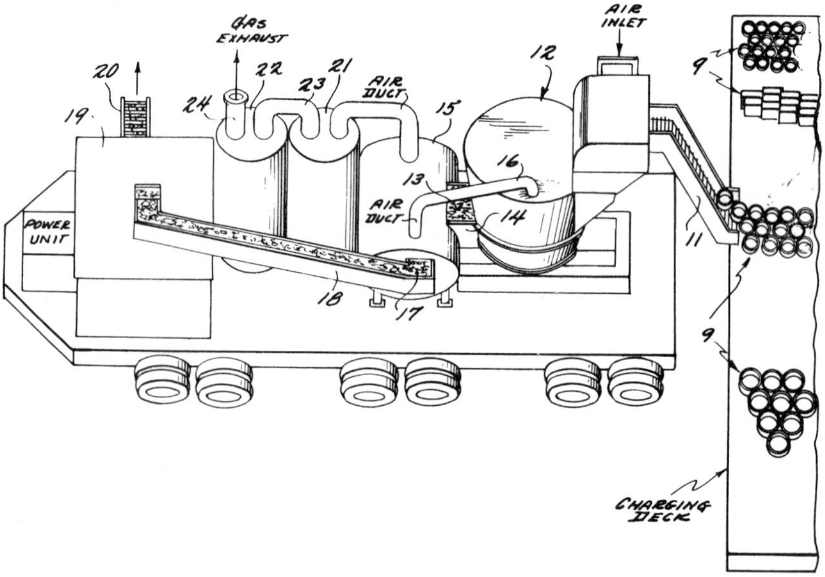

Source: J.H. McKean; U.S. Patent 3,701,355; October 31, 1972

The containers **9** to be reclaimed are placed on the charging deck or platform adjacent to a conveyor **11** for transporting the containers to a shredder **12**. A preferred shredder is of the cutting type having rotary breaker bars which tear pieces of the container and drop them by gravity into a series of grinding wheels contained in a barrel-shaped shell. As the tear pieces feed by gravity to successive lower rotating grinding wheels the distance between progressively lower wheels and the shell becomes progressively less thus causing the sheared material to become smaller in size finally attaining the size of an inch or so in diameter referred to herein as nugget-sized pieces.

To assist in this cutting and shearing action, bars are fixed inside the shell projecting into the path of the descending tear pieces. Commercial shredders are available on the market operating on this principle. However, other types of size-reduction equipment such as shears, cutters and clippers may also be employed. As the shredded pieces or nuggets **13** of the container are discharged from the size-reduction equipment they are moved by conveyor **14** to an incinerator **15** which performs several functions. Primarily the incinerator operating in the range of 1000°F to 1300°F is to volatilize and combust the residual contents of the container present on the nuggets. Since the shredder generates a rather large volume of air discharge during the shredder operation and this discharge is contaminated with the residual chemicals in the container it is advisable to conduct by duct **16** this discharge from the shredder as air supply for the incinerator together with fuel for the combustion mixture. In this way no contaminant chemicals are discharged into the atmosphere from the shredder but such chemicals as are caught up in the air discharge will be combusted in the incinerator. The incinerator also combusts the labels on the containers as well as any organic liners used to coat the inside of the containers. The resulting shredded-incinerated nuggets **17** emerge from the incinerator free from organic matter and are of high quality steel feed stock for recycle to the steel mills.

The nuggets **17** are conveyed by conveyor **18** to the storage bin **19** and therefrom by conveyor **20** to a truck, train or other suitable means for transport to steel mills. In order to prevent escape of unburned residue passing from the incinerator, an afterburner **21** is provided. This burner operating at about 1000°C serves as a back-up unit to ensure as complete combustion as possible to any residual organic matter which might pass through the incinerator incompletely combusted.

As a further precaution there is provided a gas scrubber **22** which may be of the conventional impingement baffle, venturi-impingement or venturi, cyclonic or packed tower type. The scrubber is connected by duct **23** to the discharge of the afterburner thus providing means for scrubbing all air passing through the shredder, incinerator and afterburner. An air discharge duct **24** is provided on scrubber **22** to exhaust the combustion gases.

There is depicted a mobile unit capable of being transported from one site to another such as a container area provided where the spent containers can be collected in a geographical location. It is to be understood that such equipment may also be permanently located. Such mobile unit may be the flat bed of a truck trailer or train car. The mobile unit can be equipped with a power generating source to operate the required motors for conveyors, shredder, blowers and other electrical accessories or the power may be provided at each container collection or processing point.

Recovery of Steel Drums and Barrels

A process developed by *H.W. Stratford; U.S. Patent 3,163,134; December 29, 1964* involves the reclaiming of metal barrels or drums, and more particularly safely charring or burning out combustible vapor-producing residue remaining on the interior surfaces thereof. In reclaiming barrels or drums having removable lids, it is usually a safe and economical procedure to direct a flame thereinto through the open end or otherwise heat the barrel for charring residue remaining on the inside walls prior to cleaning.

However, such burning out of barrels or drums provided with only a small opening or bung hole and containing combustible vapor-producing residue as from paint, lubricating oils, etc. is likely to be extremely dangerous. This is due to the explosive conditions produced by the mixing and confining of such vapors with oxygen which has previously entered into the substantially empty barrel from the atmosphere. This situation has rendered the restoration of such barrels so hazardous that it has often been described as an economically unjustifiable operation.

The principal objects of this process are: to provide a safe and economical method for restoring or reclaiming barrels containing combustible vapor-producing residues, which method includes burning out; to provide such a method which includes prior treatment whereby the oxygen normally contained within the barrel is displaced by a noncombustible gas prior to the burning or heating operation; to provide such a method which in a preferred form utilizes only a relatively small inexpensive piece of solidified carbon dioxide or Dry Ice for displacing oxygen in the barrel; to provide such a method wherein considerable time may elapse between the oxygen displacement step and the burning step without materially increasing the danger of explosion; and to provide such a method which requires no specialized apparatus, using only existing exterior barrel burning equipment.

Use as a Reductant for Copper Oxides

As described by S.E. Khallafalla, H.W. Kilau and L.A. Haas in *Report of Investigations No. 7301,* Washington, D.C., U.S. Bureau of Mines (October 1969), the most favorable conditions for reacting iron with Cu_2O are temperatures in excess of 560°C and a static helium or nitrogen atmosphere containing approximately 0.6% water vapor. Increasing the molar ratio of copper to iron from 1 to 3 resulted in a decrease in copper metallization. The reactivity of Cu_2O with tin cans and heavier steel scrap was variable and less than that with powdered iron. This ferrothermic process seems to involve high-temperature, solid-state cementation of copper on an iron base.

The practical application of this process on a large scale would involve continuous feeding of iron scrap and oxidic copper ore concentrate to a rotary kiln heated to the optimum reaction temperature and containing an inert atmosphere with small amounts of H_2O. Sulfidic copper concentrates could be reduced with metallic iron following their complete oxidative roasting, because the oxide would then be at a sufficiently high temperature for the reaction to proceed.

FOOD AND BEVERAGE INDUSTRY WASTES

BEVERAGE CONTAINERS

A great deal of public attention has been focused on the impact of beer and soft drink containers on the environment. This concern has centered on the aesthetic problems associated with beverage container litter, solid waste generation factors, and the overall environmental impacts associated with beverage container production and use. While these latter aspects apply to all packaging and, in fact, to all consumer goods, the litter issue renders beverage containers somewhat unique. For this reason, beverage containers have become a highly sensitive public issue, as described by the Office of Solid Waste Management Programs in their *Second Report to Congress: Resource Recovery and Source Reduction,* Washington, D.C., U.S. Environmental Protection Agency (1974).

Consumption of beer and soft drink containers has grown and continues to grow faster than population growth and the consumption of beverages themselves. Table 11 illustrates that per capita beverage container consumption rose from 87 units in 1959 to 230 units in 1969, an increase of 164%. In the same period, the per capita consumption of beer and soft drinks rose 29%. In the 1959 to 1969 period, the use of refillable bottles decreased as the average number of fillings per container declined from 3.7 to 1.8. In large part as a consequence of this decline, the total number of beverage containers consumed rose from 15.4 billion units in 1959 to 46.8 billion units in 1969.

Beer and soft drink containers form a large and highly visible segment of roadside litter. It has been estimated, for example, that at least 2.2 billion beverage containers became litter in 1969, from 20 to 32% of all roadside litter by item count. Of these containers, it is estimated that 71.3% were beer containers, 25.7% were soft drink containers, and approximately 3.0% were wine and liquor bottles. By type of container, it is estimated that 73.1% were cans, 17.0% were nonrefillable bottles, and 9.9% were refillable bottles.

A survey by the Oregon State Highway Division indicates that by volume, beverage cans and bottles form approximately 62% of the litter along the state's highways, bottles accounting for 22% and cans, 40%. Although this is but one survey, it indicates that beverage containers might represent a greater visual blight than item counts would seem to indicate. This is further borne out by a somewhat limited four-city survey that revealed that the public thinks beverage containers constitute nearly 40% of all litter. In 1972, approximately 8.8 million tons of beer and soft drink containers (6.2 million tons of glass, 2 million tons of steel, and 0.6 million tons of aluminum) were consumed in the United States. This represents approximately 20% of all packaging waste and 7% of total municipal solid waste.

TABLE 11: BEVERAGE CONTAINER CONSUMPTION*

Item	1959	1969	Growth rate (percent)
Containers consumed (billions)	15.4	46.8	204
Beverage fillings consumed (billions)	58.4	85.8	47
Average number of fillings per container	3.7	1.8	–
Container consumption per capita	87	230	164
Beverage fillings consumed per capita	325	420	29

*Bingham, T. H., and P. F. Mulligan [Research Triangle Institute]. The beverage container problem; analysis and recommendations. Washington, U.S. Government Printing Office, 1972.

Source: EPA Report SW-122.

With respect to the environmental impacts resulting from beverage container production, data are presented in Table 12 on the impacts associated with the production of five container types. These data reveal that, for the delivery of equivalent volumes of beverage, a refillable bottle (assuming usage of each bottle 10 times), as compared to any other type of beverage container considered, provides a reduction in energy consumption from 41 to 74%, a 34 to 87% reduction in waterborne waste, and a reduction in air effluents from 30 to 71%.

There are three major types of strategies that have been proposed for reversing the trend toward nonrefillable containers and curbing the beverage container portions of litter and solid waste: (1) mandatory deposit systems for all beverage containers, (2) bans on the production and sale of nonrefillable containers and (3) low taxes on beverage containers to be used for increased litter cleanup. Each of these strategies is to generate revenues analyzed in this subsection, and Table 13 summarizes the results of the analysis.

TABLE 12: COMPARISON OF FIVE DIFFERENT CONTAINERS* FOR DELIVERING 1,000 GALLONS OF BEVERAGE†

Environmental impact	10-trip returnable glass‡	All steel§	Bimetallic§	One-way glass‡	Aluminum§
Energy (10^6 Btu)	24	41	57	72	91
Virgin raw materials (lb)	1,538	2,029	1,677	7,515	578
Water volume (10^3 gal)	11	38	34	37	16
Waterborne waste (lb)	45	349	335	68	249
Atmospheric emissions (lb)	111	157	234	328	381
Post-consumer solid waste (ft^3)	12	4	3	41	3
Industrial solid waste (lb)	8	71	61	32	29

*All containers are 12-ounce beer.
†Source: Preliminary data prepared by the Midwest Research Institute for U.S. Environmental Protection Agency Contract No. 68-01-1848.
‡Capped with steel closures; solid bleached sulfate paperboard carriers are included.
§Plastic ring-type carriers are included.

Source: EPA Report SW-122

TABLE 13: PROJECTED BENEFITS AND COSTS OF CONTROL MEASURES

Impact	Mandatory deposit	Ban	Tax
Benefits:			
Litter reduction	Substantial improvement	Substantial improvement	Substantial improvement
Solid waste reduction	Some improvement	Some improvement	No change
Environmental impact reduction	Substantial improvement	Substantial improvement	No change
Costs:			
Sales volume change	Possible slight decrease	Likely decrease	No decrease
Industry dislocation	Dislocation, but less than with a ban	Substantial dislocation	No dislocation
Employment impact	Dislocation, but no net loss	Dislocation, and likely net loss	No dislocation
Tax revenue loss	Likely decrease	Likely decrease	No decrease
Beverage price	Lower on average	Lower on average	Higher on average
Consumer choice impact	Some limitation	Choice limited	No change
Consumer convenience impact	Higher price for convenience	Higher price for convenience	No change

Source: EPA Report SW-122

Mandatory Deposit: The mandatory deposit alternative selected for analysis would require the retailer to pay $0.05 for every empty container of beer and carbonated soft drinks. The retailer would be required to accept from the consumer any empty container of the kind, size, and brand sold by that retail outlet. Retailers, in turn, could return empty containers to the distributor who would also be required to pay the 0.05 refund.

Mandatory deposit legislation has been passed at both the local and state level. (Oregon and Vermont; Oberlin, Ohio; Bowie, Maryland; and Ann Arbor, Michigan, have enacted mandatory deposit legislation.) In these cases, the costs and benefits of the approach would have to be analyzed on an individual basis because of varying degrees of littering, industry intensity, and consumption. The projected costs and benefits of implementation of this strategy at the national level follow.

It is estimated that implementation of a $0.05 mandatory deposit is likely to result in a reduction in the beverage container portion of the litter. This would result partly from decreased littering and partly from increased scavenging. One estimate of the quantities of beverage container litter to be reduced is 60%. Preliminary data from the state of Oregon indicate a reduction of beverage container litter of from 70 to 75%.

Beverage container consumption in 1972 equaled 8.8 million tons. If a $0.05 mandatory deposit system resulted in the use of only refillable beverage containers (each container to be used 15 times), solid waste reductions of approximately 6 million tons would occur.

Any measure that resulted in increased usage of the refillable glass container would also result in significantly decreased environmental impact. A mandatory deposit system resulting in a predominately refillable bottle market (85 to 90% refillables), as compared to the current market mix of containers, would result in (1) a material use reduction, (2) energy savings, (3) an air pollution reduction, (4) a water pollution reduction, and (5) reduced mine waste production.

Sales of beverages may decline slightly under a mandatory deposit system. A sales decline could result from a decreased number of sales outlets (e.g., vending machines may decrease in number) or from a switch to other beverages without the required deposit (e.g., juice or wine). This decline has been estimated at from 4 to 8%. Preliminary experience in Oregon, as will be discussed later in this section, reveals no decline in sales. Current growth rate in the industry is estimated at 6% a year. Thus, the effect could range from no change in growth to 1 year of no growth and subsequent years at present growth.

A $0.05 mandatory deposit that resulted in a switch to refillable bottles would eliminate a substantial portion of metal beverage cans and would have significant impacts on that industry (e.g., a reduction of 75% in the use of beverage cans would be equivalent to a decline of $1.1 billion worth of shipments in 1971). Approximately 2% of steel production is currently related to can manufacture, and this steel use could be totally eliminated. In the aluminum sector, beverage containers represent 11% of aluminum shipments, which would be significantly affected. Major disruptions could also occur within the brewing industry, particularly for the national shipping brewers, if a switch was made from current beer distribution methods.

A mandatory deposit could result in large reductions in employment in the container industries (estimated at 60,500 jobs, mostly in skilled categories) and large additions to employment in the beverage and distribution industries (estimated at 60,800, mostly unskilled categories). The net effect could be a small increase in jobs and probable drop in labor income, accompanied by substantial disruptions in the affected industries.

Tax revenues would decline substantially during a period of transition because of employment wage decrease and tax write-offs. An estimate of the quantity of tax revenue loss is $803 million for the first year (based on the total elimination of beverage can production and an 8% beverage sales decline). This figure would be decreased if beverage sales did not decline substantially and if beverage cans continued to be sold.

The average price paid by all consumers for beer and soft drinks should decrease slightly because the higher-priced nonrefillable containers would only be sold in small quantities. Increased handling costs (estimated at $0.015 per container at retail) and costs related to equipment changes in the brewing and soft drink industries would likely be passed on to the consumer, however. These costs could result in a rise from the current price of refillable containers. It is estimated that the price per unit of beverage is likely to be on the order of $0.005 per unit lower than the current nonrefillable price.

A mandatory deposit system would raise the price of convenience. If a consumer purchased a container and did not return it for a refund, he would in effect be paying $0.05 extra for not returning it. A ban that prohibited the sale of various types of containers would limit consumer choice. A mandatory deposit does not directly prohibit the sale of any container type. However, it forces the consumer to pay a higher price—equivalent to the deposit, for the convenience of discarding a container.

Ban on Nonrefillable Containers: A ban on nonrefillable containers would act in the same way as a mandatory deposit, as bottlers of beer and soft drinks would probably place deposits on their refillable beverage containers to retrieve them for refilling. A ban on nonrefillables would, therefore, result in benefits similar to those of a mandatory deposit. The costs of a ban would be more severe than those of a deposit, however, because a ban would prohibit utilization of any container other than one that is refillable.

Specifically, a ban would completely eliminate the beverage can manufacturing industry ($1.5 billion in shipments in 1971) as well as the contract canning industry. The uses of steel and aluminum for beverage cans would also be eliminated. A ban prohibiting the sale of nonrefillables would limit consumer choice because the only containers available would be the refillable bottle. If a consumer discarded rather than returned a refillable container, he would lose the deposit and, in effect, he would be paying extra for this convenience.

Litter Tax: The low litter tax selected for analysis would require that an additional $0.005 per container be paid on the sale of each container for beer or carbonated soft drinks. The tax would be imposed at the point of purchase of the container by the beverage industry. The projected costs and benefits of implementation of this measure at the national level follow. Litter taxes can also be imposed at the state or local level (the state of Washington has enacted a low litter tax). Where implemented at the State or local level, the costs and benefits must be analyzed in relation to the characteristics of the particular area. The specific effects of a litter tax in reducing beverage container litter are difficult to predict accurately. Some qualitative comments follow.

While a low litter tax probably would not cause any change in the rate of littering, it would raise revenue to be used for litter collection. In 1972, a $0.005 per container tax would have raised approximately $278 million in revenue. If this amount were applied totally to litter cleanup activities, it would increase the frequency of litter collection by approximately five to six times. However, it is possible that once raised, the revenues might be used to substitute for present funds.

A low litter tax would have no effect on beverage container solid waste. A low litter tax would have no effect on the environmental impact of beverage container production and use. No changes are expected because a tax of $0.005 per container is not likely to affect consumption patterns. Studies of the elasticity of demand for beer and soft drinks indicate that both are relatively inelastic and that a price increase of $0.005 per container would not affect demand substantially. A low litter tax is not likely to cause dislocation in the beverage or container industries. No employment impact would be expected as a result of imposition of a low litter tax. No increase in tax write-offs would be likely as a result of a low litter tax as the tax is not likely to affect beverage or container consumption.

A low litter tax would increase the costs of all container types by the amount of the tax. If the charges were passed on to the consumer, the average price of beer and soft drinks would be increased an average of slightly less than $0.005 per filling. The price increase for refillables will be less as the tax can be amortized over several fillings. Consumer choice and convenience would be unafffected by a low litter tax.

A mandatory deposit law in Oregon has been in effect since October 1, 1972, and a similar law in Vermont went into effect September 1, 1973. Because Oregon is the only state for which detailed data are available on the impacts of beverage container legislation, the effectiveness and impacts of its law will be discussed in some detail.

Oregon's "Bottle Bill" requires a minimum $0.02 refund to purchasers on the return of "certified" containers of beer, malt beverages, and carbonated soft drinks, and a $0.05 refund on the return of all other beverage containers for those beverages. Certified containers are defined as containers that are used by, and that will be accepted for, reuse by more than one manufacturer. In addition, the law outlaws the sale of the fliptop or pulltab beverage container. The remainder of this subsection summarizes the effects of the law through June 1973.

Beverage containers in litter decreased substantially between winter and spring of 1971 to 1972 (before enactment) and winter and spring of 1972 to 1973 (after enactment). Table 14 illustrates these findings. Cans for beer and soft drinks have declined to approximately 2% of the market share, and nonrefillable glass bottles have been completely eliminated.

Sales of beer and soft drinks have not declined since the law went into effect. A price rise of up to 1.7 cents per container did occur in the spring of 1973, although it is not clear as to whether this can be attributed to the bottle bill, inflation, or other cost increases. Approximately 142 jobs were lost as a result of the bottle bill, 62 in a canning facility in Oregon, and 80 in a can manufacturing plant outside of the state.

In conclusion, it is found that beverage containers form a significant and visible portion of roadside litter and a substantial and increasing percentage of solid waste. This is due in large measure to the growth of the nonrefillable container at the expense of the refillable bottle. There is analytical evidence to indicate that replacement of the existing beverage packaging system with a refillable bottle system would result in substantive decreases in air and water pollution and energy and resource use.

Preliminary data from Oregon indicate that a mandatory deposit system on all beverage containers results in the establishment of a predominantly refillable bottle system and also results in a reduction in litter. These effects are accompanied by some unemployment and by adverse economic impacts on the manufacturers and fillers of nonrefillable containers. More conclusive data on the impacts of the Oregon law are expected in the future. An excellent source of background data is *The Beverage Container Problem: Analysis and Recommendations,* Report EPA-R2-72-059, Washington, D.C., U.S. Environmental Protection Agency (Sept 1972).

TABLE 14: A COMPARISON OF OREGON LITTER DATA BEFORE AND AFTER INSTITUTION OF THE MANDATORY DEPOSIT ACT*

	Winter (October to February)			Spring (March to June)		
Type of litter	Average number of items per mile per month, 1971-72	Average number of items per mile per month, 1972-73	Decrease (percent)	Average number of items per mile per month, 1972	Average number of items per mile per month, 1973	Decrease (percent)
Beverage containers	269	51	74	103	19	-75
Other litter	456	219	52	187	175	5
Total	728	270	56	290	195	-25
Beverage containers as a percent of total litter	37	19	42	36	10	-65

* EPA analysis of data supplied by the Oregon State Highway Department of Litter Surveys.

Source: EPA Report SW-122

BEVERAGE INDUSTRY WASTES

In the manufacture of beverages of the soft drink categories there are no by-product aqueous solutions produced which would be considered to have recoverable value. The primary loss of valuable materials in the soft drink manufacturing industry is in the form of sugar-bearing syrups or concentrates that are either spilled during process or intentionally dumped to the sewer because of their unacceptable quality characteristics.

Two factors play a part in the unrecoverable nature of these syrups or concentrates. The first is the fact that reclamation of food or beverage products would not be compatible with normal quality control and consumer acceptance standards. The second is the fact that the dollar value of the sugar lost in these syrups is relatively small and could not support the expenditure of costly capital equipment for their recovery.

Detailed studies of several soft drink manufacturing operations have indicated that the BOD loading of the waste materials from a normal bottling or concentrate producing plant are small enough that they are not of an intolerable character. Moderate amounts of sodium hydroxide from bottle washing operations are also thrown to waste but, once again, the dollar burden to the operation and the public acceptance of recovery for this purpose would not render the operation practical. Imposed upon this aspect of bottling operations is the fact that there is a strong tendency for the "one-way" or nonreturnable bottle. In this instance no bottle washing wastes are produced.

In the production of fermented alcoholic beverages from grain, substantially more wastes are generated but, once again, the use of classical desalination methods would not be practical in recovery operations. These wastes are generally in the form of spent grain or hulls. Broad-scale utilization is made of these wastes, particularly in the form of animal and livestock foods.

In most cases, however, the bulk of the grain is recovered by filtration or centrifugal separation means. It is only in the case of dilute washings and rinsings that waterborne wastes containing low-density dissolved solids are produced. Because of their dilute character and the relatively low value of the product recovered, it is not considered feasible to apply desalination techniques to their recovery. With further development of desalination methods, it is possible that in the future either ultrafiltration, reverse osmosis, or high-efficiency distillation methods could be used in this recovery according to Aqua-Chem, Inc. in Resource and Development Progress Report No. 581, Washington, D.C., Office of Saline Water (October 1970).

CHEESE INDUSTRY WASTES

Recovery of Whey

This is discussed in a subsequent part of this volume (Whey p. 79). The reader is also referred to *Whey Processing and Utilization* by M.T. Gillies, Park Ridge, N.J., Noyes Data Corporation (1974).

FATS

Recovery from Liquid Wastes

Recovery from Edible Oil Refinery Effluents: A process developed by W.C. Seng is described in Report EPA-660/2-73-015, Washington, D.C., U.S. Environmental Protection Agency (December 1973). New full-scale equipment and modification to the existing standard waste treatment equipment were installed at the Swift and Company modern, high-volume, edible fat and oil refining plant at Bradley, Illinois, complete with necessary controls and instrumentation to study methods for removing and upgrading the fatty materials for resale.

FIGURE 17: BRADLEY WASTE TREATMENT FLOW DIAGRAM

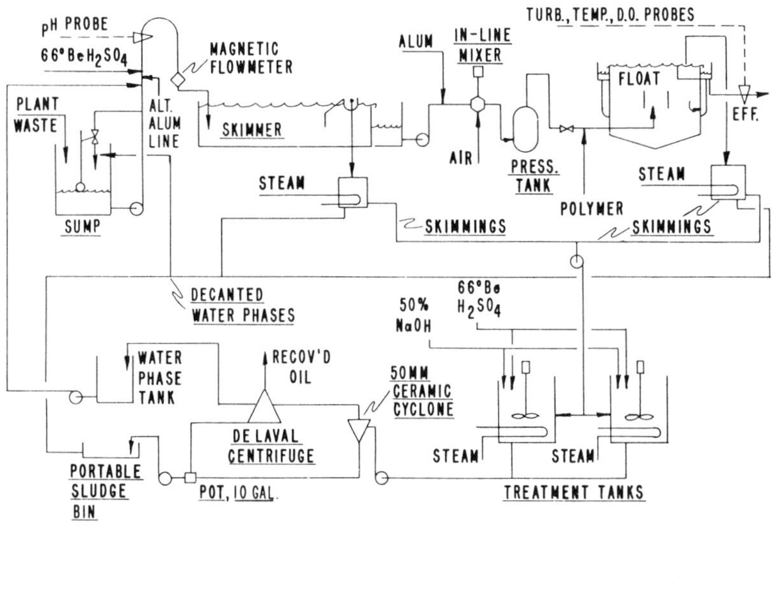

Source: Report EPA-660/2-73-015

Concurrent with the above, an in-plant survey was conducted. The waste streams were characterized as to source, flow rate, and ether solubles, suspended solids and BOD content.

Many commercially available synthetic acrylamide polymers were laboratory tested as flocculants in combination mainly with alum for plant wastewater clarification. Four were tested over several months at Bradley. All were found very effective, provided that suitable pH levels and dosage were used.

Cathodic protection devices and impressed current successfully controlled corrosion and build-up of solids on the walls of the existing skimmer and air flotation units, but statistically significant enhancement of flocculation was not shown. However, other concurrent and later work by Swift at other plants has shown impressed current is a valuable aid to wastewater clarification.

Several instruments and controls were evaluated relative to the water clarification system. Concurrently, a DeLaval PX-213 bowl-opening, disc-stack, centrifuge was tested to concentrate and upgrade the removed fatty materials after caustic and sulfuric acid treatment. The system was capable of handling in one shift all the material removed over 24 hours.

An overall economic evaluation indicated the 7,000 pounds of oil recovered (99% ether-soluble) would offset 60% of the total daily direct operating costs for the waste treatment system including the oil in reclaiming system. A flow diagram of this plant is shown in Figure 17.

Recovery from Solid Wastes

Recovery from Damaged Butter and Oleo Packages: A process developed by *A. Moustafa; U.S. Patent 3,476,570; November 4, 1969* is a process for reclaiming edible oleaginous material from damaged parchment wrappers, cartons and related plastic containers (packaging materials), commonly used in the consumer packaging of margarine and related oleaginous products.

Since these packaging materials are often imperfectly prepared, it is highly desirable to reclaim the adhering and/or absorbed oleaginous materials therefrom. This reclaiming process comprises the following five basic steps:

(1) A melting of the oleaginous fat from the packaging materials;
(2) A screening thereof. In the case of plastic containers, screening is preceded by a shredding of plastic containers;
(3) A washing of the packaging materials and further screening;
(4) A centrifuging of the resultant emulsion of Step 3 aforesaid;
(5) A vacuumizing of the reclaimed oleaginous materials resulting in recovery of pure oil.

Figure 18 is a block flow diagram of this process.

Rejected and damaged prints, wrappers, plastic containers, etc. otherwise defined as packaging materials, are introduced into the Step 1 heating and agitating tank, proceeding thereafter through the steps of:

(1) Melting and agitating;
(2) First stage screening;
(3) Washing;
(2a) Rescreening;
(3a) Centrifuging;
(4) Vacuumizing until the final reclaimed oleaginous product is collected and stored.

FIGURE 18: PROCESS FOR RECLAMATION OF FATS FROM PACKAGING MATERIALS

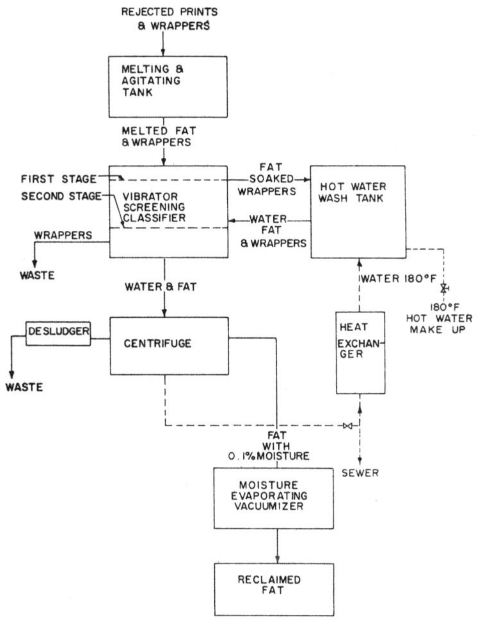

Source: A. Moustafa; U.S. Patent 3,476,570; November 4, 1969

FRUIT AND VEGETABLE INDUSTRY WASTES

Recovery from Liquid Wastes

The canning of fruits and vegetables in the United States is an industry of great proportions in which substantial volumes of wastes are generated. In most cases, these wastes do not contain materials that would be considered recoverable or valuable as by-products from the canning industry. The only minor exception to this recovery situation would be the extraction of alcohol or brandy from certain fruit juices and the recovery of tomato or strawberry skins and seeds for reuse in poultry and pet food. The wastes generated in the canning industry are generally washwaters or caustic solutions used to aid peeling of fruits and vegetables.

In total, approximately one billion gallons per day of these wastes are produced in the United States. The state of California alone has approximately 25% of the food processing and canning in the United States. Current cannery practice is toward shifting as much of the preprocessing load to field operations as possible.

This means that washing, caustic treatments, and other primary operations will be done in the field at the time of harvest rather than in the canning plant itself. The obvious ramification to this method is that the waste will then not be generated in large industrial plants typically located in urban or populated areas. In this case the wastewaters produced by these operations will very likely be discharged for irrigation disposal on fields.

The wastes now generated from canning operations typically contain three primary constituents: the first of these is insoluble dirt, soil, and grit entrained upon the food itself; the second is a small amount of soluble sugars and starches washed from the fruit and vegetables; and the last is dilute caustic solutions used to aid in peeling and skinning various fruits and vegetables. These wastes do not appear to offer recoverable advantages according to Aqua-Chem, Inc. in Research and Development Progress Report No. 581, Washington, D.C., Office of Saline Water (October 1970).

Production of Animal Feed and Molasses from Pear Wastes: A process developed by *R.P. Graham, J.H. Thompson and A.D. Shepherd; U.S. Patent 2,614,933; October 21, 1952; assigned to the U.S. Secretary of Agriculture* relates to the utilization of pear waste. In particular, it relates to methods for treating raw pear waste whereby this material can be separated into useful fractions. Thus the process contemplates separating the waste into a solid fraction and a juice fraction.

The solid fraction contains minerals and carbohydrates and can be dried and used as an animal feed. The juice fraction is a clear, straw yellow to amber solution containing sugars in such amounts that it can be used for many purposes. For example, the juice fraction can be used as a nutrient medium for the culture of yeast or other microorganisms. Further, it can be concentrated by evaporation to produce a molasses which can be used in animal feeds, human foods, or which can be utilized for preparing microbiological culture media. The molasses produced in this manner has a higher ratio of sugar to total solids than does beet or cane molasses and hence is a valuable carbohydrate source.

Pear waste is a material produced in enormous tonnages at pear canning plants and usually consists of pear peels, seeds, cores, trimmings, overripe, and cull fruit. In pear canning operations, the amount of waste is from 40% to 45% of the pear used. The disposal of this waste constitutes a serious problem. This problem is especially acute in areas where local and state authorities forbid dumping the waste directly into streams and where facilities for sewage disposal are not adequate to handle the waste. In such cases the waste must be hauled from the factory to a dumping ground, at considerable cost to the processor.

Many attempts have been made in the past to utilize pear waste, particularly to separate it into its solid and juice fractions. Such attempts have all been unsuccessful. The difficulty is that pear waste is a slimy mass containing fine particles of cellular material. When filtration or pressing is attempted to separate the juice from the solids, the fine particles of cellular material clog the pores of the filter or press cloth so that little or no passage of liquid can take place.

If any juice is obtained, it still contains so much suspended matter that if it is evaporated sufficiently to be self-preserving, it will solidify in the evaporator. Further, the filter or press cake obtained thereby still contains so much water that if it is introduced into a dryer such as a rotary kiln dryer, it forms a taffy-like mass which sticks to the walls of the dryer or forms lumps or balls which do not dry properly. If the pear waste is finely ground and subjected to filtration or pressing, much of the waste will pass through the filter or press cloth and little separation will be achieved.

Various methods have been advocated for separating the juice and solid fractions of fruit pulps. These processes involve a lime treatment or a treatment using both lime and phosphoric acid (or other polybasic acid). These methods are based on the addition of an excess of lime to the waste to cause formation of calcium salts with the organic acids in the fruit material. The formed calcium salts then precipitate occluding with them various other solid materials such as fruit pulp, fiber and so forth. In many cases, particularly where the fruit pulp is deficient in organic acids, phosphoric acid or other polybasic mineral acid is also added.

The resulting precipitate of calcium phosphate then assists in the separation as it occludes solid material as it comes out of solution. With some fruit wastes these methods have proved satisfactory. However in the case of pear waste, the known methods do not give successful results—the separation is inefficient because the material still tends to blind the filter or press cloths and the juice obtained is contaminated with pectinous materials which cause difficulties when the juice is evaporated.

The process involves, basically, a two-stage liming of the pear waste. In the first stage, which is similar to previously used liming treatments, sufficient lime is added to the waste to raise its pH to a level which activates the enzyme content of the waste whereby the pectin in the waste is largely deesterified forming low-methoxyl pectins and/or acid. These deesterified products gel as they are formed due to the presence of calcium ions and occlude the finely-divided particles suspended in the waste.

While this action results in a modification of the character of the waste which makes possible separation thereof into its solid and juice components, the product produced at this point can be further altered to make it more capable of separation, that is, the rate of filtration and porosity of the material is greatly increased. To this end the treated waste is mixed with an additional amount of lime whereby certain chemical and/or physical changes take place whereby the mass becomes much easier to separate into its solid and liquid fractions. This second liming gives a result not yielded by the prior art process involving only a single stage of liming and constitutes the principal element of this process.

Recovery of Soluble Starches and Sugars from Corn Milling Wastewaters: The process of wet corn milling is of large proportions in the United States, with the total dollar value of products approaching one billion dollars annually. By far the greatest activity in this industry is located in the Midwest and Central United States, with approximately 80% of the corn products coming from Illinois, Iowa, Missouri, Indiana, and Ohio. Shelled corn is received at the corn processing plant and converted into the several products shown on the attached flow diagram. The basic products that are taken from corn are corn oils, starches, sugars, and various animal foods.

The chemical makeup of corn is as follows: 15% moisture, 60% starch, 9% protein, 5% oil, 5% hull, 6% ash and insolubles. Starch, which makes up the largest portion of the corn kernel, is extracted for use as starches as well as through hydrolysis conversion into sugars. These various corn sugars (dextrose) are utilized either as highly purified crystal forms, water syrups, or syrup solids known as molasses.

The major portion of the oil present in corn is extracted as such and finds its use in products such as margarine and frying and food preparation oils. The protein portions of the corn are contained largely in the germ and the hull where the largest outlet is in animal and livestock feeds.

As will be observed from the flow diagram of a corn milling process in Figure 19 almost the entire product is utilized and any losses are kept to very minimal amounts. On this basis there are relatively few areas in the wet corn milling operations where recoverable products of substantial value can be collected. The primary areas which offer potential for desalination equipment in recovery operations would be reverse osmosis or ultrafiltration recovery of some of the soluble starches and sugars resulting from washing operations. In typical corn processing operations every attempt is made to recover the maximum amount of these materials through evaporative methods.

However, there are inevitable highly dilute solutions of these soluble fractions for which the cost of evaporation would be far too high to justify their recovery. On this basis, high flux reverse osmosis or ultrafiltration could have some utility in separating these materials from the large volumes of water in which they are carried. The exact character of the waste from the corn milling operations is considered highly proprietary data by the various operators and it would be almost impossible to adapt standard designs or approaches which could be used on a broad scale throughout the industry.

However, it has been estimated that the soluble wastes carried away from corn milling operations do not contain more than one pound of soluble sugar or starch per one thousand gallons. On this basis, the value of product wasted would not exceed ten cents per thousand gallons thus putting very stringent economic requirements on any recovery operations. A large corn milling operation may have several hundred thousand or even millions of gallons of wastewater being sewered each day, but it is considered that the value of the product typically being lost in these streams would not justify expensive recovery operations. High-flux ultrafiltration could be used to prevent the various starch and sugar fractions from contributing to the BOD of wastewaters, but operation on them for the full purpose of product recovery does not seem feasible according to Aqua-Chem, Inc. in Research and Development Progress Report No. 581, Washington, D.C., Office of Saline Water (October 1970).

FIGURE 19: THE CORN WET MILLING PROCESS

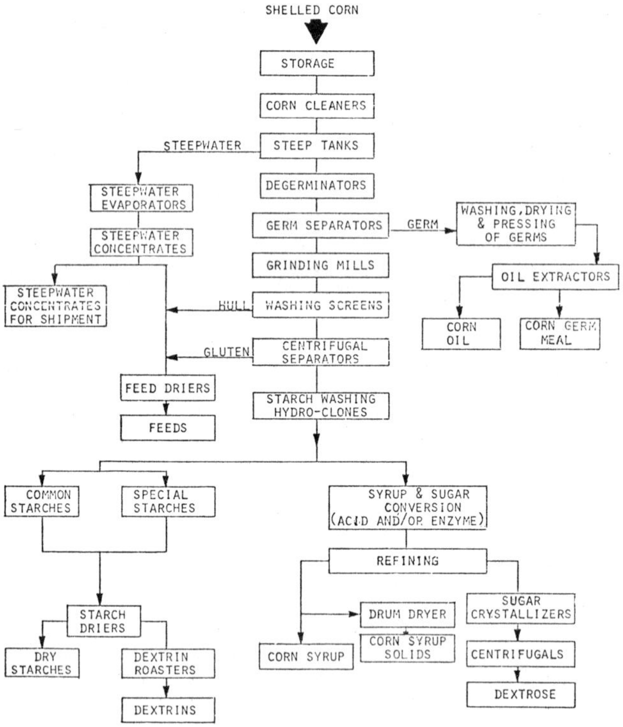

Source: OSW Research and Development Report No. 581

Recovery of Protein from Soy Whey: In the soybean processing industry, two general operations are carried out for the extraction of valuable materials from soybeans. These are the pressing of soybean for the recovery of soybean oil, and the extraction of protein from the soybean. In the first operation by which soybeans are processed for the recovery of oils and materials, as such, there are no aqueous wastes which would be amenable to treatment.

In the second operation, that is, the extraction of protein from soybeans, a certain amount of soy whey (water-soluble soy protein) is generated in washing and extraction operations. This soy whey may be a solution of about four and one-half to five percent solids by weight and may contain approximately one-half protein. As in other protein extraction operations, this solution is contaminated with a certain amount of soluble salt which must be removed prior to direct utilization. The transport depletion process is useful in the demineralization of this whey so that it can be recovered for further use in animal and human food products.

The value of transport depletion equipment for soy whey processing is comparable to equipment that would be used in cheese whey processing. In the case of the market value of the product, there is a lesser value put on soy protein, with present markets in the range of ten to twenty cents per pound. In the United States, there are approximately forty million pounds of soy protein recoverable and suitable for upgrading by transport depletion methods.

These operations are centered primarily in the Midwest, with the biggest operations being in Illinois, Iowa and Minnesota. The typical plant for the recovery of approximately three to four million pounds of soy protein per year will sell for approximately $80,000 to $100,000. On this basis, there is a total market potential for between $800,000 and one million dollars in transport depletion and electrodialysis equipment for the recovery of soy protein from soy whey, according to Aqua-Chem, Inc. in Research and Development Progress Report No. 581, Washington, D.C., Office of Saline Water (October 1970).

Recovery from Solid Wastes

Conversion of Citrus Pulps to Fertilizers: A process developed by *W.P. Moore; U.S. Patent 3,878,304; April 15, 1975; assigned to Allied Chemical Corp.* is one in which waste or by-product carbohydrate materials comprising polysaccharides are converted into slow-release nitrogen products valuable as a protein substitute in ruminant feeding and as a slow-release fertilizer for growing plants. The slow-release properties are achieved by acid-catalyzed reaction of an amide such as urea or biuret with partially degraded waste or by-product polysaccharides and by agglomeration of this combination into particles having substantial physical integrity. The waste or by-product polysaccharide materials include citrus pulps, bagasse, peanut hulls, wheat straw and wastepaper.

Enzyme Treatment of Citrus Pulps to Increase Juice Production: A process developed by *K.J.S. Villadsen and K.J. Moller; U.S. Patent 3,347,678; assigned to A.B. Grindstedvaerket, Denmark* is one in which, after initial juice extraction, the removed solids are subjected to a treatment with pectolytic enzymes for a prolonged period at room temperature, or for a shortened period at increased temperature. By the enzymatic action, a further substantial amount of juice is liberated, and suitably extracted.

The following is one specific example of the operation of the process. To 1,000 grams of orange pulp, from which the juice had been screened, and which contained 19.5% dry matter, was added 1 gram Pectolase DE 10. After standing at room temperature for 20 hours, a further yield of 254 grams of juice was obtained by pressing.

MEAT PROCESSING WASTES

Recovery from Liquid Wastes

Recovery of Protein from Animal Wastes: A process developed by *R.A. Grant, U.S. Patent 3,835,041; September 10, 1974; assigned to Tasman Vaccine Laboratory Ltd, New Zealand* involves removing protein from waste effluents containing same using ion exchange methods. The ion exchange resin is a crosslinked regenerated cellulose matrix modified by the introduction of groups capable of cation exchange or groups capable of anion exchange.

Meat works, abattoirs, fish canneries, dairies, glue works and the like industrial concerns handling animal products produce large volumes of effluent. Such effluent may vary in composition, but a typical effluent from an abattoir results from the use of large quantities of water for the washing of slaughterhouse equipment and carcasses. It thus contains appreciable amounts of blood proteins, soluble protein from muscle and other tissue, and suspended or colloidal protein material, together with a certain amount of particulate tissue fragments and fat.

Although the concentration of protein in the effluent is usually quite low (about ½ pound per 100 gallons) the solution is capable of nourishing abundant undesirable organisms. It has been estimated that in a typical slaughterhouse the material lost in the effluent constitutes about 5% of the weight of slaughtered animals, the remaining 95% of the weight being utilized in one way or another. Another source of protein containing effluents are natural rubber latex processing factories.

Some factories producing protein-containing effluents may be so sited that the effluent can be discharged directly into a river or into the sea without treatment. However, where this is not possible, it is usual to treat the effluent to reduce the concentration of proteins and other nutrients before discharging the effluent into a waterway in order to prevent the pollution effects of the undesirable organisms that may grow on the untreated effluent as well as to prevent the stench caused by the decomposing animal matter.

Effluents such as those from meat works are usually treated by biological treament in holding tanks. Such biological treatment involves both aerobic and anaerobic degradation of the organic material in the effluent. This has the double disadvantage that a large area of ground is required for the equipment and that the process destroys the potentially valuable proteins. Furthermore the expense of providing suitable holding tanks is considerable.

Where is is desired to discharge the effluent directly into a municipal sewer the authorities responsible for treating the sewage may levy a considerable charge on the meat works for disposing of the effluent because of the capital expenditure involved in providing suitable holding tanks and the running costs of operating the process.

It has been proposed (U.S. Patent 2,446,913—Erlich) to pass the slop or thin stillage remaining after the alcohol has been distilled off from a fermented mixture of 90% wheat and 10% barley malt through beds of an ion exchange resin after first having centrifuged the slop or thin stillage. The slop or thin stillage, before passage through the beds of ion exchange resin, is said to contain "proteinaceous compounds" in addition to salts and a nitrogen-free extract. The proportion of proteinaceous compounds is estimated by performing a nitrogen analysis and multiplying the result by 6.25. However, there is no unambiguous proof that the proteinaceous compounds are true proteins.

The proteinaceous compounds-containing solution is passed first through a cation-absorbing type ("C") synthetic phenol-formaldehyde resin and then through an anion-absorbing type ("A") synthetic amine-formaldehyde resin. On regenerating the resins and combining the regenerants a solid is obtained which is stated to contain 29% proteinaceous compounds in addition to various inorganic salts. However there is no proof given that the proteinaceous compounds are true protein materials, the figure of 29% being merely based upon a nitrogen analysis the result of which is multiplied by 6.25. Even ammonium salts can be classed as proteinaceous compounds on this basis.

It is also known to extract proteins on a laboratory scale from aqueous solutions thereof by passage through a column of diethylaminoethyl cellulose (DEAE-cellulose). An example of the use of DEAE-cellulose for extracting protein from solution is described in U.S. Patent 3,409,605 (Florini). However, although DEAE-cellulose is successful on a laboratory scale where very low flow rates can be tolerated, it is impracticable to use DEAE-cellulose resins for industrial purification of raw protein-containing effluents because the DEAE-cellulose resins very quickly become blocked and the beds of the DEAE-cellulose become choked so that the flow rate drops to an unacceptably low level.

It is an object of the process to provide a method for purifying a waste effluent by removal of proteins therefrom which can be practiced using equipment of low capital cost and occupying only a small ground area.

Essentially, the process for the purification of a waste effluent containing protein comprises the steps of: (a) passing the effluent through a bed of a granular ion exchange material in the form of a crosslinked regenerated cellulose modified by the introduction of cationic or anionic exchange groups to remove protein from the effluent, (b) recovering essentially protein-free effluent, and (c) subsequently regenerating the ion exchange material for use in a further cycle.

A process developed by *P. Nettli; U.S. Patent 3,842,003; October 15, 1974; assigned to A/S Apothekernes Laboratorium for Specialpraeparator, Norway* is based on the discovery that sugar esters of sulfuric acid such as saccharose sulfate, lactose sulfate, amylum sulfate, etc., are useful for the removal of proteins and any decomposition products thereof from wastewater by precipitation.

Recovery from Solid Wastes

Recovery of Tallow from Waste Animal Flesh: A continuous process developed by *E.J. Jemal; U.S. Patent 3,355,255; November 28, 1967; assigned to Mine and Smelter Supply Co.* involves an apparatus for separation and recovery of tallow from waste animal flesh in slaughterhouses, butcher shops, and the like. Further, the process provides a continuous operation wherein animal flesh to be treated may be continuously fed into the apparatus and tallow continuously recovered therefrom, as distinguished from conventional procedures which attempt the tallow recovery by treatment of a batch of material at one time.

WHEY

Recovery from Cheese Manufacturing Effluents

As described by Aqua-Chem, Inc. in Research and Development Progress Report No. 581, Washington, D.C., Office of Saline Water (October 1970), the manufacture of cheese is one of the biggest aspects of dairy operations in the United States. Approximately one and one-quarter billion pounds of cheese are produced annually in the United States, and in the process approximately one and one-quarter billion pounds of whey solids are released. These whey solids are generated from the manufacture of both natural and cottage cheeses.

A typical flow diagram for the manufacture of natural cheese by the enzyme coagulation method is shown in Figure 20. Cottage cheese is produced in a slightly different manner through the acid coagulation of milk. Whey production is divided so that approximately fourteen billion pounds of fluid are produced annually from natural cheese, while approximately half that quantity, or seven billion pounds per year, are produced from cottage cheese operations. Of the cheese marketed in the United States approximately one-third is of the cottage cheese type, while the remaining two-thirds are of the natural or processed types.

The solids present in whey are a mixture of milk protein or lactalbumen, milk sugar or lactose, and a small amount of ash or dissolved salts. As whey is produced in the cheese making process, it is collected at a concentration of about 6% by weight. In other words, every one hundred gallons of fluid whey contain approximately fifty pounds of whey solids. This quantity of whey solids is approximately equal to, or slightly less than, the weight of cheese which was extracted from the whole milk depending upon the cheese making process.

From the flow diagram of cheese making the various steps of whey processing can be seen. Whole milk has either enzyme or acid coagulated to form the product cheese. The cheese curd is skimmed from the remaining fluid portion of the milk which is referred to as whey. The whey extracted from the cheese making vat is normally about 6% weight concentration. The whey solids themselves are composed of approximately 70% lactose, 20% lactalbumens, and 10% ash. Varying amounts of acids, ash, and other constituents may be present depending upon the specific mode of cheese making operation.

FIGURE 20: FLOW DIAGRAM OF CHEESE, LACTOSE, WHEY PROCESSING OPERATIONS

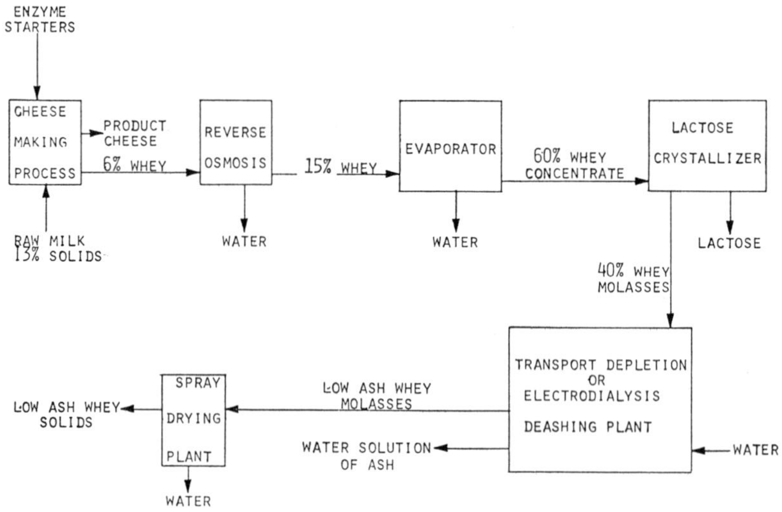

Source: OSW Research and Development Report No. 581

The first step in whey processing can be the preconcentration of whey from 6% solids to approximately 15% solids through the use of reverse osmosis. This step has been demonstrated to be feasible and is distinct in lower operating costs over the conventional types of evaporating equipment. At about 15% concentration the osmotic pressure of whey solution becomes too high for continued reverse osmosis processing. At this point conventional milk or dairy evaporating equipment is used to concentrate the whey to approximately 60% solids.

Dairy evaporators are of a special sanitary, low-temperature construction that is not found in the desalination industry. On this basis no potential is assumed for desalination-type evaporators in this industry. At the 60% solids concentration level lactose can be crystallized from the whey through cooling of a warm solution. About one-half of the lactose present is normally removed in this manner. The remaining whey molasses at approximately 50% concentration can be sent into a transport depletion or electrodialysis deashing plant.

In this equipment anywhere from 50 to 95% of the dissolved ionized salts can be removed resulting in a substantial upgrading of whey value. The final step in whey molasses processing is typically the spray drying of the fluid to a stable dry powder. In this form the product has a long shelf life and can be easily transported and marketed. In the fluid form the whey molasses must be used within approximately twenty-four hours or bacterial action may result in spoilage.

The uses of whey have only touched upon the relatively high-value specialty markets such as synthetic milk products, special cheese foods, dietary supplements, infant formulas and substitute low-fat dairy products. The further researching and application of whey protein and solids to low-cost, high-volume food products give rise to the thought that ostensibly all of the whey production could be utilized in solving food scarcity problems.

GLASS AND CERAMICS

CEMENT

As pointed out by M. Sittig in *Pollution Control in the Asbestos, Cement and Glass Industries,* Noyes Data Corp., Park Ridge, N.J. (1975), dust is a major problem in the cement industry and dust reclamation is an economic necessity. The reader is referred to that volume for details of air pollution control equipment as applied to cement plants.

T.E. Kreichelt, D.A. Kemnitz & S.T. Cuffe in Publication 999-AP-17, Cincinnati, Ohio, U.S. Public Health Service (1967) state that the economic benefits resulting from the reuse of collected dust do not completely affect the cost of the dust collection equipment. However, the esthetic value to the community plus compliance with air pollution control ordinances make dust recovery worthwhile.

FOUNDRY SAND

As pointed out by E. Aleshin, S.A. Bortz and T.B. Wade in a final report on IIR Research Institute Project No. G6027 wherein they made analyses and projections of various industrial metal and mineral wastes, foundry sand presents a major disposal problem.

Regeneration by Fluidized Treatment

A process developed by *H. Jacob; U.S. Patent 3,857,201; December 31, 1974* is one in which spent contaminant coated casting sand is mixed with fresh uncontaminated casting sand, and the thus obtained mixture is subjected to mechanical agitation, during which the fresh sand scours the contaminant coating off the spent sand and thus regenerates the sand.

GLASS

Total glass production in the U.S. is over 10 million tons, of which containers account for about 75%, the balance being flat glass and pressed and blown glass. A large proportion of the container glass plus a fair proportion of the other types end up in municipal solid waste; only a minute proportion of glass waste (almost exclusively flat glass) is associated with industrial sources. There is virtually no recovery of glass from mixed waste, but a small amount of glass is recycled through voluntary collection centers and cullet dealers. Compared to other materials, glass is among the lowest in recycling ratios, or about 4.5% of consumption, when home scrap (scrap generated in the glass manufacturer's plant) is excluded.

The best sources of quality cullet have been declining. Clear glass milk bottles and returnable glass containers rejected from bottle washing operation—major sources of cullet in the past—are gradually disappearing. Sorting, collection, and delivery costs have risen, principally because these operations are highly labor intensive. Plants have not been maintained and equipment has not been purchased because of the limited capital of the few dealers still in operation. As the quality and availability of purchased cullet have deteriorated, its use in the glass industry has also declined.

In addition to the use of purchased cullet in glass furnaces, there are several alternatives for cullet utilization. The most widely publicized alternative is in "glasphalt," a road surfacing material in which cullet replaces part of the asphalt aggregate. Initial testing results at the University of Missouri indicate that glasphalt is equal to or superior to conventional asphalt. However, cullet would have to compete economically with asphalt aggregate, which ranges in price from $1.50 to $5.00 per ton when delivered to the asphalt plant. Present cullet prices are significantly higher than this amount. Other proposed uses for cullet include construction materials, such as glass-cement blocks, and cullet-terrazzo. Experiments to determine the feasibility of cullet utilization in these products are currently underway.

A comprehensive discussion of the utilization of waste glass in secondary products has been published as a result of a symposium held at the Technology Application Center of the University of New Mexico in Albuquerque, New Mexico in January 1973. It is available as Report N73-33460, Springfield, Va., Nat. Tech. Information Service.

The glass industry has certain characteristics that make high levels of waste recycling much more favorable than in other industries. First, the manufacture of glass containers is essentially a one-step process, starting with raw materials and ending with the finished product. Second, cullet can be substituted for virgin raw materials in large percentages, provided that the cullet meets minimum specifications of color, cleanliness, and purity. From a technology standpoint, glass manufacture from 100% cullet appears possible.

There are, however, two problem areas: comparative economics and the recovery of cullet from mixed waste. With respect to economics, the cost of virgin raw materials averages $15.48 per ton batch, as compared to a range of from $16.00 to $22.50 per ton batch of cullet; both include freight charges to the plant. Processing cost differentials are not significant. The conversion of an existing plant to using increased quantities of purchased cullet would cost from $50,000 to $100,000, depending upon the plant; but the changeover could be accommodated within a framework of normal periodic plant improvements. A new plant designed to use cullet would be no more costly than a new plant designed for virgin materials.

The recovery of large quantities of cullet from municipal waste is dependent on the development of a technical process for separating and upgrading the cullet. However, the possibility of source separation of glass containers in the home for separate collections is an alternative that cannot be eliminated. Neither traditional cullet dealers nor voluntary citizen delivery of glass to recycling centers is likely to increase the cullet flow by more than a few percent.

Mechanical separation methods for removing glass from other components of municipal waste are still under development. One promising system that combines density classification and optical color sorting is currently being tested at Franklin, Ohio. Other methods, including one developed by the Bureau of Mines, are not yet ready for a comprehensive test.

Until the technology is further developed, utilization of purchased cullet on a large scale does not appear possible. Further, because glass is only a small percent of solid waste, complete glass recovery from mixed waste is not likely to occur until full-scale recovery centers concerned with all major materials are set up.

Unless source separation of glass containers is found to be feasible, utilization of purchased cullet on a large scale appears to be closely tied to development of full-scale municipal resource recovery centers. The glass coming out of such systems will not be attractive to the glass industry on a cost basis, however, unless economic incentives are provided.

Glass constituted about 10% by weight of municipal solid waste and totaled approximately 11.6 million tons in 1971. Virtually all of this glass consisted of discarded containers and packaging. Beverage containers accounted for about half the total.

According to estimates of the Glass Container Manufacturers' Institute, in 1972 nearly 225,000 tons of postconsumer glass were recycled as the result of volunteer collection efforts, particularly community recycling centers. Other glass cullet consumed by the industry was primarily in-plant manufacturing scrap and cullet from bottling operations.

In terms of market suitability and potential demand, glass is an inherently recyclable waste material. Clean, color sorted glass is an attractive raw material to the glass industry, and demand exists at prices comparable to virgin materials. No major process changes are required to utilize even very large quantities of waste glass (cullet) in glass manufacturing. Because use of cullet reduces fuel consumption and refractory wear, glass cullet is to some degree preferable to virgin materials.

Glass recycling is limited by supply. Glass is not easily removed from municipal waste. Home separation and separate collection of glass are possible and have been practiced, but only on a limited scale.

The best opportunity for removal of glass from waste is through mechanical separation in recovery facilities where waste is already undergoing shredding, air classification, or other types of separation. The separation of glass in such instances can generally be achieved, but the glass must also be color sorted to make it marketable.

The economic attractiveness of glass recovery is still not clear and may vary considerably depending on the type of processing employed in the recovery system. Recovery of glass is usually not based on an independent investment decision; separation of glass may be required to produce a clean organic waste fuel or may be undertaken to obtain aluminum, with a glass fraction occurring as a by-product. In the Franklin, Ohio, demonstration, for example, the projected economics of the combined aluminum and glass recovery look attractive, but primarily because of the value of the aluminum extracted with the glass. However, the small additional investment to color sort the glass after this initial processing also appears justified.

The proximity of markets is also an important influence on glass recovery economics. Glass recovered at a site more than 200 miles from a glass plant is unlikely to yield sufficient revenue to justify recycling.

Glass can also be used as a road paving material or a component of building products. Such new uses are now being tested by various researchers. As a rule, however, the value of glass in such applications is much lower than its value as cullet.

The most reasonable conclusion to draw at this time is that glass recovery in some form will accompany at least the larger installations of energy recovery or other mixed waste processing systems. Measures that would eliminate nonreturnable beverage bottles could considerably alter the necessity or desirability of glass recovery.

Environmental impacts occur at every step of glass manufacturing, from the mining of raw materials to final waste disposal. Changes in the amount of cullet (glass scrap) in the raw material batch are responsible for significant changes in environmental effects.

Comparing the environmental impact of glass manufacturing using 15 and 60% cullet mixes, it is clear that increased cullet usage results in reduced quantities of residual discharge. Table 15 illustrates the difference in impact for the two cullet mixes. A 60% cullet batch would result in over 50% less mining and postconsumer waste, 50% less water consumption, and up to 22% less atmospheric emissions. The energy requirements either increase by 3% or decrease by 6%, depending on the recovery system used for obtaining the cullet.

TABLE 15: SUMMARY OF CULLET-DEPENDENT ENVIRONMENTAL IMPACTS FOR 1,000 TONS OF GLASS CONTAINERS, BY IMPACT CATEGORY*

Environmental impact	15% Cullet	60% Cullet	Change+ (%)
Mining wastes	104 tons	22 tons	−79
Atmospheric emissions (all sources)	13.9 tons	13 tons §10.9 tons	−6‡ −22§
Water consumption (intake minus discharge)	200,000 gal	100,000 gal	−50
Energy use	$16,150 \times 10^6$ Btu	$16,750 \times 10^6$ Btu $15,175 \times 10^6$ Btu	3‡ −6§
Virgin raw material consumption	1,100 tons	500 tons	−54
Net postconsumer	1,000 tons	450 tons	−55¶

*Midwest Research Institute. Economic studies in support of policy formation on resource recovery. Unpublished report to the Council on Environmental Quality, 1972.

+Negative number represents a decrease in that category resulting from increased recycling.

‡Calculated for the Black-Clawson wet recovery system for recovery of cullet from municipal waste.

§Calculated for the Bureau of Mines incinerator residue recovery system for cullet recovery from municipal waste.

¶Based primarily on surveys conducted in 1967-1969.

Source: EPA Report SW-118

Cost comparisons of glass manufacture from either waste glass (cullet) or virgin raw materials depend primarily on the delivered cost to the plant of each raw material. Glass can be made from cullet in existing plants with only minor and inexpensive changes in process. Production costs are essentially the same with either raw material. Similarly, a new plant designed to use cullet would be very similar to a plant based on virgin materials and would be no more costly to construct.

Table 16 compares costs when virgin raw materials are used with the costs of using cullet. The lower end of the cullet price range reflects a transportation distance of 25 miles or less. As distance from the glass plant increases, the price obviously rises. Because most recovered glass would need to move more than 25 miles, the upper end of the range provides the best estimate.

Glass manufacturers are not likely to make even the minor process changes required to increase cullet consumption in cases where the cost of using virgin materials from well established sources, with predictable supplies and prices, is equal to or less than that of bringing in an unfamiliar, possibly contaminated substitute.

TABLE 16: COST COMPARISON FOR GLASS*

Cost component	Virgin materials ($/ton)	Cullet (waste glass) ($/ton)
Raw materials delivered	15.48	0
Cullet delivered	0	17.77–22.77
Fusion loss	2.95	0
Incremental handling costs at glass plant	0	0.50–1.00
Total	18.43	18.27–23.77

*Midwest Research Institute. Economic studies in support of policy formation on resource recovery. Unpublished report to the Council on Environmental Quality, 1972.

Source: EPA Report SW-118

Separation from Municipal Waste

A process developed by *P.G. Marsh; U.S. Patent 3,720,380; March 13, 1973; assigned to The Black Clawson Company* is one in which glass and aluminum are recovered from a mixture of solid waste, such as municipal waste, by reducing the frangible portions of the waste to a size less than a preselected maximum size, centrifugally treating the reduced portions to divide the glass, aluminum and other inorganics from the fibrous and other organic components, separating the aluminum from the glass and other inorganics, and segregating the colorless glass from the remaining, relatively opaque inorganics.

A very similar process is described by *P.G. Marsh; U.S. Patent 3,788,568; January 29, 1974; assigned to Black-Clawson Fibreclaim, Inc.* Figure 21 is a flow diagram of the process. The system will be described, for purposes of illustration, in conjunction with the treatment of refuse of the type typified by municipal waste or refuse. In *An Interim Report; 1968 National Survey of Community Solid Waste Practices,* U.S. Dept. of HEW, Environmental Control Administration, the typical composition of municipal refuse is given as follows.

Components	Wet Basis, %	Dry Basis, %
Metals	9	11
Glass	8	10
Dirt	3	3
Food waste	19	17
Plastics	4	5
Yard waste	4	4
Cloth	3	3
Wood	2	2
Paper	48	45

In a similar analysis made by The Black Clawson Company, the constituents of municipal refuse for a particular community were found to be as follows.

Components	Wet Basis, %	Dry Basis, %
Metals, glass and dirt	20	27
Food wastes, cloth and wood	30	20
Paper	50	53

Such wastes may comprise on an average 75% solids and 25% liquids, although the liquid portion has been observed to vary between 20 and 40%.

FIGURE 21: PROCESS FOR SORTING GLASS FROM MUNICIPAL WASTE

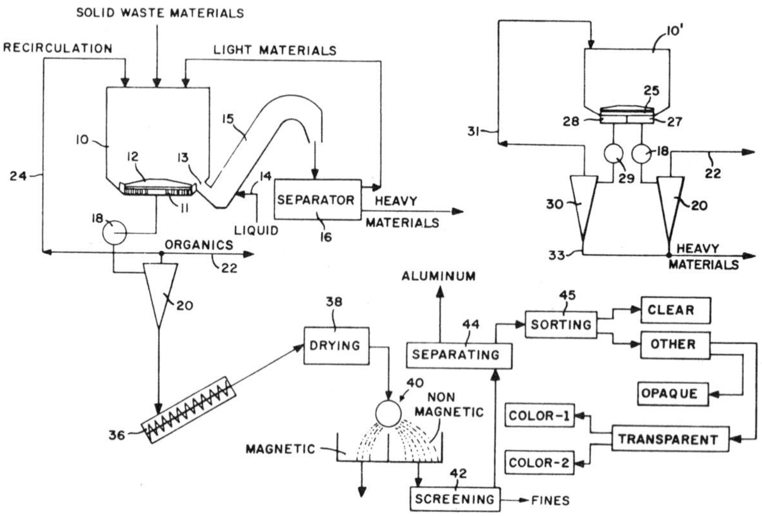

Source: P.G. Marsh; U.S. Patent 3,720,380; March 13, 1973

The refuse and an aqueous medium, such as water, are fed into a treatment vessel **10** where frangible portions of the refuse are reduced in the presence of the water or other liquid to a preselected maximum size. This is accomplished by the use of a rotor **12** rotating above a perforated bed plate **11** having perforations therethrough of the preselected maximum size so that as solids are sufficiently reduced in size, by the mechanical and hydraulic shear forces created by the rotor, they pass continuously from the treatment vessel with a portion of the water or other liquid in slurry form for further treatment.

Perforations of uniform size in a range from ⅛ to 1½ inches in diameter have been used successfully, and in one embodiment the bed plate has one section provided with substantially larger perforations than those in the balance of the bed plate. The smaller perforations are in the range of ⅛ to ¾ inch, and the large perforations in the range from ¾ to 2 inches, with a typical example being ½ and 1½ inches, respectively.

A second, larger outlet **13** from the treatment vessel **10** communicates downwardly with a junk remover **15** for the removal of relatively large, infrangible solids, such as the majority of ferrous metal pieces. An appreciable portion of the solids contained in the raw refuse deposited in the vessel **10** will usually be removed through the junk remover, a typical analysis of this portion being as follows.

Material	% Based on Portion Removed Through Junk Remover	% Based on Wet Refuse Added to Treatment Vessel
Glass	10-40	1-4
Magnetic metal	50-80	5-8
Nonmagnetic metal	1-5	0.1-0.5

(continued)

Material	% Based on Portion Removed Through Junk Remover	% Based on Wet Refuse Added to Treatment Vessel
Aluminum	0.4 - 2.0	0.04 - 0.20
Brass and copper	0.5 - 2.5	0.05 - 0.25
Lead	0.1 - 0.5	0.01 - 0.05
Other (stones, ceramics, wood, rubber, plastic, etc.)	3 - 8	0.3 - 0.8

It will be noted from the figure that the liquid added to the treatment vessel, as indicated at **14**, is preferably added through the junk remover in countercurrent to the gravity induced movement of solids through outlet **13** from the treatment vessel, it being understood that the junk remover incorporates conveying means, such as are disclosed in detail in U.S. Patent 3,549,092, for carrying such solids upwardly for dumping from the upper end of the junk remover to the separator **16**.

In addition to cleaning the solids being removed, by washing them free of fibrous and other small particle size materials tending to adhere thereto, this liquid has the important action of carrying back into the treatment vessel through the outlet **13**, the pieces of relatively light metal, particularly aluminum, which would otherwise tend to remain with the heavier ferrous metal pieces. This is especially true of aluminum which is present in the form of cans or pieces of foil such that the counterflow of liquid forces them back into the treatment vessel against the forces of gravity which separate out the ferrous metals and the like heavies.

As a result of this countercurrent flow back into the treatment vessel, the aluminum tends to remain in the treatment vessel until it has been separated from other heavier metals with which it may be associated, as in composite cans, and retained in the treatment vessel until it has been reduced to a sufficiently small size that it may be extracted from the vessel through the perforated bed plate thereof. Usually the size reduction is effected by wadding or crumpling to forms like a roughly spherical pellet, especially in the case of aluminum sheet as it occurs in cans and foil, and a substantial amount is also cut into relatively small pieces.

Some aluminum and relatively frangible materials such as glass, plaster and paper are removed from the treatment vessel by the junk remover along with the relatively infrangible materials. Preferably, therefore, the material ejected from the junk remover is fed through one or more separation steps to separate the two above noted types of materials. For example, the output from the junk remover may be fed through a magnetic separator from which nonferrous metals and glass return to the treatment vessel for further treatment.

Additionally, or as an alternate to the magnetic separation step, the materials may be subjected to an air separation step to separate relatively light materials from relatively heavy materials, and the lighter materials are recirculated to the treatment vessel. These various separating steps or combinations thereof are indicated diagrammatically at **16** in the drawing, along with the return line **17** for the glass and aluminum to vessel **10**.

The relatively frangible solids deposited in the vessel, as well as some of the liquid added thereto, are removed in slurry form through the perforated bed plate **11** after they have been reduced in size sufficiently to pass through the perforations in the bed plate. This slurry is delivered by a pump **18** to a centrifugal cleaner **20**, which operates to divide the solids in the slurry according to specific gravity. The fibrous material and other organics, which are relatively less dense than glass, metal and other inorganics, are separated from the inorganics for further processing, as indicated at **22**, particularly such as for purposes of fiber reclamation.

As already noted, it is desirable to control the size of the particles of the inorganics separated from the remainder of the slurry, both to facilitate later sorting operations and to provide, particularly in the case of reclaimed glass, a product which is well suited for reuse.

Maximum particle size is controlled by properly sizing the perforations through the bed plate of the treatment vessel, but it is also desirable to control minimum size, particularly of the glass particles and other inorganics. This is accomplished by recirculating back to the treatment vessel a substantial portion of the organics separated at the cleaner **20**, as indicated at **24**. This permits a relatively high flow rate through the treatment vessel, so that the solid materials are not retained in the treatment vessel for an extended period of time, which would cause the frangible materials to be broken down to a much smaller size than is desired.

In actual operation, it has been found that the average size of the glass particles varies more or less inversely with the length of time they are in the treatment vessel, but the fibrous particles need a greater reduction in particle size in order to facilitate their recovery at subsequent stations. A desired balance between these objectives is obtained by completely recycling the contents of the vessel at frequent intervals, such as a maximum of about 10 minutes and preferably several times faster, such as about every 2 minutes. These frequencies will result in recycling approximately 20 to 60% of the inorganics from the cleaner **20** to the treatment vessel, with an optimum value being approximately 40%.

It has also been noted that minimum particle size can be regulated through control of the speed of the rotor in the treatment vessel. Thus, in order to prevent an excess generation of fine materials, it is desirable to operate at relatively low rotor speeds and to utilize a fairly large rotor. Again, in actual operation it has been found that utilizing a 48-inch rotor at speeds of approximately 100 to 300 rpm provides sufficient control of minimum particle size. Utilizing these two control techniques, therefore, minimum particle size can be maintained above one-eighth inch for a major portion of the frangible materials, and generally in a range of one-half to three-fourths inch for the glass when the perforations in bed plate **11** are three-fourths to one inch in diameter.

An additional feature, and an unexpected result obtained, is that the particles of glass produced in the treatment vessel are quite uniform or regular in shape, with few slivers and a ratio of maximum to minimum dimensions of the glass particles of from approximately 1/1 to 2/1. While not completely explained, it is believed that this results from the fact that the glass is broken up while it is submerged in the liquid in the treatment vessel. In any event, the production of fairly regularly shaped glass particles provides a product which is easily handled and readily lends itself to sorting, and the relative freedom from slivers has the further advantage of minimizing the content and production of fine particles and their attendant disadvantages noted above.

The modified system illustrated in the inset drawing in the upper right hand corner of Figure 21 is especially effective for obtaining desired control of particle sizes, particularly with respect to the glass. The treatment vessel **10'** is equipped with a bed plate **25** having one section provided with substantially larger holes than the remainder of the bed plate, and each of these two sections has separate compartments **27** and **28** directly thereunder. The compartment **27** below the smaller sized perforation is connected as in the main figure to the centrifugal cleaner **20**.

The other compartment **28** is connected through a pump **29** with a centrifugal cleaner **30**, and the lighter materials separated by this cleaner, which includes much of the large fibrous particles, are recirculated to the treatment vessel as indicated at **31**. The heavy materials discharged at the tip of the cleaner are combined, as indicated at **33**, with the flow from the tip of the cleaner **20** for further treatment as described hereinafter.

In practice with the modified system illustrated in the inset drawing in Figure 21, the smaller sized perforations should not be greater than about three-fourths inch in diameter, and are preferably near the middle of the range from one-eighth to three-fourths inch. The larger perforations, above the compartment **28**, may be in the range of three-fourths to 2 inches in diameter, and preferably in the upper portion of this range. Under these conditions, solid materials which are reduced to a particle size sufficient to enter the compartment **27** do not return to the treatment vessel and are carried further in the system as described below. On the other hand, the organic particles which reach the compartment **28** are recirculated through the treatment vessel for further size reduction, but the rate of recirculation should be kept sufficiently high to minimize the time that any given particle remains in the treatment vessel. A recirculation rate as high as every 2 minutes is not too great, and in any event, the recirculation rate should not be slower than every 10 minutes.

The inorganics which are separated out by the centrifugal cleaner **20**, together with the output from the smaller end of the cleaner **30**, in the system shown in the inset drawing, are directed to a dewatering device **36** and thence to suitable drying apparatus **38**. The inorganics delivered to the cleaner **20** will comprise approximately the following percentages.

Material	% Based on Portion Delivered to Centrifugal Cleaner	% Based on Wet Refuse Added to Treatment Vessel
Glass	70-90	7.0-9.0
Magnetic metal	1-5	0.1-0.5
Aluminum	1-5	0.1-0.5
Ceramics	1-4	0.1-0.4
Other (bone, plastic, remaining nonmagnetic metal, etc.)	5-20	0.5-2.0

It will be seen, therefore, that the inorganics from dryer **38** comprise mainly nonmagnetic materials such as glass, aluminum, ceramics and various other materials. These can be separated as indicated at **40** by a magnetic separator with the magnetic materials being delivered downstream for further processing. While the maximum particle size is controlled by the size of the perforations through the bed plate of the vessel, and the minimum size can be controlled to a great extent through the techniques described above, there will nonetheless be some production of fine material. Therefore, the nonmagnetics from the separator **40** are then preferably screened, as at **42**, to remove the fines.

From the screens **42**, the dried inorganic particles within the desired range of particle sizes are separated as indicated at **44** to separate the heavy materials, including the glass and ceramics, from the lighter materials which will consist primarily of aluminum. This is done quite readily by means such as an air cyclone, notwithstanding the relatively close nominal specific gravities of glass and aluminum, because the pieces and pellets of aluminum exhibit an apparent specific gravity considerably lower than glass, by reason of their greater surface areas in comparison with pieces of glass of comparable overall dimensions.

Further, the separated aluminum is in very convenient condition for reclamation, and even if it does include minor amounts of other materials, the latter are easily burnt off when the aluminum is melted. Aside from this, it should be noted that it is highly desirable to remove the aluminum from the glass particles since the presence of aluminum detracts from the value of the recovered glass. Therefore, not only is valuable aluminum recovered in the recovery process, but the quality of the recovered glass is simultaneously upgraded.

The unsorted materials from the separator **44** will comprise glass of all kinds, ceramics, bone, plastics, and stones and the like, and they are initially sorted at **45** into two categories. One of these categories will be the clear or colorless glass, sometimes known as flint glass, and the other category will comprise the remaining material including colored glass. Techniques and apparatus for accomplishing this sorting already exist commercially, reference being made to U.S. Patent 3,650,396, to Sortex of North America, Inc. and the clear glass can be reused as cullet in the manufacture of new glass.

It is contemplated that techniques will soon be perfected for sorting the other category of solids, at least to the extent necessary to separate colored glass from the remaining opaque materials, and thereafter to sort this glass according to color. The reuse value of the opaque materials is primarily as aggregate or landfill. The unsorted material from the sorter **45** could be used for either of these purposes without further sorting, and the glass component actually contributes to its utility as aggregate, especially for use with asphalt in road building.

A process developed by *H.R. Rhys; U.S. Patent 3,802,558; April 9, 1974; assigned to Sortex Company of North America, Inc.* is a process for sorting refuse into its components wherein the refuse is comminuted into a mass of particles, and a fraction containing a mixture of glass particles, which are at least partially transparent, and substantially opaque particles are sorted from the remainder of the refuse. The particles in the separated fraction are passed seriatim through a transparency sorter which senses the degree of transparency of the particles and sorts the particles in accordance with the sensed transparency values.

The transparent sorted glass particles can also be sorted according to color during the transparency and color sorting process. The transparency sorter preferably comprises a strip light source with a plurality of photosensitive means, such as photodiodes, aligned in a row with the strip light source. The photosensitive means scan across the light source as the particles fall past the light source. Sorting is effected when a certain minimum number or percentage of the photosensitive means sense darkness as the particles pass the photosensitive means. Figure 22 is a flow diagram of that process.

FIGURE 22: ALTERNATIVE PROCESS FOR SORTING GLASS FROM MUNICIPAL WASTE

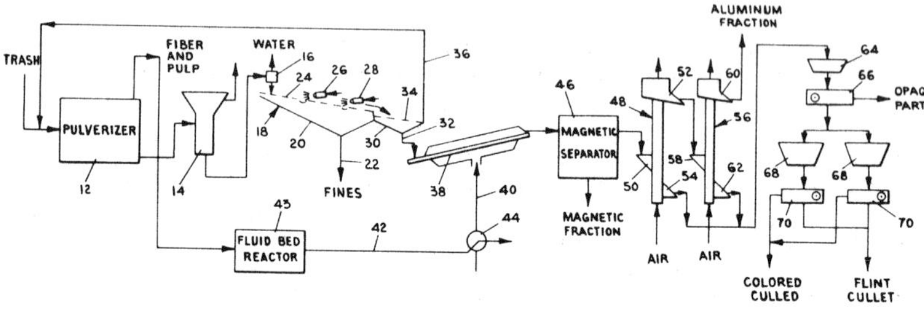

Source: H.R. Rhys; U.S. Patent 3,802,558; April 9, 1974

Domestic trash or other suitable refuse is fed to a pulverizer unit **12** which pulverizes or comminutes the feed producing a fiber and pulp fraction and a concentrated solids fraction. A suitable pulverizer is a Black Clawson Hydrapulper. The solids fraction is passed to a cyclone separator **14** which removes a water and fiber fraction from an upper portion and a more concentrated solids fraction from a bottom portion.

The dewatered solids fraction is passed by a suitable conveyor to a washer and distributor **16** where water is added thereto. The watered fraction is distributed onto a screening and washing machine **18** having a hopper **20** with an outlet **22** at the bottom thereof and a screen **24** across the top. This screen has vibratory motion imparted thereto to move the solids fraction along the top of the screen where it is subjected to a further water wash from a pair of nozzles **26** and **28** positioned above the screen. A suitable screen has ¼ inch openings to permit particles of ¼ inch or less to pass therethrough into the hopper.

These particles or fines are removed through outlet **22**; those particles larger than ¼ inch then move to the lower end of the screening and washing machine onto a second screen **34**. Vibratory motion is imparted to this screen also. This screen **34** preferably has openings of ¾ inch to permit particles smaller than ¾ inch to pass therethrough. The larger particles move to the bottom end of screen **34** where they are removed and passed through a recycle line **36** back to the pulverizer **12**. The recycle line can be a pneumatic conveyor or any suitable means for returning the oversized particles back to the pulverizer. For example, the oversize particles can even be carted back to the pulverizer in a truck. The particles collected in the hopper **30** will have a size predominately between one-quarter and three-quarters of an inch. A suitable screening and washing device is manufactured by Derrick Manufacturing Co.

The particles are removed from the bottom of the hopper through an outlet **32** and passed to a fluidized bed conveyor and dryer **38**. As the particles are passed through this fluidized bed and dryer, air is directed upwardly therethrough to remove water therefrom. This air is heated and supplied through a line **40**.

The pulp and fibrous fraction from the pulverizer **12** can be passed to a fluid bed reactor **43** after suitable water removal steps (not shown) and incinerated in the reactor **43**. Excess heat in the form of hot air, suitably scrubbed clean of noxious elements, may be bled from the fluid bed reactor, diluted with cold air if required, and admitted to the dryer either directly or through a heat exchanger **44** through line **42**.

Alternatively, the dryer may have its own provision for heat generation. Suitable machines are manufactured by the Door Oliver Co. (reactor) and the Jeffrey Manufacturing Co. (dryer). The dryer may include a cooling stage if required. Alternatively, the hot gases may be used to convey the material directly through a pneumatic conveyor (Meyer Machine Co.).

The dried particles are removed from an upper end of the dryer **38** and passed to a magnetic separator **46**. Any suitable magnetic separator can be used. An example of such a separator is a rotating drum type magnetic separator (Dings Magnetic Co.).

A magnetic fraction is separated from a nonmagnetic fraction in the separator **46**. The nonmagnetic fraction is passed to a first pneumatic separator **48**. This separator has an inlet **50**, a light fraction outlet **52**, and a heavy fraction outlet **54**. Air is supplied to a bottom portion of the separator and passed upwardly therethrough to entrain particles of lighter specific gravity. The particles of lower specific gravity are removed through outlet **52** and passed to inlet **58** of a second pneumatic separator **56**. This separator has outlet **62** for a heavy fraction and an outlet **60** for a light fraction.

Air is passed upwardly through the separator **56** to separate particles of lower specific gravity from particles of higher specific gravity. The lower specific gravity particles are removed through outlet **60**. The flow of air through separators **48** and **56** is regulated to entrain those particles whose specific gravity is lighter than glass and to permit glass particles to settle to the bottom of the separator for removal. This lighter fraction contains predominately lightweight metals such as aluminum, stones, ceramics, bones, wood, rubber and plastics. This lighter fraction can be passed to an aluminum concentrating apparatus where an aluminum rich fraction can be recovered for reuse.

The separation is achieved by regulation of the upward air velocity to split the feed according to the different settling velocities of its particulate components. This differential is a function of article shape as well as specific gravity. The split can be made in free or hindered settling conditions. A suitable air separator is manufactured by Sortex Co. of North America, Inc.

The heavier fractions separated from separators **48** and **56** comprise essentially glass and some residual stones and ceramic, and conceivably some balled up aluminum. These heavy glass fractions are combined and passed to a hopper dispenser **64** which dispenses the particles one at a time to a transparency sorting apparatus **66** wherein the particles are separated according to their ability to pass light. The opaque particles, such as the stones, crockery, and other ceramics, metals, wood, hard rubber, bone, clinker, etc. are separated out and the resulting glass fraction is passed to a pair of hopper dispensers **68** which dispense the particles one at a time to a photometric sorting apparatus **70**.

The glass particles are separated according to color in the photometric sorting apparatus producing a colored cullet fraction and a flint colored fraction. The flint colored fraction, which is worth considerably more than the colored cullet fraction, can be remelted and used for making clear bottles or other types of clear glassware. The colored fraction can be either further separated by other photometric sensing means (not shown) into different colors, or can be used as a melt for brown bottles.

The electronic color sorting of glass from urban waste has been discussed by F.J. Palumbo, M.H. Stanczyk and P.M. Sullivan in Technical Progress Report No. 45, Washington, D.C., U.S. Bureau of Mines (October 1971).

An electronic color-sorting machine was tested for recovering colorless glass concentrates from mixed glass products. The tests were made using glass recovered from unincinerated (raw) urban refuse and residues resulting from the incineration of refuse. The results of the study revealed that concentrates analyzing about 86% colorless glass could be recovered from mixed unincinerated glass containing 54% colorless glass material. Recovery of the colorless glass was about 62%. Color sorting the incinerated glass was somewhat more difficult. Multiple-stage processing was required to recover 59% of the colorless glass material in a product analyzing 89% colorless glass. The feed material analyzed 41% colorless glass.

Use in Building Brick Manufacture

A study of the fabrication of experimental building bricks from waste glass has been reported by M.E. Tyrrell, I.L. Field and J.A. Barclay in Report of Investigations No. 7605, Washington, D.C., U.S. Bureau of Mines (1972).

The Bureau of Mines produced building brick of good quality in an investigation designed to explore the possible utilization of high-glass fractions of municipal incinerator residues in the manufacture of building products. Common clays, lime, and sodium silicate were used as binders for the nonplastic residues. Test bricks were formed by dry pressing, soft mud molding, and stiff mud extrusion. A run of standard size face brick dry pressed and fired in the West Virginia University Coal Research Bureau pilot plant demonstrated that SW-grade glass-clay face brick can be produced on a commercial scale without difficulty.

Another related study involved the use of waste glass as a flux for brick clays as reported by M.E. Tyrrell and A.H. Goode in Report of Investigations No. 7701, Washington, D.C., U.S. Bureau of Mines (1972).

Work in this investigation was undertaken by the Bureau of Mines to demonstrate how waste glass might be utilized to lower the required firing temperature and firing time of structural clay products. Mixtures of waste glass and common clays were formed by stiff mud extrusion and fired to the temperature required to produce a body having 6 to 7% absorption. Substitution of glass for one-half of the clay in a red body reduced the firing temperature 500°F, which made possible a 30% increase in production; in a tan body, the firing temperature was reduced 400°F, which made possible a 23% increase in production.

GRINDING MATERIALS

Recovery from Glass Grinding

A process developed by *W.F. Brown, G.C. Mook, J.J. Jarosi and E.H. Cunningham; U.S. Patent 3,257,081; June 21, 1966; assigned to Libbey-Owens-Ford Glass Company* involves recovering the extremely finely divided and intimately associated mixture of pulverized sand and glass particles obtained as a waste product in the grinding or surfacing of plate glass.

Present day sheet glass and plate glass are generally made by charging into one end of a continuous melting tank a glass batch comprised primarily of sand, lime and soda ash, along with lesser amounts of minor ingredients to impart various desired characteristics to the finished glass. A certain amount of cullet, or broken glass, is also charged into the furnace along with the batch materials, either intimately admixed therewith or separately as in a layer or blanket upon which the batch materials are supported for movement into the melting area of the tank, and the molten glass then flows successively through melting (refining and conditioning areas) into the working area where a sheet or plate is continuously drawn or rolled therefrom.

In the production of plate glass, the relatively viscous molten glass is withdrawn from the working end of the melting tank between a pair of counterrotating forming rolls which reduce the glass to a continuous rough blank or ribbon of predetermined width and thickness. Due to the nature of its manufacture, the sheet or ribbon has relatively rough or irregular surfaces. In order to smooth the surfaces as well as to provide parallelism therebetween, the sheet is subjected to a surfacing or grinding operation with an abrasive material such as ordinary sand. The abrasive material is usually fed onto the glass sheet as a slurry where it is utilized by a series of grinding runners, generally of cast iron, as a grinding medium to wear away irregularities in the surface of the sheet.

During the course of the grinding operation, the sand particles are gradually worn down and as they are thus worn down, the particles are reclassified and used at subsequent stages in the grinding process where the sheet approaches its final smoothness until such time as they become so small as to be no longer effective in grinding. The spent grinding slurry, which also contains extremely fine particles of glass eroded from the sheet and fragments of iron abraded from the grinding runners in addition to the particles of sand, has heretofore been discarded as being of no value and, in fact, disposal of the material has constituted a major problem due to the large quantities involved.

Various uses for the material have been proposed, one of the most important of which has been as constituent in the batch materials for making additional glass. However, due to its extremely small particle size, which averages in the neighborhood of from 2 to 5 microns, the material has been difficult to collect in a usable form on an economic basis. According to this process, as shown in the process flow diagram in Figure 23, waste grinding slurry is collected from one or more discharge points in the grinding system and treated to remove any impurities, such as iron, which may be objectionable in the ultimate use of the recovered product.

FIGURE 23: PROCESS FOR RECOVERY OF WASTE GRINDING MATERIALS IN GLASS MANUFACTURE

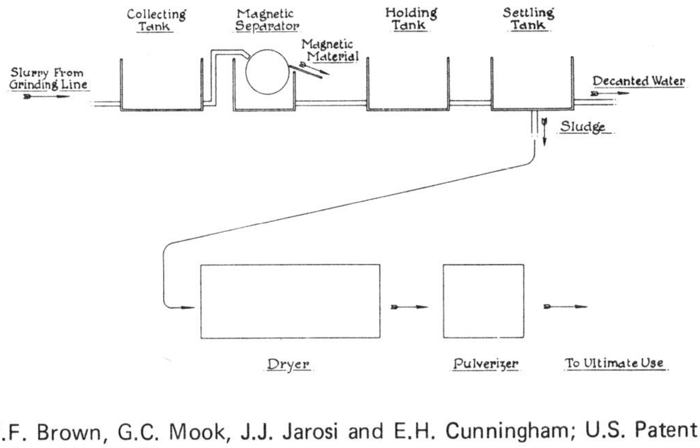

Source: W.F. Brown, G.C. Mook, J.J. Jarosi and E.H. Cunningham; U.S. Patent 3,257,081; June 21, 1966

After purification, the slurry is transferred to a settling tank where predetermined quantities of flocculating agents are added and rapidly dispersed through the slurry. Thereafter, a floc is formed which causes the suspended solids to settle, leaving relatively clear water overhead which is decanted or drawn off. The settled material, or sludge, is then dewatered to a predetermined moisture content, or even completely dried, to become a frangible, easily crushed cake. Depending upon the ultimate use to which the recovered material is to be put, the cake may thereafter be crushed to break down the agglomerated material to a desired particle size which, in the case of a glass batch constituent, is generally the original size of the sand and glass particles.

Recovery from Grinding Swarf

The reader is referred to the section bearing this heading under "Ferrous Metal: Steel" in this volume for a discussion of recovery of silicon carbide and aluminum oxide for reuse after the grinding of steel.

HEAT

The optimum solution to problems associated with waste heat disposal would be the use of rejected heat for beneficial purposes when dealing with power plant discharges, for example. However, one is confronted with immense quantities of water which are of low quality when considered as a heat source. Utilization of waste heat in this form is therefore restricted to only a very few possible applications.

One potential for such utilization is aquaculture, the farming of plants or animals in fresh or salt water. Aquaculture has been successfully practiced in Japan and elsewhere. Heated discharges could be used to enhance the environment and increase production of commercially valuable species such as pompano, catfish, shrimp, oysters, and scallops. Research, development, and pilot studies are in progress to determine the feasibility of such cultivation in American waters.

Some nuclear plants could possibly be located to provide beneficial use of heated discharges in keeping shipping lanes free from ice for extended seasons. There have been recommendations for such application to the St. Lawrence Seaway, proposing extension of the shipping season to the end of December or even January. Heated discharges may have application in irrigating and creating controlled environments for agricultural crops. In this manner, growing seasons could be lengthened in certain areas for common crops, and subtropical or tropical varieties might be produced where they are not normally adaptable as described by A.G. Christianson and B.A. Tichenor in *Industrial Waste Guide on Thermal Pollution,* PB 197 262, Springfield, Va., Nat. Tech. Information Service (Sept. 1968).

Cooling water may provide heat to warm swimming areas. Such use should be guided by the National Technical Advisory Committee on Water Quality Criteria which recommends: "In primary contact recreation waters, except where caused by natural conditions, maximum water temperature should not exceed 30°C (85°F)."

Thermal discharges may also provide heat for desalination or other processes, some of which might require higher temperature water than that normally discharged from steam plants. In such cases, it may be possible to implement a "trade-off" whereby it would be economical to alter the usual process for other benefits. For example, steam from the final turbine exhaust may be 80°F and 0.5 psia. If the steam were tapped off in front of the final turbine stage it may be 250°F and at 10 psig, which might be worth more for some other use than its final turbine passage. Marketable steam for use by other industries, for home heating, or for any other purpose may warrant trade-off situations which would help to reduce waste heat loads to streams and other water bodies.

Creation and application of improved technology for electrical energy conversion is an indirect, but effective, control measure. The advent of the fluorescent light is a classic example of getting more useful work from a watt of electricity. Other fertile areas are the heating, refrigeration, air conditioning and motor industries.

RECOVERY FROM GASEOUS WASTES

Recovery from Fertilizer Plant Waste Streams

As discussed in this volume, both under "Ammonium Nitrate: Recovery from Gaseous Wastes" and under "Urea: Recovery from Gaseous Wastes," waste process heat can be utilized and fertilizer materials simultaneously can be recovered in a process described in U.S. Patent 3,798,021.

Recovery from Styrene Plant Reactor Effluent

A process developed by *R.E. Hughes; U.S. Patent 3,691,020; September 12, 1972; assigned to The Badger Company, Inc.* is one in which low-temperature-level waste heat recovered from the dehydrogenation reactor effluent of a styrene manufacturing unit is used in the flash evaporator of a water desalination unit to produce pure water from contaminated or high-mineral-content water.

In the typical styrene manufacturing involving dehydrogenation of ethylbenzene in the presence of steam the effluent from the dehydrogenation reactor is heat exchanged with the incoming feed to recover high-temperature-level heat. Subsequently, the reactor effluent is cooled further to condense out steam and some high-boiling organic by-products at a comparatively low temperature level and then passed to a fractional distillation section for separation and recovery of unreacted ethylbenzene, styrene and tar residue.

Although the heat removed in this secondary cooling and condensing step is at a low temperature level, the quantity of heat is quite large. Accordingly it has been suggested (see British Patent 921,081 and U.S. Patent 3,294,856) that this heat be recovered and used to generate low-pressure steam or to provide a heating medium for distillation columns. However, many styrene plants are located in regions where it is necessary to treat brackish groundwater or seawater to provide water suitable for human consumption and domestic and industrial uses. Such treatment is expensive, with a major cost being the heat required for vaporizing the water so as to effect separation and removal of impurities such as salt and other minerals.

Accordingly the chief object of the present process is to provide a method and apparatus system for utilizing some of the heat content of the dehydrogenation reactor effluent in a styrene plant to produce high-purity water from a contaminated source such as salt or brackish water.

RECOVERY FROM LIQUID WASTES

Recovery from Bayer Alumina Process

A process developed by *C. Sato, Y. Yamada and Y. Takenaka; U.S. Patent 3,869,537; March 4, 1975; assigned to Showa Denko KK, Japan* involves the recovery of waste heat from the steam which is generated when bauxite-dissolved slurry is subjected to flashing in the manufacture of alumina by the Bayer process and for the removal of sand from the slurry.

The production of alumina by the Bayer process may be broadly divided into two steps. In the first step, bauxite is worked with the aqueous solution of caustic soda to dissolve alumina and obtain a slurry which contains the aqueous solution of sodium aluminate and insolubles. In the latter step, solids are removed from this slurry and then hydrated alumina is precipitated from the aqueous solution of sodium aluminate. Bauxite containing monohydrated alumina is dissolved with the aqueous solution of caustic soda at an elevated temperature under an increased pressure in a reactor called a "digester." The slurry of high temperature and pressure formed in the digester is subjected to a plurality of stages of flashing so as to be successively lowered in both temperature and pressure. It is sent through a process designed to separate the sand, namely, coarse particles predominantly of quartz, from the slurry. Then it is forwarded to the red mud thickener.

Steam is generated as the slurry is freed of high temperature and pressure in the flashing operation. The heat of this steam is recovered for heating the aqueous solution of caustic soda called spent liquor which is used for dissolving bauxite. As the slurry is passed through the successive stages of flashing, both temperature and pressure continue to fall by degrees. The steam generated in the last stage of flashing has a pressure equal to the atmospheric pressure and, therefore, does not have sufficiently high temperature. Unable to find any effective use, this steam has to date been blown off and wasted. Generally, the separation of sand is carried out subsequent to the blow-off. For the separation of sand, there are generally employed methods which utilize a liquid cyclone and other types of classifiers.

According to the conventional method, the discharge of waste steam and the separation of sand are carried out independently of each other. Lacking effective use of the steam, this method inevitably involves heavy consumption of steam. It also requires a special device designed exclusively for the removal of sand. This, accordingly, necessitates installation of pumps to effect the transfer of the slurry to these separate devices. Slurry pumps are highly liable to develop mechanical trouble and demand troblesome maintenance. It is, therefore, desirable to minimize use of slurry pumps.

The slurry which has been deprived of sand is washed in a multistage washing thickener to separate solids called red mud and produce alumina-dissolved liquid called pregnant liquor, which is forwarded to the process designed to precipitate hydrated alumina. It is the main object of this procedure to provide a method and a device for heightening the ratio of heat recovery from the waste steam and improving the operation of sand removal. The general outlines of the process are shown in Figure 24.

FIGURE 24: FLOW DIAGRAM OF BAYER ALUMINA PROCESS

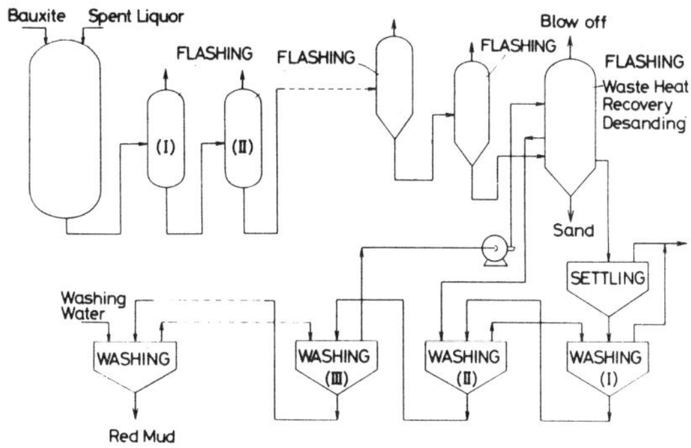

Source: C. Sato, Y. Yamada and Y. Takenaka; U.S. Patent 3,869,537; March 4, 1975

As shown there, the digester serves to cause reaction of bauxite with the aqueous solution of caustic soda under conditions of high temperature and high pressure such as, for example, 230° to 240°C and 30 to 40 kg/cm^2. The alumina component of the bauxite is dissolved by the aqueous solution of caustic soda. In conjunction with the sand (coarse particles predominantly of quartz) and the red mud composed mainly of Fe_2O_3, the dissolved alumina is discharged in the form of slurry out of the digester.

This slurry has high temperature and high pressure and, therefore, must be sent through a plurality of flashing tanks so as to be lowered successively in both temperature and pressure. While in passage through the flashing tanks, the slurry releases steam. The steam generated in consequence of the flashing operation is recovered for preheating the aqueous solution of caustic soda which is used for dissolving bauxite. The last-stage flashing tank serves not only as a flashing tank but also as the waste heat recovery and desanding device.

Recovery from Power Plant Coolants

A process developed by *A.C. Kuehn and R.M. Kuebler; U.S. Patent 3,822,504; July 9, 1974* provides a scheme for using an aquaculture housing pod and a method of culture employing heated water from a power plant in which the water temperature is cooled to near ambient for safe discharge into the ambient ecology. The aquaculture housing pods are designed to cover large areas, on the order of 150 acres each, and comprise a column and cable structure supporting overhead sprinkler systems for discharge on hydroponic beds contained under the housing pod structure. A series of aqueducts and hydroponic planting beds are arranged underneath the structure.

The method involves discharging hot water from a power plant into a first holding and distribution reservoir, withdrawing water from the reservoir and passing it through either or both the aqueducts and sprinkler system of the housing pods, collecting excess water from the housing pod, and returning it to a neutral temperature reservoir before either recycling the cooled water to the power plant or discharging it to the environment. The aquaculture housing pod is simple of construction and inexpensive, and the method provides for cooling the water and discharge to the environment without harming the ecology.

The reader is also referred to the volume entitled *Utilization of Waste Heat from Power Plants* by D. Rimberg, Park Ridge, N.J., Noyes Data Corp. (1974) for more detailed information on this general topic.

RECOVERY FROM SOLID WASTES

Recovery from Incineration of Municipal Wastes

The reader is referred to *Energy from Solid Waste* by F.R. Jackson, Park Ridge, N.J., Noyes Data Corp. (1974) for a detailed treatment of this topic. Some discussion is given in this volume under "Municipal Waste" but in the interests of holding this volume to a reasonable size, the reader is referred to the more comprehensive volume cited above.

Recovery from Incineration of Plastics and Rubber

The reader is referred to *Pollution Control in the Plastics and Rubber Industry* by M. Sittig, Park Ridge, N.J., Noyes Data Corp. (1975) for more detail on this topic. As noted there incineration accompanied by heat recovery may be a particularly attractive technique for disposing of waste rubber tires.

INORGANIC MATERIALS

AMMONIA

Recovery from Gaseous Wastes

Recovery from Fertilizer Plant Exit Gases: A process developed by *W.J. Sackett, Sr.; U.S. Patent 3,499,731; March 10, 1970* provides a fully implemented control system for preventing the discharge of obnoxious wastes from plants such as those of the ammonium phosphate fertilizer types. The problem that the wastes are of different vapors and dusts in differing combinations is solved by the use of regenerating combinations of dry and wet cyclones and scrubbers taking into account that certain effluents are at elevated temperature. These effluents in an ammonium-potassium phosphate fertilizer plant are: (a) ammonia, water vapor, and fumes and finely divided ammonium and potassium chloride particles from the ammoniator-granulator; (b) ammonia and combustion dust products from the dryer; (c) ammonia and product dust from the cooler.

A process developed by *H.J. Clausen; U.S. Patent 3,687,618; August 29, 1972; assigned to Cities Services Company* is one in which ammonia can be recovered from the exit gases of an ammonium polyphosphate plant by scrubbing the reactor exit gases with partially ammoniated superphosphoric acid having a pH above about 3, and recycling the product obtained thereby to the reactor. In this way the production of orthophosphate is minimized. In the manufacture of ammonium polyphosphates from concentrated superphosphoric acid at atmospheric pressure, the reactor exit gases, containing substantial quantities of ammonia, are usually either vented to the atmosphere or are recovered in a way which precludes their use in making additional ammonium polyphosphate fertilizer.

Recovery from Textile Treating Effluents: In recent years, with such developments as permanent-press garments and demanding industrial requirements for textiles formed into specific shapes, there has been increased emphasis on treatment of fabrics with liquid ammonia. Large quantities of ammonia find their way into such processes, so there is a consequent involvement in disposing of or reclaiming ammonia effluent from ammonia treatment chambers. Unfortunately, recovery of ammonia from treatment chambers has been frustrated. Some difficulties encountered in ammonia recovery have been explosion hazards, unreliability, interference with (and worse, disruption of) the prime system of processing textiles.

Last, but certainly not least, expense has been prohibitive. Textile mills have been able to find more probitable ways to invest their captital than in ammonia recovery systems. A clear and present need has arisen for ammonia recovery from effluents of ammonia treatment chambers, within operational and economic parameters with which the textile industry can cope.

A process developed by *G.C. Briley, J.J. Shepherd and T.A. Lyons; U.S. Patent 3,721,097; March 20, 1973; assigned to Cluett Peabody & Co., Inc.* is one in which ammonia is recovered and liquefied from a mixture of ammonia and air emitted from an ammonia treatment chamber used in processing of textiles. The processing of textiles is regarded herein as the prime system. In the recovery system, subordinated to the prime system, cooling and compression are followed by condensing then purging. Noncondensibles are oxidized in a furnace and liquid ammonia is returned for reuse to the treatment chamber of the prime system. Controlling delivery of the mixture from the treatment chamber based on pressure therein accommodates maintenance of a minimal negative pressure in the treatment chamber.

The furnace is used also to dispose of the ammonia and air mixture during startup and shutdown of the prime system. The furnace also may be used to oxidize ammonia from the treatment chamber so that the treatment chamber can function in its normal manner even in the event the recovery system is out of service. When feed to the recovery system is interrupted, a recycled stream is brought into play to load falsely the compressor so that shutdown thereof is avoided.

Recovery from Liquid Wastes

Recovery from Urea Plant Wastewater: A process developed by *I. Mavrovic; U.S. Patent 3,826,815; July 30, 1974* is one in which ammonia is recovered from a waste, liquid effluent comprising a weak ammoniacal aqueous solution containing ammonia and/or ammonium carbamate or ammonium carbonate by stripping the solution with steam and an inert gas in a fractionator at specified conditions. Substantially all of the ammonia is recovered in an overhead gaseous product and a bottoms liquid product which is essentially condensate substantially free of ammonia is also recovered.

It is known that in the synthesis of urea from NH_3 and CO_2 as starting raw materials, for each two mols of NH_3 reacted with one mol of CO_2 to form one mol of urea, one mol of process water is formed in the reactor. In a subsequent step of producing solid urea from an aqueous solution thereof obtained from a synthesis section, this stoichiometric amount of process water formed in the reactor is separated from urea either by evaporation or by crystallization and invariably contains from 0.5 to 4.0 weight percent of residual NH_3. This poses serious problems in the waste disposal system employed and financial losses through rejection of the residual NH_3.

Various methods have been suggested in the past for the recovery of this residual amount of ammonia from such waste effluents before rejection to sewer. One of these known methods involves feeding such an ammoniacal process water solution into a bubble cap or sieve tray of a rectification column operating at about 20 to 50 psig, steam heating the resulting bottoms directly or indirectly, and partially cooling the top section of the column to provide the proper refluxing of the solution through the tray section thereof. The ammonia expelled from the solution is removed overhead with water vapor in equilibrium with the refluxing solution.

The main drawback of such a method is the fact that the bottoms discharged (as to a sewer) from such a column usually contain more than 300 ppm of residual NH_3 in solution. The disposal of such a wastewater represents a problem to the plant operator due to contamination of the surroundings with ammonia.

Another such method, described in U.S. Patent 3,488,293, discloses the use of a similar apparatus, as described above, but having the purpose of reducing the water content of a gaseous phase containing about 20 to 25 weight percent of NH_3 and about 10 to 15 weight percent of CO_2 together with water vapor. Such a gaseous phase is obtained from a urea synthesis reactor effluent stream subsequent to the letdown in pressure from a urea synthesis reactor. This method has several drawbacks. One consists of the fact that from such a gaseous mixture fed to the top of the rectification column, ammonium carbamate is formed on the top trays of the rectification column with a consequent relatively high corrosion rate of the stainless steel equipment in contact with the process fluid. The second is that, in order to reduce the corrosion rate of the stainless steel equipment exposed to the process, fluid oxygen must be fed to the rectification column in order to maintain the surface of the wetted parts of the equipment passivated against corrosion; however, due to the addition of this passivating oxygen and due to the fact that the gaseous phase separated from the reactor effluent stream and fed to the rectification column invariably contains various explosive gases as H_2, CH_4, etc., the operation of such a rectifier is very precarious and must be subject to careful measuring and diluting of oxygen feed with an inert gas such as CO_2 for instance, prior to introduction into such a rectifier in order to avoid an explosion of the apparatus.

On the contrary, according to the process, a weak aqueous solution containing relatively small amounts of NH_3 (e.g., from about 0.1 to 10 weight percent) and ammonium carbamate instead of a gaseous phase rich in NH_3 and CO_2 is fed to the rectification column. Such a weak aqueous solution is substantially free of volatile explosive gases as H_2 or CH_4 because they are separated from the liquid phase together with the remaining NH_3 and CO_2 gaseous phase in a previous step not described here, and the solution inside the rectification column is practically noncorrosive because substantially all the CO_2 entering the column is in the form of ammonium carbamate or ammonium carbonate and not in the form of gaseous CO_2 which could cause corrosion on the column internals due to reaction with NH_3 to ammonium carbamate or ammonium carbonate. If desired, passivating air instead of oxygen can be fed to such a rectification column without any danger of explosions.

Inasmuch, it has been found that by adding an inert gas, preferably CO_2, to the bottom section of such a rectification column the ammonia content on the wastewater collected in the bottom section of the rectifier and rejected to sewer can be reduced well below 20 ppm without incurring corrosion problems of the rectifier internals.

Advantages attained with this method include: NH_3 is recovered very economically from weak ammoniacal aqueous solutions; corrosion problems of the rectifier internals and explosion hazards are eliminated, and the process water is rejected from the system with a practically undectectable residual ammonia content and well below 20 ppm, thus eliminating the effluent waste disposal problem.

Thus, there is provided a process wherein ammonia gas with a relatively low moisture content is recovered from such aqueous solutions by steam stripping and CO_2 stripping at medium pressures between from about 20 to 400 psig in a bubble cap or sieve tray column, and the remaining purified liquid phase with a relatively low content of total ammonia of about 4 to 5 ppm is also recovered. The NH_3 can be present in solution in one or more of the following forms: free dissolved ammonia, ammonium carbonate or bicarbonate, ammonium carbamate, urea, etc.

Figure 25 shows a suitable form of apparatus for the conduct of the process. Referring to the figure, solution in line **10** contains about 2 weight pereent NH_3, about 1 weight percent CO_2 as ammonium carbamate and/or carbonate, about 0.5 weight percent urea and about 96.5 weight percent H_2O. The solution in line **10** can be withdrawn, for example, from the hotwell of a vacuum surface condenser of a crystallizer (not shown), in which urea solution is crystallized under vacuum to form urea crystals and mother liquor. The water evaporated under vacuum in the cyrstallizer carries over small amounts of urea mist and mother liquor, together with residual ammonia and carbon dioxide originally present in the urea solution fed to the crystallizer for processing.

The solution in line **10**, at about 120°F for example, is elevated in pressure by means of pump **11** to about 150 psig and is divided into two streams, **12** and **13**, in an approximate ratio of about 1:4, respectively. Stream **13** is passed through tubes **14** of economizer **15** in indirect heat exchange with stream **16**. The latter is withdrawn from bottom section **17** of recovery stripper **18** at about 370°F, or more precisely at the temperature at which water condensate boils at the pressure prevailing in the bottom section **17** of recovery stripper **18**. In economizer **15**, stream **13** is heated to about 320°F and stream **16** is cooled to about 150°F.

FIGURE 25: PROCESS FOR RECOVERY OF RESIDUAL AMMONIA FROM SOLUTION

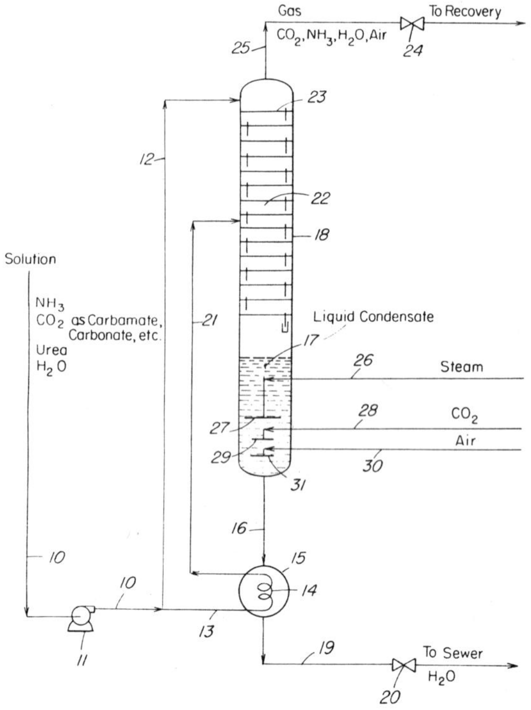

Source: I. Mavrovic; U.S. Patent 3,826,815; July 30, 1974

Stream **16** is discharged to sewer through line **19** and valve **20.** The latter serves to maintain a constant level of condensate in the bottom section **17** of recovery stripper **18.** The heated stream **13** is passed through line **21** and is introduced approximately into the midsection of the bubble cap or sieve tray section **22** of recovery stripper **18.** If economizer **15** is omitted, lines **13** and **21** are also omitted and stream **10** is not split into two streams. Instead stream **10** is fed in its entirety to top tray **23** of tray section **22** through line **12,** and the relatively hot stream **16** is discharged to sewer through line **19** without heat recovery.

The pressure inside recovery stripper **18** is maintained at about 150 psig, and is controlled by means of valve **24** located on overheat line **25.** The recovery stripper **18** can be operated within the pressure range of from about 20 to 400 psig with desired results. The solution from lines **12** and **21** is heated inside the tray section **22** of recovery stripper **18** in countercurrent contact with an uprising gaseous stream substantially containing NH_3, CO_2 and water vapor. Water vapor condenses out from the gaseous phase and provides on the trays the major part of heat required to heat the downcoming solution.

Practically all of the ammonia originally contained in stream **10** is expelled from the liquid phase in stripper **18** due to heating and countercurrent stripping with the uprising gas. The urea originally contained in stream **10** is hydrolyzed to ammonium carbamate in the tray section **22** and in the bottom section **17.** The ammonium carbamate is decomposed to gaseous NH_3 and CO_2, and these gases are stripped from the liquid phase in stripper **18.**

The equivalent amount of heat required to decompose carbamate to NH_3 and to CO_2 and to strip the gases from the liquid phase is supplied to the recovery stripper **18** by means of steam preferably passed through line **26** and distributed through steam sparger **27** immersed in the liquid phase collected in the bottom section **17**. The steam is saturated and flows through the liquid phase to function as a stripping agent for the residual NH_3 dissolved in the liquid phase of bottom section **17** and tray section **22**. Instead of feeding live steam to the bottom section **17**, heat can be supplied to recovery stripper **18** by indirectly condensing steam in a coil (not shown) immersed in the liquid phase in bottom section.

CO_2 gas is supplied to the bottom section **17** through line **28** and through spacer **29** immersed into the liquid phase of bottom section **17** and is preferably positioned under steam sparger **27**. If recovery stripper **18** and its internals exposed to the mildly corrosive action of the refluxing solution therein is made of stainless steel, a small amount of air (or oxygen) is optionally introduced into the bottom section **17** of recovery stripper **18** to passivate the stainless steel. The amount of air (or oxygen) used is equivalent to from about 0.1 to 5 mols of oxygen per 100 mols of carbon dioxide present in line **25**. The air (or oxygen) is introduced through line **30** and through sparger **31** immersed into the liquid phase. The sparger is preferably positioned below spargers **27** and **29**.

The stripped gas, containing NH_3, CO_2, water vapor and air and flowing upwardly through recovery stripper **18**, is discharged overhead from stripper **18** through line **25** for recovery in another section, not shown. The liquid discharged to sewer from recovery stripper **18** through line **19** is substantially all condensate and contains only about 4 to 5 ppm of total residual ammonia as free NH_3 and ammonia fixed either as carbamate or as urea.

The amount of inert gas so charged to the stripper through line **28** can be varied from about 0.1 to 10 mols of inert per mol of ammonia as free ammonia dissolved in water, ammonia fixed in ammonium carbamate, ammonium carbonate, ammonium bicarbonate, urea or a mixture thereof. With CO_2 as the preferred inert, preferred operation is with from about 0.5 to 2 mols of CO_2 per mol of ammonia, as defined in the charge; and with an inert other than CO_2, the preferred range is from about 3 to 6 mols.

AMMONIUM NITRATE

Recovery from Gaseous Wastes

Recovery from Fertilizer Plant Waste Streams: A process developed by *D.F. Bress and C.A. Petrarca; U.S. Patent 3,798,021; March 19, 1974; assigned to Foster Wheeler Corp.* is primarily concerned with two pollution streams, namely, a first stream from the neutralizer, containing nitric oxide, nitric acid, ammonium nitrate and water, and a second stream from the top of the prill tower containing air and entrained solid ammonium nitrate. Conventionally, the first stream is either passed into the atmosphere or is condensed and conveyed into a sewer line, and the second stream is exhausted into the atmosphere. It should be noted that ammonium nitrate can be fairly devastating to the surrounding land and in particular, adversely affects water streams in the area.

In the process, the first stream containing water, vapor and contaminants produced during the manufacture of fertilizer is passed into a heat exchange zone in indirect heat exchange relation with the second stream containing air and entrained fertilizer solids such that a portion of the water, vapor and contaminants from the first stream are condensed to increase the water-carrying capacity of the second stream. The condensed portion of the first stream is recirculated in countercurrent contact with the second stream from the heat exchange zone such that water free of contaminants is vaporized from the condensed portion of the first stream and entrained solids are removed from the second stream. Thus, substantially pure water and air can be removed and a fertilizer containing solution is produced which can be used in the manufacture of fertilizer.

Recovery from Liquid Wastes

Recovery from Fertilizer Plant Waste Streams: A process developed by *I.R. Higgins; U.S. Patent 3,775,088; November 27, 1973; assigned to Small Business Investment Company of New York, Inc.* relates to a method of treating a fertilizer plant waste stream containing ammonia and nitrate values by continuous ion exchange to recover the same as fertilizer product and concomitantly to condition the stream so that it can advantageously be employed as a recycle stream.

Figure 26 shows a suitable form of apparatus for the conduct of the process. The apparatus shown diagrammatically in the drawing comprises a cation exchange column loop system **10** in combination with an anion exchange column loop system **13** for recovering valuable fertilizer values from a fertilizer plant waste stream while simultaneously producing from the same waste stream process water suitable for recycle within the total system, thus providing an ideal overall operation from both an ecological and economical viewpoint.

The continuous cation exchange column loop system **10** has two operating ion exchange sections, i.e., a loading section **12** and a stripping section **14**. The direction of liquid flow is indicated by the solid arrows **a** and the direction of resin flow is indicated by the open arrows **b**. The loading section **12** is positioned adjacent the upper portion of one of the legs of the column loop and fertilizer plant waste enters through conduit **16** adjacent the upper portion of the loading section **12**. This loading section contains a strong acid-type cation exchange resin in the acid or salt form, and as the fertilizer plant waste stream flows downwardly therethrough the hydrogen ions of the ion exchange resin are exchanged into solution and the ammonium and metal ions such as calcium and magnesium are adsorbed on the resin.

Thus, a softened product water is withdrawn from the lower portion or region of the loading section **12** through conduit **18** and is delivered to the anion exchange column loop system **13**, as will be described in greater detail below, for further processing. Thus, the ammonium and metal-ion-rich-resin in the loading section **12** is shifted upwardly around the upper loop end **20** and is passed downwardly to the stripping section **14**. In stripping section **14** a nitric acid regenerating solution is passed upwardly through the bed of loaded resin, and the ammonium and metal ions loaded thereon are exchanged into solution with adsorption of hydrogen ions from the regenerating solution. Thus, an ammonium nitrate-containing stream is removed from the stripping section **14** via conduit **22** and is delivered to ammonium nitrate product recovery station **23**.

The aqueous stream leaving the loading section **12** of the cation exchange column loop system **10** via conduit **18** is characterized in that it has substantially reduced hardness and ammonia values. The stream is delivered, directly or indirectly via storage vessels, not shown, to the continuous anion exchange column loop system **13** which has two operating anion exchange sections, i.e., a loading section **12'** and a stripping section **14'**. Again, the direction of liquid flow is indicated by the solid arrow **a** and the direction of resin flow is indicated by the open arrow **b**. As in the cation exchange column loop system **10**, the aqueous stream in these sections of the anion exchange column loop system generally flows countercurrently to the movement of the resin.

The softened and reduced ammonia value aqueous stream from cation exchange column loop **10** is introduced into the loading section **12'** via conduit **16'**. The loading section **12'** contains a weak base anion exchange resin and as the softened aqueous stream flows therethrough, the hydroxyl ions are exchanged into solution and the nitrate anions of the softened aqueous stream are absorbed on the resin. Thus, a substantially contaminating nitrate anion and cation-free aqueous stream, i.e., a demineralized stream is taken off from the loading section **12'** via conduit **18'**, the demineralized aqueous stream advantageously being recycled within the overall operation as a rinse medium for both the regenerated cation and anion resin via conduits **38** and **38'**. The contaminating nitrate anion-rich resin in the loading zone **12'** is shifted to the stripping section **14'**. In stripping section **14'**, a regenerating agent, an aqueous ammonia solution, is passed upwardly through the resin and the nitrate anions are exchanged into solution with absorption of hydroxyl ions. Thus, an ammonium nitrate-containing stream is removed from the stripping section **14'** via conduit **22'** and is delivered to ammonium nitrate product recovery station **23**.

FIGURE 26: RECOVERY OF AMMONIUM NITRATE FROM WASTEWATER BY ION EXCHANGE

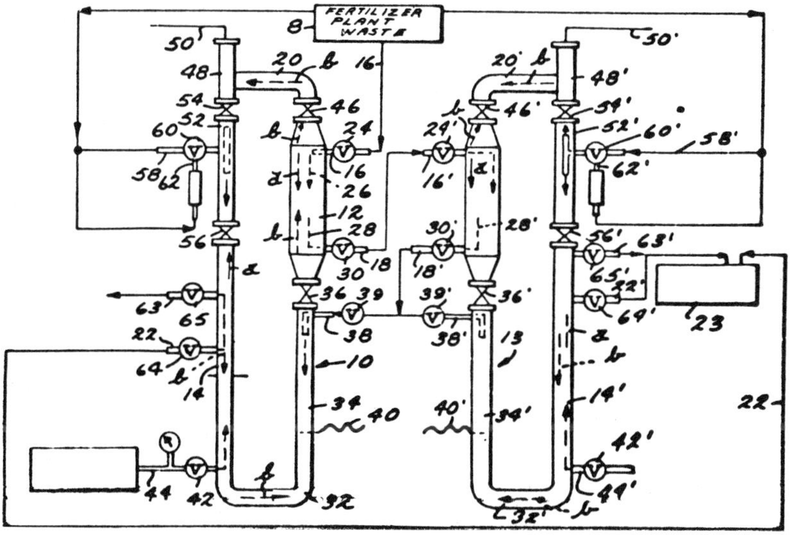

Source: I.R. Higgins; U.S. Patent 3,775,088; November 27, 1973

In addition to the recovery of valuable ammonium nitrate fertilizer values from the fertilizer plant waste stream, which from an ecological viewpoint is a most advantageous feature, another advantage of the development is the production of a demineralized aqueous stream in the anion exchange column loop system **13** for recycle within the total system as a rinse medium for both the regenerated cation and anion exchange resin.

AMMONIUM SULFATE

Recovery from Liquid Wastes

Recovery from Acrylonitrile Manufacture: A process developed by *A.T. Guttmann, M.G. Konicek and A.J. Grigsby; U.S. Patent 3,576,607; April 27, 1971; assigned to The Standard Oil Company* involves the recovery of fertilizer-grade ammonium salts in high yields from acrylonitrile wastewater process streams containing water-soluble organic tars. The removal of water and volatile organic matter from the wastewater process streams is carried out in the presence of a water-miscible solvent, recovering the solid ammonium salt from the residual ammonium salt-solvent-tar mixture and recovering the solvent for reuse.

Recovery from Battery Plate Wash Water: A process developed by *Z.L. Burke; U.S. Patent 3,882,019; May 6, 1974; assigned to Hill Brothers Chemical Co., Inc.* involves converting effluent water which has been contaminated with lead and sulfuric acid as a result of a manufacturing process into water suitable for reuse in such manufacturing process. The process involves the steps of adding lead carbonate to the effluent water in a first agitated enclosure thereby forming water, carbon dioxide, and insoluble lead sulfate, filtering the water from the lead sulfate, and conducting the filtered water back to the manufacturing process for reuse. The filtered lead sulfate is added to a solution of ammonia, ammonium sulfate, carbon dioxide, and water in a second agitated enclosure to form ammonium sulfate and lead carbonate. The ammonium sulfate liquid is filtered from the lead carbonate and conducted to a holding tank for reuse in the second agitated enclosure. The filtered lead carbonate is dried to remove residual ammonia therefrom whereupon the dried lead carbonate powder may be conducted back to the first agitated enclosure for reuse. The carbon dioxide from the first agitated enclosure and the liberated ammonia from the dryer may be conducted back to the ammonium sulfate holding tank. Finally, excess ammonium sulfate from the holding tank may be conducted to a crystallizer for generation of ammonium sulfate crystals.

Recovery from Caprolactam Manufacture: In the well-known Beckmann rearrangement, cyclohexanone oxime is rearranged to obtain caprolactam which is the monomer for making nylon-6. In commercial practice, following completion of the rearrangement, the reaction mixture is neutralized and the lactam separated from the mixture by solvent extraction or by other appropriate means. The most common neutralizing agent used is aqueous ammonium hydroxide. Thus, when sulfuric acid is used as the rearrangement catalyst, the by-product of the neutralization is ammonium sulfate which cannot be reused in the process. The ammonium sulfate therefore must be sold as for fertilizer use; however, this material is often in over-supply on the market and thus may bring only low prices.

Moreover, about three pounds of ammonium sulfate is produced per pound of caprolactam, so that disposal of the by-product becomes a severe problem as the quantity of caprolactam produced is increased. Besides leading to a low priced by-product, this neutralization process also involves large volumes of water which are required since the neutralization is exothermic and the heat released must be dissipated to control the process. The large amount of neutralization medium involved keeps high the cost of separating the lactam and recovering the by-product, ammonium sulfate.

Neutralization of the sulfuric acid with other bases leads to materials which are of less value or used in lesser quantity. Calcium hydroxide, for example, which is a low-cost base would lead to calcium sulfate which, in addition to having a low market value, is insoluble and would clog up the reaction system. Thus, the desirable alternative is not to recover ammonium sulfate (or other sulfates) but is to develop a process which avoids this problem.

A discussion of the manufacture of caprolactam without concurrent ammonium sulfate production has been presented by R. Mattone, G. Sioli and L. Giuffre of Snia Viscosa in *Hydrocarbon Processing,* January 1975, pp 83-87.

Recovery from Solid Wastes

Recovery from Scrap Storage Batteries: See U.S. Patent 3,883,348; May 13, 1975 by F.A. Acoveno and T.W. Freudinger, assignors to R.S.R. Corp. under Lead: Recovery from Used Storage Batteries, elsewhere in this volume.

CALCIUM CHLORIDE

Recovery from Liquid Wastes

Calcium chloride presents a serious pollution problem as described in *Chemical Week,* August 3, 1970, pp 22-24. It is a by-product of the chlorohydrin process for propylene oxide manufacture and is also a by-product of the Solvay soda ash process. By-product calcium chloride from the Solvay and chlorohydrin processes thus presents a massive disposal problem. Although some waste calcium chloride is reclaimed, Wyandotte, for example, sells Solvay by-product $CaCl_2$, reclamation of all that is dumped would be impossible even if it could be sold. The current market for calcium chloride is about 1.3 million tons annually. Soda ash producers alone get about 6 million tons of calcium chloride annually as by-product. Currently less than 10% of this amount is recovered; the rest is waste. In selling the recovered calcium chloride, soda ash producers msut compete with production from natural brines, which currently totals about 700,000 tons. So total recovery of calcium chloride from the Solvay process alone would produce an amount about five times the size of the available market. Recovery of by-product calcium chloride from the chlorohydrin process would add another 1 million or so tons. Some estimates of quantities produced and of recoveries obtained are as follows:

	Million tons, 1969
Recoverable $CaCl_2$ by-product	
From Solvay soda ash production	6.0
From chlorohydrin propylene oxide production	1.1
$CaCl_2$ recovered for market	
As Solvay by-product	0.58[a]
From natural brines	0.70[a]

a Solid and liquid.

CHLORINE

Recovery from Sniff Gases from Chlorine Cells

A process developed by *C.J. Howard and E.B. Port; U.S. Patent 3,357,796; December 12, 1967; assigned to Allied Chemical Corporation* is one in which chlorine in purified and concentrated form is recovered from dilute chlorine gas by a process of passing the dilute chlorine gas in contact with a calcium hydroxide slurry to effect reaction of the chlorine and calcium hydroxide to produce an aqueous solution of calcium hypochlorite and calcium chloride. Figure 27 shows a suitable form of apparatus for the conduct of the process.

The first step of the operation involves the absorption of chlorine from tail and sniff gases from a chlorine plant by passing the gas countercurrent to a falling stream of calcium hydroxide slurry. The lime slurry is chlorinated until its chlorine absorption capacity is nearly exhausted at which point it can be referred to as spent lime slurry. Referring to the drawing the lime slurry in which the amount of calcium hydroxide may vary from about 2% to in excess of about 10% by weight of the slurry is passed through line **1** into gas scrubber **2** which may be any suitable vessel equipped with baffles or plates or filled with packing to effect intimate contact between the gas and lime slurry.

Dilute chlorine gas containing minor amounts of chlorine of the order of a few percent down to as low as a fraction of a percent chlorine is passed through line **3** into the bottom of gas scrubber **2**, then upwardly through the gas scrubber countercurrent to the down flowing lime slurry and the scrubbed gas exits into the atmosphere via vent **4**. The amount of lime slurry fed to the gas scrubber is controlled so that only a small excess of lime remains in the slurry leaving the gas scrubber via line **5** to settler **6**. The absorption of chlorine from dilute chlorine gases by passing the gases in contact with calcium hydroxide slurry in gas scrubber **2** may be illustrated by the following equation:

$$2Ca(OH)_2 + 2Cl_2 \longrightarrow Ca(OCl)_2 + CaCl_2 + 2H_2O$$

Insoluble matter settles to the bottom of settler **6** and is withdrawn via line **7**. Relatively clear liquor containing calcium hypochlorite, calcium chloride and a minor amount of calcium hydroxide overflows from settler **6** through line **8** to reactor **9**. The overflow through line **8** constitutes generally well over 90% of the liquor with the remaining few percent being discharged through line **7**. Reactor **9** is agitated preferably with a paddle-type agitator **11**, so that a fairly uniform suspension of calcium sulfate dihydrate is maintained therein. Sulfuric acid aqueous solution is fed to reactor **9** through line **12**. Desirably the entry of line **12** is placed so that it is well removed from the entry point of line **8** to avoid precipitation of poorly settling $CaSO_4 \cdot 2H_2O$ which often results when these two streams are introduced at the same point in the reactor.

FIGURE 27: APPARATUS FOR RECOVERY OF CHLORINE FROM WASTE GASES

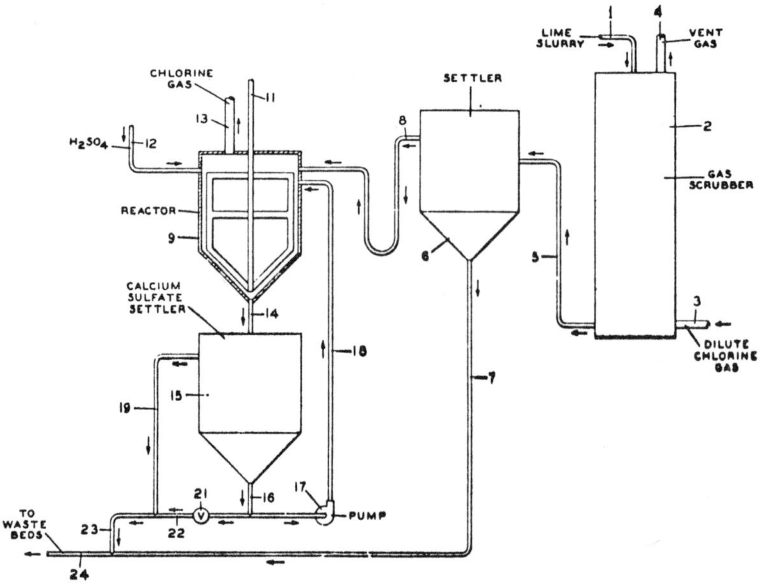

Source: C.J. Howard and E.B. Port; U.S. Patent 3,357,796; December 12, 1967

The concentration of solids in the slurry is kept at a level such that at least 20% solids are present in any given part of the slurry volume, with overall solids in suspension kept in the range of 20 to 40%. Substantially lower concentrations give rise to poorer crystals and a slurry difficult to settle, while substantially higher concentrations produce slurries that are difficult to pump and agitate. The acid addition is controlled so that free chlorine is liberated from the solution which is maintained at a pH below 2, preferably about 1.5 (measured in liquor cooled to room temperature). Chlorine leaves the reactor via line **13** which is of a sufficient purity and concentration that it generally can be combined directly with the chlorine stream produced in the plant, e.g., line **13** can be connected directly to the header drawing off chlorine from the electrolysis cells.

The temperature of the reactor is kept below 90°C and may be operated as low as about 10°C but is preferably operated within the range of 30° to 60°C. The slurry from reactor **9** passes through line **14** to calcium sulfate settler **15**. Here the slurry is thickened so that a slurry containing about 30 to 50% suspended solid $CaSO_4 \cdot 2H_2O$ is collected at the bottom of the tank. This is withdrawn through line **16** to pump **17** and returned through line **18** to reactor **9**. The thin slurry from settler **15** overflows through lines **19, 23** and **24** to waste beds. This slurry may be combined with the settled solids from calcium hypochlorite settler **6** which has sufficient alkali to raise the pH of the combined streams to above 7.

Excess thickened slurry from settler **15** can be drawn through valve **21** and line **22** and sent to waste beds via lines **23** and **24**. The reaction in reactor **9** between the clear liquor containing calcium hypochlorite and calcium chloride to produce chlorine gas which is recovered and calcium sulfate precipitate may be illustrated by the following equation:

$$Ca(OCl)_2 + CaCl_2 + 2H_2SO_4 \longrightarrow 2Cl_2 + 2CaSO_4 + 2H_2O$$

In some plants there is available waste lime and waste sulfurice acid which products often present a disposal problem. The process permits the use of waste lime and waste sulfuric acid to recover a valuable product present in dilute form in a waste gas and converts sulfuric acid and waste lime into water and an insoluble solid composed primarily of calcium sulfate dihydrate. Such a procedure reduces pollution of the air and water table in the area surrounding the plant.

CLAY

Recovery from Paper Mill Waste Sludges

See under Titanium Dioxide.

CYANIDES

Recovery from Used Plating Solutions

A process developed by *W. Zabban; U.S. Patent 2,845,330; July 29, 1958* relates to the recovery of cyanides from industrial wastes. Metal cyanides are used in various manufacturing operations. Such cyanides are used in solution for the electroplating of metal, and are also used, in molten form, as an ingredient of molten metallurgical salt baths. After periods of use, the sorptions or baths become contaminated by the introduction of foreign materials, by chemical breakdown, or both, so as to require replacement.

Cyanide compounds are violently poisonous to all animal life, even in very small concentrations, and, therefore, cannot be readily disposed of, even though highly diluted. Most metal cyanides, as they exist in industrial wastes, are water-soluble to a substantial extent, and thus cannot safely be discharged into rivers and streams. This process involves recovering the cyanide content of industrial waste such as sodium and potassium cyanide solutions, copper, brass, bronze, zinc, and silver-plating solutions, the recovered cyanide being in such a form that it may readily be reused, the recovery being so complete that the waste solution remaining contains not more than 0.5 ppm of cyanide (as CN) and a very small content, on the order of 2 ppm of metallic ions left in solution. Thus, the remaining solution, after use of the preferred method, requires very little or no dilution before it may safely be discharged into public waterways.

The cyanide-containing industrial waste, if in the form of an exhausted electroplating solution, may immediately be treated to recover metallic cyanide therefrom. If the cyanide-containing waste is in the form of a solid salt from a used metallurgical salt bath, the waste is dissolved in water prior to treatment. Usually, the cyanide-containing portion of the waste goes into solution before most of the other constituents are dissolved. The undissolved solids are removed before the solution is treated to precipitate metallic cyanides in accordance with the process.

The metallic cyanides are precipitated from the waste cyanide solution by adding thereto a mixture of copper sulfate and sodium sulfite (or equivalent in sulfur dioxide), the sodium sulfite added being sufficient to completely reduce the copper sulfate to cuprous sulfate, the copper sulfate being just sufficient to produce sufficient cuprous sulfate to completely react with the cyanide content of the waste cyanide solution to form an insoluble metallic cyanide or a mixture of insoluble metallic cyanides. The insoluble reaction product may then be separated from the remaining waste solution.

FELDSPAR

Recovery from Mica Mill Tailings

The selective flotation of minerals from North Carolina mica tailings has been reported by W.H. Eddy, J.S. Browning and J.E. Hardemon in Report of Investigations No. 7319, Washington, D.C., U.S. Bureau of Mines (November 1969).

Laboratory batch and small-scale continuous flotation tests were conducted by the Bureau of Mines on a mica waste tailing from Kings Mountain, North Carolina, to determine the feasibility of producing high-quality mica, feldspar, and glass sand. Mica was recovered by two methods; one was by flotation in an alkaline circuit, which produced a mica concentrate assaying 92.0% mica with a recovery of 87.1%, and the other by flotation in an acid circuit, which produced a mica concentrate assaying 91.0% mica with a recovery of 79.4%.

Continuous flotation processing yielded a feldspar concentrate assaying 1.10% Na_2O and 13.50% K_2O, with recoveries of 66.4 and 80.6%, respectively. The feldspar tailing was high-quality quartz glass sand analyzing 98.6% SiO_2 and 0.02% Fe_2O_3.

FLUORIDES

Recovery from Gaseous Wastes

Recovery from Aluminum Cell Exit Gases: A process developed by *A.W. Kielback; U.S. Patent 3,122,594; February 25, 1964; assigned to Aluminium Laboratories Limited, Canada* utilizes a special floating bed scurbber for fluoride removal. By way of specific example, the procedure and apparatus have been employed for scrubbing the gases from aluminum reduction cells (of the electrolytic type using molten baths), or the gases from the buildings, i.e., pot-rooms, where such cells or pots are operated. The scrubbing is effected with water, preferably water having a suitable chemical content to promote the desired absorption of gaseous or other material from the waste gases. That is to say, in such scrubbing operations, it is usually desired to recover fluorine values, present as gaseous hydrogen fluoride and particles of solid fluorine compounds (e.g., fluorides).

The gases withdrawn from the pots or pot-rooms, as by suitable pumping, contain fluorine in various amounts, as for example from 2 to 40 mg/cu ft, a relatively minor proportion of the total gas. These gases also contain other solid particles, such as alumina, tars, carbon and the like, which it is desired to remove.

As stated, the fluorine in such gases is present in part as hydrogen fluoride, but also in part as solid particulate fluorine compounds, which may account for as much as 30 to 50% of the total fluorine loss from the reduction cells. At least a major portion of this solid material in the gas results from volatilization of electrolyte constituents, which condense to extremely small fume particles. Removal of these very fine particles, understood to have sizes of a fraction of a micron, has created difficulty in getting a high efficiency of recovery. Whereas hydrogen fluoride gas alone can be removed efficiently with equipment having a low pressure drop (e.g., 0.5 to 1 inch of water), such as open spray towers, satisfactory removal of fluoride fumes requires expenditure of more energy, as in higher pressure drop apparatus. This device is highly effective for such purpose, and is specially advantageous in that it does not become clogged with accumulations of solid material.

A process developed by *L.L. Knapp and C.C. Cook; U.S. Patent 3,503,184; March 31, 1970; assigned to Aluminum Company of America* is one in which hydrogen fluoride and finely divided solids are removed from gases evolved in the electrolytic production of aluminum, by means of a bed of alumina particles fluidized by a stream of such gases. Alumina particles combined with the hydrogen fluoride, and finely divided solids entrapped in the bed, discharge from the bed and are recycled to the process. Figure 28 shows a suitable form of apparatus for the conduct of the process. That apparatus includes an elongated chamber or vessel having a blower associated therewith for blowing into the chamber a continuous stream of the gases evolved in electrolytic cells for the production of aluminum; a bed of alumina particles fluidized in the elongated chamber by the gas stream; means for feeding additional alumina particles into the fluidized bed, and an overflow outlet for solids in the bed. Filtering means is provided directly above the fluidized bed for removal of entrained solids from the gas stream and return of such solids to the bed.

FIGURE 28: APPARATUS FOR FLUIDIZED ALUMINA TREATMENT OF FLUORIDE GASES FROM ALUMINUM CELLS

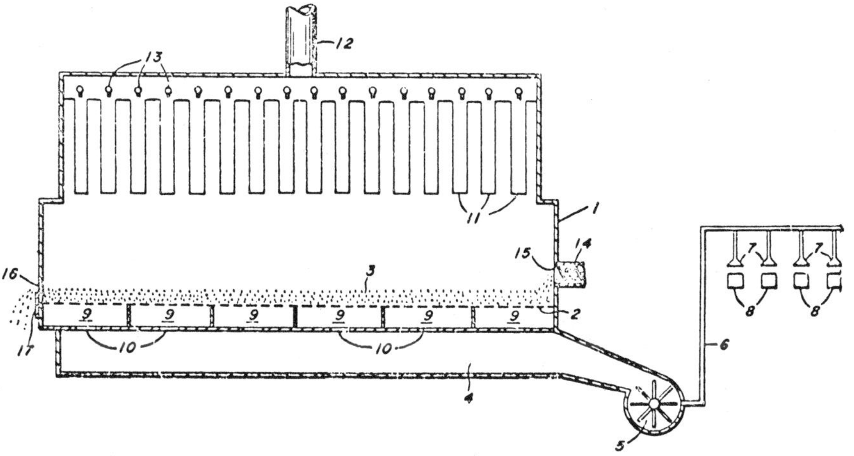

Source: L.L. Knapp and C.C. Cook; U.S. Patent 3,503,184; March 31, 1970

Referring to the drawing, in the elongated chamber **1** a horizontally disposed, perforated plate **2** supports a dense, turbulent fluidized bed **3** consisting primarily of finely divided alumina particles. At the base of the chamber **1** is a plenum chamber **4** extending substantially the entire length of the chamber **1** and having a blower unit **5** connected thereto. A conduit **6** connects the blower unit **5** with gas collecting hoods **7** over conventional electrolytic cells **8** (indicated diagrammatically) for the production of aluminum by reduction of alumina dissolved in molten cryolite. Gas distribution chambers **9** below the plate **2** communicate with the chamber **4** through spaced openings **10** in the top of chamber **4**.

Conventional bag filters **11** made of porous fabric extend across the upper portion of chamber **1**, above the bed **3**. At the top of chamber **1** is a gas discharge conduit **12**. Air jets **13** are provided for periodically dislodging deposited solids from the exterior of the filter bags by blowing a reverse air stream through the bags. At one end of chamber **1** is a conveyor **14** for feeding finely divided alumina into the bed **3** through an inlet opening **15** in an end wall of the chamber. At the opposite end of the chamber **1** is an outlet opening **16** through which solids from the bed **3** can overflow to a suitable conveyor (not shown). An adjustable weir **17** at the opening **16** permits control of the depth of the bed **3** to assure adequate contact of the gas stream with the bed for the desired purposes.

In carrying out the process in such apparatus, a stream of gases from the electrolytic cells **8** is blown through the plenum chamber **4** and the plate **2** by the blower **5**, at a sufficient velocity to keep the bed **3** fluidized. That bed has a depth of only 2 to 12 inches, whereby the pressure drop through the bed is low (e.g., 3 to 6 inches of water) with attendant low cost of operation of the blower **5**. However, 50 to 150 pounds of alumina is kept in the bed per pound of gaseous hydrogen fluoride present in the gas stream, thereby providing a large ratio of alumina to hydrogen fluoride. The gas stream, which is at a temperature of 150° to 280°F, passes through the bed **3** in from 0.25 to 1.5 seconds, depending on the bed depth and the velocity of the gas stream. Solids in the gas stream are entrapped in the bed **3**. The gas stream leaving the bed **3** flows through the bag filters **11**, depositing on the exterior of the bags remaining finely divided solids from the electrolytic cells, as well as alumina particles small enough to be carried upward out of the bed. Periodically sharp jets of air are blown into the bags **11** by the air jets **13**, dislodging the solids that had been deposited on the bags' exterior surfaces, whereupon the dislodged solids fall back into the bed **3**.

As additional alumina particles are fed into the bed **3** by the conveyor **14**, solids adjacent the outlet **16** flow automatically out of the bed through that outlet. As a result of the substantial distance between the inlet **15** and the outlet **16**, and the heterogeneous, irregular path of movement of particles in a fluidized bed, the average time of passage of a particle of alumina across and out of the bed is two to fourteen hours, thereby insuring ample opportunity for contact of the particles with gaseous hydrogen fluoride in the bed. Such average time of passage of the particles through the bed is referred to herein as average residence time.

The alumina particles fed into the bed **3** are of the same grade of alumina as is normally used for production of aluminum in electrolytic cells. That grade of alumina is of low porosity, but nevertheless gaseous hydrogen fluoride in the bed is taken up by the alumina particles, apparently by adsorption. The materials leaving the bed consist of solids which have been removed from the gas stream and entrapped in the bed, as well as alumina particles carrying hydrogen fluoride. The latter particles are referred to herein as alumina particles containing hydrogen fluoride, although it may be that hydrogen fluoride initially sorbed on the alumina has reacted, at least in part, with the alumina by the time the alumina particles leave the bed.

The mixture leaving the bed can then be fed into the fluoride bath in one or more of the cells **8**, thus returning the recovered fluorine values to the cells and thereby providing a portion of the fluoride values needed in the fluoride bath of the cells as well as supplying alumina to the cell for conversion to aluminum. By procedures described above, 99 to 100% of the total hydrogen fluoride and other fluorides in the gas stream can be removed, over a wide range of total fluoride content in the gas stream, and even though the percentage of fluorides present in the gas stream is quite low, e.g., less than 35 mg/cu ft. Moreover, the process is highly economical in view of the relatively inexpensive equipment and operating conditions employed, and the very hgih efficiency of recovery of fluoride.

A process developed by *H.H. Predikant; U.S. Patent 3,709,978; January 9, 1973; assigned to Universal Oil Products Company* provides a multistage method for scrubbing a waste gas stream containing HF and SO_2 such as from the aluminum production industry where aluminum dust can also be present. Control of wash streams is maintained to preclude too much HF reaching the second-stage washing zone cause sodium floride to precipitate and then block valves and piping. A combined effluent from each wash stage is passed to a neutralization zone where an alkaline stream such as caustic soda will cause precipitation of sodium fluoride and sodium sulfate along with entrained dust or aluminum particulates.

Figure 29 is a schematic flow diagram of the process. In the flow sheet, the two scrubbers connected in a row have the reference numerals **1** and **2**. Each of these scrubbers contains perforated trays penetrable for gases and liquids, and between the trays a fluidized bed of spherical packing **4** is provided. Above the perforated trays **3**, a liquid spray diffuser **5** and above the liquid spray diffusers drop separators **6** are provided. Through a blower **7**, the waste gases to be purified, for example at a temperature of 90°C are introduced through conduit **8** into the lower part of the scrubber **1** and flow countercurrently to the washing liquid through the perforated trays **3**, the fluidized beds of packing **4** and the drop or mist separator **6** and then leave the scrubber **1** at the other end thereof. Within the scrubber **1**, an adiabetic refrigeration of the waste gases to be purified occurs, whereby a portion of the water of the washing liquid is evaporated and leaves the other end of the scrubber **1** together with prepurified waste gases. The waste gases prepurified in this manner enter through conduit **9**, the lower section of the scrubber **2** and flow again countercurrently to the washing liquid through the perforated trays **3**, the fluidized beds of packings **4** and the drop separator **6** and leave the scrubber **2** through conduit **10**, from where they are delivered to the atmosphere or introduced into a not-shown postpurifying unit for removing sulfur dioxide.

FIGURE 29: APPARATUS FOR ALKALINE SCRUBBING OF WASTE GASES FROM ALUMINUM CELLS

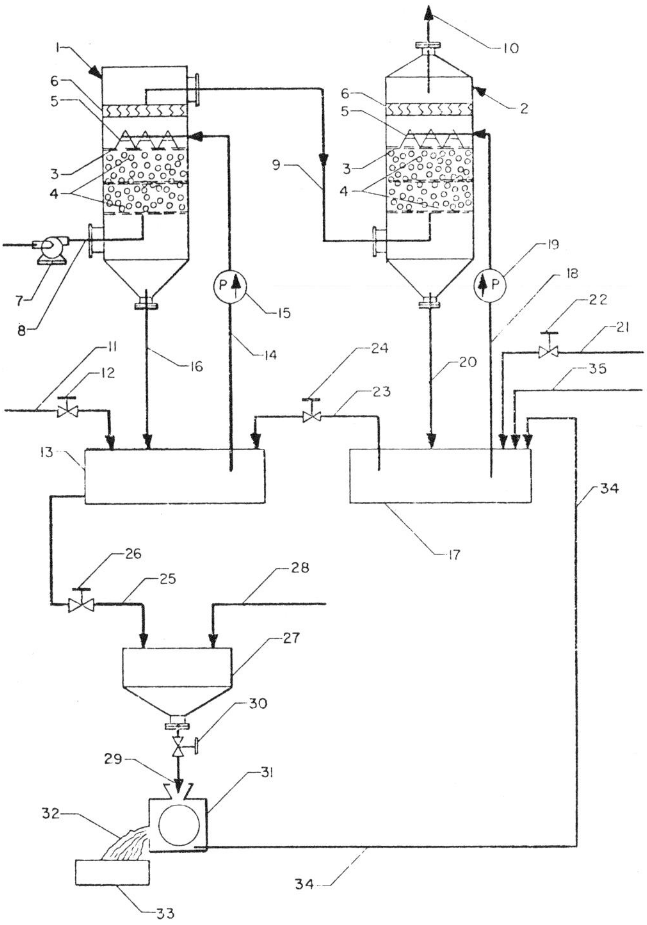

Source: H.H. Predikant; U.S. Patent 3,709,978; January 9, 1973

Through conduit **11** and valve **12** fresh water is introduced into the collecting container **13** of the scrubber **1**, and through conduit **14** and pump **15** the washing liquid is recycled from the collecting container **13** to the liquid spray diffusers of the scrubber **1** and constantly recycled through the members and conduit **16**. In the same manner, the washing liquid within the collecting container **17** of the scrubber **2** is recycled through the conduit

18 and pump **19** to the liquid spray diffusers of the scrubber **2** and is constantly recycled through the scrubber and conduit **20**. Through conduit **21** and valve **22** also to the collecting container **17** continuously fresh water is added, and through conduit **23** and valve **24** continuously a portion of the washing liquid of the collecting container **17** of the scrubber **2** is transferred to the collecting container **13** of the scrubber **1**.

Preferably, continuously a portion of the washing liquid is transferred from the collecting container **13** through conduit **25** and valve **26** to the neutralization container **27** and combined there with alkaline solution from conduit **28**, for example, with aqueous sodium hydroxide or sodium carbonate. Thereby, the hydrogen fluoride within the washing liquid is converted to sodium fluoride having a concentration that the solubility product is exceeded and the sodium fluoride is precipitated. Through conduit **29** and valve **30** the neutralization mixture is introduced into a centrifuge or another separator unit **31**, from where the solid sodium fluoride mixed with dust and little sodium sulfate is withdrawn as solid **32** and introduced into a unit **33** to convert it to cryolite. The mother liquor of the centrifuge **31** is introduced through conduit **34** into the collecting container **17** of the scrubber **2**.

Preferably, there is added through conduit **28** a stoichiometric excess of alkaline solution to the neutralization container **27**, so that the mother liquor introduced through container **34** into the collecting container **17** still contains alkali. Additionally, or if to the neutralization container **27** only stoichiometric amounts of alkali are added, through conduit **35** additional aqueous sodium hydroxide or aqueous solution of sodium carbonate can be added to the collecting container **17**.

By control of the valves **12, 22, 24** abd **26** as well as the circulating speed through the pumps **15** and **19**, the concentration of hydrogen fluoride within the washing liquids of the scrubbers **1** and **2** is adjusted to a special value within the claimed ranges. Preferably, the concentration of hydrogen fluoride within the washing liquid cycled through the scrubber **1** is within the range of from 1.3 to 1.5% by weight, and the concentration within the washing liquid cycled through the scrubber **2** is preferably within the range of from 0.05 to 0.2% by weight.

A process developed by *W.M. Fish; U.S. Patent 3,760,565; September 25, 1973; assigned to Aluminum Company of America* is based on the discovery that particles of alumina can be successfully used for removal of fluoride, particularly hydrogen fluoride, from gas containing same, especially effluent from electrolytic reduction cells, by contacting the alumina with a stream of the gas moving or flowing concurrently therewith for no longer than a total contact time of about 60 seconds, preferably less than about 20 seconds. Maximum chemisorption of the fluoride on the particles of alumina, which is believed to be in a substantially monomolecular layer on the surface of the particles, is sometimes accomplished in as little as 2 seocnds or less of total contact time between a particle of alumina and the concurrently moving stream of gas flowing upwardly in a confined zone.

A somewhat different process using alumina has been developed by *A.J. Teller; U.S. Patent 3,773,633; November 20, 1973; assigned to Wellman-Lord, Inc.* It is a process in which hydrogen fluoride is recovered from gases by contacting the gases with alumina pretreated wtth NaOH, KOH or their mixture.

Another process using alumina particles has been developed by *W. Muhlrad; U.S. Patent 3,780,497; December 25, 1973; assigned to Air Industrie, France*. In this process the alumina particles are introduced at the base of an adsorption column through which the stream of gas to be scrubbed flows, the column terminating in a separator where the particles charged with fluorine are recovered. The method is particularly suitable for scrubbing of gas from electrolytic baths in an aluminum-producing plant.

A process developed by *F. Steineke; U.S. Patent 3,813,852; June 4, 1974; assigned to Elkem-Spigerverket AS, Norway* is one in which fluorine from waste gases is recovered by contacting the gases with aluminum oxide in the presence of steam. The steam increases the adsorption rate of the aluminum oxide thereby improving fluorine recovery.

Recovery from Aluminum Smelter Exit Gases: A process developed by *T.B. Nix; U.S. Patents 3,790,143; February 5, 1974; and 3,876,394; April 8, 1975; both assigned to the Fuller Company* involves removing gaseous fluorides from gases evolved from an aluminum smelting furnace. Basically, the fluorides are removed by achieving intimate contact between finely divided particles of alumina and the fluoride-containing gases. This is achieved by feeding the particles of alumina into a stream of gases collected from the aluminum smelting furnaces countercurrent to the flow of gases. The particles of alumina adsorb the fluorides. The particles of alumina and adsorbed fluorides are then separated in a gas-solids separator, such as cyclone and the alumina may then be fed to the smelting furnaces.

Preferably, two stages are used. The gases supplied to the second or last cyclone in the direction of gas flow are supplied with new alumina countercurrent to the flow of gases. The alumina particles removed in this second cyclone are conveyed to the gas stream supplied to the first cyclone in the direction of gas flow. In a two-stage apparatus, these gases will be those evolved from the smelting furnace. The alumina separated from the first cyclone may be either conveyed to the smelting furnace or recirculated or a portion may be recirculated and a portion conveyed to the smelting furnace. The gases separated in the first cyclone are conveyed to the second cyclone.

Recovery from Fertilizer Manufacture: A process developed by *W.J. Sackett, Sr.; U.S. Patent 3,494,107; February 10, 1970* employs a dust-fume control system having a series of dry cyclones to remove a major portion of the dust, a wet impingement-type separator wherein the gas impinges on the surface of a liquid to entrap any particles of dust remaining and to remove at least some of the fumes, and a tangential wet-wall-type scrubber to remove any residual dust and fume from the gas issuing from the wet impingement-type separator.

Figure 30 shows a suitable form of apparatus for the conduct of the process. Reference numeral **12** indicates generally a duct conveying incoming, heavily polluted air to a dust-fume system **10**. This air is typically that which arises from the vats, dens, drums, mixers, reactors, and dryers of industrial plants of the chemical fertilizer type, for example. Such air is usually laden with dust and chemical fumes. A typical example is that from a superphosphate granulation process plant where crushed phosphate rock is treated with acids, and the duct **12** carries 25 pounds of dust, and as fumes, 1.25 pounds of fluorine, 1.92 pounds of silicon tetrafluoride per minute. The duct **12** leads into a first set of two paralleled dry cyclones **14**. A duct **16** connects the first set of cyclones **14** to a similar second set of paralleled dry cyclones **18**.

The source of propulsion for this air is a centrifugal fan **22** which has its intake connection attached to the outlet duct **20** of the second set of cyclones **18**. A duct **24** leads from the blower **22** into an impingement scrubber **30** where its exit end **26** terminates in an adjustable sleeve **28**. The scrubber **30** contains, in its lower portion, a funnel **32** which retains water whose surface level **W.L.** is just beneath the adjustable sleeve **28**. The upwardly open receiving end **36** of a duct **34** leads from near the top of scrubber **30** through a sidewall to terminate in a tangential discharge end **38** in the sidewall of a packed wet cyclonic scrubber **40**. The exhaust duct **42** of this scrubber **40** is given a negative pressure by a centrifugal fan **44** which discharges through a duct **46** into a stack **48**, exhausting to the atmosphere.

The dust collected to the sets of cyclones **14** and **18** is removed by a screw conveyor **50** as a dry product for return through a duct **52** to process. In the example given, 23 of the 25 pounds per minute of dust is thus recovered. In the impingement scrubber **30**, the remaining dust particles are forecefully driven into the water with the result that a slurry is gradually established. This slurry is recirculated through a circuit consisting of a pipe **54** connected at the bottom of funnel **32**, a valve **56**, a recirculating pump **58**, and a return pipe **60** back to the mouth of the funnel **32**.

When the slurry builds up to a sufficient density in this recirculating system, it is tapped off from the pipe **60** by a valve **62** and returns over a pipe **64** to process. Fresh water as indicated is released into the impingement scrubber **30** to replenish the funnel **32**. Practically all the dust is removed. A typical figure in the example is 1.8 pounds withdrawn per minute. For this reason, the following packed wet cyclonic scrubber **40** operates at high efficiency to react the vapors with relatively small requirement for fresh water replenishment except to maintain a sump with a water line **W.L.** as shown.

FIGURE 30: CYCLONE AND SCRUBBER DUST CONTROL AND RECOVERY SYSTEM

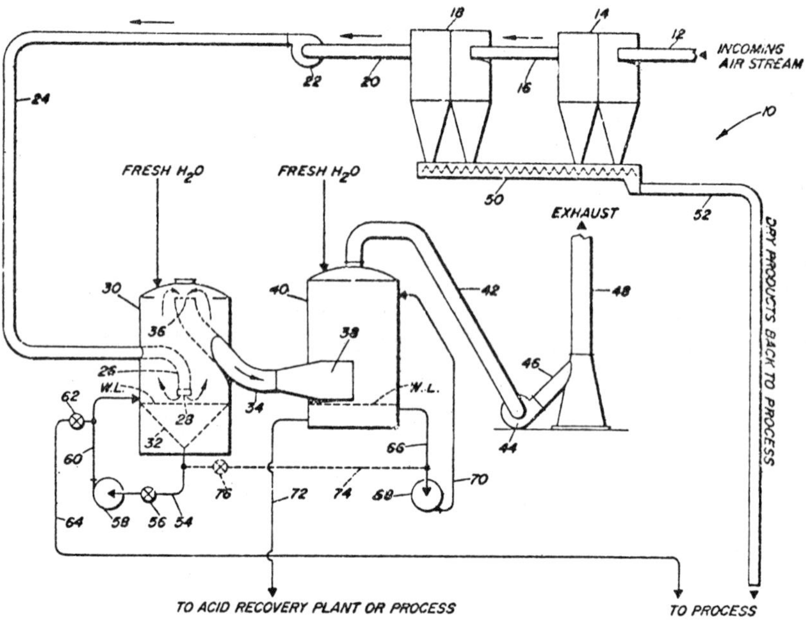

Source: W.J. Sackett, Sr.; U.S. Patent 3,494,107; February 10, 1970

A circulating pump **68** withdraws liquid from this sump through a pipe **66** and reintroduces it through a pipe **70** as spray within the scrubber **40**. A pipe **72** leads to an acid recovery plant, not shown, or if desired to process. Typically here the saving is 0.14 pound of dust (in suspension), 0.37 pound of silica, and 1.15 pounds of fluorine as hydrosilicofluoric acid in 8.25 pounds per minute of water. An optional cross connection of a pipe **74** and valve **76** between the liquid circulating systems of the two scrubbers **30** and **40** is installed when it is unnecessary to keep separate the two effluents.

While this process has been described in connection with a phosphate fertilizer plant other uses will suggest themselves wherever an air laden with a large volume of dry dust and water-soluble or water-reactant vapors is to be cleansed before release to the atmosphere and where it is important not to release an effluent to a stream or drain.

Removal from Waste Gases by Water Scrubbing and Calcium Precipitation: A process developed by *N.E. West; U.S. Patent 3,743,704; July 3, 1973; assigned to E.I. du Pont de Nemours and Company* is one in which HF is removed from an HF-containing gas by scrubbing the gas with water, whereby the HF from the gas enters the aqueous phase to form an aqueous solution of HF, and the gas phase is vented as virtually HF-free gas. Precipitation of the fluoride ion in the aqueous solution as CaF_2 is carried out by adding $Ca(OH)_2$,

$CaCO_3$, or a mixture thereof to the solution. Settling of the precipitate is improved by adding $CaSO_4$ to the aqueous solution during the reaction forming the precipitate. The resultant precipitate is readily filterable from the aqueous phase. The aqueous phase which remains is virtually HF-free and can be reused for scrubbing HF-containing gas.

Recovery from Liquid Wastes

Recovery by Stripping and Absorption: A process developed by *R.G. Hartig; U.S. Patent 3,711,596; January 16, 1973* is one in which a gas or liquid stream containing fluorine is first subjected to treatments to provide a gas stream containing fluorine in the form of HF. These treatments will vary depending upon the form of the gas or liquid stream to be processed. After all, or substantially all, of the fluorine components have been converted to HF, the HF is removed by absorption in a fluid bed constituted by powdered, granular, or pelletized sodium fluoride (NaF). The sodium fluoride combines with HF and is converted to the bifluoride ($NaHF_2$). HF is recovered by decomposition, or desorption, of the HF from the bifluoride. The sodium fluoride may be recycled for subsequent reuse. The separated HF, in substantially uncontaminated condition, is liquefied by condensation and is a high-grade product.

Recovery of CaF_2 as Concrete Additives: A process developed by C.J. Staebler, Jr. is described in EPA-660/2-73-024, Washington, D.C., U.S. Environmental Protection Agency (March 1974). This report presents the development and successful demonstration of laboratory and pilot-scale fluoride treatment techniques for selected aerospace and metal working industry chemical processing solutions and rinse waters. It includes laboratory-scale, lime treatment parameters for chemical processing solutions such as temperature, retention time, pH, slurry concentration and fluoride influent and effluent levels, and ion-exchange treatment to reduce the fluoride concentration of rinse waters to levels less than 3 ppm.

Pilot studies of centrifugal techniques to separate lime-precipitated sludges from titanium chemical milling, titanium descaling and aluminum deoxidizing solutions show that lime precipitation can give final effluents having fluoride concentrations less than 3 ppm. Aluminum conversion coating solutions, however, require secondary treatment with aluminum sulfate to give final effluents having fluoride concentrations less than 3 ppm. Chemical and mechanical property tests show that it is potentially feasible to use calcium fluoride sludge as a strength-maintaining additive for concrete.

FLY ASH

Production of Fertilizers by Combination with Manure

See Manure under Natural Products in this volume, page 153.

Production of Lightweight Aggregates

A process developed by *L.F. Gelbman; U.S. Patent 2,544,752; March 13, 1951* involves making lightweight aggregate using fragile nodules, the steps of subjecting fine sinterable aggregate body material containing fuel needed for combustion in sintering to the action of a nodulizing device and introducing a water spray into the nodulizing device to form and bond the fine material into fragile nodules, screening the fragile nodules to sizes ranging from $\frac{1}{3}$" to $\frac{1}{2}$", and sintering the nodules with a stationary forced draft sintering system. In the past great difficulty was experienced in making lightweight aggregate for use in concrete by the sintering process using a forced draft when very fine sinterable aggregate body material was used. These fine materials of 100 screen, and smaller, such as fly ash, clay, etc., have a tendency of blocking the air passages so that the sintering mass burns down a short distance and then the flame goes out.

To avoid this condition it has been a practice to add coarser material to the fine material so as to provide air passages. Various systems were devised for filling the sintering chambers with different layers of material of different coarseness. However, in all prior methods the operation was highly unreliable, depending upon chance. If the fine material accidentally assumed certain positions the passages were totally blocked and the flame would go out. As shown in Figure 31, the process includes subjecting fine sinterable aggregate body material containing the fuel needed for sintering to the action of a continuous nodulizing device **10**, such as a rotary mixer or vibrating screen, and introducing water with a spray **11** into the device **10** to form and bond the fine material into fragile nodules. The discharge end **12** of the device **10** is provided with ½" screen material **13** through which the nodules may discharge.

FIGURE 31: APPARATUS FOR PRODUCTION OF LIGHTWEIGHT AGGREGATES FROM FLY ASH

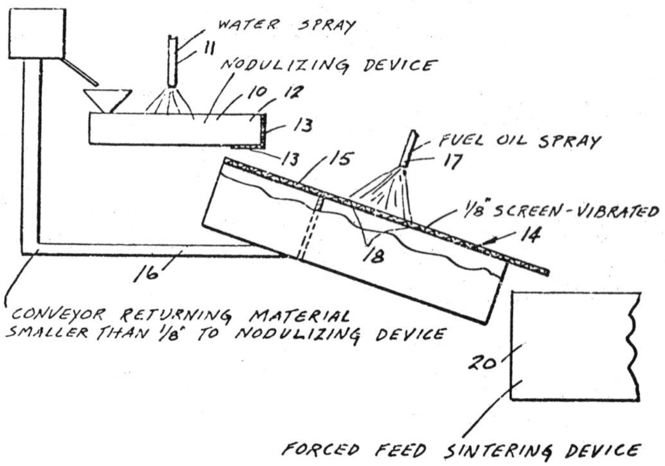

Source: L.F. Gelbman; U.S. Patent 2,544,752; March 13, 1951

The discharging nodules are then screened to remove material smaller than ⅛". This is done with a ⅛" vibrating screen **14** positioned beneath ½" discharge of the device **10**. The nodules of material of smaller size than ⅛" pass through the front portion **15** of the screen **14** and are returned by a conveyor **16** to the continuous nodulizing device **10** to be reformed into nodules of the proper size.

At the central portion of the vibrating screen **14** the nodules are strengthened by a spray **17**, preferably discharging fuel oil or a saline solution. This spray **17** is confined to a central area **18** of the vibrating screen **14** which is past the front portion **15** in which the material is returned to the device **10**. The purpose of this arrangement is to avoid the fuel oil from reaching the material being returned to the device **10** because merely surface spraying of the nodules is desired. It is pointed out that experiments have shown the inadvisability of allowing fuel oil sprayed nodules to be returned to the device **10** as the fuel oil will then be mixed throughout the entire body of the nodules.

Because the screen **14** vibrates, the fuel oil or saline solution will be discharged on most of the surfaces of the nodules passing through the spray area **18**. The nodules from the vibrating screen **14** are then discharged into a forced feed sintering apparatus **20**. It is absolutely essential to this process that the sintering device **20** be of the forced feed stationary type, that is the type in which the nodules rest upon a stationary grate. The nodules are quite fragile and would be broken up in a rotary-type kiln. This process is therefore limited to the stationary forced feed type of sintering device. Experiments have shown that if the fuel oil is saturated in the nodules the smoke discharged during the sintering steps will be excessive and highly objectionable. Various fine sinterable aggregate body material may be used in this process. Materials screening as fine as 100 to 200 mesh, or finer, may be used. Heretofore, these materials could not be used by themselves without the addition of larger materials to produce porosity. The fine materials capable of use in this process are fly ash, clay, etc. When fly ash is used no additional fuel need be added since it contains sufficient fuel for supporting combustion in the sintering step.

A process developed by *H.T. Stirling; U.S. Patent 3,381,948; May 7, 1968; assigned to Stirling Sintering Company* involves the production of valuable material by sintering or heat-hardening pellets derived from such fines as fly ash or the dust obtained from the precipitators or dust collectors of any ferrous or nonferrous metallurgical process.

During many industrial processes a substantial amount of fine material is obtained in the form of a fine dust-like material. Heretofore it has not been possible to utilize these fines and accordingly, since they could not be disposed of, the fines were stockpiled. In this country alone, it is conservatively estimated that there are millions of tons of such fine waste material which have been stockpiled in the past and these deposits are rapidly increasing. The cost of maintaining such stockpiles has been a long-standing problem in industry and most of this material would be willingly given away at no cost merely to reduce the size and cost of the stockpiles. However, there has been no method heretofore which could reclaim this material and therefore industry could not dispose of it.

It has been found that these costly, troublesome by-products can be converted into valuable and useful products. It is possible to convert into useful products such fine waste materials as ferromanganese blast furnace flue dust, basic oxygen furnace dust, red-dog, phosphate rock fines, bauxite, chrome ore fines, iron ore fines, iron concentrate, certain ceramic fines, copper fines, high-chrome and nickel fines, stainless steel grindings, iron sulfate, glass grinding fines, coal refuse, shale, clay, slag particles, fly ash, the iron concentrate from fly ash, the nonmagnetic (tailings) fraction of fly ash, and many other waste materials which at this time are being stockpiled at great cost to industry in the United States.

In a preferred embodiment of the process, fly ash, which has both a magnetic and a nonmagnetic portion, is pretreated to separate the magnetic portion thereof from the nonmagnetic portion or tailings. This can advantageously be accomplished by passing fly ash particles through a magnetic separator and storing the magnetic and nonmagnetic portions separately. In this way, it is possible to treat either of the fly ash fractions as desired by feeding the desired fraction from its storage bin to the processing equipment. In some instances it may be necessary to pass the fly ash two or more times through the magnetic separator to recover a high-grade magnetic concentrate.

Treated material is transferred from the treating device to a plurality of storage bins for the different fractions of the original feed material. In some instances, one or more bins discharges its fraction as waste. In other cases, however, such as when fly ash is separated into magnetic and nonmagnetic fractions, both portions may be further treated to form valuable products.

Material from a storage bin is charged onto a suitable feeder such as a table feeder or a proportionating weigh belt-type feeder. The feeder meters the flow of feed material to conform to the required amount. A suitable recording computer is operatively associated with the feeder to record the quantity of feed material delivered. From the feeder, the material is fed to a wetting drum where the particles are moistened to a predetermined moisture content. The amount of water added in the wetting drum is automatically controlled by a flow controller, known per se, which is responsive to a signal from the feed recorder. The use of a wetting drum increases the capacity of the plant in a manner described more fully hereinafter. The wetted material is passed from the wetting drum to a pelletizing drum where the fines are transformed into agglomerates. The agglomerates are heated by passage through a heated hood, and then the feed conveyor deposits the pellets evenly on a sinter strand. The pellets on the sinter strand are subjected to intense heat in an ignition furnace, and the sintering operation is completed by drawing air through the sinter strand. The hot gases, after passage through the material being sintered and the strand itself, are passed through a dust collector to discharge.

The dust collector preferably operates on the principle of passing the gas to be cleaned through one or more cascading streams of water. The water from the dust collector is recycled to the wetting drum whereby the wet dust particles from the dust collector are formed with the initial feed material into pellets. This step accomplishes two results, namely the elimination of dust, and the recovery of the valuable contents present in the dust. The sinter formed on the sinter strand is processed in a sizing roll crusher and deposited upon a vibrating hot screen or its equivalent and then cooled on suitable apparatus such as a wire mesh conveyor belt. The roll crusher or equivalent apparatus is preferably of an adjustable type which can be set at the desired maximum size for crushing the sinter cake before screening.

A process developed by *J.T. Pennachetti and J.F. Boux; U.S. Patent 3,769,054; October 30, 1973; assigned to Enercon International Limited, Canada* is one in which fly ash is subjected to a plurality of operations to recover therefrom a multiplicity of valuable products, including an iron concentrate, an improved pozzolanic material, a high carbon content product and an agglomerable material, suitable for formation into a controlled sintered aggregate product.

Use to Fill Underground Cavities

The use of fly ash for remote filling of underground cavities and passageways has been described by E.M. Murphy, M.O. Magnuson, P. Suder, Jr. and J. Nagy in Report of Investigations No. 7214, Washington, D.C., U.S. Bureau of Mines (December 1968). Methods are described in the above-cited report for filling underground cavities remotely with dry material pneumatically injected through a borehole to seal openings, prevent air movement, and reduce subsidence; this often brings underground fires under control by limiting the access of air.

Large-scale tests were made in the Bureau of Mines Experimental Coal Mine, Bruceton, Pa., in an operating mine, and in an abandoned Pennyslvania mine. Supplemental trials were made above ground in galleries and in the laboratory. The parameters studied include horizontal and inclined passageways, obstructions in the entry, water leakage into the borehole, dry material, and grout and slurry mixes. The materials studied were fly ash, crushed and pulverized limestone, and sand.

Best results were obtained with dry fly ash, though satisfactory seals were obtained with the other materials. When a borehole is dry and ends at an open void, casing is not necessary. If more than 5 gpm of water flow in the borehole, if the void contains loose debris, or if coarse material is injected, a material feedpipe should be inserted through the borehole. Fly ash was most suitable for remote filling because of its low angle of repose (8°) and its good flow characteristics. The depth of borehole through which the fly ash is injected has no appreciable effect on the filling operation.

Use to Reclaim Acidic Coal Mine Spoil

Experiments in reclaiming acidic surface-mined coal lands with raw fly ash from bituminous-coal-fired power plants were conducted by the Bureau of Mines at two sites in northern West Virginia. Plots at both sites were treated with varying tonnages of fly ash from three area power plants and were planted with a variety of grasses, legumes, trees, and shrubs. Eight grasses and seven legumes were investigated. Greatest promise for growth under harsh soil conditions was shown by Kentucky 31 fescue, rye and red top grasses, and birdsfoot trefoil, a legume. Survival of trees and shrubs was negligible. Addition of fly ash to the spoil increased the pH to a range tolerable to some types of plant growth, improved the texture of the soil, and increased the available water of the resulting mixture according to L.M.

Adams, J.P. Capp and E. Eisentrout in Report of Investigation No. 7504, Washington, D.C., U.S. Bureau of Mines (April 1971).

HYDROFLUOSILICIC ACID

Recovery from Phosphoric Acid Plant Pond Water

A process developed by *R.G. Smaltz, W.A. Satterwhite and P.R. Roberts; U.S. Patent 3,498,746; March 3, 1970; assigned to United States Steel Corporation* is one in which wastewater from the manufacture of wet-process phosphoric acid from phosphate rock and containing P_2O_5 fluoride and silica compounds is normally passed into a pond exposed to weather conditions leading to evaporation of water and the concentration of the P_2O_5 and fluoride values. It is utilized together with a recycle stream from the process which also contains P_2O_5 and fluoride values to dilute concentrated sulfuric acid so that the dilute acid containing P_2O_5 may be used in the digestion of phosphate rock for the recovery of the P_2O_5 while under the heat of dilution fluoride and silica compound vapors are evolved and condensed by contact with water and recovered as fluosilicic acid product. Figure 32 is a flow diagram of the process.

FIGURE 32: PROCESS FOR RECOVERY OF FLUOSILICIC ACID

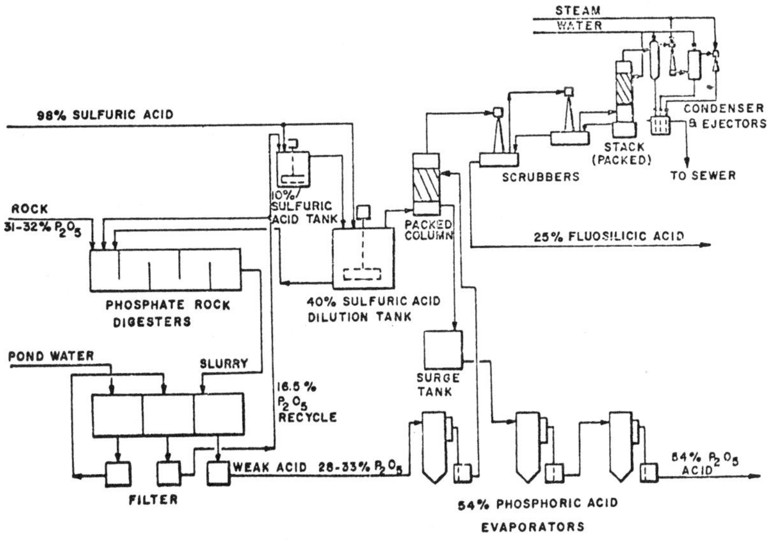

Source: R.G. Smaltz, W.A. Satterwhite and P.R. Roberts; U.S. Patent 3,498,746; March 3, 1970

Preliminary studies of the recovery of HF from waste hydrofluosilicic acid have also been reported by P.C. Good and J.E. Tress in Report of Investigation No. 7213, Washington, D.C., U.S. Bureau of Mines (December 1968). Preliminary studies showed that direct hydrolysis of H_2SiF_6 to SiO_2 and HF at elevated temperatures was subject to many mechanical difficulties, and that corrosion of construction materials was a serious problem. Elimination of SiO_2 from the HF product was generally poor. An alternative procedure, hydrolysis of lime-neutralized H_2SiF_6 by steam at 950° to 1100°C, yielded a product very low in SiO_2. Condensates containing up to 300 g/l HF were prepared.

HYDROGEN CHLORIDE

Recovery from Gaseous Wastes

Recovery from Incineration of Organochlorides: A process developed by *S. Ezaki; U.S. Patent 3,589,864; June 29, 1971; assigned to Yawata Chemical Engineering Co., Ltd., Japan* is an economical process for recovering hydrogen chloride from a spent organo-chlorine compound. It comprises carrying out a combustion treatment of spent liquor or waste gas containing organo-chlorine compound, the discharge of which in the untreated form is undesirable from the viewpoint of minimizing public nuisance, converting almost all the chlorine contained in the organic spent liquor or waste gas to hydrogen chloride, recovering the thus converted hydrogen chloride as hydrogen chloride gas or hydrochloric acid having a concentration of higher than the azeotropic composition of $HCl\text{-}H_2O$ system, i.e., approximately 20% by weight, and at the same time, discharging other components into the atmosphere in the form of completely harmless and odorless gases.

A stream having a spent organo-chlorine compound is burned so as to leave almost no soot or an unburned material, and to such an extent as to allow the chlorine contained in the stream to be converted completely to hydrogen chloride vapor. In order to assure the best possible results from this combustion, the ratio of combustion air plays an important role. In general, the use of 100 to 120% of the theoretical air requirement is preferable. The combustion temperature of higher than 1000°C is considered appropriate. If a great amount of water is present in the stream or if the spent gas is of low-calorie content, the combustion of an auxiliary fuel is occasionally required.

Any spent product stream containing organo-chlorine compound can be treated in a method of this process, for example, chloromethane, chloroethane, chloropropane, chloroethylene, chlorobutylene, chlorobenzene, chloroacetone, chloroacetaldehyde, monochloropropionaldehyde, slurry containing vinyl chloride resin powder and the like. As shown in the flow diagram in Figure 33, the spent organic compound to be subjected to combustion treatment and the combustion air are both led to burner **3**, where the spent organic compound is burned, via lines **1** and **2** respectively. The hot combustion gas containing hydrogen chloride is injected into an extracting agent liquid bath (extracting agent solution concentrator) **4**, where the gas is cooled by the direct heat exchange between the gas and the liquid, and simultaneously, the water in the extracting agent liquid bath is caused to evaporate by the combustion gas and, thus the concentration of the extracting agent solution is carried out.

The method by which a liquid is heated and caused to evaporate by contacting with a hot combustion gas injected directly into the liquid stored in a tank is generally called submerged combustion method. The concentrated extracting agent solution is transported to an extractive distillation step. The gas cooled by heat exchange is introduced to the lower part of absorber **5** via line **14**, and is contacted countercurrently with water provided via line **7** that joins the upper part of the column. The structure of the absorber does not play an essential role in the method and therefore, any suitable contacting apparatus can be used. Some of the column types usable therefore include plate tower such as bubble cap plate and perforated plate, and packed tower packed with Raschig ring, Lessing ring, cross-partition ring, Pall ring, Berl saddle, Intalox saddle, Tellerette, spiral ring, etc.

Corrosive materials are involved in the operation, so a packed tower is preferable, since it is easier to maintain. The hydrogen chloride vapor in the gas is absorbed by contacting, directly and countercurrently with a water spray, the water preferably being sprayed from the upper part of the column. As a result, a 16 to 18% hydrogen chloride solution, and in some cases, a solution approximating the azeotropic composition can be obtained from the bottom of the column. In case the concentration obtained is low, it is possible to recirculate a part of the hydrogen chloride solution obtained from the bottom of absorber **5** to the top. In case a packed tower is used as an absorber, the recirculation is required in some cases in order to secure the liquid load necessary for providing appropriate gas-liquid contact. The gas released from the upper part of the absorber **5** consisting mainly of nitrogen, carbon dioxide and steam can be transported to the lower part of scrubber **6** via line **15** in order to eliminate any remaining harmful materials. In scrubber **6**, the gas is countercurrently contacted with a dilute alkaline solution which is introduced via line **9**, and circulated through the column. The dilute alkaline solution contains, for example, sodium carbonate, sodium bicarbonate, sodium hydroxide and calcium hydroxide or the like. The small amount of unrecovered hydrogen chloride can be removed in column **6** through neutralization. Thus, the gas is made completely harmless and odorless and is released into the atmosphere from the top via line **8**.

FIGURE 33: PLANT FOR INCINERATION OF ORGANOCHLORIDES AND RECOVERY OF RESULTANT HYDROGEN CHLORIDE

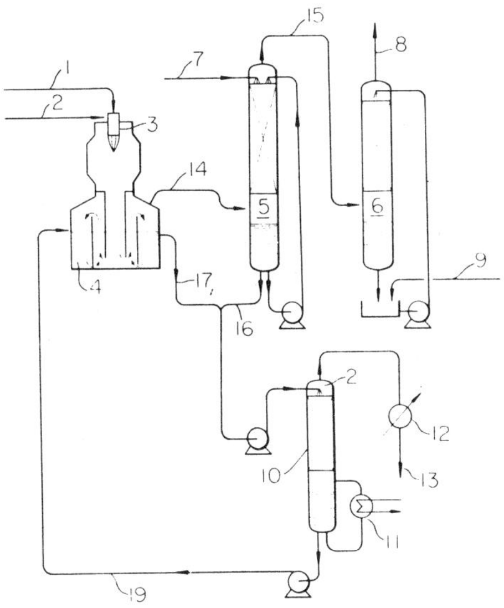

Source: S. Ezaki; U.S. Patent 3,589,864; June 29, 1971

Thereafter, in order to concentrate the dilute hydrogen chloride solution withdrawn from the bottom of absorber **5** via line **16** to a concentration of higher than the azeotropic composition, the dilute hydrogen chloride solution is subjected to a continuous mixing with the concentrated extracting agent solution passed from the liquid bath **4** via line **17**, and is provided to the top of the extractive distillation column **10**. Reboiler **11** provided at the bottom of the extractive distillation column supplies the heat necessary for effecting the distillation, and thus the extractive ditillation is carried out. The vapor withdrawn from the top of the distilltion column is condensed as it is in condenser **12**, and is recovered in the form of hydrogen chloride solution having a concentration higher than the azeotropic composition and high purity via line **13**.

As can easily be understood to those in the art, when it is required, a part of the aforementioned hydrogen chloride solution can also be returned to the extractive distillation column as a reflux stream. As a result of the extractive distillation, the extracting agent solution diluted with water is recirculated to the extracting agent liquid bath **4** via line **19** from the lower part of the column, where the extracting agent solution is reused after effecting a direct heat exchange with the combustion gas and carrying out concentration as has been mentioned in the foregoing. In the extractive distillation step, in some cases, it is also possible to obtain an anhydrous hydrogen chloride gas by the combined use of a partial condenser. As the extracting agent solution, an aqueous solution of an inorganic compound having high hygroscopicity, i.e., sulfuric acid, calcium chloride, lithium chloride and ferric chloride or the like is preferable, and particularly preferable is the use of an aqueous solution of calcium chloride or sulfuric acid. The amount of the extracting agent solution to be circulated and the concentration thereof can easily be adjusted depending on the concentration of the hydrogen chloride to be recovered.

Recovery from $MgCl_2$ Process Evaporators: A process developed by *W.R. McCormick; U.S. Patent 3,810,970; May 14, 1974; assigned to NL Industries, Inc.* is one in which hot waste gases containing small amounts of hydrogen chloride plus additional combustion products including carbon dioxide, nitrogen, oxygen and water vapor are scrubbed substantially free of the hydrogen chloride by contacting the hot gases with a fluid scrubbing medium comprising a slurry of oolitic sand and calcium chloride brine at a pH less than 7, whereby the hydrogen chloride in the gases is converted to additional calcium chloride brine which is separated and recovered from the fluid scrubbing medium, the latter being fortified by a minor portion of the recovered brine and recycled to scrub additional hot waste gases.

The system employed comprises a scrub tower in which the fluid scrubbing medium and hot waste gases are brought into intimate contact, preferably by countercurrent flow, and suitable means for separating and recovering the newly formed calcium chloride brine from the fluid scrubbing medium including means for recovering the oolitic sand and mixing it with make-up brine to form fresh fluid scrubbing medium which is recycled to the scrub tower.

Another process developed by *W.R. McCormick; U.S. Patent 3,826,816; July 30, 1974; assigned to NL Industries, Inc.* is one in which hot waste gases containing small amounts of hydrogen chloride plus additional combustion products including carbon dioxide, and water vapor are scrubbed substantially free of the hydrogen chloride by contacting the hot gases with a fluid scrubbing medium comprising an aqueous suspension of calcium hydroxide (milk of lime) and calcium chloride brine whereby the hydrogen chloride in the gases is converted to additional calcium chloride brine which is separated and recovered from the fluid scrubbing medium as a calcium chloride concentrate.

Recovery from Organic Chlorination Processes: A process developed by *W. Kunzer and F. Leutert; U.S. Patent 3,036,418; May 29, 1962; assigned to Badische Anilin- & Soda-Fabrik AG, Germany* involves working up waste gases containing hydrogen chloride and impurities and originating from organic chlorination reactions. It is known to recover hydrochloric acid from waste gases which contain hydrogen chloride and which have been obtained from organic chlorinations, by allowing the waste gas to flow upwardly through a tower filled with a suitable packing and to which an aqueous absorption liquid is supplied at the top. When so working there is the disadvantage that the waste gas passing to the further working up is in general saturated with vapors of impurities contained therein and originating from the organic chlorination reaction and that there is a risk of these vapors being condensed upon the entry of the waste gas into the absorption zone of the tower.

Even when only small amounts of liquid or solid condensate are thereby formed, it is not always certain that at the speed at which the hydrochloric acid flows away from the tower the condensate will completely vaporize again before the hydrochloric acid enters the cooler through which it passes before leaving the tower. The hydrochloric acid obtained may therefore be substantially contaminated especially by substances which evaporate only slowly or of which the boiling point is so high that they are only volatile azeotropically with steam.

It has been found that the disadvantage can be avoided by heating the contaminated gas containing hydrogen chloride, prior to the absorption, to a temperature which is above the dew point of the impurities. This heating above the dew point of the impurities may advantageously be effected by leading the gas coming from the chlorination apparatus in heat exchange relation to the hot acid flowing from the absorption zone of the tower. For this purpose it is thus only necessary to install in the absorption zone a suitable device for heating up the gases to be worked up, through which the acid produced flows downwardly.

In normal operation there is sufficient heat set free by the absorption for the gas supplied to the absorption zone to be heated above the dew point of the impurities without any appreciable change of temperature taking place in the absorption zone. The separation of the liquid droplets or solid particles entrained with the gases containing hydrogen chloride, from these gases prior to the supply of the gases to the absorption zone is preferably also effected in the device serving for the heating up of the gases. Figure 34 shows a suitable form of apparatus for use in the process.

FIGURE 34: TOWER FOR HCl RECOVERY FROM ORGANIC CHLORINATION PROCESS

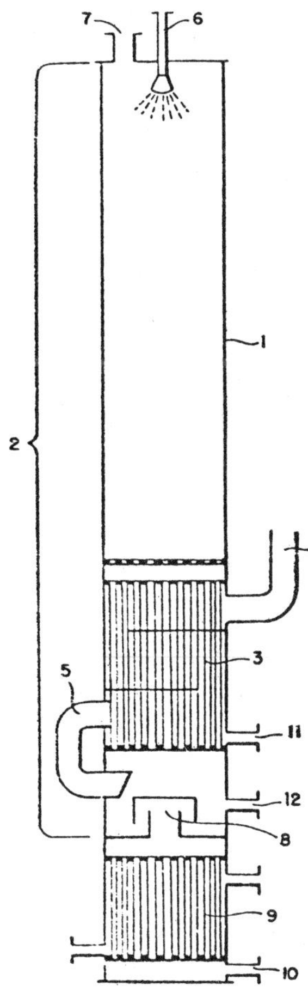

Source: W. Kunzer and F. Leutert; U.S. Patent 3,036,418; May 29, 1962

A tower **1** has an absorption zone **2** charged with packing in which there is also installed a heat-exchanger **3** to which is supplied through a pipe **4** the contaminated gas containing hydrogen chloride and in which this gas is heated up, in heat exchange with the hot hydrochloric acid formed in the absorption zone, to a temperature above the dew point of the impurities contained therein. This gas then passes through a pipe **5** into the absorption zone of the tower, first completely saturates the hot acid flowing down in the heat-exchanger and is then absorbed in the packed portion of the absorption zone above the heat-exchanger by the water sprayed in at **6** with the formation of hydrochloric acid. The residual gas, which contains steam produced by the heat of absorption and organic substances, leaves the tower through a pipe **7** and is supplied to plant for the recovery of the organic substances. The hydrochloric acid flowing away from the heat exchanger **3** passes a dip seal **8** and is then cooled in a heat exchanger **9** and led away from the tower through a pipe **10** as concentrated hydrochloric acid which is practically free from organic impurities and chlorine. Condensate entrained by the gases containing hydrogen chloride can be withdrawn through a pipe **11**.

In order to heat up the gas supplied to the tower when the latter is set in operation, steam is led in at **12** until the heat exchanger **3** has reached a temperature of about 80°C. When the absorption is proceeding, the temperature is maintained by the absorption heat formed so that it is not necessary to supply steam for a prolonged period. The heat exchanger for the heat exchange between the gases containing hydrogen chloride to be supplied to the absorption zone and the hydrochloric acid formed in the absorption zone can also be arranged in the space between the heat exchanger **9** and the dip seal **8**. The heat exchanger for the heating up of the gases containing hydrogen chloride may also be installed outside the tower in the pipe **4** and brought to the desired temperature by means of a heating medium such as steam, water, fuel, oil, diphenyl or the like.

Whereas according to the methods hitherto known there is obtained from the waste gas containing hydrogen chloride contaminated with benzene from a chlorination plant, a 30% hydrochloric acid which contains 0.2 gram of benzene per liter, the working up of the same gas according to the process makes possible the recovery of a 30% hydrochloric acid of which the content of benzene amounts to only 0.0015%. In the working up of waste gases which contain carbon tetrachloride in addition to hydrogen chloride, it is possible according to the process to obtain a 30% hydrochloric acid of which the content of carbon tetrachloride amounts to only 0.006% as compared with a content of 0.01% in 30% acid recovered by the known methods.

Recovery from Oxychlorination Off-Gases: A process developed by *N. Kafes; U.S. Patent 3,816,599; June 11, 1974; assigned to The Lummus Company* is one in which an off-gas from an oxychlorinator containing water vapor and hydrogen chloride is subjected to high pressure-low pressure azeotropic distillation to recover hydrogen chloride therefrom. An oxygen-containing gas is used as a stripping medium in the high-pressure column and the hydrogen chloride of the off-gas is recovered with the oxygen-containing gas for introduction into the oxychlorinator.

Recovery from Liquid Wastes

Recovery from Spent HCl Pickle Liquor: A process developed by *R.A. Taylor and F.G. Krikau; U.S. Patent 3,434,797; March 25, 1969; assigned to E.I. du Pont de Nemours and Company and Interlake Steel Corporation* is a process for removal of iron solids from hydrochloric acid waste pickle liquor by maintaining the system pH between 8 to 10 while the waste pickle liquor and a lime slurry, each at a temperature of from 180° to 210°F, are simultaneously introduced into a reaction vessel. The pH and the temperature are maintained for at least 5 hours. The iron is precipitated out as iron oxide crystals (Fe_3O_4) and is separated from the system by suitable means. The clear liquor contains dissolved calcium chloride.

Waste hydrochloric acid liquors from pickling operations in the iron and steel industry typically contain 20 to 30% by weight of $FeCl_2$ and ½ to 1½% acidity as HCl. These liquors may contain small amounts of other metal chlorides, lubricants, inhibitors, hydrocarbons and other organics and impurities. The removal of the iron from the liquor has been a problem of long standing in the industry. The process affords an effective method for removal and recovery of the iron oxide while affording a useful liquor residue of $CaCl_2$. This residue liquor can be concentrated and used for control of ice and snow on highways, for dust control on secondary roads, for freeze-proofing coal and other ores, and for an addition in concrete mixes to give a quicker initial set. The recovered iron oxides, being essentially pure, can be utilized in the metal industry as a raw material.

A process developed by *L.J. Lefevre; U.S. Patent 3,522,002; July 28, 1970; assigned to The Dow Chemical Company* is one in which an ion exchange process is provided for the recovery of hydrochloric acid from spent hydrochloric acid pickle liquor. The process includes regeneration of the ion exchange material with a complex sulfonic acid, and treatment of the regenerant effluent to recover the complex sulfonic acid regenerant and to produce iron oxide suitable for use in steel making. Figure 35 shows a suitable form of apparatus for the conduct of the process.

A countercurrent, continuous ion exchange reactor of the type described by Higgins in U.S. Patent 2,815,322 is particularly suitable for use in this process and is schematically shown in the drawing. The continuous reactor, generally indicated at **10**, includes an ion exchange zone **12**, a regeneration zone **14**, water rinse zones **16** and **18** immediately above the ion exchange and regeneration zones, respectively, and a recycling path **20**.

As shown in the drawing, spent hydrochloric acid pickle liquor stream **22** is initially contacted with the ion exchange material in zone **12** of continuous reactor **10**. The liquor contains an aqueous solution of ferrous chloride formed in the pickling treatment of steel and some hydrochloric acid. Broadly stated, the purpose of the ion exchange in zone **12** is to remove the ferrous ions in the liquor and replace them with hydrogen ions which form hydrochloric acid with the chloride ions present in the liquor. The hydrochloric acid is, of course, then available for further use as a pickling agent in the steel pickling process.

The ion exchange material is a cation exchange resin in the hydrogen form, capable of providing hydrogen ions for exchange with the ferrous cations in the spent pickle liquor. Exemplary of cation exchange resins which may be used in the process are strong acid resins such as the sulfonated styrene-divinylbenzene copolymers known as Dowex 50W-X8, Amberlite IR-120, Ionac C-240, and the like. A preferred cation exchange resin for use in the process is Dowex-50W-X8 resin. The spent pickle liquor to be treated is passed through the resin in continuous reactor **10** to remove its ferrous ions. The reaction that occurs in ion exchange zone **12** can be exemplified by the following equation in which RSO_3^- represents the cation exchanger resin:

$$FeCl_2 + 2RSO_3H \longrightarrow Fe(RSO_3)_2 + 2HCl$$

The ferrous chloride in spent pickle liquor **22** is thus converted to its acid component producing an effluent **24** containing hydrochloric acid. Spent cation exchange resin from zone **12**, containing ferrous ions, is then passed through rinse zone **16** where the resin is washed with water to prevent cross-contamination between the hydrochloric acid formed in ion exchange zone **12** and the acid used in regeneration zone **14**. While it is not essential to rinse the spent resin following ion exchange, such a rinse step is desirable because the presence of chloride ion in the resin would present a serious corrosion problem in the subsequent regenerant recovery step.

The spent, and desirably rinsed, cation exchange resin containing ionically bound ferrous cation is regenerated in regeneration zone **14** by contacting the resin with a complex sulfonic acid solution such as described by Wilson in U.S. Patent 3,248,278. This sulfonic acid solution is prepared by adsorbing sulfur dioxide in an aqueous solution of a suitable water-soluble aldehyde or ketone to produce a strong, complex sulfonic acid by the reaction:

$$H_2O + R\overset{\overset{\displaystyle O}{\|}}{C}R' + SO_2 \rightleftharpoons \begin{matrix} R & & OH \\ & \diagdown\;\diagup & \\ & C & \\ & \diagup\;\diagdown & \\ R' & & SO_3H \end{matrix}$$

where RCOR' is the aldehyde or ketone. This complex sulfonic acid (ZSO_3H) is a highly effective resin regenerant in the system. Exemplary of carbonyl compounds suitable for use in the preparation of the complex sulfonic acid are water-soluble C_1 to C_8 aldehydes and ketones, such as acetone, methyl ethyl ketone, acetaldehyde, formaldehyde, furfuraldehyde, isobutyraldehyde, cyclohexanone, benzaldehyde, methyl isobutyl ketone, mesityl oxide, and salicylaldehyde. Acetone is preferred for use in the process and reacts with sulfur dioxide and water to produce 2-hydroxypropane-2-sulfonic acid.

At room temperature maximum ionization of the complex sulfonic acid can be most economically achieved in the production of the complex sulfonic acid using a weight ratio of about 7 parts water to 1 part of acetone. Increasing the ratio of acetone to water above 7 to 1 only insignificantly increases the solubility of sulfur dioxide in the mixture at 20° to 25°C. An optimum sulfonic acid composition should utilize a minimum amount of acetone or other carbonyl compound, both because the carbonyl compound is the most expensive reagent in the system and because the solution should be as aqueous as possible to maximize ionization of the acid. At a weight ratio of 7 parts water to 1 part acetone, the complex sulfonic acid solution contains about 11% acetone, 12% sulfur dioxide, and 77% water, and has a normality roughly equivalent to a 2 N acid.

FIGURE 35: PLANT FOR RECOVERY OF HCl FROM SPENT PICKLE LIQUOR

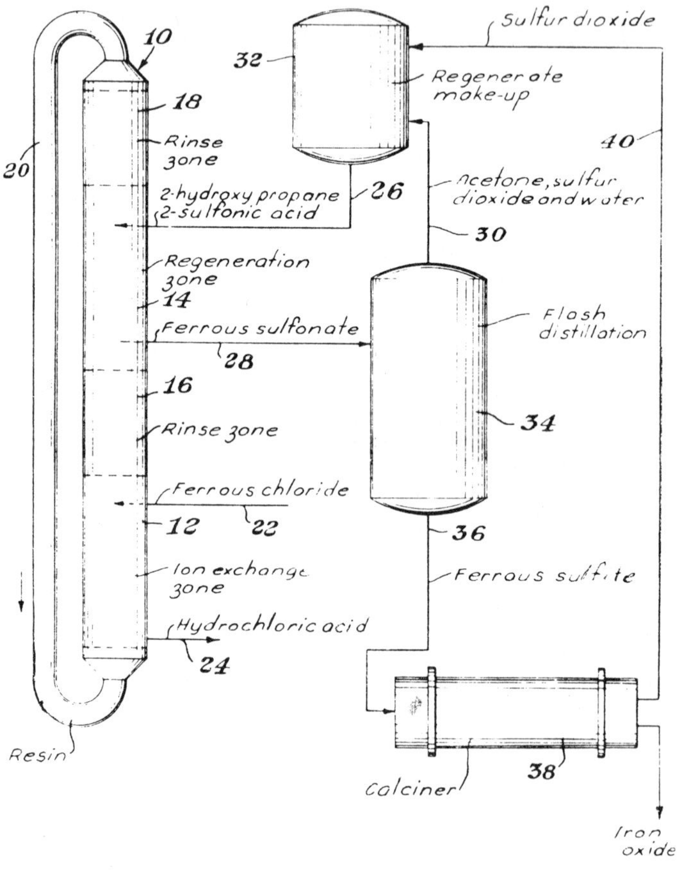

Source: L.J. Lefevre; U.S. Patent 3,522,002; July 28, 1970

Complex sulfonic acid solution **26** containing H+ cations and ZSO_3^- anions is brought into contact with the spent cation exchange resin in regeneration zone **14** of continuous reactor **10**. In this regeneration reaction, the hydrogen ions of the sulfonic acid solution are exchanged for the ferrous ions in the cation exchange resin to produce an effluent **28** containing ferrous sulfonate. The cation regeneration reaction can be exemplified by the following equation:

$$Fe(RSO_3)_2 + 2H^+ + 2ZSO_3^- \longrightarrow 2RSO_3H + Fe(ZSO_3)_2$$

The ferrous cations are thus removed from the resin as sulfonates and the resin is returned to the hydrogen form for further use in the initial treatment of spent pickle liquor. Ferrous sulfonate is relatively soluble in water and thus regenerant effluent **28** is an aqueous solution containing dissolved ferrous sulfonate. The regenerated ion exchange resin from regeneration zone **14** of continuous reactor **10** is then passed through rinse zone **18** where it is washed with water to avoid loss of residual sulfonic acid in the resin following regeneration. While it is not essential to rinse the regenerated resin, the rinsing step is preferred because the carryover of sulfonic acid in the regenerated resin represents a loss of sulfur dioxide and the carbonyl compound. The regenerated, and desirably rinsed, cation exchange resin is recycled from rinse zone **18** through passage **20** for reuse in ion exchange zone **12** of reactor **10**.

In accordance with a preferred embodiment of this process, regenerant effluent **28** is treated to recover its carbonyl and sulfur dioxide components for further use in the regeneration of cation exchange resin. This recovery procedure is preferably carried out by thermal distillation of the regenerant effluent at about 100°C in a suitable vessel **34**. Such distillation effectively strips the carbonyl compound, a portion of the sulfur dioxide, and some water from effluent **28** as distillate **30**. Since all of distillate **30** is to be recombined to form the complex sulfonic acid regenerant, the distillate can be immediately passed to regenerant make-up tank **32** without the need for separating it into its various components. Distillation of regenerant effluent **28** also produces an insoluble precipitate of ferrous sulfite, as can be exemplified by the following equation:

$$Fe(ZSO_3)_2 \xrightarrow{\Delta} FeSO_3\downarrow + 2Z'\uparrow + SO_2\uparrow + H_2O$$

where Z' is the aldehyde or ketone orginally used to form the sulfonic acid regenerant. While ferrous sulfonate can be decomposed in this equation at a temperature as low as about 100°C, the decomposition reaction is preferably carried out at about 250°C to substantially remove all water from the product as water vapor, leaving an essentially dry ferrous sulfite product **36**.

In accordance with a preferred embodiment of this process, the dry ferrous sulfite product is calcined in rotary kiln **38** at a temperature of about 450°C to decompose it and form sulfur dioxide and ferrous oxide. Decomposition of ferrous sulfite can be exemplified by the following equation:

$$FeSO_3 \xrightarrow{\Delta} FeO + SO_2$$

The sulfur dioxide given off by decomposition of ferrous sulfite is recycled at **40** to regenerant make-up vessel **32**, thus providing for the substantial recovery of all of the sulfur dioxide used in preparing the cation regenerant solution. The iron oxide produced in calciner **38** may contain trace amounts of elemental sulfur, but it is not contaminated with chloride and therefore can be used in the steel making process.

It will be apparent from the foregoing description that this process provides a simple, economical, and effective method for the recovery of hydrochloric acid from spent hydrochloric acid pickle liquor having several important advantages not heretofore realized in prior art processes. Such advantages include elimination of the problem of disposing of pickle liquor wastes, since the hydrochloric acid and iron oxide recovered by the process are both useful in the steel making process. Further, the chemical agent used to regenerate the ion exchange resin in this process is recovered substantially in its original form, and thus little or no cost is encountered in the replenishment of the regenerant once the process is on stream.

IRON OXIDE

Recovery from Gaseous Wastes

Recovery from Open Hearth Furnace Fumes: A technique developed by *N.F. Tisdale; U.S. Patent 3,365,340; January 23, 1968* involves the economical recovery of iron oxide particles suspended in fumes produced in the melting, refining and production of steel, sintering of ores, or blast furnace operations. It also relates to the recovery of iron from iron oxide scale, produced in the heating operation of iron and steel production. The oxygen-type furnace, which is being widely adopted, is a producer of copious iron oxide fumes. The open hearth furnace, with or without the oxygen lance, also produces these iron oxide laden fumes. Several other melting and refining processes all produce similar fumes.

Communities which have grown up around such melting plants have suddenly enacted laws to force the reduction of these fumes and their solids from going into the air because of their adverse effect on health. Since there is approximately 1½% of the charge volatized into the air as iron oxide during the melting operation, this becomes a substantial economic loss unless recovered and a large amount of iron oxide is put into the atmosphere. Present methods to recapture the particles from these fumes require high initial expense, substantial operating cost and an elaborate cooling system to cool the fumes. There are several methods which are used singly or in combination, in an attempt to do an efficient job.

All of the basic oxygen furnaces come equipped with dust precipitators which are essentially water scrubbers, electrostatic precipitators and bag houses. This kind of equipment costs from 18 to 25% of the total cost of installing the oxygen furnace installation, whereas this system will cost only 6 to 12% of the total cost. This system essentially takes the fumes and has them impinged in a heated liquid bath of silica slag which has been altered just a small amount by materials to make it more fluid. The furnace is so constructed that there may be one or more baffles along its length with the result of forcing impingement of the gas on the surface of the slag. This can be augmented by pressure or by a pump to force this material to hit the slag. Figure 36 shows the essential elements of the process.

FIGURE 36: SCHEME FOR IRON OXIDE RECOVERY FROM OPEN HEARTH FURNACE FUMES

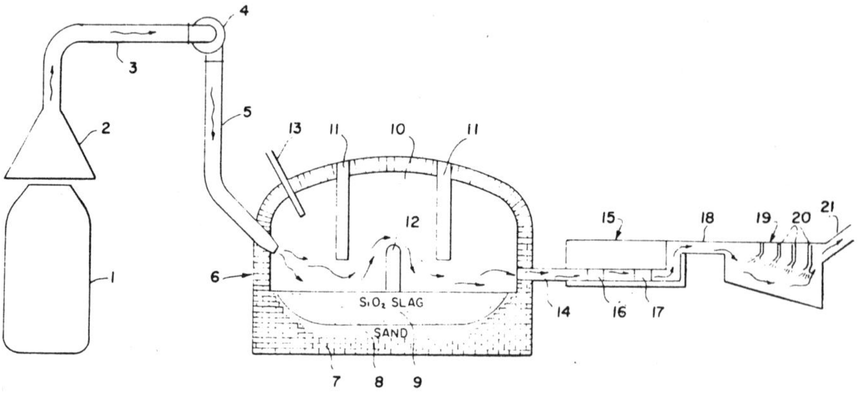

Source: N.F. Tisdale; U.S. Patent 3,365,340; January 23, 1968

Referring to the drawing, numeral **1** denotes a furnace, such as for example, a basic oxygen furnace which emits fumes containing iron oxide, which fumes are directed into a hood **2** and through a duct **3**. A pump **4** may be provided between ducts **3** and **5** to increase the velocity of flow of the fumes as they impinge the SiO_2 slag denoted by numeral **9** contained within an acid air furnace **6**, which slag rests upon a bed of sand **8** contained within the basic brick work **7** of the furnace. Fuel is introduced through tube **13** into the furnace. Preferably one or more baffles such as **11** are provided to more effectively deflect the fumes onto the slag and thus prevent the fumes from bypassing the slag and going directly through the furnace. A sand separator **12** may also be used, if desired, which acts similarly to baffles **11**.

As shown by the arrows, after the fumes impinge on slag **9** they flow through ducts **14, 16** and **17** in a waste heat chamber **15** and then through duct **18** into a spray chamber **19** where a plurality of sprays **20** will cool the fumes and clean them of the last traces of dust contained therein before passing through exit **21** into the air.

Recovery from Liquid Wastes

Recovery from Slurries from Mill Fume Wet Scrubbing: A process developed by *M. Carignani, M. Conti, D. Berte, G. Rossi, E. Pasero and H. Zmölnig; U.S. Patent 3,770,415; November 6, 1973; assigned to Italsider SpA, Italy* has as its object to recover steel mill and basic oxygen furnace gas purification plant slurries and slimes and to transform same into a solid material which may be easily carried by the usual carrying means used in steel mills, and has a controlled granulometric distribution. Such a material further may be sent together with the other iron ores to the installations for preparing the steel mill burden (homogenizing, sintering and the like), or it may be treated by particular devices which render it immediately usable in steel mills. Figure 37 is a flow diagram of this process.

FIGURE 37: SCHEME FOR RECOVERY OF IRON OXIDE FROM WET SCRUBBING EFFLUENTS IN STEEL MILL

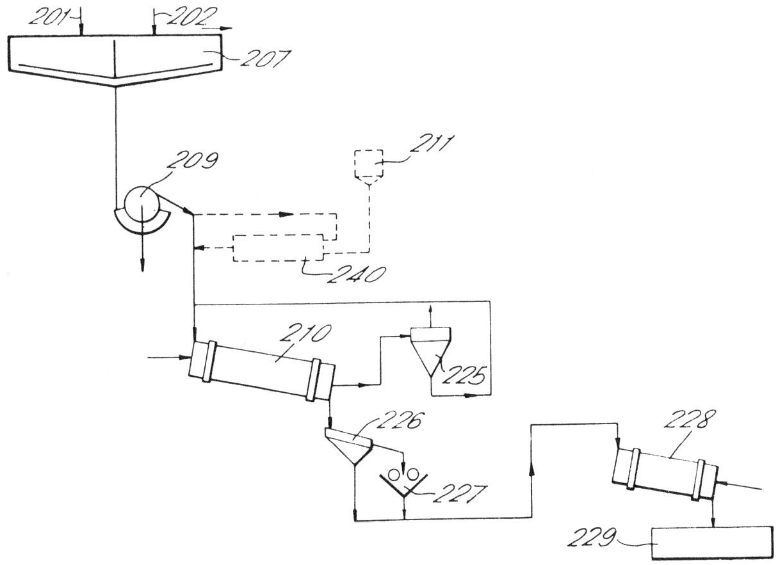

Source: M. Carignani, M. Conti, D. Berte, G. Rossi, E. Pasero and H. Zmölnig; U.S. Patent 3,770,415; November 6, 1973

The washing waters **201, 202** coming from the wet purification devices for the blast furnace fumes and those for the basic oxygen furnace fumes are all conveyed to a single decantation and mixing tank **207**, from the bottom of which a slurry having a solid content between 250 and 600 g/l is continuously removed; such a slurry is fed to a rotating vacuum filter **209** (Oliver filter) which discharges a slime filtered with an average content of humidity of 25 to 40%. Such a filtered slime, possibly added with a quantity of lime, between 5 and 10% of the slime weight coming from **211**, is fed to the drying and granulating device **210** after possible mixing in **240**.

The granular product coming from **210**, after sieving in **226** and possible grinding when required in **227**, is conveyed to a rotating baking furnace **228** at high speed (1000° to 1200°C) provided with a preheating grate, so as to be subjected to a sintering process making it suitable to the possible direct charging into a blast furnace **229**. The fumes coming out of drum **210** are purified in cyclone **225**, and the fumes so collected are recycled to drum **210**.

Recovery from Spent HCl Pickle Liquor: The reader is referred to the section of this volume on Hydrogen Chloride, Recovery from Liquid Wastes, Recovery from Spent HCl Pickle Liquor.

Recovery from Spent H_2SO_4 Pickle Liquor: A process developed by *R.K. Rathmell; U.S. Patent 3,261,665; July 19, 1966; assigned to E.I. du Pont de Nemours and Company* is a process for the treatment of waste sulfuric acid pickle liquor by adding to a reaction vessel a lime slurry and the liquor, mixing the reaction mass at a temperature in the range of from 180°F to the normal boiling point of the mass while bubbling air therethrough, and subsequently separating the aqueous content and the solids content composed essentially of calcium sulfate and ferrosoferric oxide particles, of the reaction mass from each other.

A process developed by *M.C. Irani; U.S. Patent 3,849,535; November 19, 1974* involves recovering from aqueous solutions dissolved elements and compounds which form gelatinous solids when precipitated. The following is a specific example of the conduct of this process.

Example: To 100 grams of spent sulfate pickle liquor from steel processing containing 7 grams of free sulfuric acid (H_2SO_4), 22 grams of soda ash (Na_2CO_3) was added at 100°C. The green precipitate formed was allowed to settle overnight and separated from the supernatant liquid by decantation. To the decanted precipitate enough soda ash was added with constant stirring to form a pasty mass. The paste was spread out over a flat surface and allowed to dry overnight. It was then stirred with water and the stirred product allowed to settle. The insoluble precipitate of iron oxide settled down rapidly. It was separated from the supernatant liquid by decantation and repeated washing. The washed precipitate was filtered and dried at atmospheric temperature. The dried iron oxide obtained was granular and free-flowing material.

IRON SULFATES

Recovery from Copper Mine Tailing Dumps as Fertilizer

A process developed by *B.F. Williams; U.S. Patent 3,135,595; June 2, 1964* involves providing fertilizer from a metal mine waste dump containing sulfates of copper, iron, manganese, zinc and magnesium comprising:

(1) circulating water through the weathered and oxidized material of the waste dump to extract the water-soluble contents thereof and thereby provide a pregnant liquid having a pH of about 1 and a SG of about 26° Baumé and containing copper sulfate, ferric sulfate and ferrous sulfate,

(2) transferring the pregnant liquid to a processing tank

(3) adding metallic iron to the pregnant liquid in the processing tank thereby precipitating therefrom most but not all of the copper as metallic copper

and reducing the ferric sulfate to ferrous sulfate,

(4) adjusting the pH of the liquid in the processing tank to the neighborhood of 2.5 by adding sulfuric acid thereto, and

(5) withdrawing the supernatant liquid from the tank for use as fertilizer, the liquid having a chemical analysis on the basis of pounds per one thousand gallons, approximately as follows:

Iron as ferrous sulfate	1,474.00
Copper as copper sulfate	1.80
Manganese as manganese sulfate	11.70
Zinc as zinc sulfate	5.80
Magnesium as magnesium sulfate	98.50
Sulfur in excess of that contained in above combinations calculated as SO_4	846.00

Recovery from Electrolytic Zinc Refining Residues

A process developed by *D.A. Jackson; U.S. Patent 3,871,859; March 18, 1975* is one in which the jarosite reisdues produced in the precipitation of complex basic iron sulfate from solutions of iron and zinc and other metal sulfates in the presence of ammonium, sodium and/or potassium ions in the electrolytic refining of zinc, heretofore dumped in copious quantities as waste, are treated for use as at least a component of fertilizer. The residue solids are redissolved by acidification and a substantial part of the iron sulfate is removed by fractional crystallization leaving a mother liquor containing the plant fertilizer materials. The iron separated from the jarosite residues may be recovered as sulfate, oxide or metal.

Mineral phosphate, such as rock or bone, may be treated with sulfuric acid to form superphosphate which may then be admixed with the mother liquor residues to produce a more potent fertilizer. Iron may be recovered as oxide by drying and roasting the separated iron sulfate to drive off volatile materials and metallic iron may be produced by reducing this oxide.

LIME

Recovery from Limestone Mud

A process developed by *H.B. Nielsen, E.J. Bonner and E.A. Lado; U.S. Patent 3,796,791; March 12, 1974; assigned to Nichols Engineering & Research Corporation* is a process for reclaiming limestone mud wherein a supply of the mud in a wet condition is passed to a spray-type dryer and dried to a bone dry free-flowing powder by the use of hot drying exhaust gases from a calciner furnace. The dry powder together with the exhaust gases are then passed to a dust collector for removing substantially all of the fines from the gases. The spent lime powder may then be back-mixed with some of the original spent mud in proportions designed to yield a calciner feed having the desired moisture content for optimum physical condition of both feed and calcined product.

The pugged mixture is then conveyed to the top of the calciner furnace and passed downwardly through successive zones including a heating zone, a drying zone, a decomposition zone wherein the spent lime is decomposed into calcined lime and gaseous carbon dioxide, and a second heating zone to increase the density without loss of activity thereof, while passing hot gases upwardly in countercurrent flow with respect to the downwardly flowing spent lime. The calcined lime is discharged at the bottom of the furnace and conveyed to a cooler wherein the temperature is reduced to usable levels for recycling in the system.

Figure 38 is a flow diagram of this process. There are many industrial processes which utilize calcined lime, CaO, to aid in the precipitation of undesirable impurities and absorb carbon dioxide, CO_2, which is used in decomposition, such as sewage treatment, water treatment and sugar refineries, for example. In view of the large quantities required, the spent lime used in the process is usually reclaimed and recycled through the system. For example, in the refining of sugar beets using the so-called Steffen process, calcined lime is required for saccharate formation. In addition, lime aids in the precipitation of undesirable impurities. The sludge or mud produced by the lime is removed from the system by such means as rotary vacuum filters which produce a filter cake containing between about 50 and 55% moisture. Usually, this is reslurried and pumped to a centrifuge located adjacent the reclaiming station. A centrifuge operating on a lime-mud slurry produces a cake of about 40% moisture content.

FIGURE 38: APPARATUS FOR RECLAIMING LIMESTONE MUD

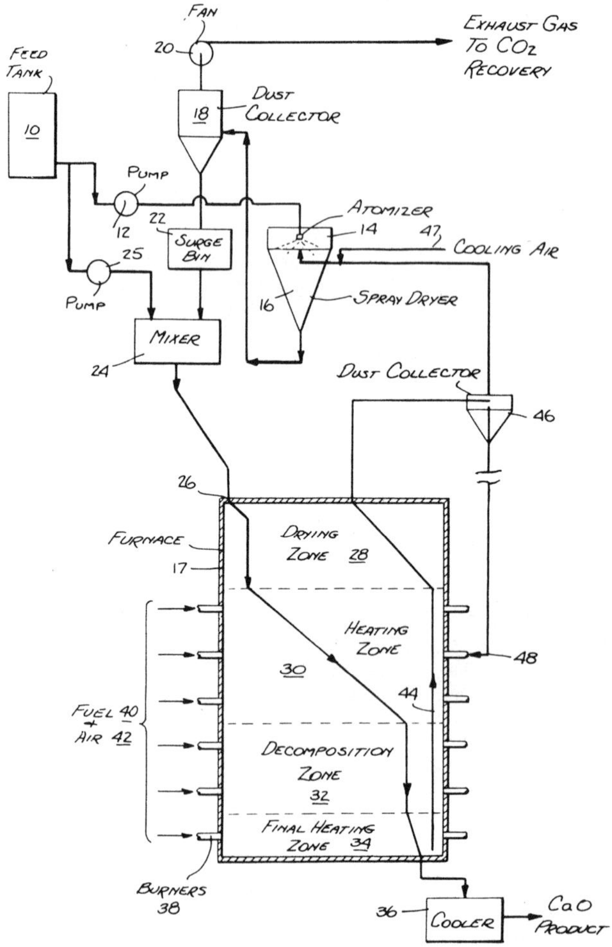

Source: H.B. Nielsen, E.J. Bonner and E.A. Lado; U.S. Patent 3,796,791; March 12, 1974

A typical lime-mud usually includes about 80% $CaCO_3$; 12.2% inert inorganics and about 7.8% organics on a dry weight basis. It is desirable that the CaO product from the reclaiming process be light-burned to yield maximum activity for saccharate formation in a sugar refining process such as the Steffen process. It is also desirable that the maximum use be made of the available heat content in the fuel fired, as the economics of lime-mud recovery depend primarily on fuel economy. In addition, it is desirable that the exhaust gases from the system contain as high a percentage of CO_2 as possible for efficient use in the recarbonation system. Moreover, the nature of beet sugar factory operation requires dependable and substantially maintenance-free, rugged equipment to insure continuous plant operation.

Referring to the drawing, the system comprises a feed tank **10** containing a supply of spent lime-mud which is to be reclaimed. This material is essentially a precipitated form of calcium carbonate, $CaCO_3$, containing some organics and inert impurities which will be disregarded herein for simplifying the description. The lime-mud is in the form of a filter cake or slurry containing free water in the range of between about 40 to 60% on a wet basis.

In the illustrated example, the filter cake contains about 40% free water, and is at approximately 60°F. The system is designated to reclaim about 150 tons per day of CaO product from the lime-mud. A positive displacement-type pump **12** takes suction from the base of the tank and delivers the spent mud directly to an atomizer **14** of a spray dryer **16**. A spray dryer that is known to be suitable for this use is the 26-ft diameter Nichols/Niro Model. In the dryer **16** the spent mud enters a drying chamber of the spray dryer at the top through a centrifugal atomizer wheel wherein it is converted into a spray of uniform droplets. Hot drying air is generally utilized in the spray dryer which in this case is actually the exhaust gas, at a temperature of about 1100°F, from a calciner furnace **17**, which will be described more fully hereinafter. This air enters the drying chamber at a location just below the atomizer wheel through an adjustable-vane air disperser, thereby accomplishing true cocurrent flow of drying air and spray.

The lime-mud is dried to a bone-dry free-flowing powder in the drying chamber and the dry powder exits from the bottom of this chamber along with exhaust gases at a temperature of about 250°F and enters a bag-house-type dust collector **18**. The bag-house collector **18** removes more than about 99.9% of the fine powder from the gases, which are then returned by a fan **20** to a recarbonating station (not shown). These gases are clean and have a temperature of about 250°F. The CO_2 content in the gases is about 23.3% on a dry basis. The dry spent lime powder is collected at the bottom of the dust collector and mechanically conveyed to a surge bin **22**.

Dry powder is drawn from this bin at a controlled rate and mechanically conveyed to a mixer **24** or pug mill wherein the bone-dry powder is mixed with a sufficient quantity of fresh mud, having a moisture content of about 40%, received directly from the feed tank **10**, through pump **25**, to produce a uniform nondusty feed for the calciner furnace **17** at about 16.6% moisture on a wet basis, and at about 125°F.

A furnace that is known to be suitable for use in accordance with this process is described in U.S. Patent 3,153,633 issued to C.F. Von Dreusche, Jr. on October 20, 1964. This furnace comprises a series of circular hearths, placed one above the other and enclosed in a refractory-lined steel shell. A vertical rotating shaft through the center of the furnace carries arms with rabble blades which stir the charge and move it in a spiral path across each hearth. Material is fed to the hearth and rabbled across it to pass through drop holes to the next hearth below. It passes in this way over and across each hearth to the bottom of the furnace whereat the product is discharged through one or more ports. Heated gases flow countercurrently to heat the charge to the desired temperature.

The spent lime enters the furnace, as indicated at **26**, and is heated to and dried at about 212°F in a drying zone **28**. From the drying zone the spent lime feed passes to a heating zone **30** wherein the dry spent lime is heated to about 1660°F before it passes into a decomposition zone **32**. In this zone, at a temperature of about 1660°F the calcium carbonate, $CaCO_3$, is decomposed into calcined lime CaO and gaseous carbon dioxide CO_2, as it moves down through the calciner by contact with gases between about 1850° and 1900°F passing upwardly through the units in countercurrent flow. In addition, combustion of the organics in the feed is completed in this zone. The so-decomposed calcined lime is finally heated to a temperature of about 1800°F in the final heating zone **34**. This provides a product of good density and activity. Hot calcined lime is then discharged from the bottom hearth and is conveyed to a cooler **36** wherein the temperature is reduced to about 150°F and then discharged for recycling through the sugar refining process.

In the calciner furnace **17**, multiple burners **38** are disposed along the fired hearths which burn natural gas, as indicated by the arrows **40**, with some excess air as indicated by the arrows **42**. Air for the combustion of the organics in the spent lime is supplied through the burners. The hot gases at a temperature of between about 1850° and 1900°F flow upwardly through the furnace, as indicated at **44**, to provide heat in the final heating zone **34**, decomposition zone **32**, heating zone **30** and drying zone **28**. The exhaust gas leaves the calciner furnace at the top hearth at about 1100°F and passes through a cyclone-type dust collector **46** to remove entrained fines prior to recycling of these gases to the spray dryer **16** wherein they are used in predrying, as was discussed more fully hereinbefore.

Minor amounts of cooling air, as indicated by arrow **47**, may be added to the spray dryer for purposes of internal cooling of exposed surfaces and protection seals. The fines collected inside the cyclone are recycled back into one of the hearths of the calciner furnace, as indicated by arrow **48** in the drawing.

NITRIC ACID

Recovery from Spent Nitric Acid Leach Liquors

A process developed by *D.W. Bailey; U.S. Patent 3,851,047; November 26, 1974; assigned to Brunswick Corporation* is a process for continuously processing nitric acid etching solutions including, e.g., leaching and pickling solutions containing ferric nitrate by flash evaporation, recovering substantially pure nitric acid through an overhead, e.g., by utilizing a condenser in the overhead line, and separating and recovering ferric hydroxide from the bottoms of the distillation. In a more limited version, single or plural serial hydrolysis stages are utilized after flash evaporation in order to further increase the yield of nitric acid and make the ferric hydroxide separation step even less difficult. Figure 39 is a flow diagram of the process.

FIGURE 39: APPARATUS FOR RECOVERY OF HNO_3 FROM SPENT ETCHING OR PICKLING SOLUTIONS

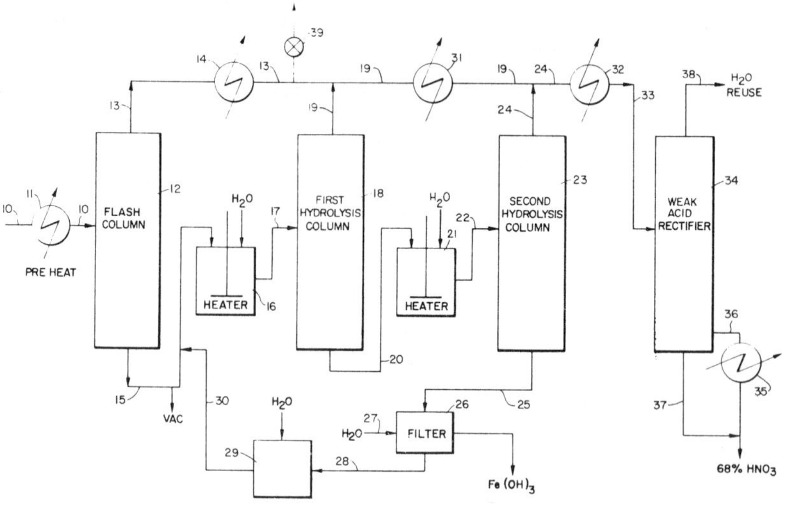

Source: D.W. Bailey: U.S. Patent 3,851,047; November 26, 1974

Referring to the drawing, the spent nitric acid solution enters through conduit **10**. It is heated by heat exchanger **11** and flashed in distillation column **12**. The overheads are taken up through conduit **13** as a nitric acid solution, cooled by heat exchanger **14** and eventually pass to the rectification column **34** or are removed via conduit **39**. The bottoms pass out of column **12** by conduit **15** and into heater **16**. Conduit **15** may also contain a take-off line to a vacuum pump, not shown. This bottoms solution containing ferric nitrate, other nitrates, water and possibly some nitric acid as well as some ferric hydroxide, is stirred and heated since it is a slurry in tank **16**. Also, water is added as needed to tank **16** to provide, e.g., the water needed for hydrolysis in column **18** and adjust the specific gravity of the slurry. Then the slurry is passed to the first-stage hydrolysis column **18** by conduit **17**.

In the first stage hydrolysis column **18**, some of the ferric nitrate is hydrolyzed to ferric hydroxide, and a nitric acid solution is taken off by conduit **19** as overhead. The more concentrated ferric hydroxide-ferric nitrate slurry is then removed from the bottoms of the column by conduit **20** and passed into heated stirring tank **21** where additional water is added as needed. Then it is passed to the second-stage hydrolysis column **23** by conduit **22**. Second-stage hydrolysis column **23** hydrolyzes substantially all of the ferric nitrate remaining in the slurry and the nitric acid is removed as a weak acid solution via conduit **24**. The bottoms containing only a small amount of ferric nitrate is removed via conduit **25**, and filtered at filter **26** as discussed above. Water is fed into the filter **26** by conduit **25**, and the ferric hydroxide removed to further processing as desired.

The filtrate then passes into holding tank **29** by conduit **28**, is optionally further diluted with water, and then it passes by conduit **30** into conduit **15** which contains the bottoms from the flash distillation column **12**. The three overheads, conduits **13, 19** and **24**, each contain condensers **14, 31** and **32** to provide for the liquefaction of the vapors. These eventually combine into conduit **33**. However, part of the liquid nitric acid solution from the flash distillation column can be recycled to the leaching or pickling operations if desired.

The mixed acids in conduit **33** are passed into rectification column **34** wherein reboiler **35**, acting in conjunction with conduits **36** and **37**, provides for the production of nitric acid of up to 68% by weight concentrations. Relatively pure water is the overhead from this column and is removed by conduit **38**. However, the acid conduits are not necessarily combined; that is, the three limes, or as many as desired, can be fed into the rectifying column at different points if desired.

Further, if the hydrolysis units are not desired, the bottoms from the flash evaporator, along with additional water, can be recycled to the flash evaporator after removal of at least a portion of the ferric hydroxide. The acid, as discussed above, can be utilized in the leaching process after removal by conduit **39**. Also, the recycle of the wash liquor from the filter and the holding tank can be recycled to either of the hydrolysis columns, although it is shown to the first. The mixed nitric acid material in conduit **33** may be utilized with suitable dilution in the leaching process as well. However, the production of the concentrated nitric acid results in a saleable product or in the alternative, results in smaller storage problems since the water produced is clean and thus is in condition for removal from the system without causing any environmental problems. Finally, the columns utilized in the process may be of any design desired, including, e.g., packed columns and columns containing sieve trays or bubble cap trays.

PHOSPHATES

Recovery from Liquid Wastes

Recovery from Fertilizer Plant Pond Water: In the manufacture of wet-process phosphoric acids and other phosphatic materials, gases and vapors are evolved and these are scrubbed with water to remove fluorine, silica and small amounts of P_2O_5 therein, and such water, which is regarded as waste material, is passed into a pond. The pond water not only is considered waste material but also presents a hazard in that during exposure to the sun and atmospheric conditions, there is some evolution of fluorine, and further, in the case of heavy rains, some of the pond water may overflow into streams.

By rerouting this wastewater back to use in the processes, the requirements of fresh water are reduced and better used. The volume of wastewater can be restricted to the capacity of the pond. Otherwise an excess volume often results which cannot be retained in the pond and must be chemically treated before discharge to public waters. This is expensive, presenting a further hazard to animal and vegetable life and a loss of valuable products.

A process developed by *R.G. Smaltz and W.A. Satterwhite; U.S. Patent 3,498,747; March 30, 1970; assigned to United States Steel Corporation* is one in which pond water is reacted in a container with concentrated sulfuric acid whereby, under the heat of dilution, fluorine and silica compounds along with steam are evolved. The compounds are scrubbed with water, and the stream filtered to remove silica, the final product being fluosilicic acid. Nearly all the fluorine was evolved, with almost none retained in the sulfuric acid body. The sulfuric acid retains the P_2O_5 values and is a valuable product preferably used in digesting phosphate rock, but useful in other ways whereby the retained P_2O_5 is recovered as product.

A process developed by *I.R. Higgins; U.S. Patent 3,579,322; May 18, 1971; assigned to Chemical Separations Corporation* involves contacting the pond water stream with a cation exchange resin bed and then contacting the cation-free aqueous stream with an anion exchange resin bed to eliminate the phosphate values therefrom and then removing the phosphate values loaded on the anion exchange resin with a regenerating agent. In another embodiment, the waste aqueous stream bypasses the cation exchange resin and is contacted only with a weak-base anion exchange resin which is, thereafter, contacted with a regenerating agent, such as ammonium hydroxide to liberate the phosphate values therefrom.

Recovery from Solid Wastes

Recovery from Phosphating Sludge by Reduction Sinter Process: A process developed by R.F. Waters, H.E. Powell and A.A. Cochran and described in Report of Investigations No. 7533, Washington, D.C., U.S. Bureau of Mines (July 1971) involves the recovery of sodium phosphates and metal values from waste sludges generated by the phosphate treatment of automobile and appliance components.

After treatment of the sludge with NaOH, followed by coke or hydrogen reduction and sintering, zinc and lead were recovered as fume. The sinter was then separated by leaching into a metallic product, chiefly iron and nickel, and high-purity disodium or trisodium phosphate.

Recovery from Phosphating Sludge by Solvent Extraction: A process developed by H.E. Powell, L.L. Smith and A.A. Cochran is described in Report of Investigations No. 7662, Washington, D.C., U.S. Bureau of Mines (August 1972). It reports laboratory-scale development of a solvent extraction procedure for recovering trisodium phosphate, zinc and iron from waste phosphate sludge as separate products.

The sludge was dissolved in hydrochloric acid and iron was extracted as ferric chloride by isopropyl ether. Zinc was extracted using di-2-ethylhexyl phosphoric acid in kerosene. The efficiencies of the one- and two-stage extractions ranged from 98 to nearly 100%. Trisodium phosphate was recovered by crystallization from the raffinate.

PHOSPHORUS

Recovery from Wastewater from Electrothermal Process

A process developed by *F. Muller, K.-H. Stendenbach and H.-H. Weizenkorn; U.S. Patent 3,684,461; August 15, 1972* involves the continual work-up of wastewater having phosphorus sludge therein, such as that obtained in the electrothermal production of phosphorus. The wastewater is first filtered in a filtration zone, the resulting filter cake is predried in a preliminary drying zone, the predried filter cake is conveyed through at least two additional drying zones maintained at temperatures between about 100° and 380°C, gaseous and vaporous matter issuing from the additional drying zones is delivered to a condensation zone, the phosphorus and water are condensed therein and separately removed therefrom.

PIGMENTS

Recovery from Pigmented Paper Process Sludge

A process developed by R.P. Labrecque and T.E. Weymouth, is described in Report EPA 12040 FES 07/71, Washington, D.C., U.S. Environmental Protection Agency (July 1971). Disposal of the sludge that results from primary treatment of wastewaters generated during the manufacture of coated and/or filled papers has caused many problems.

A process has been developed and tested in a full-scale mill trial wherein the pigment present in the waste sludge was reclaimed through incineration of the volatile components of the sludge. In essence, the process consisted of operations in which the sludge resulting from the primary treatment process was diluted to less than 0.75% solids, centricleaned, dewatered to 30% solids, shredded, dried, and burned in a rotary kiln. The ash that resulted, the pigment, was then pulverized and used as filler pigment in the papermaking process. During the trial, it was found that a pigment of acceptable abrasiveness and a GE brightness of 84 to 85% could be produced provided that sludge centricleaning was practiced and the temperature in the kiln was kept below 1600°F.

SODIUM FLUORIDE

Recovery from Spent Aluminum Reduction Cell Linings

A process developed by *H.-K.H. Lam and N.J. Samawi; U.S. Patent 3,808,322; April 30, 1974; assigned to Kaiser Aluminum & Chemical Corporation* is one in which sodium fluoferrate (Na_3FeF_6), a fluorinating agent for converting chloroalkanes to chlorofluoroalkanes, is prepared in high purity from spent, carbonaceous fluorine-containing linings of aluminum reduction cells. The spent lining, is, after comminution, subjected to controlled calcination and the residue is then selectively leached to recover substantially pure NaF solution. Adjustment of the pH precipitates residual Na_3AlF_6 impurities and the filtrate is directly used for the preparation of Na_3FeF_6 by addition of $FeCl_3$ and NaCl.

SODIUM SULFATE

Recovery from Solutions from Pyrite Roasting

As described by W. Teworte in *Chemistry & Industry,* May 1969, pp 565-574, sodium sulfate is recovered in considerable amounts from waste solutions in many industrial countries. The mother liquor of the potash industry is of considerable importance. Besides this, Glauber's salt is produced during rayon manufacture by thorough cooling during the purification of the effluent. In West Germany, the recovery of Na_2SO_4 from the end liquors of the chloridizing roasting of copper-containing pyrites cinders is of considerable importance as a source of sodium sulfate.

The recovery of sodium sulfate is based on a physical process involving cooling to about 0°C (32°F). One maintains the concentration of the solution at 120 to 160 g/l to produce the decahydrate $Na_2SO_4 \cdot 10H_2O$ which is known as Glauber's salt. There is, however, virtually no market for this intermediate product. It contains 56% water of crystallization and melts at 32.4°C (90°F). The conversion into the water-free salt occurs by melting in its own water of crystallization. The resulting mother liquor is concentrated by cooling or by evaporation. Here the process technology has brought important advantages in the last 30 years especially with the introduction of corrosion-resisting steels.

SODIUM SULFITE

Recovery from Sodium-Base Sulfite Pulping Liquors

A process developed by *R.J. Lurie; U.S. Patent 3,061,408; October 30, 1962; assigned to Arthur D. Little, Inc.* involves recovering the sodium and sulfur components from sodium-base sulfite spent liquors resulting from chemical digestion of a lignocellulose-containing material. This process comprises the steps of concentrating the spent liquors and combusting the resulting concentrated liquors, the combusting step being carried out in the presence of an additive selected from the group consisting of silica-rich and alumina-rich materials to react the additive with the sodium, and subsequently converting the resulting sodium reaction products to sodium sulfites in the presence of an acidic reactant selected from the group consisting of $SO_2 \cdot H_2SO_3$ and $NaHSO_3$.

A process developed by *N.C.S. Chari and L.D. Keller; U.S. Patent 3,392,004; July 9, 1968; assigned to Owens-Illinois, Inc.* involves the fluid-bed oxidation of a sodium sulfide smelt from a spent pulping liquor recovery process to give sodium sulfite which can be used to prepare fresh cooking liquor.

SULFUR

Recovery from Gaseous Wastes as Elemental Sulfur

A process developed by *V.B. Sefton; U.S. Patent 3,579,302; May 18, 1971; assigned to Ontario Research Foundation, Canada* is one in which hydrogen sulfide-containing gases are used to react with sulfur dioxide in a waste effluent gas stream to thereby produce sulfur. The hydrogen sulfide-containing gases are generated by reaction between sulfur, a hydrocarbon fuel and steam at high temperatures. The sulfur used to form the hydrogen sulfide may be that formed in the reaction of the sulfur dioxide and the hydrogen sulfide-containing gases.

A process developed by *R.J.J. Hamblin; U.S. Patent 3,594,125; July 20, 1971; assigned to Universal Oil Products Company* is one in which H_2S is removed from a gas stream and sulfur is produced by the steps of:

(1) sequentially scrubbing the gas stream with a first recycle water stream containing NH_4OH and then with a second recycle water stream which is substantially free of NH_4OH to produce a treated gas stream which is reduced in H_2S content and is substantially free of NH_3 and a bottom water stream containing NH_4HS and NH_4OH;

(2) catalytically treating the bottom water stream with an air stream to produce an effluent stream containing ammonium polysulfide;

(3) treating the ammonium polysulfide-containing stream to recover sulfur and to produce a water stream which is substantially free of NH_4OH and a water stream containing NH_4OH; and,

(4) separately recycling at least a portion of these last two streams to the scrubbing step.

Key feature of the resulting process is the production in the regeneration section of two separate recycle water streams and the judicious use of these streams in the scrubbing step to remove H_2S from a gas stream and minimize the amount of NH_3 contained in the treated gas stream.

A process developed by *C.H. Watkins; U.S. Patent 3,687,629; August 29, 1972; assigned to Universal Oil Products Company* is a process for effecting the desulfurization of a hydrocarbonaceous charge stock accompanied by the recovery of sulfur therefrom. Catalyst regeneration facilities are integrated within the process, and the sulfur dioxide resulting therefrom is reacted with hydrogen sulfide, resulting from the conversion of the sulfurous compounds in the charge stock, to produce elemental sulfur.

A process developed by *C.S. Goddin, Jr. and N.R. Montgomery; U.S. Patent 3,758,676; September 11, 1973; assigned to Amoco Production Company* involves the recovery of free sulfur from sour gas streams by the catalytic conversion of hydrogen sulfide under conditions such that unreacted sulfur compounds discharged to the atmosphere are held to a minimum.

Attempts have been made in the past to increase the catalytic conversion of hydrogen sulfide gaseous streams to free sulfur by conducting a portion of the reaction at temperatures below the sulfur dew point in an effort to remove additional sulfur compounds from the stack gases prior to their discharge. To accomplish such an operation required not only many large, expensive valves but also complicated piping arrangements.

Other procedures taking advantage of the increased level of conversion of hydrogen sulfide at the lower temperatures have been proposed. However, they employed an inert gas to regenerate catalyst beds that had become relatively unreactive as a result of sulfur deposition. The disadvantage encountered in this procedure was that under the regeneration conditions the reverse Claus reaction occurred, i.e., the free sulfur on the catalyst reacted with water vapor in the regeneration gas to convert the sulfur to hydrogen sulfide and sulfur dioxide, thus lowering the efficiency of the process.

The process takes advantage of the favorable equilibrium that exists for the catalytic reaction of H_2S with SO_2 at temperatures of the order of 270° to about 300°F while at the same time offering considerable flexibility in the manner in which the various streams can be handled. Because of the fact that one of the reactors in a multireactor plant can always be operated at an effluent temperature of at least 600°F, the process assures good clean-up of feed streams containing CS_2 and COS. Also, valving and piping are simplified by allowing one reactor to operate as a conventional Claus reactor while the other two reactors operate alternately at temperatures above and below the sulfur dew point.

A process developed by *H.W. Strong, J.E. Radway and H.A. Cook; U.S. Patent 3,784,680; January 8, 1974; assigned to Basic Incorporated* is a process for removing sulfur dioxide from gases and recovering elemental sulfur comprising scrubbing the gases with a slurry containing a metal sulfide under conditions to react the sulfur dioxide with the sulfide to form a precipitate which is predominantly metal sulfite and elemental sulfur. A portion of the slurry is withdrawn and the precipitate is separated therefrom. The remaining slurry is recycled for reaction with additional sulfur dioxide and the precipitate is heated under conditions to remove water therefrom and convert any metal thiosulfate therein to additional sulfite and elemental sulfur. Some metal sulfate may be present. The elemental sulfur is separated, and the sulfite and sulfate are then reduced to the sulfide which is recycled to the scrubbing slurry. Preferably, the metal sulfide is an alkali sulfide or alkaline earth sulfide.

A process developed by *A. Deschamps, C. Dezael, S. Franckowiack and P. Renault; U.S. Patent 3,882,222; May 6, 1975; assigned to Institut Francais du Petrole, des Carburants et Lubrifiants, France* is one in which sulfur dioxide is removed from a gas by contacting the latter with ammonia and water, concentrating and then vaporizing the resulting ammonium sulfite solution, passing the vapors through a heated liquid phase, recovering sulfur and ammonia and recycling ammonia.

SULFUR DIOXIDE

An excellent review of the control of sulfur dioxide emissions in copper, lead and zinc smelting together with various recovery options (as SO_2, H_2SO_4 or elemental sulfur) has been presented by the staff of the Bureau of Mines in Information Circular No. 8527, Washington, D.C., U.S. Bureau of Mines (1971).

Recovery as Sulfite Pulping Liquor

A process developed by *K. Mattern; U.S. Patent 3,635,665; January 18, 1972; assigned to Metallgesellschaft AG, Germany* involves recovering MgO and SO_2 from the flue gases of a combustion process and producing a digestion liquor for use in a pulp plant wherein the gases containing MgO and SO_2 are passed upwardly through outwardly dispersed atomized sprays of liquid. The liquid droplets are collected outwardly of the atomizing disks and passed to successively lower atomizing units where the process is repeated. Water introduced above the lower (wet precipitator) set of atomizing disks strips the MgO fly ash from the gas to form a suspension which is used in the upper stages to react with SO_2 (absorption stage) and produce the digestion liquor.

A process developed by *N.C.S. Chari, K.E. Olson and P.E. Shick; U.S. Patent 3,647,363; March 7, 1972; assigned to Owens-Illinois, Inc.* involves the oxidation of neutral sulfite green liquor with air or oxygen under heat and pressure to form an alkaline solution containing sodium sulfate, sodium thiosulfate, sodium sulfite, sodium hydroxide and sodium carbonate, suitable for recovery of sulfur dioxide from flue gases to obtain a product liquor suitable for use in a pulping process.

Recovery as Sulfur Dioxide

A process developed by *R.R. Cantrell and F.P. Wiley; U.S. Patent 3,538,681; November 10, 1970; assigned to W.R. Grace & Company* involves removing and recovering SO_2 from gaseous mixtures such as power plant flue gases by absorbing the SO_2 in dimethyl sulfoxide (DMSO), and subsequently desorbing the SO_2 from the DMSO/SO_2 mixture by contacting the mixture with alumina, or other suitable adsorbent.

A process developed by *T. Owaki; U.S. Patent 3,556,722; January 19, 1971; assigned to Furukawa Mining Company, Ltd., Japan* is a process which comprises treating exhaust gas containing sulfurous and/or sulfuric acid gas with a solution or suspension of the hydroxide or carbonate of calcium, to produce the corresponding sulfite and/or sulfate, adding waste sulfuric acid to the liquid so obtained to convert the sulfite to the corresponding sulfate simultaneously with the evolution of sulfurous acid gas and degassing the liquid thus obtained by blowing the air thereinto, to recover the sulfate and sulfurous acid gas from the exhaust gas and consequentially eliminate a public nuisance otherwise caused thereby.

A process developed by *I.S. Shah; U.S. Patent 3,617,212; November 2, 1971; assigned to Chemical Construction Corporation* is a process designed to economically absorb and recover sulfur dioxide gas, and thus eliminate air pollution and produce a useful sulfur-containing product. Sulfur dioxide is recovered from a waste gas stream, such as the flue gas from a power boiler which burns sulfur-containing coal or other fuel, by scrubbing the waste gas stream with an aqueous solution of magnesium bisulfite-magnesium sulfite. Sulfur dioxide is absorbed into the aqueous scrubbing solution by reacting with magnesium sulfite to form magnesium bisulfite in solution. The resulting solution is divided into two portions. Magnesium hydroxide is added to the first portion at a controlled rate, to convert magnesium bisulfite to magnesium sulfite.

The first portion is then recycled for further waste gas scrubbing. The magnesium sulfite and magnesium bisulfite content of the second solution portion are processed at elevated temperature to produce magnesium oxide and a gas stream of high sulfur dioxide content, usually about 10% by volume. The magnesium oxide is slaked with water to form magnesium hydroxide which is added to the first solution portion, and the gas stream of high sulfur dioxide content is utilized to produce a sulfur-containing product.

A process developed by *Y. Eguchi; U.S. Patent 3,667,910; June 6, 1972; assigned to Takeda Chemical Industries, Ltd., Japan* is a process for removing sulfur oxides from waste gases with employment of activated carbon as the adsorbent. The chemical exhaustion of activated carbon and the lowering of its ability for adsorbing sulfur oxides in the regeneration step can be remarkably prevented by regenerating the sulfur oxides-adsorbing activated carbon with the desorbent containing carbon monoxide gas and/or hydrogen gas in a concentration at least about 40% at a temperature from about 230° to 450°C. The most advantageous results can be attained by employing vanadium oxide-supporting activated carbon as the adsorbent.

A process developed by *J.D. Terrana and L.A. Miller; U.S. Patent 3,687,623; August 29, 1972; assigned to Wellman-Lord, Inc.* involves the reaction between SO_2 in a gas with a sulfite to produce the corresponding bisulfite. Examples of sulfites include potassium, cesium, and rubidium sulfites. In the process, bisulfite crystals are separated from a solution containing the same and SO_2-partial pressure-lowering materials and subsequently placed in either solution or slurry form and passed into a decomposition zone wherein the bisulfite is heated to drive the SO_2 out of the solution converting the bisulfite to the sulfite which can be recycled to absorb additional SO_2.

A process developed by *F.J. Ball; U.S. Patent 3,789,110; January 29, 1974; assigned to Westvaco Corporation* involves removing sulfur dioxide from waste gas mixtures by adsorbing sulfur dioxide onto activated carbon as sulfuric acid at a temperature between ambient temperature and 350°F, subsequently reacting the sulfuric acid-laden activated carbon with the correct proportion of hydrogen sulfide at a temperature below 350°F, preferably 200° to 350°F, to homogeneously reduce all the hydrogen sulfide and a portion of the adsorbed sulfuric acid to elemental sulfur, then raising the temperature to between 350° to 1000°F to react the elemental sulfur and remaining sulfuric acid to form concentrated sulfur dioxide without reaction with the activated carbon, recovering the sulfur dioxide from the activated carbon, and recycling the completely regenerated activated carbon.

The recovered sulfur dioxide may be used in a variety of ways, for instance, it may be reacted with methane to produce hydrogen sulfide, it may be reacted with hydrogen sulfide as in a Claus unit to produce elemental sulfur, or it may be converted to sulfuric acid.

Recovery as Sulfuric Acid

A process developed by *E.J. Parsi; U.S. Patent 3,524,801; August 18, 1970; assigned to Ionics, Incorporated* is a cyclic process for the removal of SO_2 from an SO_2-containing waste gas with subsequent recovery of at least some of the SO_2 in the form of sulfuric acid. A three-compartment diaphragm-type electrolytic cell is employed to convert a feed solution of an alkali metal sulfate salt into its corresponding acidic and basic components. The basic caustic solution will absorb the SO_2 from the waste gas forming primarily an alkali metal bisulfite which is at least partially oxidized into bisulfate and the resulting liquid mixture evaporated to separate out alkali metal sulfate crystals. The crystals are redissolved in water and cycled back to feed to the electrolytic cell. The mother liquid of sulfuric acid remaining from the crystallization step is collected as a valuable product.

A process developed by *L.L. Fornoff, J.J. Collins and R.D. Taylor; U.S. Patent 3,829,560; August 13, 1974; assigned to Union Carbide Corporation* is one in which sulfur dioxide in the effluent from the acid absorber of a contact process sulfuric acid plant is adsorbed in a fixed bed of molecular sieve adsorbent, purged from the adsorption bed with hot dry oxygen-containing gas and recycled to a catalytic converter in admixture with the oxygen-containing purge gas for conversion to sulfur trioxide and consequent reaction with water to form sulfuric acid.

SULFURIC ACID

In the sulfate process for the production of titanium dioxide pigment, titaniferous ores or slags are digested with 90 to 95% sulfuric acid. After hydrolysis of the digestion black liquor and separation of the titanium dioxide hydrate produced therein, large quantities of a liquid stream containing about 20% free sulfuric acid and 10 to 20% by weight of dissolved salts remain. Depending upon the type and origin of the ore or slag, the dissolved salts include iron, aluminum, magnesium, calcium, and titanium sulfates plus minor quantities of the sulfates of vanadium, chromium and manganese.

In excess of 1.5 million tons of sulfuric acid are used each year in the manufacture of titanium dioxide pigment. Almost all of this acid reports in the 20% acid stream. Another 1 million tons of sulfuric acid are used each year in the cleaning of metals, from which a somewhat similar waste acid solution results. The dilution inherent in each of these processes makes for a waste acid stream totaling in excess of 10 million tons per year.

Recovery from Waste Pickle Liquor as H_2SO_4

A process developed by *W.L. Johnson and W.R. Massey; U.S. Patent 3,816,593; June 11, 1974* is a process for regenerating aqueous sulfuric acid pickling liquor that contains residual iron-sulfur compounds characterized by the steps of (1) oxidizing the iron-sulfur compounds in at least a portion of the liquor; (2) partially neutralizing the portion of the liquor by raising its pH to at least 2 with the addition of a soluble caustic material, preferably one that contains a cation whose reaction product after the two steps forms a precipitate at a pH of 2 and higher; (3) separating the precipitates from the regenerated pickling liquor; and (4) returning the regenerated pickling liquor to pickling service.

A process developed by R.E. Lacey of Southern Research Institute and detailed in Report 12010 EQF 03/71, Washington, D.C., U.S. Environmental Protection Agency (March 1971) is an electromembrane process for regenerating acid from spent sulfuric acid pickle liquor. These studies have indicated that the process is technically feasible. The studies have shown that the iron ions in spent pickle liquor can be removed and replaced by hydrogen ions to regenerate H_2SO_4 in electromembrane cells.

A method of removing iron from spent liquor that involves the formation of insoluble iron hydroxides is preferable to plating iron metal onto cathodes. Estimated treatment costs were \$0.045 ± 0.002 per gallon, whereas the combined costs of purchasing acid and disposing of spent liquor by existing methods were in the range of \$0.015 to \$0.06 per gallon of spent liquor. A determination of the long-term performance of the ion-exchange membranes when treating actual pickle liquors that contain organic pickling aids is needed, however, the report concludes.

Recovery from Waste Pickle Liquor as Fertilizer

A process developed by *R.B. Bahme; U.S. Patent 3,232,738; February 1, 1966* involves making fertilizers from waste pickling liquors of the steel and galvanizing industries. In some forms, it also utilizes waste ore tailings from mining operations, waste pyrites fines as used in the manufacture of sulfuric acid from pyrites ores, or waste ash and wood slag from the burning of wood scraps, sawdust, and the like in the preparation of such fertilizer products.

The process results in formation of ammine (i.e. the diammine, triammine, tetraammine, hexammine, and other ammonia complex) compounds of iron, zinc, and other metals. These ammine compounds are solids at standard temperature and pressure, and are soluble in water. Substantially all concentrations of these compounds in water may be made, and water solutions thereof may be evaporated to dryness to get the solid compound. These metallic ammine compounds are made by reacting anhydrous or aqueous ammonia with waste pickling liquors. Typical reactions that occur when sulfates of these liquors are reacted with anhydrous ammonia are as follows:

$$FeSO_4 + 4NH_4OH \longrightarrow Fe(NH_3)_4SO_4{\cdot}4H_2O$$

$$ZnSO_4 + 4NH_4OH \longrightarrow Zn(NH_3)_4SO_4{\cdot}2H_2O + 2H_2O$$

$$2FeSO_4 + 2NH_4OH \longrightarrow (NH_4)_2Fe(SO_4)_2 + Fe(OH)_2$$

The ammines dissociate in the following manner:

$$Fe(NH_3)_4SO_4 \longrightarrow Fe(NH_3)_4^{++} + SO_4^{=}$$
$$Zn(NH_3)_4SO_4 \longrightarrow Zn(NH_3)_4^{++} + SO_4^{=}$$
$$(NH_4)_2Fe(SO_4)_2 \longrightarrow NH_4^{+} + NH_4Fe(SO_4)_2^{-}$$

These compounds make outstanding fertilizers. For example:

$Fe(NH_3)_4SO_4$ contains about 25% nitrogen, 25% iron and 15% sulfur;
$Cu(NH_3)_4SO_4$ contains about 25% nitrogen, 28% copper and 14% sulfur;
$Zn(NH_3)_4SO_4$ contains about 25% nitrogen, 29% zinc and 14% sulfur;
$(NH_4)_2Fe(SO_4)_2$ contains about 10% nitrogen, 20% iron and 22% sulfur.

The water solubility of these metallic ammine sulfates makes them ideally suited for use as liquid fertilizers. Simple aqueous solutions of these compounds contain plant nutrients in a form readily available to plant life. Where desirable, additional fertilizer components may be combined with these metallic ammine sulfates to provide a fortified product. For instance, ammonia, urea, ammonium nitrate, ammonium sulfate, and other similar ammonia compounds, singly or in combination, may be dissolved in the water solutions of the sulfates to increase the nitrogen content. Of course, other suitable compositions and compounds, such as phosphoric acid and/or potassium salts, also may be added to these solutions.

Waste pickling liquor from the steel and galvanizing industry includes dilute sulfuric acid and metal sulfates. The metallic sulfates can be precipitated by concentration, and then reacted with ammonia to form the particular metallic ammine and ammonium sulfates, which are valuable soil conditioners and fertilizers. This pickling acid treatment also can be used with tailings and fines which have not been previously leached with ammonia or ammonium sulfide. Thus, waste materials, i.e., mine ore tailings, pyrites fines from sulfuric acid manufacturing, and the pickling liquor from the steel and galvanizing industry are transformed into valuable raw materials for manufacturing fertilizers in a less expensive manner.

Another plentiful supply of inexpensive raw material for manufacturing fertilizers is the ash resulting from burning waste wood scraps, sawdust, and other organic matter in the lumber milling industry. At practically every lumber mill and plywood plant, and at many other wood product manufacturing establishments, much waste wood scraps, chips, sawdust, bark, leaves, etc. are produced, and they are burned as the easiest method of disposal.

The ash resulting from this burning is very alkaline, usually around 11.5 pH or higher, but, depending upon what type of plant matter has been burned, it contains substantially every element except nitrogen and sulfur, essential to plant life including potassium, copper, phosphorus, manganese, magnesium, aluminum, calcium, iron, and zinc, mostly as basic oxides and/or carbonates.

If this waste ash, which sometimes occurs as a slag-like material, is neutralized with the waste pickling liquor from the steel and galvanizing industries, the resulting liquid contains many metallic sulfates and other essential compounds and elements in a form easily and readily assimilated by plants. This liquid can be applied directly to the soil without further modification, or other fertilizer components, such as ammonia and other nitrogen-containing compounds, can be added to it, thus providing nitrogen and increased amounts of other elements.

Alternatively, the pickling liquor can be modified by adding to it various compounds especially nitric and/or phosphoric acids, before combining it with the ash. Also, the pickling liquor can be neutralized partly by ash and partly by ammonia, either separately or by combining the ash and ammonia before reacting with the liquor.

A process developed by *S.J. Shirley; U.S. Patent 3,888,653; June 10, 1975; assigned to Shamrock Chemicals Limited, Canada* involves producing a potassium sulfate fertilizer from spent sulfuric acid having impurities therein comprising the steps of: (a) selecting as a first feed material spent sulfuric acid; (b) selecting, as a second feed material, one consisting essentially of potassium chloride; (c) mixing the first and second feed materials to achieve a feed mix; (d) heating the feed mix at an elevated temperature in excess of 300°F whereby the potassium sulfate fertilizer is created in fine crystalline powder; (e) mixing the recovered crystalline potassium sulfate with granulating agents and small amounts of water until the potassium sulfate congeals into small pellets; and (f) grading the pellets into a predetermined range of sizes.

Recovery from TiO_2 Pigment Plant Wastes

A process developed by *C.H. Umstead; U.S. Patent 3,713,786; January 30, 1973; assigned to The New Jersey Zinc Company* is one in which sulfuric acid in concentrated form is recovered from dilute, waste sulfuric acid liquors containing metal sulfates, such as those resulting from the leaching of mineral values from ores and the like or from the cleaning of metals such as iron and steel. Separation of the sulfate salts in substantially dry form is effected by total evaporation of the water and of the sulfuric acid; condensation of these vapors is controlled so that the resulting acid is of increased concentration. Figure 40 is a flow diagram of the process.

FIGURE 40: APPARATUS FOR EVAPORATIVE SULFURIC ACID RECOVERY FROM SULFURIC ACIDS CONTAINING SULFATES

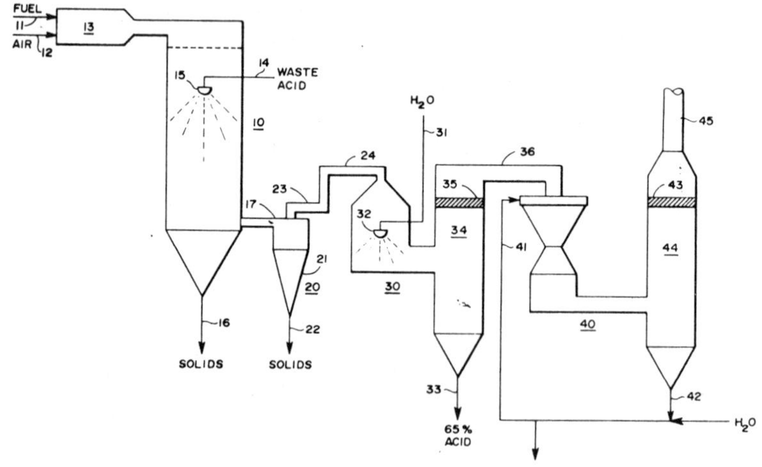

Source: C.H. Umstead; U.S. Patent 3,713,786; January 30, 1973

The process centers on the spray evaporator **10**. Fuel and air are fed through lines **11** and **12** respectively to burner **13**. Waste acid liquor enters the spray evaporator through line **14** which terminates in a spray device **15**, to provide for the introduction of the liquor in a finely divided state. In order to avoid decomposition of the sulfuric acid, it is essential that intimate contact between the hot gases from burner **13** at a temperature of about 600° to 1000°C and the finely divided liquor issuing from spray device **15** at or somewhat above ambient temperature, be effected as promptly as possible. The temperature of the mixture of water and sulfuric acid vapors, and combustion gases leaving the spray evaporator **10** through outlet **17**, should not exceed about 350°C. The spray evaporator **10** has two outlets therefrom; an outlet **16**, at the base, which is coupled to a closed container to receive the discharge of dry sulfate solids and an outlet **17**, at the side, through which the vapors and gases are transferred to cyclone separator **21**. Any solids which remain in the gas stream exiting the spray evaporator, are caused to drop down the conical sides **21** of the separator to the bottom outlet **22** and then discharged to a sealed receptacle (not shown). Substantially solids-free gases and vapors exit the cyclone separator through conduit **24** and enter the condensation unit **30** wherein they are contacted and cooled by evaporation of liquid water introduced via line **31** through spray device **32**. The temperature in the condensation unit is controlled such that substantially all of the sulfuric acid vapor and less than all of the water vapor are condensed. With a feed liquor containing about 20% sulfuric acid, a temperature of about 150°C in condensation unit **30**, will result in a product having a sulfuric acid concentration of about 65 to 70%. Control of the temperature is effected by the amount of water evaporated.

Vapor-liquid separation chamber **34** provides for disengagement of liquid droplets from the noncondensable gases, i.e., principally those from the combustion in burner **13**, and for coalescence of the liquid droplets. The condensation product is collected in the base and discharged through outlet **33** to a storage tank (not shown). Mist eliminator **35** which may consist of commercially available demisting pads, functions to impede escape of the fine droplets of sulfuric acid and to allow for their collection with the product acid. The noncondensable gases pass through the mist eliminator **35** via a conduit **36** to a Venturi scrubber **40**.

Venturi scrubber **40** is provided to assure that any last traces of sulfuric acid are removed from the gas stream before discharge through the stack **45**. Water is recirculated from outlet **42** of the vapor-liquid separator **44** through line **41** to the scrubber. Another mist eliminator **43** insures that the discharged gases will consist of water vapor and the noncondensables of combustion.

TITANIUM DIOXIDE

Recovery from Paper Mill Waste Sludges

A process developed by *C.A. Hoffman; U.S. Patent 3,876,497; April 8, 1975; assigned to Sterling Drug, Inc.* is one in which paper mill waste sludges are subjected to wet air oxidation whereby organic materials are converted to innocuous oxidation products, and the inorganic filler materials, e.g., titanium dioxide, clay and the like, are recovered in a form suitable for reuse.

Inorganic filler materials are added to cellulose fiber in the papermaking process in order to modify certain physical properties, especially to enhance the whiteness and opacity of the paper. Such filler materials, also called pigments, include aluminum silicate (China clay), calcium carbonate, zinc oxide, aluminum hydroxide, zinc sulfide, titanium dioxide, calcium sulfate, calcium sulfite, barium sulfate (blanc fixe), hydrated magnesium silicate (talc), silica, and mixtures of two or more of the foregoing. The preferred and most commonly used fillers are aluminum silicate, titanium dioxide and calcium carbonate.

In conventional papermaking processes the pulp fiber and fillers are dispersed in water and run on to a paper machine wire. A certian amount of the solid material is not retained but passes through the wire along with the water used for suspending the fibers and fillers. This effluent is known as white water and, although it can be reused to a certain extent, it must eventually be disposed of. Generally, the white water is passed through a clarifier or other type of sedimentation device before being discharged. In the clarifier a sludge accumulates which contains waste fiber, filler and other organic material. This sludge, unless processed, represents a loss to the paper mill and also presents a waste disposal problem.

The filler in the waste sludge is not reusable as such since it is contaminated with waste fiber and other organic material. The organic fraction must be removed before the fillers can be used again. Conventional incineration of the sludge destroys the fibers and other organic matter but results in undesirable chemical and physical changes in the filler materials. For example, upon conventional burning, clay becomes abrasive and, if reused, results in excessive wear of the wire of the papermaking machines. Furthermore, if calcium carbonate is present, incineration converts it to calcium oxide which is not useful per se as a filler. Incineration also produces objectionable air pollution and losses of the filler material. In the process, the inorganic filler materials are recovered substantially completely from paper mill waste sludges, suitable for reuse, by heating the waste sludges with an oxygen-containing gas (pure oxygen or oxygen mixed with inert gas, e.g., air) at a temperature between 160° and 375°C and at a corresponding pressure which will maintain most of the water content of the sludge in the liquid phase, between about 500 and 3,500 psig, directly proportional to the temperature used, and then separating the gas, liquid and solid phases, the last being the recovered filler material suitable for reuse.

If pure oxygen is used, pressures in the lower part of the abovementioned range are used. A preferred temperature range for the oxidation process when air is used is from 225° to 320°C. The oxidation is carried out for a time necessary to effect maximum reduction of COD (chemical oxygen demand) under the conditions used and to produce a recovered filler having a brightness of at least about 80% GE (General Electric measurement). The oxidation time varies from about 0.25 to 5 hours, inversely proportional to the temperature and pressure conditions. The wet air oxidation of the paper mill sludge converts the waste fiber and other organic materials to carbon dioxide or to low-molecular-weight organic compounds which are water-soluble and readily removed by simple washing procedures. Figure 41 is a flow diagram of the process.

FIGURE 41: PROCESS FOR WET AIR OXIDATION OF PAPER MILL SLUDGE TO PERMIT FILLER RECOVERY

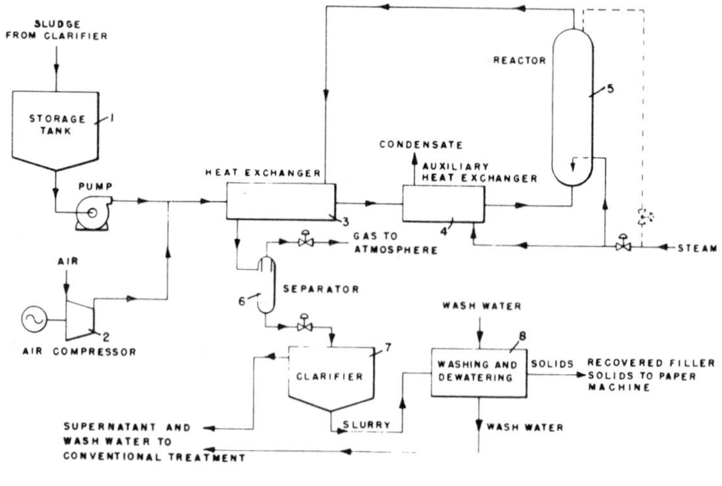

Source: C.A. Hoffman; U.S. Patent 3,876,497; April 8, 1975

Paper mill sludge from a clarifier is transferred to a storage tank **1**. From the storage tank, the sludge is pumped into the system where it is mixed with air provided by a compressor **2**. The mixture of air and sludge passes through one or more heat exchangers **3** and **4** and into a reactor **5**. Steam for start-up or continuous use may be added indirectly to the process by means of heat exchanger **4** as shown or may be injected directly into the reactor. Also, steam may be injected into the reactor as a means of rapid start-up and thereafter utilized in the auxiliary heat exchanger, thus conserving the condensate. Oxidized material exiting from the reactor **5** passes through the heat exchanger **3** and into a separator **6**.

From the separator, the gaseous phase passes into the atmosphere, preferably by way of a purifying device such as a scrubber, catalyst, activated carbon tower or the like. The liquid phase containing suspended filler material is transferred to a clarifier **7** where the solid material settles and is withdrawn from the bottom of the clarifier as a slurry. The slurry is transferred to a washing and dewatering vessel **8** to afford the recovered filler suitable for reuse as an additive for fresh cellulose fiber in the papermaking machine. The supernatant liquor from the clarifier **7** and wastewater from vessel **8** are combined for conventional treatment and disposal.

TUNGSTEN CARBIDE

Recovery from Used Grinding Tools

As described in *Chemical Week* for April 1, 1970, page 63, the tool industry uses $250 million worth of cemented carbides every year. When the tools are no longer usable, the tungsten carbide may be virtually intact.

A process developed by the U.S. Bureau of Mines permits separation and reuse of the tungsten carbide by mixing the scrap with molten zinc. This removes the carbide from the cobalt cementing agent by forming a zinc-cobalt alloy from which the zinc is eventually recovered by distillation.

NATURAL PRODUCTS

BAGASSE

Bagasse is the wood stem of the sugar cane. It is a by-product of the sugar industry present in large amounts at low costs in areas where sugar is produced. The abundance of this by-product material has attracted the attention of many, and through the years proposals for utilization of the material have been frequent. A use commonly suggested is to carbonize the bagasse to provide it in a form suitable for use as fuel, and various of these proposals have been tried.

A problem incidental to such use of bagasse is with respect to bulk density of the carbonized product. Thus, absent special precautions carbonized bagasse has a very low bulk density so that storage and transportation are complicated. Further, the low bulk density gives rise to various difficulties in the utilization of the material as fuel. One suggestion to obviate the bulk density problem, in the prior art, is to briquette the bagasse before subjecting it to carbonization. While this does provide a solution to the problem, the cost of briquetting is substantial.

Recovery in the Form of Charcoal

A process developed by *M.F. Leonor; U.S. Patent 3,436,312; April 1, 1969* relates to the carbonization of bagasse to provide the carbon values in the bagasse in a form suitable for use, for example, as a heating medium.

In the process, bagasse is carbonized in a process which comprises the step of contacting the bagasse with sulfuric acid to reduce its weight by more than about 20%, followed by the further step of separating the bagasse residue from the contacting medium, and thereafter carbonizing the separated residue.

The following is one specific example of the conduct of the process. 100 lb of green bagasse shredded to particles ranging in length up to ¾" is mixed in a digestor with 25 lb of 100% H_2SO_4 and 6 parts H_2O. The mixture is agitated, heated to 150°F, and steam at atmospheric pressure is then injected at the rate of 5 lb/hr. Contacting at these conditions is continued for twenty minutes. The digestion is then complete, and the digested bagasse and liquor are removed from the digestor and separated by filtering. The bagasse can then be washed with water, and the wet bagasse, which is partially carbonized, can be subjected to a low temperature carbonization in a kiln.

The carbonization is preferably a low temperature carbonization. It can be at a temperature of about 450° to 700°F. The carbonization can be carried out by known means, and is preferably carried out in a rotary kiln. The carbonization product can be about 50% by weight of the green bagasse starting material.

Another process developed by *M.F. Leonor; U.S. Patent 3,436,314; April 1, 1969* is one for converting moist plant matter, such as bagasse, into charcoal, wherein the matter in particulate form is introduced in a suspended state into a drying tube supplied with a flue gas to dry the particles, the dried particles being introduced in a suspended state into a preheating tube supplied with a flue gas for heating the particles to a temperature effecting partial carbonization thereof to produce heated, partially carbonized particles which are then introduced in a suspended state into a carbonization tube supplied with a flue gas heating the particles to a temperature effecting complete carbonization thereof.

BANANAS

Recovery of Waste as Cattle Feed

As described in *Chem. & Eng. News,* June 9, 1975, p. 50, some 200 million pounds of bananas go to waste annually in Panama. The bananas, bruised and cut during harvesting and handling are left in dumps to rot. Now protein-rich feed for beef cattle may be produced by a fermentation process developed by professors from M.I.T. and the University of Missouri.

The proposed process would consist essentially of chopping up the rejected bananas and feeding them to fodder yeast or bakers' yeast in an aerobic fermenter. Both strains have been used before to feed cattle, and both multiply rapidly, doubling in weight every four to eight hours. After fermentation, undigested material would be filtered out and the yeast dried and mixed with nutrient salts, vitamins, and a source of fiber, such as chopped banana leaves. The mixture would be pelletized for use.

CASTOR BEANS

Production of Fertilizer from Castor Bean Residues

A process developed by *J.R. Spies, E.J. Coulson and P.A. Wells; U.S. Patent 3,101,266; August 20, 1963; assigned to the U.S. Secretary of Agriculture* relates to the deallergenization and detoxification of castor bean pomace.

Castor beans contain two principal harmful components: ricin, an extremely poisonous albumin, and an exceptionally stable allergenic component. In processing the castor bean for castor oil these components are retained in the pomace, the ground, defatted bean residue. The toxicity of ricin is known to be destroyed when heated in water to the boiling point or even to the coagulation temperature of the protein.

Despite intensive efforts to inactivate the allergen, the presence of this factor in castor bean pomace is still a serious problem in handling castor bean pomace or products into which the pomace is incorporated. This is because the allergen has unusually potent sensitizing capacity for those exposed to the dust of castor bean pomace. Even trace amounts provoke severe asthma in hypersensitive individuals. Although processes have been proposed for deactivating the allergen, none of these currently reported suffice to yield a product which can be safely handled by all people.

An object of the process is to render castor bean pomace nonallergenic and nontoxic. A further object is to provide a castor bean product having both organic and inorganic nutritive values as a fertilizer. Still another object is to provide a fertilizer containing castor bean pomace which can be used without endangering the health of handlers and consumers.

According to the process, castor bean pomace, water and calcium hydroxide are combined to give a mixture having a pH in the range of about 9.5 to 12.5, and the alkaline mixture is heated at about 100° to 120°C for a time interval of about eight to sixty minutes. This treatment inactivates the allergen and destroys the toxic components. The pomace product is used as an alkaline fertilizer, or preferably the reaction mixture is partially or completely neutralized to give a pomace-calcium phosphate solid, a new combination fertilizer.

Various alternatives are available for handling the products as fertilizers. For local use, the wet slurry may be directly applied to fields, or solids may be collected from the slurry by conventional means, such as centrifuging or filtering, and the product used as a wet sludge. In practice, it is anticipated that a considerable portion or all of the slurry or the collected pomace product will be dried, thus facilitating storage, shipping and consumer use.

CORNCOBS

The commercial utilization of corncobs has been reviewed by L.K. Arnold of the Department of Chemical Engineering at Iowa State University in *Solid Wastes,* C.L. Mantell, Ed., New York, John Wiley & Sons (1975). The reader is also referred to the section on "Furfural" under "Organic Materials" for a reference on the production of furfural from corncobs.

Production of Stock or Poultry Litter

A process developed by *R.R. Yoerger; U.S. Patent 2,766,795; October 16, 1956; assigned to Iowa State College Research Foundation, Inc.* utilizes a crusher adapted to crush corncobs, etc. for the forming of a stock or poultry litter and other products.

In the crushing of corncobs, etc. with mechanisms available, it is found that a very substantial amount of dust is formed, which dust is not only undesirable in the handling and transporting of the product, but also, when the product is delivered and placed within a stock or poultry building. Further, the dust itself is largely lost and becomes ineffective as a litter. The devices available for treating corncobs also are of sharply limited capacity, and efforts to operate such devices at a higher speed result in a greater formation of dust.

An object of the device is to provide a machine for crushing corncobs, etc. in such a manner as to avoid dust formation while at the same time producing corncob segments which are usable as an effective litter quantitatively. A further object is to provide mechanism in which corncobs are split first in longitudinal segments and then cut transversely to produce sizable chunks or pieces of corncob having relatively no dust therein and which are employed as an entirety in the forming of litter.

A still further object is to provide an adjustable mechanism effective for separating corncobs into long segments while at the same time delivering the segments automatically to a desired destination and in substantially uniform pieces. A still further object is to provide mechanism for carrying corncobs to a zone in which the cobs are broken into substantially uniform segments while delivering the segments therefrom to a desired destination. A still further object is to provide a method and means for breaking corncobs into sizable and relatively uniform chunks highly desirable for litter while substantially eliminating the formation of dust.

CORNSTALKS

An excellent review of the commercial utilization of cornstalks has been presented by L.K. Arnold of the Department of Chemical Engineering of Iowa State University in *Solid Wastes,* C.L. Mantell, Ed., New York, John Wiley & Sons (1975). Topics covered by Arnold include the following.

(1) Paper from cornstalks
(2) Building board from cornstalks
    (a) Insulating board
    (b) Acoustical board
    (c) Hard board
    (d) Molded fiber products
(3) Miscellaneous products from cornstalks
    (a) Xylose
    (b) Maizolith
    (c) Fuel gas

LEATHER

Recovery from Leather Scrap

A process developed by *H. Okamura; U.S. Patent 3,441,470; April 29, 1969; assigned one-half to Fuji Spinning Company Limited, Japan* involves utilizing defective leather scrap effectively for fabricating woven or nonwoven leather goods by retaining collagen fibers from the defective leather scrap.

The process involves the steps of dipping a leather raw material into warm water containing a nonionic active agent, rotating the treated leather, dipping the leather again into warm water, and detanning the leather. The treated leather is rotated in a dilute alkaline solution and then washed. The leather is then treated in a urea solution and washed, and then dehydrated partially. Sulfonated olive oil emulsion is added to the treated leather, and the leather is then beaten and pressed against a card clothing roll for disintegration. The collagen fibers are obtained by affixation to a rotating wooden roll on which is provided a card clothing.

MANURE

Production of Cattle Feed

A process developed by *D.W. Seckler and J.M. Harper; U.S. Patent 3,875,319; April 1, 1975; assigned to Ceres Ecology Corporation* involves recovering valuable feed products from animal manure by adding water to the manure to form a slurry from which mineral matter, grain and fiber particles, proteinous material and water are recovered by successive slurry-forming and liquid/solid separation stages. The recovered water is reused in the forming of the initial manure slurry and in subsequent steps of the process. The recovered grain and fiber particles and proteinous material are useful as feed products which can be refed to the animals from which the manure is collected or, of course, other animals.

Figure 42 is a flow diagram of the process. Raw manure recovered from an animal feedlot is ground by a hammermill or other suitable device **10** and then conveyed in any suitable fashion to a first mixing tank **12**. The device shreds the raw manure to divide lumps of manure into constituent particles, without substantial grinding of the particles, and thus aids in the subsequent slurrying steps. Preferably, the feedlot will have a concrete floor to minimize contamination but the process may be used with any type of feedlot or other confinement.

The size of the first mixing tank depends upon the amount of manure to be processed and, to some extent, the manner in which the manure is delivered to the tank. For example, if the manure is delivered to the tank in truckload batches, a somewhat larger tank would be required than if the manure were delivered in a steady stream such as, for example, by auger, chain drive or conveyor belt.

FIGURE 42: PROCESS FOR PRODUCTION OF CATTLE FEED FROM MANURE

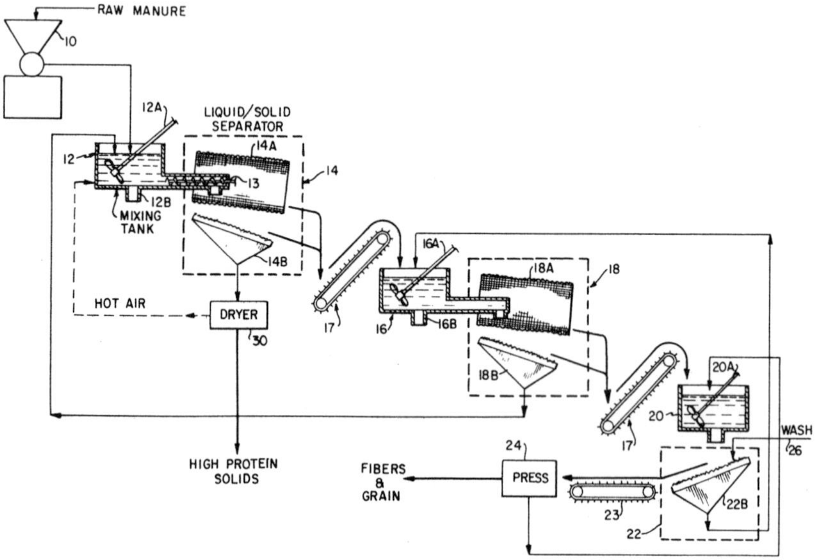

Source: D.W. Seckler and J.M. Harper; U.S. Patent 3,875,319; April 1, 1975

A slurry is formed in the first mixing tank **12** as described below, the slurry comprising about 15 to 30% dry matter and about 85 to 70% water. These percentages may vary widely, with more water being desirable with freshly collected manure because of the presence of mucus. The amount of water in the raw manure collected from the feedlot may vary substantially (for example, from about 50 to about 80%) and, therefore, the amount of water to be added to the tank will vary accordingly. For example, the amount of water added to the slurry tank may vary from none to about 3 lb of water for each pound of raw manure.

In addition to water, a small amount of chlorine may be introduced into the slurry tank in order to kill some of the bacteria and reduce or eliminate the smell of the slurry. Conventional apparatus may be used for this purpose. The amount of chlorine added is ordinarily less than twenty parts of chlorine per million parts of slurry. A suitable detergent (such as a surfactant) may also be added to the slurry in the tank to aid in the dissolving process. Also, the slurry in the tank may be heated (e.g., by introduction of steam) to reduce its viscosity and thus facilitate subsequent liquid/solid separation stages. If the slurry is heated to about 120°F some bacteria will be killed, which may be desirable.

The mixing tank may be provided with a standard agitator **12A** which serves to mix the slurry to an approximately uniform consistency. The agitator may be of the propeller type or any other conventional type used in analogous applications. The tank may include a trap **12B** toward which rocks and sand will tend to sink for periodic removal as required.

The slurry from the tank is delivered to a first liquid-solid separator **14** which separates the slurry into liquid and solid fractions. In an experimental embodiment, the separator consisted of a coarse rotary screen **14A** which rotated above a pitched finer mesh vibrating screen **14B**. The slurry was fed axially into one end of the rotary screen and the coarse wet solids were conveyed axially through the screen **14A** by appropriate vanes or the like while finer solids and the liquid fraction fall through to vibrating screen **14B**. These finer solids are separated by screen **14B** and combined with the coarse solids from the rotary screen to yield the solid fraction from the separator **14**.

The particular mesh size selected for the screens will depend in part upon the characteristics of the manure which, in turn, may depend upon the type of feed eaten by the animals from which the manure is collected. For example, if the manure is collected from cattle which have been fed whole kernel corn, the manure will contain a certain proportion of partially digested whole kernels.

In such a case, the coarse rotating screen may have openings of about ½" across so that only the larger kernels and coarse bits of fiber or straw are retained on the screen **14B** while the remainder of the manure slurry washes through the rotating screen to the vibrating screen. On the other hand, if the manure is collected from cattle that have been fed ground grain, the rotating screen may have openings of about $^3/_{16}$" across or less so as to retain the larger undigested bits of ground grain and coarse bits of fiber or straw.

Instead of helping to separate the solid fraction from tank **12**, the rotary screen may be used to further shred the manure into its constituent particles and to screen rocks and other undesired objects from the slurry. In such a case, the mesh size will be selected to allow the solid and liquid fractions of the slurry to pass through to screen **14B**. An augur or the like may be used to remove the undesired material from the exit end of the screen **14A**. The vibrating screen has a finer mesh, about 20 mesh for example, so as to retain the finer undigested feed and fiber particles. Very fine particles such as mineral matter and suspended solids (which comprise most of the valuable protein) pass through the second vibrating screen.

The solid fraction from the separator is then fed to a second mixing tank **16**, which may be essentially the same as tank **12**, including an agitator **16A** to mix the slurry to a uniform consistency and a trap **16B** to collect rocks and other debris. A chain drag **17** may be used to deliver the solid fraction from the separator to tank **16** though any other suitable conveying or pumping means may be used.

The slurry from the second tank is coupled to a second liquid-solid separator **18** which may be the same as separator **14** including, for example, a coarse rotary screen **18A** and a finer vibrating screen **18B** for separating the liquid and solid fractions of the second slurry. Because the slurry in tank **16** is relatively thin (e.g., about 15% dry matter), the head in the tank will ordinarily be sufficient to allow gravity feed to separator **18** although an augur or similar device may be used as an auxiliary pumping means.

According to the preferred embodiment, the solid fraction from separator **18** is fed to a third mixing tank **20** which may be essentially the same as tanks **12** and **16** (although ordinarily a rock trap will not be required). The slurry from tank **20** is fed to a third liquid-solid separator **22** to separate the liquid and solid fractions of the third slurry. The third separator may include only a single vibrating screen **22B** similar or identical to vibrating screens **14B** and **18B**.

The wet solids from the third separator are then conveyed by a chain drag **23** or the like to a press **24** which compresses these wet solids to remove excess water. The solid matter remaining is composed principally of fiber and grain particles and is one of the two basic products recovered from the manure in accordance with the recycling method of the process. It consists of about 45% dry matter and chemical analyses in feed trials with sheep and cattle have shown this product to be the equivalent of dried beet pulp or low-grade barley in feed value on a dry weight basis.

The liquid fractions from the separators **18** and **22** and the press (which is also a separator) are recycled to the preceding slurry tanks **12, 16** and **20**, respectively, to be mixed with the manure in those tanks. That is, liquid from separator **18** is returned to tank **12**, liquid from separator **22** is returned to tank **16**, and the liquid drawn off by press **24** is fed to tank **20**. Suitable pumping means and valves (not shown) may be included in the various lines to control the liquid flows to the mixing tanks to maintain the respective slurries at their preferred consistencies. In the preferred embodiment, the only water introduced into the system (excluding the water contained in the raw manure) is wash water shown diagrammatically on line **26** which may be used to clean the fiber and grain matter on the vibrating screen **22B**.

The use of a multistage slurrying and separating process provides more complete separation of fiber and grain material from the nutrients contained in the soluble matter of the manure and the fine particles which contain or constitute most of the recoverable protein. The solid fractions from the respective separators will have progressively higher concentrations of dry matter (i.e., grain and fiber) whereas the liquid fraction will become progressively clearer. As a result of the countercurrent flow of the recycled liquid fractions as described, substantially less water is required for separation and the liquid fraction from separator **14** (i.e., the wash water from screen **14B**) will contain an extremely high concentration (as high as 40%) of valuable proteinous particles held in a colloidal type suspension.

Previously, economic recovery of these particles has proven difficult and substantial proteinous matter has been lost. However, the concentration of these particles may be sufficiently great to allow separation by evaporation in a dryer **30** or similar device to which the liquid fraction from separator **14** is pumped or otherwise conveyed. Hence, the process provides a liquid fraction having a higher concentration of proteinous material from which, consequently, removal of water is simpler. Moreover, because of this higher concentration, evaporation may be used to recover substantially all of such material and other solids.

The basic feed product recovered from the press may be further processed in several ways before being fed back to the animals in the feedlot. For example, the feed product from the press may be tumbled through an open flame to burn off the animal hair present in the product. After the hair has been burned off, the remaining product which is about 60% dry matter (because of partial dehydration) may be passed through an air separator, screen or other suitable device to separate the grain particles from the fiber particles. The grain particles constitute a high-energy feed and may be fed back directly to the animals in the feedlot. The fiber particles may be mixed with various feed supplements and then fed back to the animals.

Alternatively, after the hair has been removed, the product may be mixed with molasses or another carbohydrate such as sugar and then fermented to improve protein content, palatability, storability and shelf life in the feed bunks. Alternatively, the fermented product may be dehydrated, mixed with other feed ingredients and pelletized prior to being fed back to the animals.

A process developed by *S. Hashimoto; U.S. Patent 3,878,303; April 15, 1975; assigned to Union Oil Company of California* facilitates the bacterial decomposition of animal waste without many of the economic and operational hindrances of conventional waste treatment. According to the process, an aqueous medium containing the animal waste is prepared and a foam stabilizer such as a surface active agent or a protective colloid is added to the resulting aqueous medium. The aqueous medium is then foamed into a stable foam with air in the discontinuous phase surrounded by the continuous aqueous medium. Preferably, the medium is inoculated with a bacterial culture before foaming.

The foam is then discharged into the fermentation vessels where it is retained for a sufficient period of time, e.g., from about 3 to about 48 hours at a temperature from about 15° to about 60°C to permit substantially complete bacterial decomposition of the animal waste products. Upon completion of the fermentation, the foam composition is removed from the fermentation vessel and a clear, aqueous effluent is separated for recycle in the process or for discharge. The bacterial cells can be concentrated from the aqueous medium by various solid-liquid separators and finally dried to recover cellular matter containing approximately 60% protein.

Production of Fertilizers by Combination with Fly Ash

A process developed by *G. Carlberg; U.S. Patent 3,877,920; April 15, 1975* is one in which organic waste material such as animal manure, spent mushroom compost, composted municipal wastes, sewage plant solids, and agricultural processing wastes are reclaimed and reusable by homogeneously mixing one or more of the organic wastes with a fly ash binder in weight ratios between about 20:1 and about 1:9. The reclaimed wastes are deodorized and useful as fertilizers and as support material for plants.

Production of Fertilizers by Combustion Process

A process described by E.G. Davis, I.L. Feld and J.H. Brown in Technical Report No. 46, Wash., D.C., U.S. Bureau of Mines (Jan. 1972) is one in which agricultural manure wastes were combusted in a fluid-bed reactor or a small rotary kiln as a method for disposal of this waste material. As much as 90% weight reduction and 85% volume reduction was obtained by burning manures.

Dry manure burning in the fluid-bed reactor was self-sustaining, whereas wet manure was both dried and burned in the heated rotary kiln. Heat balance estimations indicate that preheating would be required to dry the wet manure prior to burning in the fluid-bed reactor. However, the estimation indicated that no extra heat was needed in the process if the wet manure was predried with exhausted combustion gases before being fed to the fluid bed. The burned residues were pelletized and found suitable for use both as a potassium and phosphorus fertilizer and as a lime soil conditioner.

Production of Soil Additives and Food Supplements

A method developed by *C.K. Megronigle; U.S. Patent 3,732,089; May 8, 1973* for the processing of the excrement of lactating cows is one having a digestion phase, a fermentation phase and an algae culture phase. In the digestion stage a slurry of excrement of lactating cows is reacted with brewer's yeast, sodium chloride, an inoculant and water to provide a first phase supernatant for use, separated from wastes, in the second phase. In the second or fermentation phase the first supernatant is filtered, concentrated and fermented, wastes separated, and a second supernatant drawn off for use in the third or algae culture stage.

In the latter stage nutrients and fresh water are added to the second supernatant which is held in pools with withdrawal of the wastes and separation of a culture useful in the treatment of animal wastes, sewage and wastewaters and products useful for a variety of purposes including reduction of odors, increased oxidation in sewage effluent, degreasing of sewer conduits and plant equipment, accelerated settlement of sewage solids, foaming reduction and control, nitrogen and phosphorus level reductions in sewage plant effluent, accelerated cellulose fiber degradation, organic soil additives, and animal feed supplement and/or additives.

OAT HULLS

Production of Animal Feed

A process developed by *B. Kviesitis and W.E. Rogerson; U.S. Patent 3,395,019; July 30, 1968; assigned to Vylactos Laboratories, Inc.* is a process of preparing an animal feed wherein ground oat hulls are mixed with water and a treating substance is added thereto. The mass is then baked to modify its starch and protein characteristics to dry the same and to expand the pores of the material. A water and molasses solution is then mixed into the baked mass and the mass is then dried.

Production of Furfural

The reader is referred to the section of the volume on Furfural under Organic Materials.

NONFERROUS METALS

The history of the nonferrous metals recycling industry can be traced back to antiquity. Records of ancient peoples give evidence of crude melting techniques that existed at the dawn of civilization. Archaeologists believe the rudimentary recovery of iron, lead, and copper occurred in earliest times.

But, recycling, as it is known today, is the development of the past 100 years. During this period, scientific and sophisticated technology has been developed to transform waste material into valuable products. More significantly, the volume of scrap expanded with the growth of industrialization, particularly in the United States, which, today, has the largest and most effective recycling industry in the world.

More than three million tons of nonferrous scrap metals are recovered annually by secondary smelters, refiners, ingot manufacturers, fabricators, foundries, and other industrial consumers in the United States. The growth potential of the secondary industry is also substantial.

The companies that make up the secondary nonferrous metals industry are generally small and generally are not well equipped to deal with the development and application of new technology for the control of pollution. They lack the needed technical capabilities and financial resources. However, they are concerned about the environment and are making efforts to reduce the pollution potential of the industry, as well as its capabilities for resource recovery.

As described by H. Nack, K. Murthy, E. Stambaugh, H. Carlton and G.R. Smithson, Jr. in PB 233 646, Springfield, Va., Nat. Tech. Information Service (May 1974), the secondary nonferrous metals industry consists of about 20 segments. The major segments in regard to volume of metal recovered from scrap are: copper segment, brass and bronze segment, lead segment, aluminum segment, and zinc segment. The minor segments are: tin, nickel, cobalt, magnesium, mercury, antimony, precious metals, titanium, selenium, cadmium, germanium, hafnium, zirconium, indium, and beryllium.

The secondary nonferrous metals industry encompasses a host of companies. For example, in 1973, there were 125 companies producing secondary aluminum, 130 companies producing secondary lead, and 44 companies producing secondary copper and brass and bronze products. The size of these companies ranges from small family-owned facilities to large corporations.

Manufacturing operations fall into three basic categories: (1) scrap pretreatment, (2) smelting/refining, and (3) casting (product formation). Raw materials for the industry are divided into three broad categories as follows.

(1) Old Scrap: This is the kind most commonly associated with the scrap business though not necessarily the largest in volume. Old scrap is discarded, dismantled, worn out metallic elements. Typical examples are: a radiator removed from a wrecked car; copper pipes taken from an old building; high strength, heat resistant rotor blades once used in a jet engine, appliances discarded by homeowners.

(2) New Scrap: This is metal which has never been made into or used as an end product. It comes from industrial sources and is the by-product of some part of the manufacturing process. Common examples include: a manufacturer of aluminum cans buys huge coils of sheet aluminum and when the can lids (tops) are made on stamping (blanking) machines which punch the lids out of sheets of original material, the remaining skeleton is scrapped and usually described as clippings; or, when household faucets are produced in foundries which make the body of the faucet in a sand mold to produce a casting; or, when pouring molten metal into the mold some will spill over the mold opening or into the sand rather than into the mold cavity and these spillings are scrapped because they pick up impurities or become mixed with dirt and sand. These are usually called foundry spillings or residues. This new scrap differs from old scrap in that it is generally available in larger and more uniform lots; and its metallic constituents are more readily identifiable.

(3) Obsolete Scrap: This category consists of generally new, unused but technologically obsolete parts, over-runs from the production process, inventory that is no longer required, repossessed merchandise or auctioned material. Some examples: The U.S. Government sells for scrap, spare parts for a radar network which is no longer needed for national security; or a railroad has in stock a type of journal bearing but decides to standardize its entire fleet of cars with roller bearings, making the solid bearings available as scrap.

Products from the secondary nonferrous metals industry are generally ingots of refined or partially refined metals which are shipped to the fabricator for processing into a finished product or shipped to the primary industry for further refining.

The secondary nonferrous metals industry is one of the sources of environmental pollution. The pollution may be in the form of atmospheric emissions, liquid wastes, and solid wastes. Atmospheric emissions generally contain particulate and gaseous materials, both of which are sources of pollution and potential health hazards.

The composition of the atmospheric emissions depends on the segment of the industry from which the emissions are derived. However, it has been established that the industry as a whole emits such metallic materials as lead, zinc, mercury, arsenic, cadmium, antimony, precious metals, magnesium, nickel, copper, manganese, and aluminum.

Emissions factors for atmospheric emissions for every segment of the industry are not available although it can be said that, in many cases, significant quantities are emitted. For example, the raw emissions factor for metallic zinc sweating process with zinc cholride was reported to be 10.8 lb of particulates per ton of scrap, whereas the raw emission factor for sweating residual scrap was reported at 24.5 lb/ton of scrap processed.

Morphology of the particles is dependent on the type of scrap, segment from which the particulates are generated, and many other factors. In most cases, the particulates' sizes range from less than 0.1 to several microns. Particle shape will vary from spherical to irregular shapes.

Gaseous emissions also depend on the segment of the industry and the processing step. In general, atmospheric emissions may contain, in addition to the combustion products, harmful materials like sulfur oxides, halides, inorganic acids, hydrocarbons, and ammonia.

A substantial portion of the pollution from the industry comes from the burning of fuels to provide energy for the processes. Quantitatively, data are not available on the industry demand for this energy. However, the primary sources of energy for the secondary nonferrous metals industry are: (1) natural gas, (2) fuel oil, (3) coke, and (4) electricity. Data on quantities of energy consumed by this industry are not readily available.

The secondary nonferrous metals industry can be a significant source of liquid wastes as water is used for many applications throughout the industry. For example, in the aluminum segment, aqueous wastes are generated principally in these operations: cooling molten aluminum alloy, wet scrubbing of fumes during chemical magnesium removal, and the wet milling of aluminum melt residues such as dross and slag. Ingots and shot are cooled with water by direct contact with the mold and metal. Magnesium content in aluminum alloys is adjusted by the chemical removal of magnesium using either chlorine or aluminum fluoride. Wastewaters containing very large levels of suspended and dissolved solids are produced during the wet milling of residues containing aluminum. Another example is the copper segment, where wastewater is generated in 15 to 25 processes.

Composition of these aqueous wastes depends on the segment of the industry generating the wastes. Some of the wastes are simply cooling water and can be recycled with a minimum of treatment. Other wastes contain a wide variety of soluble and insoluble metallic and nonmetallic materials. An example of the contaminants in the aqueous wastes for the secondary aluminum industry is shown as follows: sulfate, chloride, fluoride, aluminum, calcium, copper, magnesium, nickel, sodium, zinc, cadmium, lead, manganese, oil, grease, and phenols.

Also, metal values such as arsenic, mercury, and barium are found in the aqueous wastes. While many of these heavy metal values may be harmless, metals such as cadmium, mercury, and arsenic are known to be toxic. In summary, the secondary nonferrous metals industry is a potential source of aqueous wastes to the environment. If these wastes are not controlled by treatment before discharging or if they are not recycled, serious pollution problems could arise.

Solid wastes are generated in this industry by the nature of the numerous processes. Slag from the smelting of copper and drosses from the aluminum industry are examples. Drosses from the treatment of aqueous wastes generated in the production of electrolytic copper is another example. Another source of solid wastes common to this and other industries is the dust collected in control equipment such as baghouses. Solid wastes such as slurries (mixtures of solids and liquids) and fine particulate materials are disposed of in lagoons, in landfills, or, in many cases, become a source of raw material for the secondary industry.

As with the other wastes from this industry, the composition of this waste is dependent on the process and processing steps in each segment. These wastes are known to contain metallic and nonmetallic materials found also in atmospheric emissions, aqueous wastes, and the raw materials to this industry. Many of them are toxic.

In summary, this industry generates substantial quantities of solid wastes. In many cases, serious pollution problems can result if these wastes are not disposed of in a safe and adequate manner.

A study of the use of cryogens to reclaim nonferrous scrap metals has been made by E.G. Valdez, K.C. Dean and W.J. Wilson and the results presented in Report of Investigation No. 7716, Washington, D.C., U.S. Bureau of Mines (1973).

Three cryogenic systems that utilized liquid nitrogen, dry ice, and methanol with dry ice were investigated in conjunction with crushing and classifying techniques. The systems were used to aid in separating and reclaiming the metallic components contained in insulated wires, shredded automobile nonferrous metal concentrates, small motors, generators, and rubber tires.

Wire strands, 2 to 6 inches long and insulated with polyvinyl chloride and neoprene, were chilled at -60° and -195°C, then roll-crushed and processed by water elutriation; this procedure resulted in two products, a sink fraction that was 99% metallic and a float fraction that was 99% nonmetallic.

Excellent separation of zinc die-casting alloys from copper and aluminum contained in shredded automobile nonferrous metal concentrates was attained by chilling at -72°C for one minute, crushing in a grateless hammermill, and screening. From the screened products, 97.2 and 100% of the copper and aluminum, respectively, were recovered in the plus one inch fraction, and 100% of the zinc was recovered in the minus one inch fraction of over 97% zinc die-cast purity.

ALUMINUM

Recovery from Liquid Wastes

Recovery from Anodizing Process Waste Liquor: A process developed by *C.J. Huffmann; U.S. Patent 3,798,160; March 19, 1974; assigned to American Metal Climax, Inc.,* permits recovering aluminum values from aluminum waste liquor and comprises adjusting the pH of the liquor to 6 to 8.5, thus causing precipitation of aluminum hydroxide, adding sufficient polyacrylamide resin solution to produce a concentration of 1 to 5 ppm of resin in the liquor, allowing the aluminum hydroxide precipitate to settle in a thickener, withdrawing the supernatant liquid from the thickener, removing the precipitate from the thickener by filtration, recycling the filtrate to the thickener, and collecting the aluminum hydroxide filter cake from the filter.

Figure 43 is a block flow diagram of this process. It is a flow sheet for the treatment of aluminum anodizing waste liquors, having pH values between 4 and 12. These liquors are passed through a neutralizing tank **1** where the pH is adjusted to 7.5 to 8.5 with acid or alkali, as required. If the waste liquors are acidic (e.g., from the anodizing rinse) they are preferably treated with alkaline liquors from the alkali etching rinse, and if they are alkaline, an acid liquor is preferably used. Conventional automatic control equipment (not shown) controls the addition of spent acidic or alkaline liquors to adjust the pH to the desired range. In either procedure the pH is brought to a value near 7.8, at which point aluminum hydroxide precipitates, forming a suspension ordinarily containing about 800 parts of solids per million parts of water.

The aluminum hydroxide suspension is pumped to a circular thickener or clarifier **2** where a solution of high molecular weight anionic resin is added to facilitate settling of solids. The quantity of resin added (usually as 0.020 to 0.050% aqueous solution) is selected to produce a final resin concentration of 1 to 4 parts per million part of liquor.

The solids settle to the bottom of the thickener **2** and the clear effluent overflows to the sewage system, control of this effluent being conducted by means of a turbidimeter **4** and a pH recorder **5**. The underflow from the thickener consists of 10 to 15% of the feed volume and is a soupy suspension containing 2 to 4% solids. It is fed to a drum-type vacuum or pressure filter **6**. A paste consisting of 16 to 20% solids builds up on the filter drum and is continuously removed by a knife as the drum rotates. The paste (or filter cake) is collected and removed. The filtrate is returned to the center well of the thickener **2** for preprocessing.

In a typical operation of the type described above, 350 gallons per minute of aluminum rinse liquor enters a 10,000 gallon neutralizing tank where it is brought to pH 7.5 to 8.5 with spent acidic or alkaline waste liquor, forming a suspension of aluminum hydroxide (800 ppm). It is then pumped to the thickener where a 0.025% solution of high molecular weight polyacrylamide resin is added. Flocculation occurs and the suspension is rapidly thickened and settles as a sludge in the thickener from which it is removed as an underflow at the rate of 40 gallons per minute. The overflow of 310 gallons per minute is discharged

FIGURE 43: PROCESS FOR ALUMINUM RECOVERY FROM ANODIZING WASTE LIQUORS

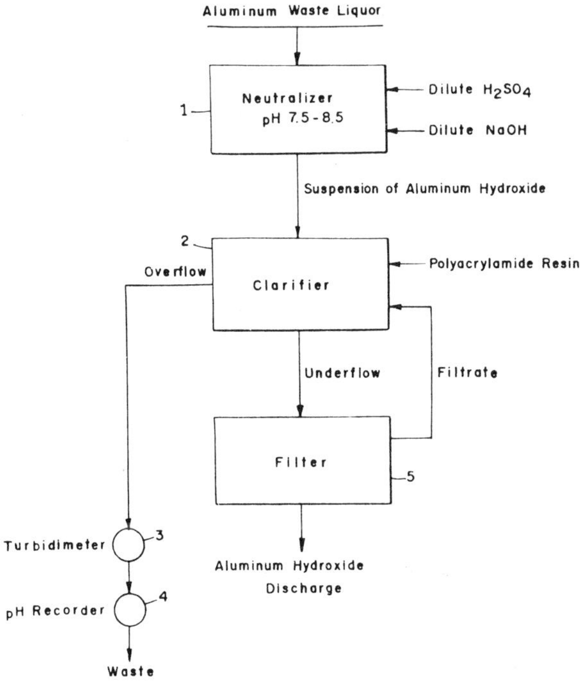

Source: C.J. Huffman; U.S. Patent 3,798,160; March 19, 1974

to sewer or river at pH 6.5 to 8.6 and maximum suspended solids of 20 ppm. The underflow (2 to 4% solids) is fed to a drum-type vacuum filter coated with a filter precoat of diatomaceous earth. A filter cake consisting of 16 to 25% solids builds up on the drum and is removed with a knife, collected in a hopper and disposed of as waste, or treated with sulfuric acid to produce a 28% aluminum sulfate solution suitable as a flocculent for other water treatment. The diatomaceous earth is insoluble in the aluminum sulfate solution and can be removed by filtration. A roll discharge vacuum filter is also feasible. If a pressure filter is used instead of a vacuum filter, it is possible to produce a filter cake with 20 to 50% solids. With either filter, about 3 tons per day of a cake is produced which is substantially all aluminum hydroxide and which can be used as a commercial source of aluminum values.

Recovery from Wastewater from AlF_3 Manufacturing: A process developed by *A. Schmidt; U.S. Patent 3,816,591; June 11, 1974; assigned to Osterreichische Stickstoffwerke AG, Austria* is a process for the recovery of aluminum and fluorine values from dilute aqueous solutions containing aluminum fluoride and silicofluoride whereby in the solution bound fluorine is present in excess with respect to aluminum by adding an ammoniacal medium in order to adjust the pH of the solution between 6 and 6.5 and separating the precipitated ammonium fluoroaluminate.

If aluminum fluoride trihydrate is manufactured by crystallization from its aqueous solutions, the mother liquor which remains contains a residue of about 2% by weight of aluminum fluoride. Further reduction of this content is no longer possible through a longer reaction time or by changing the temperature during the crystallization, so that considerable amounts of aluminum and fluoride are lost.

Apart from the aluminum fluoride, the mother liquors also still contain silicofluoride. If attempts are made to precipitate the aluminum as hydroxide by adding a caustic alkali solution, the coprecipitated silicic acid contaminates the aluminum hydroxide precipitate. This silicic acid is of a jelly-like consistency so that it is practically impossible to remove the precipitate by filtration. As a result of the hydrolysis, almost the entire fluorine content furthermore remains in the liquid phase and is therefore lost.

Suprisingly, it has been found that the aluminum content of the mother liquor may be recovered practically completely if the pH value of the mother liquor which must contain an excess of bound fluorine with respect to the aluminum being present is adjusted, at a temperature between 15° and 35°C to a certain pH range by adding ammonia, ammonium carbonate or ammonium bicarbonate. Hereupon, a crystalline substance precipitates, the composition of which approximately corresponds to the formula $(NH_4)_5Al_3F_{14}$, and which thus represents an ammonium compound approximately analogous to chiolite. It is contaminated with only small amounts of silicic acid (0.2 to 2.5% of SiO_2) and is easily filtrable.

Recovery from Solid Waste

Tonnage-wise, the aluminum segment constitutes the second largest segment of the secondary nonferrous metals industry after copper. In 1974 some 1,250,000 short tons of secondary aluminum was produced.

Sources of raw materials to this segment include both new and old scrap. The new scrap, that which comes from a fabricator who does not choose or it is not equipped to recycle the scrap, accounts for about 75% of the scrap. Old scrap, a product of obsolescence, becomes available to this segment when consumer products have reached the end of their economic life and have been discarded, and accounts for approximately 25% of the domestic scrap consumed in the U.S. Basically, the scrap can be divided into 6 categories: (1) sheet and castings, (2) new clippings, (3) borings and turnings, (4) high aluminum-iron alloy scrap, (5) drosses or skimmings with fluxing salts, and (6) drosses or skimmings without fluxing salts.

The choice of scrap which a secondary smelter purchases for use in its furnaces is determined by the equipment or flow sheet of the individual smelter. Obviously, a smelter with no burning or drying equipment for processing oily borings and turnings cannot use this material. Likewise, smelters without crushing or residue processing equipment are similarly restricted in the types of scrap they can process. Air pollution codes in many cities also have limited the processing of heavily painted material such as painted siding or Venetian blinds. Hence, smelters having limited processing equipment must be selective in scrap purchasing, while more sophisticated smelters have greater latitude in scrap purchasing.

The products from the secondary aluminum segment are: (1) aluminum alloy castings, (2) aluminum shot, (3) hot metal, (4) aluminum fines, and (5) hardeners. The aluminum alloy castings may be in the form of ingots (15 to 30 lb), billets (approximately 1,000 lb), notched bars (2 to 5 lb), and hardeners of different weights. Hot metal is molten aluminum alloy which is tapped directly from the furnaces into preheated crucibles with capacities up to 15,000 lb and transported directly up to distances of 300 to 400 miles to the consumer (foundries). Aluminum shot is small beads of quenched aluminum metal. Aluminum fines are the undersize (–20 microns) material from screening of the aluminum scrap treated by the burning/drying process.

The recovery of aluminum from scrap normally involves two manufacturing operations: (1) scrap pretreatment, and (2) smelting-refining. Aluminum scrap is pretreated prior to smelting to remove contaminants and physically prepare the material for further processing. Three types of pretreating (mechanical, pyrometallurgical, and hydrometallurgical) are used. The pretreatment process employed depends on the type of scrap. The types of pretreatment processes used are as follows: (1) shredding/classifying process, (2) baling process, (3) crushing/screening process, (4) burning/drying process, (5) sweating process, and (6) leaching process. Some additional information on each of these scrap pretreatment processes is given in the following paragraphs.

(1) Shredding/Classifying Process: The iron core and neoprene or plastic insulation is removed by the shredding/classifying process from electrical conductor scrap containing superpure aluminum. This treated aluminum scrap is used to produce, among other things, hardeners. The processing steps are: (a) shredding of the scrap into small pieces to separate the iron core and plastic coating from the aluminum alloy, (b) magnetically treating to remove the ferrous portion of the scrap, and (c) separating the neoprene or plastic insulation from the high-purity aluminum by an air classification system. Energy demand for this process is that necessary to drive the equipment.

The shredding/classifying process is a minor source of atmospheric emissions. However, significant quantities of solid wastes—ferrous scrap and neoprene or plastic waste—are generated. The scrap is recycled and plastic waste is disposed of in a landfill. This process has no serious potential pollution problems of any kind.

(2) Baling Process: Densification or compaction of sheet, castings, and clippings is achieved by baling in specially designed baling machines. Baling is normally conducted by the scrap dealer or at the source of scrap generation. Occasionally, the baled scrap is further processed by crushing and screening prior to smelting. Energy demand is limited to that needed to drive the equipment.

Atmospheric emissions from this process are composed of suspended particulate matter containing primarily dirt from the scrap and alumina resulting from the oxidation of the aluminum dust. Solid wastes generated are scrap iron, magnesium, and other scrap found mixed with the scrap aluminum. The quantity of pollutants generated by baling is low; therefore, the process has a low pollution potential.

(3) Crushing/Screening Process: Borings and turnings which result from machining and drilling are treated by the crushing/screening process to densify the aluminum values and remove contaminants. This process entails three process steps: (a) crushing in a hammermill to densify the scrap, (b) deoil, (c) dry, (d) screening to remove aluminum fines, and then (e) passing over a magnetic separator to remove the tramp iron. Energy demand for this process is that needed to drive the equipment and oil/gas for the dryer.

The process produces essentially no atmospheric emissions and no aqueous wastes. The quantity of solid wastes such as tramp iron generated is expected to be low as the majority of the iron is removed prior to crushing. The process has no potential for serious pollution problems.

(4) Burning/Drying Process: Borings, turnings, and other organic-contaminated scrap are, in some cases, smelted without removal of the organic contaminants and water. However, the scrap is generally treated to remove the water and organic matter such as machining oils by the burning/drying process. This process involves several processing steps: (a) crushing in a hammermill to densify the scrap, (b) heating the crushed material in a gas-oil fired rotary dryer to remove organic contaminants and water, (c) screening the dried material to remove the aluminum fines which are sold for explosive or pyrotechnic purposes and (d) magnetically treating the remainder to remove the tramp iron. Energy demands for this process are: (a) gas or oil to supply the heat to burn the organic compounds and evaporate water from the scrap, and (b) that required to drive the equipment.

The burning/drying process is a source of atmospheric emissions and solid wastes. The atmospheric emissions consist of gaseous products primarily carbon dioxide from combustion of the fuels and the organic contaminants, and particulate matter entrained in the gases. The gases may also contain some chlorides, possibly fluorides and sulfur oxides, depending on the composition of the organic contaminants. The particulate matter is primarily alumina resulting from oxidation of the small particulates of aluminum swept from the kiln by the combustion gases. The solid waste is primarily tramp iron removed in the magnetic treatment of the crushed scrap.

Thus, the burning/drying process presents serious potential pollution problems if the emissions are not controlled properly, especially in those cases where the organic contaminants contain significant quantities of halides, where significant quantities of alumina are emitted, and where oil is used in place of gas as the fuel.

Normally, the atmospheric effluent from the burning/drying process is treated with an afterburner located in the stack. However, wet scrubbers are beginning to replace the afterburner, in which case the air pollution problem can become a water pollution problem. The solid wastes generated are not a serious pollution problem.

(5) Sweating Process: The contaminant-iron is removed from high iron scrap by the sweating process, employing a sloping hearth or grate-type furnace. The processing steps employed are: (a) heating the scrap to melt the aluminum and other low melting constituents which flow by gravity into a collecting pot or ingot mold, and (b) casting the melt into pigs (30 lb ingots) or sows (up to 1,000 lb ingots) in aircooled molds. The furnace is heated to approximately 1400°F to separate the low melting aluminum along with other low melting materials from the scrap. Energy demand for this process is natural gas and that needed to drive the equipment.

Emissions from this process are: (a) atmospheric emissions composed of the combustion products and particulate matter, primarily aluminum trioxide, and (b) solid wastes containing iron contaminated with aluminum and other nonferrous elements such as zinc, magnesium, and lead. Gaseous emissions from the furnace are generally passed through an afterburner before being emitted to the atmosphere. The solid wastes are discarded.

The process can cause serious atmospheric pollution problems, if not properly controlled. According to a survey, raw (uncontrolled) particulate emission factor from the sweating process is reported to be 14.5 lb/ton (7.25 kg/MT) of metal processed. This factor does not include gaseous emissions. Composition of the emissions, both gases and particulate matter, will need to be determined by source testing. A baghouse is usually used to control dust emissions. The solid wastes should cause no pollution problems.

(6) Leaching Process: Leaching removes the contaminants such as fluxing salts and other water-soluble components from drosses, skimmings, and slags to produce an aluminum scrap suitable for the smelter. The processing steps are: (a) wet milling of the scrap in an attrition mill or a ball mill to densify the scrap, (b) screening to remove the fines and dissolved salts, (c) drying to remove the water, and finally (d) treating with a magnetic separator to remove the ferrous portion. Energy required is that needed to operate the equipment and to dry the crushed scrap.

The process produces no atmospheric emissions except possibly in the drying step. However, significant quantities of water and solid wastes are produced. The water wastes containing dissolved and suspended solids are disposed of via a series of settling ponds. The solid wastes result from the settling pond and the undersize material from the screening operation.

The process does not present an atmospheric pollution problem; however, there is a water pollution problem as well as a problem relating to disposal of the solid wastes. From a recent survey of the secondary aluminum industry by Battelle, the water pollution problem essentially can be handled by disposing of the water wastes in settling ponds. The solid wastes can be placed in a landfill.

A variety of smelting-refining processes may be used in the secondary aluminum industry as follows: (1) reverberatory (chlorine) smelting-refining, (2) reverberatory (fluoride) smelting-refining, (3) crucible smelting-refining, (4) induction smelting-refining, and (5) rotary furnace smelting-refining. Some additional information on each of these processes is given in the following paragraphs.

(1) Reverberatory (Chlorine) Smelting-Refining Process: Alloy ingots, shot, and hot metal are produced from treated aluminum scrap, sweated pigs, and, in some cases, untreated scrap by this process. The process steps are: (a) charging of the scrap to the furnace; (b) melting of the scrap; (c) fluxing to remove certain contaminants; (d) alloying, i.e., the addition of certain metals to produce a melt of the desired composition; (e) mixing of the melt to produce a homogeneous material; (f) demagging with gaseous chlorine; (g) degassing with an inert gas such as nitrogen or nitrogen-chlorine mixture to remove entrapped hydrogen and other gases; and (h) skimming to remove the slag or dross containing the impurities.

After slag removal, the melt is (1) poured into molds and cast into alloy ingots, billets, and notched bars; or (2) poured onto a vibrating feeder and quenched in water to produce shot; or (3) poured into heated crucibles and transported to the consumer as hot metal, i.e., molten metal. Any given smelter may not incorporate all of these steps into the process, but may use a combination of the steps.

The reverberatory furnace, a rectangular box usually ranging in capacity from 15 to 90 tons is used in this process to smelt and refine the aluminum. The furnace is direct fired; furnace fuel may be either natural gas or fuel oil. Heat input ranges from 2,000 to 2,500 Btu per pound of alloy. Additional energy is required to operate the auxiliary equipment.

Emissions from this process are atmospheric emissions (gases and particulate matter), aqueous wastes, and solid wastes. In addition to the combustion products, the gases may contain chlorine, hydrogen chloride, zinc chloride, magnesium chloride, aluminum chloride, and aluminum oxide, along with minute quantities of a wide variety of other metals contained in the scrap and added during the smelting-refining. Likewise, the aqueous and solid wastes contain a wide variety of metal values depending on the scrap.

All processing steps are sources of atmospheric emissions. The addition of scrap to the furnace and melting of the scrap is a major source of atmospheric emissions depending on the quality of the scrap. Quantitative data are not available for the secondary aluminum segment. However, for the brass and bronze segment, raw emission factors for the charging step are reported to range from 11.7 to 59.0 lb of particulate matter per ton of feed. Comparable emission factors may be expected for the charging step for the secondary aluminum segment, especially at those smelters charging untreated scrap and scrap containing significant quantity of volatile impurities such as zinc.

Fluxing is practiced to some extent at most secondary aluminum smelters. Some of the components volatize during the smelting-refining while others, along with impurities or contaminants in the scrap aluminum, are entrained in the furnace exit gases. Composition of the atmospheric emissions will vary depending on the scrap and the flux. However, these emissions have been found to contain such elements as sodium, aluminum, magnesium, calcium, iron, lead, manganese, potassium, chromium, zinc, and nickel, in addition to the combustion gases and hydrogen halides.

Alloying produces minor amounts of fumes and dust which are carried from the furnace by entrainment in the combustion gases and other gases formed during this processing step.

Mixing of the melt to insure uniform composition and to mix the fluxes into the melt is generally accomplished by injecting nitrogen gas below the surface of the melt. As a result, mixing becomes a significant source of atmospheric emissions. The gaseous portion of the emissions is composed of nitrogen, hydrogen halides, and other volatile components. Composition of the particulate matter is similar to the emissions from the fluxing processing step.

Removal of magnesium (demagging) is accomplished by chlorination. This entails lancing the molten aluminum with chlorine gas which reacts with the aluminum to form aluminum chloride. The aluminum chloride then reacts with the magnesium to form magnesium chloride which is carried along with other metallic and nonmetallic impurities and gaseous contaminants to the surface of the melt. Some of these are trapped in the flux on the surface of the melt, while the remainder is volatized and becomes part of the atmospheric emissions.

According to H. Nack, K. Murthy, E. Stambaugh, H. Carlton and G.R. Smithson, Jr. in PB 233 646, Springfield, Va., Nat. Tech. Information Service (May 1974), the demagging of aluminum with gaseous chlorine results in the evolution of large quantities of chlorine, primarily as aluminum chloride. The use of solid demagging agents such as aluminum fluoride or aluminum chloride instead of chlorine reduces the degree of severity of the problem and is being practiced by the industry. A new approach involves the use of a precoated baghouse (chromatographic baghouse) which removes both the halides and particulate matter. This technique is being tested on a commercial scale in Canada, and several U.S. companies are contemplating building commercial-size units.

The demagging step is still, however, a major source of atmospheric emissions in the secondary aluminum segment. Raw emission factor for this processing step is reported to be 1,000 lb of particulate matter per ton of chlorine used (500 kg/t). Through the use of a baghouse as the control device (which does not remove the gases), the particulate emissions factor is reduced to 50 lb of particulate matter per ton of chlorine used (25 kg/t).

In order to remove the gases and the particulate matter, a wet scrubber using a 10% caustic solution as the scrubbing medium in series with a baghouse is employed. In one installation, two of the scrubbers were arranged in parallel. After scrubbing, the fumes were sent to a five-compartment baghouse for particulate removal. In this installation, the scrubber removed virtually all of the hydrogen chloride gas and greater than 90% of the chlorine gas. Wet scrubbing also hydrolyzes and removes the majority of the chlorides such as aluminum chloride as the corresponding oxides or hydroxides.

Composition of the atmospheric emissions from the demagging step depends somewhat on the composition of melt and fluxes employed in the fluxing step. However, in addition to the combustion gases, the gaseous portion of the atmospheric emissions is composed of chlorine, hydrogen halides such as hydrogen chloride, along with chlorides of the volatile metals. The particulate emissions are composed of such metal values as calcium, copper, magnesium, nickel, zinc, cadmium, and aluminum.

Degassing of the molten aluminum prior to casting is necessary to remove hydrogen absorbed from the atmosphere or other sources of moisture or water vapor. The metal is degasified by lancing with dry nitrogen, chlorine, or mixtures of the two gases. If a smelter does not have adequate ventilation and air pollution abatement equipment, nitrogen is used for degassing, because degassing with chlorine or mixtures of chlorine and nitrogen results in severe fuming. Composition of these may vary from smelter to smelter, but, in general, is similar to those emissions from the demagging processing step. Skimming to remove the dross or slag collected on the surface of the melt and casting of the ingots are minor sources of atmospheric emissions.

Another potential source of atmospheric emissions results from cooling and from storage of the slag (drosses and skimmings) while en route to the processor. If allowed to come in contact with water, the nitrides and carbides of aluminum react with the water to form hydrocarbons and ammonia which are liberated to the atmosphere.

Aqueous wastes are generated during ingot and shot cooling or quenching, and from scrubbing of the emissions from the reverberatory furnace. Also, if the drosses and skimmings are stored in the open, runoff water may become a source of aqueous wastes. The waste cooling water is disposed of by vaporization, discharging directly to a municipal sanitary sewer or stream, or a pond. In a survey of 50 secondary aluminum plants, 24 discharged to sewers or streams, 4 discharged to a pond, 13 recycled continuously, 7 discharged after recycle, and 2 vaporized the water. This wastewater contains significant quantities of heavy metals, phenol, and other contaminants and is, therefore, a potential source of water pollution.

The other source of wastewater from the reverberatory (chlorine) smelting-refining process is that from the wet scrubbers. This waste which may or may not be neutralized is discharged to ponds. That which has not been neutralized has a pH of about 1.5 and contains hydrolyzed metal chlorides of aluminum, magnesium, zinc, manganese, cadmium, copper, nickel, and lead. The neutralized scrubbing liquor with a pH of 9 to 11 contains sodium, potassium, and calcium and lesser quantities of the heavy metals, aluminum and magnesium.

Solid wastes generated by this process include: (1) drosses and skimmings from the process itself, and (2) wastes from the air pollution control equipment. The drosses and skimmings are recycled, while the other are probably disposed of by routine methods such as landfill. If not properly prepared, the leachate from the landfill can enter the ground water and cause pollution problems.

Thus, the reverberatory (chlorine) smelting-refining process is a major source of atmospheric emissions from the aluminum segment of the secondary metals industry. These emissions containing gaseous and fine particulate matter present serious pollution problems if emitted to the atmosphere. Raw emission factors exclusive of the demagging processing step for this process have been reported to be 4.3 lb of particulate matter per ton (2.15 kg/t) of metal processed. Using a baghouse or an electrostatic precipitator as the pollution control device reduced the emission factor to 1.3 lb of particulate matter per ton (0.65 kg/t) of metal processed. Raw emission factor for the demagging processing is reported at 1,000 pounds of particulate matter per ton of chlorine (500 kg/t) used. The emission factor is reduced to 50 lb of particulate matter per ton of chlorine (25 kg/MT) used by employing a baghouse as the control device.

Currently, most of the secondary smelters using chlorine in the demagging processing step employ wet scrubbers alone or in combination with baghouses to reduce the atmospheric emissions. Although emission factor data are not available, it is expected that this combination of pollution abatement equipment will remove the gaseous emissions such as chlorine and hydrogen chloride as well as the particulate matter more efficiently than either the baghouse or electrostatic precipitator. In one installation, the scrubber removed virtually all of the hydrogen chloride gas, greater than 90% of the chlorine, and greater than 80% of the aluminum chloride.

Particle morphology of the particulate emissions, i.e., particle size and shape, is somewhat dependent upon the composition of the scrap and the processing parameters. However, in general, the particulate matter is composed of particles ranging in size from a few hundredths of a micron (approximately 0.05 micron) to several microns (approximately 2 to 5 microns). In one case, the fume from salt-cryolite mixtures was found to be composed of particles ranging in size from less than 2 microns to less than 0.1 micron. In another case the fume from a demagging operation was found to be composed of a mixture of particles with 90 to 95% less than 1 micron in diameter. This small particle size makes it difficult to collect the emissions in baghouses and electrostatic precipitators.

The emissions are generally composed of particles exhibiting both regular and irregular particle morphology. The morphology will range from acicular shapes (long needle shape particles) to spherical particles. Solid wastes and aqueous wastes, if not disposed of properly, can present pollution problems.

(2) Reverberatory (Fluoride) Smelting-Refining Process: This process is essentially the same [i.e., the same processing steps are employed as in reverberatory (chlorine) smelting-refining] except that aluminum fluoride (AlF_3) is used in place of gaseous chlorine in the demagging processing step. The Mg is removed by adding to the molten metal solid AlF_3 which reacts with the Mg to give Al metal and magnesium fluoride which floats or is carried

by the mixing gas to the surface of the molten metal and is removed as part of the dross or skimmings. Conducting the demagging with AlF_3 results in a significant reduction in the quantity of atmospheric emissions. Consequently, the problems associated with atmospheric pollution are significantly reduced, yet not completely eliminated as significant quantities of atmospheric pollutants are discharged. Instead of large quantities of chlorine gas, the emissions contain fluorides as gaseous fluorides or fluoride dusts. If not properly controlled, the fluorides as well as the other particulate matter and gases in the effluent can cause serious pollution problems. The atmospheric emissions may be controlled by the same dry or wet methods used in the chlorine process.

Aqueous waste generated in this process consists of waste cooling water generated during ingot and shot formation. This wastewater is disposed of either by discharging to sewers, streams, or ponds; recycled continuously; recycled partially; or evaporated (see discussion of waste cooling water under the chlorine process as same treatment and disposal methods are employed). The wastewater from the fume scrubber (which is recycled) is first treated with caustic to remove the fluoride. The solids formed during neutralization are separated in settling tanks and the effluent is recycled to the scrubber system.

Solid wastes generated by this process are: (1) drosses and skimmings from the process itself, and (2) solids from the air pollution control equipment. The dross and skimmings are recycled. Solid waste (sludge) from the wet scrubber is dewatered and placed in a landfill. If not properly prepared, the leachate from the landfill can enter the underground water system and cause pollution problems. Those smelters using the dry collection system also dispose of the fume dust and bag coating substances in landfills. Although the fluorides are considered to be insoluble, the solubility may be great enough to contaminate ground water.

Because of the toxic and hazardous nature of the pollutants generated by this process, the process presents serious pollution problems if these pollutants are not collected and disposed of in a proper manner.

(3) Crucible Smelting-Refining Process: For producing small quantities of aluminum alloy castings, i.e., up to approximately 1,000 lb, the crucible smelting-refining process is used. The processing steps are: (a) charging of the melt to the furnace, (b) melting of the charge, (c) fluxing to remove the contaminants if necessary, (d) alloying to produce a melt of the desired composition, (e) mixing to homogenize the molten metal, (f) demagging, (g) degassing, and (h) skimming. Afterwards, the melt is poured and cast into the desired shapes. Any given smelter may not incorporate all of these steps into the process, but may use a combination of the steps depending on the source of scrap.

Energy demand is limited to that required to operate the equipment and to heat the furnace. The furnace is heated indirectly with gas or fuel oil as the source of fuel. However, electricity may also be used as the source of energy.

Potential environmental pollutants from this process are: atmospheric emissions which are composed of gases and particulate matter, aqueous wastes containing dissolved and suspended solids and solid wastes. Since this process is quite similar to reverberatory smelting-refining except for the size of operation, see discussions in the previous sections on the chlorine and fluorine processes for data on composition of pollutants, processing steps generating the wastes, and disposal of the pollutants. This process has the potential for generating significant pollution problems.

(4) Induction Smelting-Refining: This process produces hardeners, alloys of known composition, which are used to introduce precise amounts of particular metals into a melt to meet predetermined specifications. The most common hardening agents are titanium, boron, and chromium. The hardeners are produced by melting and blending electrical conductor scrap (superpure aluminum) with the hardening agent in an induction furnace. The processing steps are: (a) charging of the scrap to the furnace, (b) melting of the scrap, (c) blending the molten metal with the alloying agent, (d) skimming of the melt to remove

the oxides on the surface, (e) pouring of the melt, and (f) casting into notched bars. Energy demand is limited to that needed to run the equipment and the electricity to melt the scrap.

The process produces small quantities of atmospheric emissions and aqueous and solid wastes. The atmospheric emissions contain, in addition to aluminum, small amounts of the alloying agent and other metals found in the scrap. These emissions are generated in all of the processing steps. The aqueous wastes are generated during the casting of the notched bars, whereas the solid wastes are drosses (skimmings from the surface of the melt). These wastes can be collected and disposed of by the methods employed in the chlorine process. This process has a low potential for the production of pollutants.

(5) Rotary Furnace Smelting-Refining: This process is employed to recover the aluminum values from such raw materials as drosses to produce high aluminum ingots which are refined in the reverberatory furnace. The processing steps are: (a) charging of the scrap and flux, (b) melting of the charge, (c) pouring, and (d) casting of the ingots. The energy demand for this process is that needed to drive the equipment and natural gas or fuel oil to heat and melt the charge.

The process produces atmospheric emissions composed of gases such as the combustion products, entrained air, and volatile metal values and particulate matter such as the flux and particles of the charge. Solid wastes are the slag containing impurities extracted from the aluminum and the sludge from the wet scrubber. Aqueous waste consists of primarily the cooling water used in casting the ingots. These wastes may be disposed of by methods used for disposal of similar wastes from other processes. The process can cause serious pollution problems, if not properly controlled.

Recovery from Cables: A process developed by *H. Lenz, J. Mehlbeer, H. Macura and O. Weidner; U.S. Patent 3,582,004; June 1, 1971; assigned to Eisen und Metall AG, Germany* provides a method for recovering metals from discarded cables and the like wherein the discarded cables are comminuted in a hammermill and, after removing light components by suction and iron components by magnetic action, may be additionally comminuted by a cutting action. The comminuted material is successively treated in belt sink separators individually associated with draining screens and conveyors operatively connected therewith for removing floats from the first separator or separators and sinks from the last separator. The several sink separators are operated at different turbidities.

Discarded cables, as ground cables and the like, are usually composed of more than 50% by volume of insulating materials, such as synthetic substances, rubber and paper, of the conductor proper, usually copper or aluminum, or armor and cladding of steel and lead. Such cables can be decomposed into their components by means of a hammermill or by so-called cutting mills. Mechanical sorting of the product of comminution is very difficult because this product, as a whole, is likely to mat and, particularly, the metals which are in the form of more or less long wires and strips tend to bunch.

In this process insulating material, aluminum, copper and lead may be received by the first belt sink separator. The insulating material is separated as floats at a turbidity of about 2.0 grams per cubic centimeter. Aluminum floats up in the second belt sink separator at a turbidity of 2.8 to 3 grams per cubic centimeter. The remaining sinks still contain copper and lead which may be separated by hand on a subsequent picking belt. In lieu of belt sink separators, other sink separators, for example drum sink float separators may be used.

Figure 44 is a flow diagram of the process. As shown there, a feeder belt **1** transports raw material to a hammermill **2**. The material comminuted in the hammermill is placed on a conveyor belt **4**. A large portion of the fine and light components is sucked off by a suction device **3**. A lifting magnet **5** is placed at the end of the conveyor belt **4** for extracting iron pieces. The material which is now free of dust and iron parts is supplied to a first belt sink separator **6**, the turbidity of which is so adjusted that the insulating material floats up. It is rinsed off on a draining screen **12** and removed on a conveyor

FIGURE 44: FLOW SCHEME FOR ALUMINUM RECOVERY FROM DISCARDED CABLES

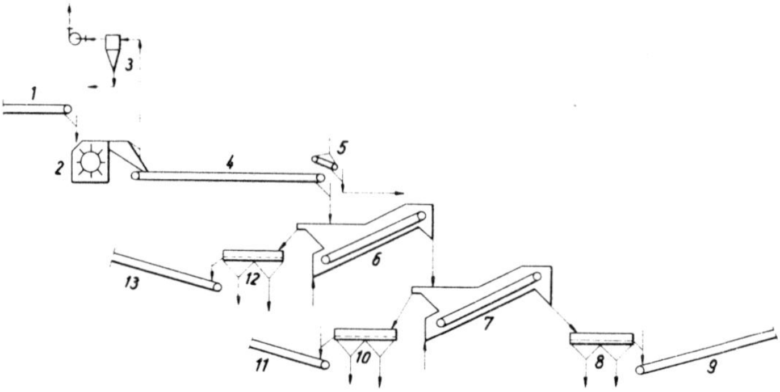

Source: H. Lenz, J. Mehlbeer, H. Macura and O. Weidner; U.S. Patent 3,582,004; June 1, 1971

belt **13**. The sinks of the first separator **6** are divided into aluminum and copper in a second belt sink separator **7**. The floating aluminum arrives at a conveyor belt **11** by way of draining screen **10**; the heavy copper at belt **9** by draining screen **8**.

A process developed by *F.J. Graveman; U.S. Patent 3,612,412; October 12, 1971; assigned to Aluminum Converter Sales & Research, Inc.* is one in which cable comprising aluminum strands wound about a steel core is reduced to its separate metal components by cutting substantially through the aluminum strands at closely spaced intervals, but leaving the steel core intact. This frees small aluminum strand segments from the core and these segments may be reduced still further in a reduction mill. The core is wound into a roll after the segments are cut away from it.

The aluminum strands are severed into the segments by blades having notches therein which are sized to fit around the core but not around the strands about the core. The blades may be mounted on revolving wheels or they may reciprocate.

Recovery from Electronic Scrap: As pointed out by R.O. Dannenberg, J.M. Maurice and G.M. Potter in *Recovery of Precious Metals from Electronic Scrap,* Report of Investigations 7683, Washington, D.C., U.S. Bureau of Mines (1972), aluminum melts at a lower temperature than most other metals and can be separated as a molten liquid in a process commonly called sweating. Recovery of aluminum from electronic scrap by sweating is complicated by the high solubility of precious metals in molten aluminum. Tests showed that as much as 60% of the precious metal content of complex scrap dissolves in the aluminum. Table 17 shows the assay of an ingot of aluminum sweated from electronic scrap by a commercial secondary metal smelter.

Fused-salt electrolysis may be a feasible method for separating precious metals from aluminum. Aluminum may be eliminated by leaching the scrap in aqueous sodium hydroxide. A 5 N NaOH solution will load to a concentration of approximately 100 grams of aluminum per liter in 6 hours at 80° to 85°C while leaching typical incinerated electronic scrap including aluminum chassis and housings. The loaded caustic may be either discarded or treated to recover a portion of the sodium hydroxide. If the loaded sodium aluminate solution were discarded, sodium hydroxide requirements for leaching one ton of scrap containing 34.4% aluminum would be 1,370 pounds.

TABLE 17: ANALYSIS OF SWEATED ALUMINUM

Metal	Weight-percent	Troy oz/ton
Aluminum	69.7	-
Copper	20.0	-
Zinc	4.02	-
Silicon	2.59	-
Tin	1.19	-
Lead	1.07	-
Iron	.54	-
Nickel	.22	-
Manganese	.20	-
Magnesium	.13	-
Silver	-	190
Gold	-	11.7

Source: Bureau of Mines R.I. 7683

A portion of the sodium hydroxide can be regenerated in a manner similar to that used to recover aluminum from clay.

Aluminum, base metals and precious metals may all be recovered from electronic scrap by a technique described by R.A. Sullivan, R.L. De Beauchamp and E.L. Singleton in Report of Investigation 7617, Washington, D.C., U.S. Bureau of Mines (1972).

The processing of sweated aluminum electronic scrap was investigated to develop methods for recovering the aluminum and concentrating the other metals, including copper, lead, gold, and silver, into a product that can be separated by known methods. Two molten-salt electrorefining processes were developed and tested.

The first process utilized a three-layer refining cell with a $BaCl_2$-NaF-AlF_3 electrolyte, which operated at 750° to 850°C. The second process used an all-chloride electrolyte of NaCl-KCl-$AlCl_3$ at 750° to 800°C, with separate anode and cathode compartments. High-quality aluminum was recovered that averaged 99.8% pure in the three-layer cell and 99.6% pure in the compartment cell. Over 94% of the aluminum was recovered with cathode current efficiencies of approximately 90% in both methods.

The copper, lead, tin, silver, gold, and other metals were concentrated threefold in an anode product. Separation of lead and part of the tin by liquation from this product seemed feasible. The anode product was smelted to prepare a 96% copper bullion that contained 690 ounces of silver and 65 ounces of gold per ton, with 98% recovery of these values.

Recovery from Furnace Skim: A process developed by *J.W. Lowry, W.S. Peterson and T.W.F. Foster; U.S. Patent 3,043,678; July 10, 1962; assigned to Kaiser Aluminum & Chemical Corp.* relates to the recovering of the free metal entrained in the dross or skimmings obtained from melts of aluminum and aluminum base alloys.

In accordance with this process, the aluminous dross is placed inside a rotary furnace. Next, a predetermined amount of salt flux in solid form, preferably granular, comprising a mixture of sodium chloride, potassium chloride and a fluoride-containing compound or constituent is placed in the rotary furnace on top of the dross. The furnace is then rotated at a suitable ratio of speed to obtain a violent tumbling or cascading action of the mixture of dross and solid salt flux in order to break up large lumps of dross. Heat is applied, such as by a suitable burner, to the mixture of dross and salt flux in order to liquefy the flux.

The application of the external heat may be started at any time during the rotation to give the tumbling or cascading action, however it is preferred to start the application of heat at the beginning of the tumbling action, that is, the burner is positioned in the rotary furnace after the charging of the flux. After the flux is liquid, the mixture is subjected to a gentle rolling action to allow maximum opportunity for separation of droplets of metal from the dross and to allow the droplets of metal to agglomerate in a pool of metal. The liquid flux, which is intimately mixed with the dross particles, selectively wets the oxides, nitrides and nonmetallic impurities, thereby promoting the separation of the metal from the dross. After rotating the furnace and heating the material for a suitable period of time, the molten metal is drained from the furnace followed by dumping of the dross and flux residue.

A process developed by *P.H. Floyd and G.J. Binczewski; U.S. Patent 3,770,424; Nov. 6, 1973; assigned to Kaiser Aluminum & Chemical Corp.* relates to the recovery of metal values from furnace skim which is formed during the melting of light metals such as aluminum, magnesium and the like.

During the course of melting operations, such as remelting and refining of metals, oxides, nitrides and other nonmetallic impurities accumulate on the surface of the molten metal and in the trade are referred to as furnace skim. Normally, prior to tapping of the molten metal, these nonmetallics are removed or skimmed from the surface of the melt. Substantial quantities of metal will be entrained unavoidably with the nonmetallics and be removed along with these nonmetallics. The amount of free metal in this furnace skim can vary up to 95% (the normal range would be 50 to 80%) of the total weight of the furnace skim, depending of a number of factors such as the composition of the alloy being melted, the melting procedure followed, and the care with which the furnace skim is removed from the melt.

Usually, when the skim is removed from the top of the melt, the skim forms a heterogeneous mass which may be loosely packed as well as dense. The size of these masses varies from microscopic size to as coarse as 1½ feet in diameter. If the hot skim removed from the melt is allowed to stand, some free metal will accumulate at the bottom of the mass, but a large part of the free metal will remain intimately mixed with the nonmetallics in the form of globules or small particles, which cannot be readily separated from the nonmetallic portion. The separation of the free metal from the nonmetallic portion of the dross has been a difficult problem particularly with aluminum.

Several methods have been proposed and used for effecting the separation. One method is to cool the dross to room temperature as quickly as possible, thoroughly dry grinding the cooled dross in a ball mill and then screening the comminuted material. In practice, this method did not provide an efficient separation because a large portion of the aluminum particles were very fine and could not readily be separated for reclaiming from the oxide dross.

Furthermore, the metallic particles were not readily wetted by molten metal. Another method is described in U.S. Patent 3,043,678 wherein the dross is mixed with a salt flux comprising a mixture of sodium chloride, potassium chloride, and a fluoride-containing compound in a rotating barrel or furnace. Although good metal recoveries could be obtained by that process as set forth, the process presents a substantial problem in the disposal of the resulting salt dross. Because of the increased government regulations related to environmental pollution, the disposal of this untreated salt dross is severely restricted if not completely prohibited. The treatment of skim with the salt flux process would be economically unacceptable when resulting dross is disposed in a manner acceptable to the governmental bodies.

In a Bureau of Mines report, RI 3874 by O.C. Garst, et al, a process is described for recovery of metallic aluminum from dross, or furnace skim, which comprises hand sorting large metallics from the skim, crushing the resultant skim in a jaw crusher, and screening on a ¾ inch vibrating screen. The undersized material is passed to a Williams mill (a hammermill) followed by an air classifier. The oversized material is further reduced through a roll crusher and then subsequently screened on a ¾ inch vibrating screen. The screen oversized from the roll grinding operation is considered as metallics; the screen undersized from the roll crushing is further ground in a Williams mill and passed to an air classifier, the fines being considered nonmetallics, and the coarse produce being considered dirty metallics. The dirty metallics are subsequently screened on a 35 mesh vibratory screen. The oversized is further ground on a roll crusher, again screened on a 35 mesh screen, the oversized being considered metallics and the undersized being discarded as nonmetallics. All of the metallics are then mixed with a flux material to separate the residual oxide material from the metallics in a standard crucible melting furnace. The total metal recovery from this rather complex process was about 67%, which is too low to justify the capital expense and operating costs involved. Furthermore, a substantial quantity of salt flux is required, which presents a dross disposal problem.

This process is directed to an advantageous solution to the problem of recovering metal values from furnace skim which is formed during the melting of light metals such as aluminum, magnesium and the like. Further, this process provides for recovering up to 90% or more of the metallic material in the skim, yet maintaining the metal content of the product recovered greater than 85%.

The process essentially comprises the controlled aqueous comminution of furnace skim at a reduced work input level and then separating the slurried product into coarse and fine fractions. The skim, along with sufficient water to effect a mixture of about 40 to 60% by weight solids, is fed to a ball mill having a ball charge from 5 to 20% of the internal mill volume. The work input is maintained from 3 to 8 net horsepower hours per ton of skim feed, preferably from 4 to 7 net horsepower hours per ton. Preferably, the comminuted slurried product is separated with a screen having a mesh opening from 10 to 65 mesh (Tyler). Figure 45 illustrates the operation of this process in somewhat more detail.

There is shown a skim barrel 10 containing hot skim material from the furnace. Molten metal that will readily separate from the oxide skim can be decanted into a sow mold (not shown). The skim remaining in the skim barrel after decantation, which includes as much as 30 to 80% of the original total metal, is dumped into quench launder 11. The quench launder is provided with a high volume of water, as much as 100 times as much water to weight of skim to quickly quench the hot skim to an ambient temperature and to convey the quenched skim through the launder to the dewatering system comprising a screen 12 and spiral classifier 13.

The dewatering screen 12 scalps off the coarse solids fraction (such as plus ⅜ inch material) and delivers the undersized to the dewatering spiral classifier 13. The spiral classifier separates dewatered fine solids which, along with the previously dewatered coarse solids, are conveyed by any suitable means to storage bin 14. The aqueous fraction from the dewatering system is conveyed by pump 25 to the quench launder water. However, a bypass valve 26 is provided to bleed off a proper fraction of the slurry to maintain the recycle water solids content at a reasonable level, preferably 1% or less. The bypass line discharges directly to the thickener 20.

The surge bin 14 holds the quenched skim prior to attrition-peening beneficiation. Several bins may be used to accumulate and store several alloy types, thus providing alloy segregation. The surge bin is equipped with a feeder 15 that draws a controlled feed for the attrition-peening beneficiation which follows. The skim is conveyed to the intake spout of the attrition-peening mill 16 where a proper quantity of water is mixed to produce a slurry between 40 and 60% solids.

The attrition-peening mill 16, with a ball charge of 5 to 20%, selectively comminutes the skim slurry in such a way that the brittle oxides and other wastes are selectively comminuted to a much finer fraction. The malleable metal sustains a minimum of size reduction and is simultaneously densified and washed to remove the brittle waste materials. The finished slurry is discharged from the mill onto the dewatering mechanical screen 17.

The mechanical screen 17 is supplied with water sprays 18 to facilitate the separation of fine waste from the coarser metal as dewatering and sizing take place. The screen under-

FIGURE 45: PROCESS FOR RECOVERY OF ALUMINUM FROM FURNACE SKIM

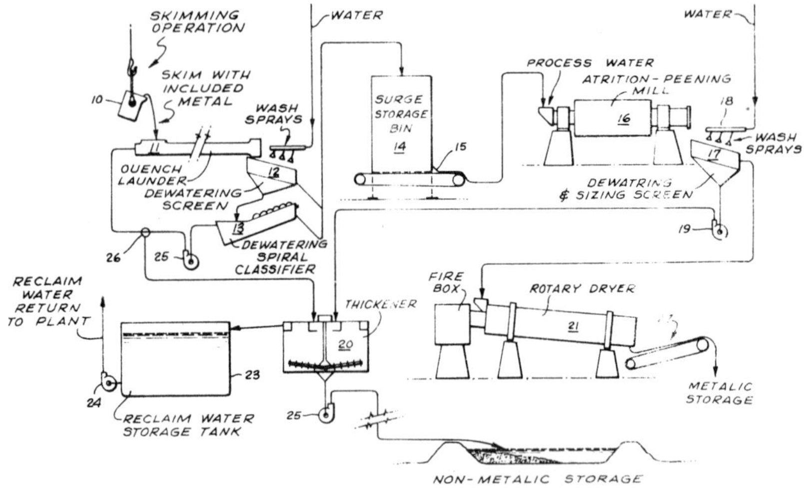

Source: P.H. Floyd and G.J. Binczewski; U.S. Patent 3,770,424; November 6, 1973

size product contains as much as 90% or more of the total nonmetallic material in the mill feed. The undersized product, a slurry, is transferred by pump **19** to the thickener **20**. The screen oversize is fed to the rotary dryer **21** and the dried product is conveyed by inclined belt conveyer **27** to metallic storage.

The rotary dryer lowers the entrained water in the coarse metallic product from approximately 20 to 0.5% moisture. The dryer discharge constitutes a finished product that may discharge to storage pending remelting operations or be directly transported to remelting operations.

The thickener **20** receives the process waste slurries for separation of a clean water effluent that overflows to reclaim water storage tank **23** for return by pump **24** to the process water system. The thickener underflow consists essentially of the nonmetallic wastes concentrated into a 40 to 60% solids slurry that is discharged to pump **25**. The pump delivers the process waste solids to a bermed storage area where it is further consolidated and dewatered pending future usage.

The following is a typical example of the operation of this process. Twenty tons per hour of quenched skim from an aluminum melting furnace containing 65% metallics are fed to a 6' x 18' ball mill loaded 5% with 3 inch steel balls, and rotated at 60% of the critical speed. Along with the mill feed was added 4,790 gallons per hour of water to effect a slurried product containing 50% by weight solids. The net horsepower input to the mill was 5 horsepower hours per ton of feed. The slurried product is passed to a 28 mesh vibrating screen for dewatering and sizing. The dewatered +28 mesh material amounted to 13.4 tons per hour containing 90% metallics. This represents a 93% recovery of the metallics which entered the mill. The -28 mesh material amounted to 6.6 tons per hour and was 86% nonmetallics.

Recovery from Furnace Pot Linings: A study conducted by R.S. McClain, G.V. Sullivan and W.A. Stickney, Report of Investigations 5777, Washington, D.C., U.S. Bureau of Mines (1961) was one in which the U.S. Bureau of Mines investigated residues from aluminum plants, to determine if carbon could be removed by flotation while recovering fluorine and aluminum compounds for recycling to the reduction process. This publication summarizes studies on samples representing three types of plant waste products, supplied by three aluminum-producing companies.

Samples of flue dust, pot skimmings, and pot linings from the various plants all responded to a simple turpentine flotation scheme to reject 90 to 97% of the carbon. Aluminum compound recoveries ranged from 42 to 94% on the flue dust, were about 95% on the pot skimmings, and ranged from 75 to 89% on the pot linings. Fluorine recoveries from the same samples ranged from 35 to 63, 92 to 95, and 77 to 86%, respectively. The pot linings contained soluble salts of aluminum and fluorine which were recovered by precipitation with sodium aluminate and carbon dioxide gas.

Another study by R.S. McClain and G.V. Sullivan, Report of Investigation 6219, Washington, D.C., U.S. Bureau of Mines (1963) was one in which continuous-circuit flotation and batch-scale precipitation tests were made by the Bureau of Mines to determine rejection of carbon and recovery of fluorine and aluminum compounds from aluminum reduction plant residues. About 40 to 60 pounds each of aluminum fluoride and cryolite are lost for each ton of aluminum produced. Gasoline and turpentine were satisfactory for floating carbon from both fresh water and saturated solution systems. Soluble fluorine and aluminum compounds were recovered by precipitation with sodium aluminate and carbon dioxide.

Over 99% of the carbon was removed by flotation. Fluorine and aluminum recoveries up to 77 and 63%, respectively, were obtained in a combined precipitate and nonfloat product. In the fresh water system the combined product was mostly cryolite with some gibbsite. In the saturated solution system the product was mostly sodium fluoride, some cryolite, and minor amounts of gibbsite.

As described by P.C. Good and W.G. Gruzensky, Report of Investigation 7264, Washington, D.C., U.S. Bureau of Mines (June 1969), the Bureau of Mines conducted a laboratory investigation of a sinter-leach method to recover aluminum and fluorine from waste carbon potlining residues from secondary cryolite recovery operations in aluminum reduction plants. After removal of carbon by burning, the residue was sintered at 900° to 1100°C with silica and Na_2CO_3, then crushed and leached with dilute NaOH solution, dilute Na_2CO_3 solution, water, or water containing residual Na_2O which remained in the sinter after reaction and decomposition of Na_2CO_3. Over 90% of the contained aluminum and up to 87% of the contained fluorine were extracted from this waste product in the form of a mixed alkaline solution.

Treatment of the alkaline solution with CO_2 yielded a white precipitate containing 89% of the original aluminum and 61% of the original fluorine as a mixture of synthetic cryolite and alumina, together with a variable excess of soda.

Recovery from Mixed Scrap Metal: A process developed by *B.E. Anderson and W.C. Larson; U.S. Patent 2,942,792; June 28, 1960; assigned to American Smelting and Refining Company* involves sorting nonferrous composite scrap metal, and especially the sorting of composite aluminum scrap.

The term composite scrap metal comprehends scrap metal comprising physically contaminated mixtures of metal castings or wrought shapes. Automobile engine pistons, foil, pipes, pipe fittings, boxes, washing machine tubs, toys, and utensils of various kinds such as kitchen utensils, tools and the like, which have been wrought from sheet metal or cast as such, are typical of the physical condition of such scrap and the sizes and shapes encountered therein. Such scrap may arise as rejects or discards during manufacture or as worn out or obsolete parts.

Although it is comparatively easy to hand-sort the composite scrap encountered in the secondary metals industry into segregated shapes and sizes, it has not been possible heretofore to sort such scrap satisfactorily according to composition. At best, it could be sorted only roughly according to composition by hand or by the other methods heretofore available, and this was particularly true of scrap which is a physical mixture of different metals or alloys having the same or similar color. Such difficulty of sorting according to composition is especially true of light colored or so-called white colored metals, particularly aluminum scrap.

Consequently, composite scrap purchased by a secondary metals smelter as being scrap of a certain metal or alloy, for example composite aluminum scrap, varies from batch to batch as to the size and shape of the scrap in a batch as well as the amounts of other metals and other constituents which are either or both physically mixed or physically attached to the metal intended to be purchased. Moreover, as indicated above, such purchased composite scrap comprises relatively large shapes, most of which, if not all, are larger and usually substantially larger than 8 inches in maximum dimension.

The problem of sorting composite scrap is further illustrated by composite aluminum scrap of the aforementioned typical commercial shapes. Aluminum engine pistons, for example, may have such foreign constituents attached thereto as either or both steel piston rings or a steel piston rod pin. Cast or wrought aluminum washing machine tubs may have attached thereto either or both a steel drive shaft or iron or stainless steel nuts and bolts attached to a tub flange or flanges. Wrought or cast aluminum pots or pans may have wooden handles attached thereto by an ordinary steel or stainless steel member.

Other aluminum utensils, toys and tools may likewise be assemblies of aluminum parts with other parts of different composition such as magnesium, steel, zinc, solder, brass, bronze, plastic, rubber, wood and the like. In addition, even though the scrap had been previously sorted by the procedures, such as hand sorting, which are available to the industry, the aluminum scrap contained physically mixed therewith shapes which contained no aluminum, for example, nuts, bolts, nails, zinc-die cast toys, steel pots and pans in the case of utensil scrap, rubber, wood, dirt, etc.

Thus, in addition to aluminum and aluminum alloys, the composite aluminum scrap purchased by the secondary metals smelter is physically contaminated with other metal and other constituents, some of which are heavier and some of which are lighter than the intended aluminum or aluminum metal base.

In the past, many attempts have been made without notable success to enhance the quality of the secondary metal recovered from composite scrap. Thus, for example, for want of a better procedure in the case of composite aluminum scrap, it is the practice on the part of the secondary metal smelter to further hand sort the scrap and to pass it through a magnetic separator to remove the loose nonmagnetic and magnetic contaminants physically mixed in the scrap. Regardless of the degree of care exercised, it is not possible adequately to remove such loose physical contaminants and especially the loose nonmagnetic contaminants. Moreover, such procedures are completely ineffective where the contaminants are physically attached to the aluminum.

After thus sorting the scrap, it is then melted down in a so-called sweating process in which an attempt is made to selectively melt the aluminum away from the metallic contaminants still mixed therewith. While the sweating procedure is somewhat effective in separating the higher melting contaminants physically present in the scrap, it is not sufficiently effective to produce molten aluminum which has a sufficiently low content of such physically unremoved contaminants, particularly, zinc, lead and tin. Although it is possible to purify the thus recovered molten aluminum of at least some of such contaminants by metallurgical procedures, such procedures are in general prohibitive as to cost. Consequently, in most instances the molten aluminum obtained from composite scrap is diluted with relatively pure aluminum to produce therefrom an aluminum product having a desired or acceptable composition.

This process comprises in combination a hammermill, which for best results possesses impact and cutting members, for reducing the scrap to a mixture comprised of fractured and torn fragments of reduced size and shape, and gravity separation means to separate the thus prepared fragments into a plurality of fractions of different specific gravity.

It was discovered that when the composite aluminum scrap was passed through a hammermill provided with impact and cutting members which reduced the scrap to a mixture of fractured and torn fragments, substantially all of which were less than about 8 inches in their maximum dimension, the thus prepared scrap when subjected to gravity separation could be separated into a fraction which was lighter than aluminum, a fraction which was heavier than aluminum, and an aluminum fraction which was substantially free of physical contaminants.

A device developed by *B.R. Ankersen; U.S. Patent 3,215,421; November 2, 1965* is a rotary barrel salt bath furnace especially adapted to melting oxidizable materials such as aluminum, particularly aluminum containing supporting members of other metals. For example, it is frequently desired to melt aluminum chips, filings, turnings and the like scrap. Conventional melting methods are either too slow, too costly or the losses in aluminum through oxidation are too high to be economically feasible.

It is also frequently necessary to melt aluminum, zinc or other light relatively low melting metal scrap containing supporting elements, or wearing elements or the like without melting such other metal elements and contaminating the molten metal. For example, aluminum pistons have steel supporting rods in the body of the piston for reinforcement, zinc die cast journals have bronze bearings cast into them. In every such case, it is important to melt away the low melting metal and remove the higher melting metal before it is attacked. Various attempts to solve such problems have been made but without any real success.

This furnace structure will overcome these problems. By the use of this furnace, it is possible to continuously melt aluminum or like chips and similar fine scrap without excessive losses and without the expensive techniques heretofore proposed or to melt aluminum, zinc or similar lower melting metals away from higher melting metals.

Recovery from Municipal Waste: Aluminum constitutes less than 1% by weight of municipal solid waste but comprises the bulk of nonferrous metals. The estimated quantity of aluminum in municipal waste is 800,000 tons a year. Current recovery of aluminum is estimated to be 80,000 tons. Most of this is being recovered in industry programs of separate collection through recycling centers.

Aluminum, like glass, is a valuable and desirable material for recycling from the standpoint of the user, in this case secondary aluminum smelters. The market price for secondary aluminum is $200 per ton, 10 times that of many other materials in mixed municipal waste.

The major constraint to aluminum recycling is supply. Recovery through consumer separation efforts has been successful in some instances and after some years of such practice, it appears that these efforts are becoming institutionalized and will continue. Aluminum recovery has been considered in most large-scale recovery systems under development, but the actual separation process (heavy media separation in most instances) has not yet been demonstrated.

Aluminum recovery will be tested as a part of some of the resource recovery systems EPA is sponsoring. Aluminum recovery could also be significantly affected by beverage container measures eliminating the beverage can or the pulltab top.

A study of *Aluminum as a Component of Solid Waste and a Recoverable Resource* has been prepared by the National Environmental Research Center, Cincinnati, Ohio and is available as PB 235 770, Springfield, Va., Nat. Tech. Information Service (Sept. 1974).

Some aspects of aluminum recovery from mixed municipal waste, particularly from admixtures with glass, have been discussed by W. Herbert and W.A. Flower in Public Works 102, (8), 70, 110, 112 (Aug. 1971). See also the reference to U.S. Patent 3,720,380 in this volume under "Glass: Separation from Municipal Waste."

Recovery from Plastic Composites: A process developed by *G. Haveman; U.S. Patent 3,516,841; June 23, 1970; assigned to The Dow Chemical Company* involves four variations in converting waste portions of metallic-plastic laminate structures into reusable and valuable raw materials. These are (1) separation of the metallic component from the plastic component and individually recovering each essentially free of contamination of the other; (2) separating the metallic component from the plastic component in such a manner that a random mixture of the two components itself can be reused, and when aluminum is the metallic component of the laminate structure; (3) recovering the plastic component completely free of aluminum for reuse in other processes and products; and (4) converting an aluminum-plastic laminate structure into an aluminum oxide-filled plastic raw material which can be used for fabricating other products.

Figure 46 shows the form of apparatus which may be employed in the conduct of the process, with particular reference to the first variation described above. A separator, generally designated by the reference numeral **10**, comprises essentially two zones, one of which is an agitation zone **11** and the second is the settling and separation zone **12**. Separator **10** is beneficially a closed tank (which may have an open top) having a generally downwardly sloping bottom from the agitation zone towards the settling and separating zone. Agitation zone **11** is not totally separated from settling and separating zone **12** but is generally so defined by baffles **13** and **14**.

The cut, sheared and ground laminate segments, that is the laminate structure which has been cut into relatively minute segments, and after being subjected to a shearing and grinding operation to further reduce the size of the segments and to free the metallic component from the plastic component, is fed into the separator through hopper **15**. Upon entering the separator, it is dispersed in liquid medium **16**, which has a density between the densities of the metallic and plastic components, by the action of agitator **17**.

FIGURE 46: DEVICE FOR SEPARATING ALUMINUM FROM WASTE ALUMINUM-PLASTIC LAMINATES

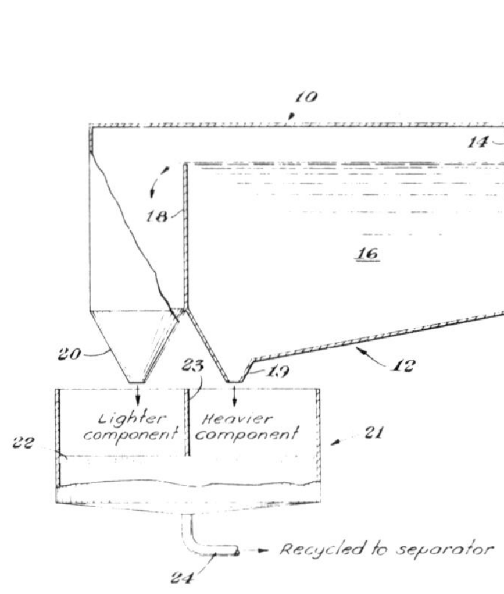

Source: G. Haveman; U.S. Patent 3,516,841; June 23, 1970

The liquid medium containing the dispersed laminate segments is allowed to flow by gravity towards the settling and separating zone, during which time the heavier component (which is in almost all cases the metallic component) settles to the bottom of the liquid medium and the lighter component (which in almost all cases is the plastic component) floats to the surface of the liquid medium.

The heavier component together with some of the liquid medium flows out from the settling and separation zone through outlet **19** and the lighter component together with some of the liquid medium flows over baffle **18** and out of the settling and separation zone through outlet **20**. Both components together wtih some of the liquid medium flow into collector **21** which is separated into two compartments in its upper half by baffle **23.** The liquid medium containing the component then flows down upon filter **22** whereat the components are retained on top of the filter, and the liquid medium passes on through the filter and is subsequently withdrawn through outlet **24** and recycled back into settler **10**. The recycled liquid medium can be introduced through a separate inlet into the top of settler **10** or it can be added together with the laminate segments through hopper **15** which beneficially aids in facilitating dispersion of the laminate segments in the liquid medium. The plastic and metallic components are recovered from the top of the filter from time to time or they can be recovered continuously by known methods.

The laminate waste material that is fed through the separator is conveniently prepared by precutting the material in a suitable cutter or breaker to about 5-cm or 2-in square chips or pieces and then subsequently further grinding the pieces in another suitable crush-cut mill provided with a sieve having openings on the order of about $1/128$ inch or about 0.2 millimeter. Although conveniently, this operation is performed in two separate steps, the cutting to the required size and applying the shearing force to the laminate structure to free the metal from the plastic can be accomplished in one operation. Any suitable equipment can be used for this purpose that is adapted to cut and apply the necessary shearing force or action to the laminate structure to free essentially all of the metal from the plastic.

Ordinarily, any such action which provides a crushing or shearing effect will free the metal component including such means as ball and roll mills and the like. Other suitable means for freeing the metallic component include gear crushers wherein the laminate particles are passed between counterrotating gear-like teeth or by agitating the segments with blades rotating at high speed. The following is one specific example of the operation of this process.

Laminate waste material collected from sheets, edge trimmings and the like and fabricated from a thin sheet of aluminum securely interlaminated between two thin sheets of cellulose acetate butyrate was cut into pieces having a maximum dimension of two inches. These pieces were then forwarded through a Peppink crush-cut mill having a sieve with $1/128$ inch opening. The shearing and grinding action freed, on the average, about 99% of the aluminum from the butyrate.

The resulting mixture of aluminum and butyrate was then fed to a separator according to the procedure discussed in connection with the drawing. The aqueous medium employed was an aqueous solution of calcium chloride having a specific gravity of about 1.33. The aluminum (specific gravity about 2.7) settled to the bottom, while the butyrate (specific gravity about 1.3) floated to the top of the solution and both the aluminum and butyrate were recovered essentially completely free of each other in accordance with the procedure outlined with reference to the drawing.

A process developed by *E.A. Mathis; U.S. Patent 3,650,830; March 21, 1972; assigned to Nichols Engineering & Research* is one in which aluminum is recovered from scrap by volatilization of organics from the scrap in a nonoxidizing furnace having multiple hearths, and thereafter separating the aluminum from the ash by mechanical means.

A considerable amount of such scrap is generated at plants where printing is done on aluminum foil. The foil is usually laminated with an organic backing material such as wax, plastic, lacquer, paper or paperboard. The metallic alluminum may constitute anywhere from 7 to 55% of the total weight of the scrap.

Scrap aluminum is generally collected and baled for storage or for shipment. It is processed according to prior art techniques by breaking open the bales and shredding the material, and thereafter blowing the shredded material through a continuous combustion unit. The material becomes ignited in the combustion unit and then passes into one end of an elongated combustion trough. The bottom of the trough is made of a flat steel plate on which rides a drag chain conveyor with steel flights, so as to move the burning scrap continuously through the trough. An insulated hood having several smokestacks is positioned over the trough and combustion air is admitted into the trough just under the hood. At the other end of the trough the burned material is milled or hammered and subjected to mechanical separation. The aluminum portion is then passed through a rotary kiln at about 950°F where the carbon is mostly burned off.

The continuous combustion technique described above is subject to several serious defects. The most serious of these defects involves the occurrence of thermite type reactions whereby some of the aluminum reacts very violently with oxygen to produce temperatures in the region of 2300°F. These high temperatures melt the surrounding aluminum and cause it to adhere to and clog the interior of the trough. The high temperatures also have a destructive effect on the insulation material which is used to line the trough. As a result, the operation of the system is intermittent rather than continuous. Also, the system becomes very expensive to operate and maintain.

An additional difficulty associated with the continuous combustion trough is that it produces an exorbitant amount of heavy black smoke which presents a nuisance. This smoke has been measured consistently at No. 4 on the Ringleman scale, and very rarely at No. 3. The Ringleman scale extends from No. 0 to No. 5, where No. 5 is smoke of maximum blackness and maximum density.

This process overcomes the above described problems of the continuous combustion system for removal of organic materials from aluminum scrap. Scrap may be continually processed in relatively compact and inexpensive equipment. Further, the aluminum values are recovered at a high degree of purity, and no danger of thermite type reaction is presented. The equipment is never subject to temperatures higher than about 1100°F so that melting does not occur; nor is there any destructive effect produced on the insulative materials in the equipment.

In general this process achieves its superior results by heating the dispersed aluminum scrap in a nonoxidizing (e.g., carbonaceous) atmosphere up to a temperature (e.g., 700° to 1100°F) sufficient to evaporate or to drive off the volatile components of the backing material. This leaves a product comprising aluminum, carbon and ash, from which the aluminum may be readily separated by mechanical means. This process is shown schematically in Figure 47.

Bales **10** of aluminum scrap are moved along a belt conveyor **12** and are dropped into a hopper **14** of a shredding machine **16**. The shredding machine separates the baled material into individual fragments which then pass out through a conveyor tube **18**. The conveyor tube **18** opens into an air conveyor **20** which is powered by a blower **22**. The blower serves to convey the shredded aluminum scrap to a cyclone **24** where the scrap fragments are separated from the air stream. The scrap passes downwardly from the cyclone **24** and through an airtight lock **26** into a controlled atmosphere furnace **28**.

The interior of the furnace **28** is maintained at a temperature above 700°F, and preferably between 900° and 1100°F. Also the composition of the furnace atmosphere is kept under close control so that no oxygen will come into contact with the metallic aluminum in the scrap material. As a result of this, the volatile organic materials forming the major portion of the nonaluminum values are vaporized and separate from the solid scrap. These materials are then discharged, or as will be described, are burned and their products of combustion are discharged out through a stack **29**.

The remainder of the scrap material, i.e., the solid residue composed chiefly of metallic aluminum values and solid, loosely adherent, carbon and ash are passed through a mechanical separator **30** where these materials are crushed and separated to provide high grade aluminum.

FIGURE 47: APPARATUS FOR ALUMINUM RECOVERY BY VOLATILIZATION OF ORGANICS FROM SCRAP CONTAINING ORGANICS

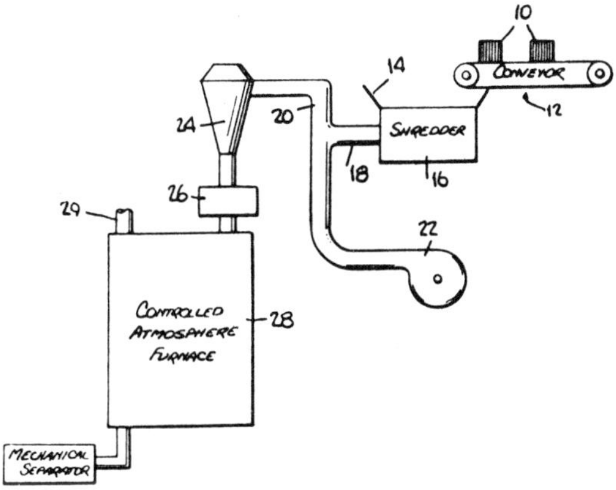

Source: E.A. Mathis; U.S. Patent 3,650,830; March 21, 1972

Recovery from Scrap Aluminum: A process developed by *B.J. Racunas and N. Jarrett; U.S. Patent 3,846,123; November 5, 1974; assigned to Aluminum Company of America* involves reclaiming aluminum values from secondary aluminum materials containing nonmetallic solids such as oxides, nitrides and carbides of aluminum and alloy constituents thereof. The method includes the steps of feeding aluminum materials into a pool of molten salt flux maintained at a temperature greater than the melting point of aluminum.

The pool is stirred to mix the aluminum materials therein, the flux being effective to separate nonmetallic solids and fines from the aluminum material thereby releasing molten aluminum globules or droplets which settle to the bottom of the pool while nonmetallic fines are suspended therein by the stirring action. The aluminum globules coalesce to form a layer or pad of molten aluminum beneath the pool of flux from which the molten aluminum is removed.

This enriches the flux with nonmetallic solids, which enriched flux is directed to a location within the pool where the solids are removed therefrom, and the flux returned thereto. The feed of aluminum materials to the pool of flux is correlated with the removal of the nonmetallic fines to maintain in the pool a nonmetallic solids content not exceeding 25% by weight of the bath therein.

Use of Aluminizing Dross in Steel Deoxidation: The feasibility of using waste aluminizing dross to deoxidize steel has been described by H.V. Makar and P.J. Gallagher in Technical Progress Report No. 55, Washington, D.C., U.S. Bureau of Mines (July 1972).

Aluminizing dross in the form of large blocks or castings, is the waste produced during an aluminum-coating process for steel products known as aluminizing. Small quantities of the material can be used in some secondary aluminum alloys, but widespread use is severely limited by its high iron content. Consequently, this dross is accumulating in large quantities and is becoming a severe disposal problem. Accurate data on total quantities generated are not available, but discussions with industry representatives indicate an order of magnitude of several hundred tons per month. Analysis on the dross reportedly ranges as follows (in weight-percent): aluminum 55 to 65; silicon 5 to 9; iron 10 to 25; and zinc up to 6.

The aluminizing dross was tested for possible use as a steel deoxidizer. This material, consisting mainly of aluminum, silicon, and iron, was used to deoxidize laboratory-scale melts ranging from 7 to 100 pounds. Oxygen analysis showed that the aluminizing dross was effective in reducing oxygen levels in steel. A direct comparison with aluminum wire showed the dross material to be at least as effective. Under similar test conditions with 7 lb melts, 3 grams of aluminum reduced the oxygen concentration from 210 to 175 ppm, while 4 grams of aluminizing dross (equivalent to approximately 2 grams aluminum) reduced the oxygen concentration from 250 to 170 ppm. An additional 4 grams of aluminizing dross produced a final oxygen concentration of 140 ppm.

ANTIMONY

Recovery from Liquid Wastes

Recovery from Spent Catalyst Solutions: A process developed by *E. Joerchel and R. Kohlhaas; U.S. Patent 3,784,671; January 8, 1974; assigned to Farbwerke Hoechst AG, Germany* involves recovering antimony pentachloride from catalyst solutions used for the fluorination of chlorine containing hydrocarbons. The process is based upon a thermal decomposition of antimony pentachloride into antimony trichloride and chlorine at a temperature of 60° to 300°C.

The organic impurities are extracted from the precipitated antimony trichloride by means of halogenated hydrocarbons. The remaining antimony trichloride is reoxidized with chlorine, which may be recycled from the step of the thermal decomposition, at a temperature of from 60° to 150°C and a pressure of zero to five atmospheres to yield antimony pentachloride which is refined by distillation.

Recovery from Solid Wastes

As described by H. Nack, K, Murthy, E. Stambaugh, H. Carlton and G.R. Smithson, Jr. of Battelle-Columbus Laboratories, the secondary antimony segment recovers antimony as an alloy with lead rather than as pure metal because of the far greater demand for the alloy. Also, primary lead refineries recover the metal as antimonial lead. The bulk of the antimonial lead is used in storage battery grids and contains about 3% antimony. Statistics for the secondary antimony industry for 1971 are presented below.

Kind of Scrap	Production, Short Tons	Form of Recovery	Production, Short Tons
New scrap	3,342	As antimonial lead	15,839
Old scrap	17,575	In other lead alloys	5,078*
Total scrap	20,917	Total recovery	20,917

* Includes 11 short tons in tin-base alloys.

Further, in 1971, by-product antimonial lead produced at primary lead refineries amounted to 19,686 tons containing 1,191 tons of antimony (6%), according to Nack et al in Report PB 233 646, Springfield, Va., Nat. Tech. Information Service (May 1974).

Recovery from Scrap: New scrap consisting of residues and drosses resulting from manufacturing and casting amounted to 16% of the total scrap processed in 1971. Old scrap, obtained mostly from used lead-acid type battery grids, amounted to the remaining 84%. Greater than 99% of both old and new scrap was in the form of lead-base alloys while less than 1% was comprised in tin-base alloys.

Although the increased useful life of batteries tended to reduce the quantity of available battery scrap, the trend was soon counterbalanced by the increase in the automobile population, thus insuring a continued supply of used battery grids as scrap. The raw materials to this industry, therefore, include: old battery grids; residue and drosses from manufacturing and casting, and scrap from the machining and manufacturing industry.

Products from the antimony scrap recovery industry include: antimonial lead alloys used in storage battery grids containing about 3 to 4% antimony (grid metal), body solder used in automobile bodies; type metal used in printing, having 4 to 23% antimony, 17 to 30% tin, and the remaining is lead; master alloys containing large amounts of antimony (15% or more) and used in blending to manufacture other alloys; and pure lead as a by-product.

The recovery of antimony as alloy from scrap involves three manufacturing operations: scrap pretreatment, smelting, and refining/casting.

New scrap generated in machine shops is not pretreated. Old scrap from type and babbitt metals is not only insignificant in quantity, but is generally not pretreated before smelting. However, since 70% of the old scrap is used battery scrap, pretreatment of these used batteries will be described. This process which separates the metallic parts of the old battery from the plastic and rubber portions is conducted in one of two ways. The more common method, adopted by about 75% of the industry, consists of the following steps: mechanical sawing of the top of the battery, manually separating the bottom case from the acid and grids by dumping, cleaning grids by water spray, crushing the battery tops and separating the metals on a vibratory table, crushing the bottom case in a hammermill for disposal in landfills, and neutralizing the acid effluent from dump tanks with lime in a settling pond.

The second method, used in about 25% of the industry, consists of the following process steps: crush the whole battery (as received) in a hammermill, drain and treat the acid, and conduct heavy media separation to separate plastics from the lead oxide and lead metal. This process produces a richer feed to smelters. The smelting operation consists of two processes which are described below.

Reverberatory Smelting Process — This process melts, purifies, and separates the metallic portions of the feed scrap generated in the pretreatment steps. The process steps are: charging the furnace, melting the charge, pouring out the lead metal, and removing the slag.

Blast Furnace Smelting Process — The blast furnace produces lead metal with a high antimony content and also reduces the oxides of lead and antimony to their respective metals. The feed to this furnace includes slag from the reverberatory furnace. The process steps are: continuous charging of coke, slag, and treated battery scrap; heating with blasts of air; and withdrawing the metal and slag.

In the refining/casting step, the metal products from the reverberatory and blast furnaces are molded after adjusting to required alloy compositions. Air blowing is done if removal of all antimony is desired. The process steps are: charging the furnace, melting the charge, blowing if necessary, pouring the molten metal into the molds and casting into ingots.

ARSENIC

Recovery from Copper Smelting Flue Dusts

As described by A. Patrick, Jr., H.J. Bennett, K.E. Starch and R.C. Weisner in *The Economics of By-Product Metals,* Report I.C. 8569, Washington, D.C., U.S. Bureau of Mines, arsenic is produced as a by-product from flue dusts from copper smelters, as shown in Figure 48.

The arsenic refining process begins with roasting of the flue dusts and their collection in the form of crude arsenic trioxide in an operation consisting of roasting furnaces and collecting chambers connected in series. Furnaces are operated at 650°F. Higher temperatures would cause fusion in the condensers. The roasting furnaces have a capacity of 50 tons per hour of charge.

Refining furnaces of the reverberatory type are of refractory brick construction with cast iron hearths, each handling 30 tons per day. High-grade material is dropped into the furnace on each of three shifts, and residues are removed once every 2 to 3 days. The condensing chambers for the arsenic trioxide consist of a series of kitchens or subchambers.

The first of these operate at a temperature above the condensing temperature of white arsenic (218°C), and thus it collects the nonarsenical dust. The other kitchens near the furnace precipitate black amorphous impure arsenic (95% As_2O_3) which is re-refined. The succeeding kitchens precipitate white crystalline arsenic trioxide of 99.9% purity which is packaged and marketed.

FIGURE 48: ARSENIC PRODUCTION FROM FLUE DUSTS

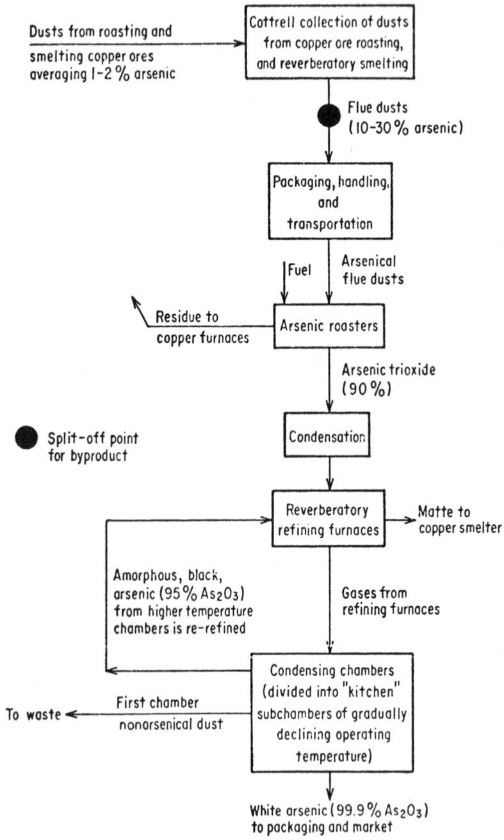

Source: Bureau of Mines I.C. 8569

BERYLLIUM

Recovery from Liquid Waste

Recovery from Spent Pickle Acids: See this same subheading under copper for reference to a process which involves beryllium recovery.

Recovery from Solid Waste

Recovery from Processing Scrap: Beryllium, like indium, is a very expensive metal with a current market value of about $6 per ounce. Total U.S. production of beryllium can be estimated at less than 500 tons per year. Exact production figures are a trade secret and no data on secondary production are available. In fact, there is no distinct secondary segment because most of the scrap is generated in-house. Tonnagewise, scrap availability is estimated at less than 100 tons per year by H. Nack, K. Murthy, E. Stambaugh, H. Carlton and G.R. Smithson, Jr. in Report PB 233 646, Springfield, Va., Nat. Tech. Information Service (May 1974).

Scrap beryllium is treated by pyrometallurgical methods to produce pure metal. Typically, vacuum casting followed by hot pressing and grinding are used. Excellent hooding and bag fillers where necessary are used to prevent escape of beryllium. Beryllium inhalation produces effects akin to silicosis and, therefore, extreme care is taken to minimize the indoor and ambient concentration of this metal.

BRASS AND BRONZE

Recovery from Scrap

The brass and bronze segment constitutes a large portion of the secondary nonferrous metals industry. Since World War II, the production of brass and bronze ingots in the United States has been fairly constant, averaging slightly over 300,000 tons annually. Potential environmental pollutants from this industry are atmospheric emissions, both gases and particulate matter, liquid wastes, and solid wastes, which if not controlled and properly disposed of can result in serious pollution problems. Emission potential of particulates alone has been estimated at 10,000 tons annually, based on an annual production of 300,000 tons of ingots.

Obsolete domestic and industrial copper-bearing scrap is the basic raw material of the brass and bronze segment. About two-thirds of the scrap is in the form of brasses and bronzes; the other one-third is received in the form of copper scrap. Of the many hundreds of copper-base alloys that become available for reuse through scrap recovery channels, 54 primary types of copper-bearing scrap are now included in the standards published by the National Association of Secondary Materials Industries. These are listed in Table 18.

Scrap as received is frequently not clean and may contain any number of undesirable metallic and nonmetallic impurities that contribute nothing to the composition of the ingot but increase the problems in producing high quality products. Among these undesirable constituents are oil, grease, paint, insulation, rubber, and antifreeze.

TABLE 18: TYPES OF COPPER-BEARING SCRAP

Number	Designation
1	No. 1 copper wire
2	No. 2 copper wire
3	No. 1 heavy copper
4	Mixed heavy copper
5	Light copper
6	Composition or red brass

(continued)

TABLE 18: (continued)

Number	Designation
7	Red brass composition turnings
8	Genuine babbitt-lined brass bushings
9	High-grade, low-lead bronze solids
10	Bronze papermill wire cloth
11	High-lead bronze solids and borings
12	Machinery or hard red brass solids
13	Unlined standard red car boxes (clean journals)
14	Lined standard red car boxes (lined journals)
15	Cocks and faucets
16	Mixed brass screens
17	Yellow brass scrap
18	Yellow brass castings
19	Old rolled brass
20	New brass clippings
21	Brass shell cases without primers
22	Brass shell cases with primers
23	Brass small arms and rifle shells, clean fired
24	Brass small arms and rifle shells, clean muffled (popped)
25	Yellow brass primer
26	Brass pipe
27	Yellow brass rod turnings
28	Yellow brass rod ends
29	Yellow brass turnings
30	Mixed unsweated auto radiators
31	Admiralty brass condenser tubes
32	Aluminum brass condenser tubes
33	Muntz metal tubes
34	Plated rolled brass
35	Manganese bronze solids
36	New cupro-nickel clippings and solids
37	Old cupro-nickel solids
38	Soldered cupro-nickel solids
39	Cupro-nickel turnings and borings
40	Miscellaneous nickel copper and nickel-copper-iron scrap
41	New monel clippings and solids
42	Monel rods and forgings
43	Old monel sheet and solids
44	Soldered monel sheet and solids
45	Soldered monel wire, screen, and cloth
46	New monel wire, screen, and cloth
47	Monel castings
48	Monel turnings and borings
49	Mixed nickel silver clippings
50	New nickel silver clippings and solids
51	New segregated nickel silver clippings
52	Old nickel silver
53	Nickel silver castings
54	Nickel silver turnings

Source: Report PB 233 646

Problems encountered in the removal of organics such as cutting oils from scrap brass and bronze turnings, shavings and borings are: high incidence of fires in the baghouses, heavy emissions of zinc oxide and other metallic fumes, and severe fire hazard during the charging of the scrap to the melting pot.

One specific solution to the problem of fire in baghouses in current use is described as follows: a fine powder is injected into the off gases from the brass and bronze melting furnaces. This fine powder reduces the incidence of fires in the baghouses, apparently by reaction with the fumes, dusts, and/or gases to form products which do not ignite as readily as the original material. Other companies are de-oiling the scrap before melt-down by burning to prevent fire. The company using the fine powder claims that de-oiling is unnecessary, according to H. Nack, K. Murthy, E. Stambaugh, H. Carlton and G.R. Smithson, Jr. in

Report PB 233 646, Springfield, Va., Nat. Tech. Information Service (May 1974).

The products from the brass and bronze segment of the secondary nonferrous metals industry are hardeners and brass and bronze alloy ingots. Brass and bronze are copper-base alloys with zinc and tin, respectively, as the largest secondary component. Other alloy agents may include such elements as lead, iron, aluminum, nickel, silicon, or manganese. Members of the Brass and Bronze Ingot Institute produce 31 standard copper-base alloys as shown in Table 19.

TABLE 19: NOMINAL CHEMICAL SPECIFICATIONS FOR BBII STANDARD ALLOYS

Alloy No.	Classification	Cu, %	Sn, %	Pb, %	Zn, %	Fe, %	Al, %	Ni, %	Si, %	Mn, %
1A	Tin bronze	88.0	10.0	--	2.0	--	--	--	--	--
1B	Tin bronze	88.0	8.0	--	4.0	--	--	--	--	--
2A	Leaded tin bronze	88.0	6.0	1.5	4.5	--	--	--	--	--
2B	Leaded tin bronze	87.0	8.0	1.0	4.0	--	--	--	--	--
2C	Leaded tin bronze	87.0	10.0	1.0	2.0	--	--	--	--	--
3A	High-lead tin bronze	80.0	10.0	10.0	--	--	--	--	--	--
3B	High-lead tin bronze	83.0	7.0	7.0	3.0	--	--	--	--	--
3C	High-lead tin bronze	85.0	5.0	9.0	1.0	--	--	--	--	--
3D	High-lead tin bronze	78.0	7.0	15.0	--	--	--	--	--	--
3E	High-lead tin bronze	71.0	5.0	24.0	--	--	--	--	--	--
4A	Leaded red brass	85.0	5.0	5.0	5.0	--	--	--	--	--
4B	Leaded red brass	83.0	4.0	6.0	7.0	--	--	--	--	--
5A	Leaded semired brass	81.0	3.0	7.0	9.0	--	--	--	--	--
5B	Leaded semired brass	76.0	2.5	6.5	15.0	--	--	--	--	--
6A	Leaded yellow brass	72.0	1.0	3.0	24.0	--	--	--	--	--
6B	Leaded yellow brass	67.0	1.0	3.0	29.0	--	--	--	--	--
6C	Leaded yellow brass	61.0	1.0	1.0	37.0	--	--	--	--	--
7A	Manganese bronze	59.0	1.0	1.0	37.0	1.0	0.6	--	--	0.5
8A	Hi-strength Mn bronze	57.5	--	--	39.0	1.0	1.0	--	--	1.5
8B	Hi-strength Mn bronze	64.0	--	--	24.0	3.0	5.0	--	--	3.5
8C	Hi-strength Mn bronze	64.0	--	--	24.0	3.0	5.0	--	--	3.5
9A	Aluminum bronze	88.0	--	--	--	3.0	9.0	--	--	--
9B	Aluminum bronze	89.0	--	--	--	1.0	10.0	--	--	--
9C	Aluminum bronze	85.0	--	--	--	4.0	11.0	--	--	--
9D	Aluminum bronze	81.0	--	--	--	4.0	11.0	4.0	--	--
10A	Leaded nickel brass	57.0	2.0	9.0	20.0	--	--	12.0	--	--
10B	Leaded nickel brass	60.0	3.0	5.0	16.0	--	--	16.0	--	--
11A	Leaded nickel bronze	64.0	4.0	4.0	8.0	--	--	20.0	--	--
11B	Leaded nickel bronze	66.5	5.0	1.5	2.0	--	--	25.0	--	--
12A	Silicon bronze	88.0	--	--	5.0	1.5	--	--	4.0	1.5
12B	Silicon brass	82.0	--	--	14.0	--	--	--	4.0	--

Source: Report PB 233 646

The production of brass and bronze products using scrap as the basic source of raw material entails two manufacturing operations: scrap pretreatment and smelting-refining.

Before the scrap is blended in a furnace to produce the alloy of a desired composition, removal of some of the metallic and nonmetallic impurities or contaminants and densification of the scrap are conducted to produce a material more suitable for subsequent smelting-refining.

The scrap pretreatment step may take the form of one of the following individual processes: stripping, briquetting, shredding, magnetizing, sweating, burning, drying, cupolaing and gravity concentration. Some additional detail is given on each of these scrap pretreatment processes in the paragraphs which follow.

Stripping Process: Insulation is removed from electrical conductors such as cables by specially designed stripping machines or by hand. Energy demand is that needed to drive the equipment.

Essentially no atmospheric emissions or liquid wastes are generated by this process. However, significant quantites of solid wastes are produced. These wastes consist of primarily organic materials such as plastics and other materials used as protective coverings on copper scrap.

The pollutants generated are expected to cause no atmospheric pollution problems, unless the solid wastes are disposed of by burning. Disposal of the solid wastes in landfills may possibly cause water pollution if the landfill is not properly prepared.

Briquetting Process: Compressing the bulky scrap such as borings, turnings, and wire into small bales densifies the scrap, permits more compact storage, and makes for easier handling and faster melting. Briquetting is carried out by compacting the scrap with hydraulic presses. Energy required is that to drive the equipment. Essentially no atmospheric emissions and no liquid or solid wastes are generated. The process has little or no potential for the production of pollution problems.

Shredding Process: The shredding process also achieves separation of the insulation from the copper-bearing scrap. The process steps are: introducing the insulated copper scrap such as copper wire into a hammermill where the scrap is cut into small pieces and the insulation is broken loose from the metal, and removing the insulation by air classification. This process is used in larger plants because of high capital costs. Energy demand is that required to drive the equipment.

Potential environmental pollutants from this process are atmospheric emissions (gases and particulate matter) and solid wastes. The atmospheric emissions are the fine particulate matter composed of the insulation and fine metal particles and the gas, generally air, used in the classification. The air is not a pollutant; however, the particulate matter is a potential pollutant if not collected and disposed of properly. Cyclones are used as the pollution abatement device. Solid wastes are the insulation and metal particles removed from the scrap by the air classification.

The process has the potential for creating environmental pollution problems. The fine particulates may cause atmospheric pollution problems; the solid wastes may cause water pollution problems if not disposed of in an approved landfill.

Magnetizing Process: The scrap may be a mixture of ferrous as well as nonferrous components. The ferrous component of the scrap is removed by the magnetizing process which entails passing the scrap over magnetized pulleys. Scrap treated in this manner is brass borings and small items. Energy required is limited to that to drive the equipment.

The only wastes generated by this process are solid wastes, scrap iron which may be contaminated with other metals and organic compounds such as the machining oils. These are disposed of by either landfill or selling to a ferrous scrap dealer. The process has essentially no potential for the production of a pollution problem.

Sweating Process: Much of the scrap such as radiators contains low melting components such as lead, solder, and babbitt metal. These are removed from the scrap by the sweating process which entails heating the scrap in a furnace. The process steps are: charging the furnace, melting the low-melting components, collecting the melt, and removing the treated scrap from the furnace for further processing. The collected metal may be made into white alloys, used for lead and tin addition to the copper base alloys, or sold as produced to a refinery.

Energy required includes the fuel to fire the furnace and electricity to drive the equipment. Several types of furnaces, rotary kiln, tunnel furnace, pot furnace, or reverberatory furnace, may be used. Both gas and oil are potential sources of fuel, depending on the availability and type of furnace.

Environmental pollutants generated by this process are: atmospheric emissions, solid wastes, and possibly aqueous wastes. Atmospheric emissions contain primarily fumes and combustion products of antifreeze residues, soldering salts, hose connections, and fuel. Metal content of the emissions is expected to be low because of the low sweating temperature. Solid wastes are those from the pollution abatement equipment. If wet scrubbers are used to reduce atmosphere pollution, liquid wastes are generated.

The process is a potential source of serious atmospheric pollution and presents potential pollution problems resulting from disposal of the solid wastes and liquid wastes, if generated.

Burning Process: Much of the scrap is covered with insulation such as polyethylene, polypropylene, or polyvinyl chloride. These contaminants are removed from the scrap by the burning process using such furnaces as muffles and rotary kilns. The process steps are: charging the furnace, and burning off the organic contaminants. Whereupon, the pretreated scrap is removed from the furnace for further processing. Energy required is the fuel, gas or oil, to heat the furnaces and electricity to drive the equipment.

The burning process is a potential source of serious pollution problems. Burning of the organic contaminants in the scrap results in atmospheric pollution problems. In addition to the combustion products such as carbon dioxide and water, the emissions may contain such gases as phthalic anhydride and hydrogen chloride from the burning of, for example, polyvinyl chloride. Flurorcarbon insulation releases hydrogen fluoride when burned. Many of these gases are highly toxic and corrosive.

Little information is available on these emissions. Source tests in Los Angeles indicate that uncontrolled emissions can be dense black smoke containing particulate matter in concentrations as large as 29 grains/scf at 12% carbon dioxide. These particulates must have been primarily carbon, because burning the emissions at 2000°F (probably in an afterburner) reduced the particulate matter level to 0.16 grains/scf at 12% carbon dioxide.

Solid and liquid wastes are not generated by the process; however, both may be generated by the pollution control devices. Disposal of these materials is probably by landfill and settling ponds, respectively.

Drying Process: Borings, turnings, and chips from machining are covered with cutting fluids, oils, and greases. These contaminants are removed in the drying process, where the scrap is heated in, for example, a rotary kiln to vaporize the contaminants. Energy required is the fuel to heat the furnace and drive the equipment. Fuel used in this case may be oil or gas.

Drying results in the evolution of considerable quantities of hydrocarbons depending on the amount present in the scrap. In addition, the oils, greases, and cutting fluids contain sulfonated and chlorinated hydrocarbons. Therefore, the gaseous emissions are composed of sulfur oxides, hydrogen chloride, hydrocarbons, and other combustion products. Particulate matter in the atmospheric emissions is soot and possible metallic fumes. Essentially no solid or liquid wastes are generated by the process.

The atmospheric emissions are controlled by burning the vaporized fumes in afterburners to oxidize the hydrocarbons to carbon dioxide and water. However, this technique does not remove the sulfur oxides and chloride emissions. Wet scrubbing is required. This process can present atmospheric pollution problems, if the emissions are not controlled.

Cupolaing Process: This process densifies bulky scrap and recovers metal values from slags, skimmings, and other low copper scrap to produce cupola melt (black copper) which is refined to produce the finished alloy. The process steps are: charging of the cupola, melting the charge, and removing the slag. Afterwards, the cupola melt is poured and cast. The charge is made up of coke, flux, and the copper-containing scrap. Air is introduced through tuyeres around the bottom of the shaft. Coke is used both as a fuel and as a reducing agent. Limestone or other materials is used for fluxing. Slag and concentrated alloy (cupola melt) are tapped near the bottom of the furnace. Energy required for this process is coke which is used both as a fuel and as a reducing agent.

The process produces: atmospheric emissions containing both gases and particulate matter (fume and dust), liquid wastes, and solid wastes. The composition of the particulate matter and gases from the cupola are variable because of the wide variety of scrap, slag, and skimmings added as the source of alloy and the various fluxes used. However, it is expected that the particulate matter will contain such metal values as zinc, lead, tin, copper, silicon, manganese, and some unburned coke, with the majority of the emissions being zinc because of its high vapor pressure at the cupolaing temperatures. The gases are expected to contain, in addition to the combustion products, such materials as sulfur oxides and halides.

Emission rates also vary because of the wide variation in the charge composition. For example, raw emission factors (furnace emission factors) were reported to range from 16.7 to 73.2 lb of particulate matter per ton of feed. Using a baghouse with a collection efficiency of 96.4% as the pollution control device, emission factors were reduced to within 0.9 to 4.1 lb of dust per ton of feed.

Since zinc oxide is normally the major component in the fume, particle shape is generally acicular, i.e., long needles with a large length to width ratio (~5 to 1). This shape is characteristic of zinc oxide. In some cases, the particle resembles a wheel with the rim removed. Particle size ranges from 0.03 to 0.3 micron which makes collection very difficult. The dust particles have an irregular shape and are expected to be much larger, possibly in the 5 to 10 micron range.

Liquid wastes are generated primarily as cooling wastewater from water-quenching of the slag and from cooling of the cupola. Solid wastes generated by this process are slags, fumes and dusts collected in the baghouse, spills, and sweepings. The wastewater is disposed of in settling ponds, recycled, or sent to sewers or nearby streams. Some of the solid wastes are recycled; other are discarded. Atmospheric emissions and liquid and solid wastes generated by the cupola can result in serious pollution problems.

Gravity Concentration Process: Metallic values are recovered from slags, drosses, skimmings, spills, and sweepings by the gravity concentration process, whereby the heavy (more dense) particles settle faster than the lighter particles in a water medium. The process steps are: grinding, screening, and gravity separation in a water medium. Energy required is that necessary to drive the equipment.

Only minor quantities of atmospheric emissions are expected from the gravity separation process. These are generated in the crushing and screening steps. However, large quantities of liquid wastes containing both suspended and dissolved solids are generated in the gravity separation step. The liquid wastes are probably treated in settling ponds and the water recycled or discharged to sewers or streams. This process does not pose an atmospheric pollution problem; however, if not controlled, liquid and solid wastes could cause water pollution problems.

Pretreatment of the scrap removes a portion of the metallic and nonmetallic impurities found in scrap and physically prepares the material for further processing by the smelting-refining manufacturing operation. In this operation, additional metallic and nonmetallic impurities are removed and the composition of the alloys adjusted to produce alloy ingots of the desired specifications. The processes employed to produce the alloy ingots are as follows: (1) reverberatory smelting-refining, (2) rotary smelting-refining, (3) crucible smelting-refining, and (4) electric crucible smelting-refining.

(1) Reverberatory Smelting-Refining Process: Brass and bronze alloys of desired specifications are produced by the reverberatory smelting-refining process. This is achieved by the removal of metallic and nonmetallic impurities from pretreated scrap and by the addition of alloying agents. The process steps are: (a) charging of the reverberatory furnace, (b) melting of the charge, (c) fluxing to remove the impurities, (d) alloying to adjust the com-

position of the melt, (e) pouring of the molten alloy, and (f) casting of the molten alloy into ingots.

This process produces large quantities of atmospheric emissions and liquid and solid wastes. The atmospheric emissions are composed of gases and particulate matter. The particulate matter includes such materials as carbon particles, entrained dust, unburned fuel, and metallic fumes. The gases in the atmospheric emissions include carbon dioxide and carbon monoxide from combustion of the fuel, sulfur oxides, halogens, and nitrogen oxides. Each processing step is a source of atmospheric emissions.

Charging is a major source of atmospheric emissions from the reverberatory smelting-refining process. The type and condition of the scrap are factors that affect the quantity and composition of the emissions. If the scrap is oily, the emissions will contain, in addition to the combustion products, unburned hydrocarbons and dust particles. When the scrap contains large quantities of highly volatile constituents such as zinc, heavy evolution of fumes will occur.

Emissions are also dependent upon factors such as location of the charging doors and methods of charging. Overhead charging results in heavy evolution of emissions containing both gases and particulate matter. End and side charging results in lower emissions. Batch charging produces large bursts of emissions that are almost impossible to control.

Melting of the charge is another major source of atmospheric emissions from this process. These emissions contain larger proportions of metallic fumes than those from the charging step. Emissions data are not available for each individual step. However, data collected in an earlier survey of industry revealed that particulate emission factors for charging-melting ranges from 11.7 to 59.0 lb/ton of feed charged to the furnace. These particulate emissions contained, in some cases, 60 to 90% zinc, probably as zinc oxide; in other cases, the particulate emissions contained 33.7 to 73.6% nonmetallic materials, 16.7 to 52.8% zinc, probably as zinc oxide, and 0.38 to 4.62% copper. No doubt, the particulate emissions also contained trace quantities of other metallic materials such as lead and cadmium.

Emission factors are not available for the gaseous emissions. However, large volumes of gases are evolved from combustion of the fuel and other sources. Reported particle size also varied over a wide range, possibly because the extremely fine particles may have passed through the filter medium used to collect the samples. In any event, particle sizes were reported to be less than 20 microns in some cases and in others, less than 1 micron.

The fluxing process step is a purification or smelting-refining step, whereby impurities are removed from the melt. These materials are removed by the addition of gaseous, liquid, or solid fluxes. Consequently, fluxing can be a major source of atmospheric emissions, especially in those cases where gaseous fluxes are used. These emissions may contain a wide variety of metallic constituents such as iron, manganese, silicon, aluminum, copper, zinc, and lead, along with those materials from the flux. These emissions may be volatilized from the melt as fume or carried from the melt by entrainment in the gaseous emissions. Emission factors for this step are not available.

In the alloying step, virgin metal or specialized scrap is added to adjust the composition of the melt. Normally, this step is not a major source of atmospheric emissions, unless high zinc alloys are being produced. In this case, large quantities of zinc fumes may be emitted.

Metal oxide fumes are produced as the hot metal is poured through the air into the molds. Other emissions are also produced, depending on the type of linings or coverings associated with the mold as it is filled with hot metal. Thus, the pouring step can be a significant source of atmospheric emissions and will vary depending on such factors as composition of the alloy, pouring temperature, pouring rate, and type of molds. Raw particulate emission factors collected in an earlier survey were noted to range from 0.4 to 10.9 lb of dust per ton of product. These emissions contained a high zinc content; in one case, 60 to 90%. Casting is a minor source of atmospheric emissions.

Thus, the reverberatory smelting-refining process for the production of brass and bronze ingots is a major source of atmospheric emissions. Raw particulate emissions factors (including both fumes and dusts) range from 125 to 140 lb/ton of feed. Composition will vary but, in general, the emissions will contain both metallic and nonmetallic constituents. The metallic emissions are composed of, in addition to a high concentration of zinc, varying concentrations of lead, tin, copper, silicon, and other metal values.

In one case, the dust collected in a brass and bronze smelter baghouse contained 45.0 to 77.0% zinc, 1.0 to 12.0% lead, 0.3 to 2.0% tin, 0.05 to 1.0% copper, 0.5 to 1.5% chlorine, and 0.1 to 0.7% sulfur. Gaseous emissions contain, in addition to the combustion products, volatilized organic materials such as oil and grease, sulfur oxides, nitrogen oxides, and hydrogen halides such as hydrogen chloride. The baghouse is the most widely used pollution abatement device. Liquid wastes generated by this process are primarily cooling wastewater. One source is cooling of the ingots by a water spray.

Significant quantities of solid wastes, slags, flue dusts, and baghouse dusts, are generated by this process. In one case, 10,000 lb of slag and 4,000 lb of flue dust were generated in the production of 178,000 lb of brass ingot product. Losses designated as gases, dusts, and others were estimated at 1,600 lb.

Thus, the reverberatory smelting-refining process produces large quantities of potential environmental pollutants. Therefore, this process has the potential for the production of serious pollution problems.

Rotary Smelting-Refining Process: This process, like the reverberatory smelting-refining process, produces brass and bronze ingots from the various types of scrap. Capacity of the rotary furnace used in this process ranges from several tons to approximately 50 tons of nonferrous metals. The process steps are: charging of the scrap through ports on the side of the cylinder, melting of the charge, fluxing to remove impurities, and alloying to adjust composition of the melt. Afterwards, the melt is poured and cast into ingots.

Energy required is the fuel used in the various process steps and electricity to drive the equipment. The rotary furnace is direct-fired using oil or gas as the fuel.

The process produces significant quantities of atmospheric emissions, liquid wastes, and solid wastes. The atmospheric emissions are generated in each of the process steps. Since the rotary smelting-refining process is similar to the reverberatory smelting-refining process in that both are direct-fired and the same process steps are employed, the quantity, composition, and particle morphology of emissions are comparable. For example, the raw emissions factor for the rotary smelting-refining process was reported to be 147 lb of particulate matter per ton of feed, whereas the factor for the reverberatory smelting-refining process was reported to by 156.9 lb of particulate matter per ton of feed.

Liquid wastes generated by this process are the wastewater from cooling of the ingots during casting. Solid wastes include wastes such as slag and baghouse dusts. Thus, the process has a high potential for the production of serious pollution problems if emissions are not controlled.

Crucible Smelting-Refining: The crucible smelting-refining process is employed for producing brass and bronze alloys from clean, well segregated scrap and for refining specialized alloys. The process steps are the same for this process as for the other smelting-refining processes.

Energy required is the fuel necessary to fire the furnace and the electricity to drive the equipment. The furnace is indirect fired using gas or oil as the source of fuel; thus, the charge never comes in contact with the flame.

The process produces a modest quantity of potential environmental pollutants. Atmospheric emissions are generated in each processing step but at a rate signigicantly less than for the reverberatory smelting-refining process. For example, furnace emissions factor for crucible-smelting-refining was reported to range from 6.5 to 12.3 lb of fume per ton of feed, whereas furnace emissions factor for reverberatory smelting-refining was reported to range from 125 to 140 lb of dust per ton of feed. Using baghouses with collection efficiencies of 93.7 and 96.2%, respectively, the atmospheric emissions were lowered to 0.24 pound per ton of feed for the low zinc alloy and 0.80 lb per ton of feed for the high zinc alloy. Liquid wastes are generated in the casting of the ingots. Solid wastes generated by this process include slags and baghouse dusts. Composition and particle morphology of atmospheric emissions and composition of the liquid and solid wastes should be comparable to those from the other smelting-refining processes.

Although this process produces significantly less potential environmental pollutants than the above smelting-refining processes, improper disposal of the pollutants and/or emitting them to the environment can result in atmospheric and water pollution problems.

Electric Crucible Smelting-Refining Process: This process is used to produce special high-grade brass and bronze alloys from treated scrap. The process involves the same process steps as employed in the other smelting-refining processes. Energy required is the electricity to drive the equipment and process the alloy.

Electric crucible smelting-refining produces significant quantities of atmospheric emissions, solid wastes, and liquid wastes. These emissions and wastes are generated in the same processing steps as in the other processes. Also, the composition will be similar. The primary difference is the atmospheric emissions do not contain combustion products since the source of heat for the electric smelting-refining process is electricity rather than fossil fuels. Thus, the quantity of atmospheric emissions, including both gases and particulate matter, are significantly reduced. Data are not available for the gaseous emissions. However, furnace particulate emission factor was reported to be approximately 2.6 lb per ton of feed. The baghouse removed 96.0% of the particulates. Other wastes generated by the electric crucible smelting-refining process include liquid wastes from cooling of the ingots during casting and solid wastes such as slags and baghouse dusts.

Although this process emits significantly less particulate matter than other smelting-refining processes, there is the potential for pollution of air and water if atmospheric emissions and liquid and solid wastes are not disposed of properly.

CADMIUM

Recovery from Fume from Roasting of Zinc Concentrates

Cadmium-bearing materials obtained as by-products by zinc, lead, and copper processing include the dusts from roasters, sintering machines, furnaces, fuming, and Waelz kilns. They also include the blue powder from the distillation of zinc, the cadmium fraction in the rectification of zinc, and the copper-cadmium filter cakes derived from purification of sulfate by cementation. The hydrometallurgical process shown in Figure 49 may be used on any of these materials.

The hydrometallurgical process for cadmium involves leaching the cadmium-bearing materials with sulfuric acid to transfer the cadmium to solution. Subsequently, the cadmium sulfate solution is treated to remove impurities such as arsenic, antimony, and iron. The cadmium sponge is then precipitated, usually with zinc, or it is electrodeposited from the solution. The sponge is then further treated by electrolytic or distillation methods, yielding a final product of high-purity cadmium. Special high-purity cadmium may be obtained by vacuum distillation, zone recrystallization, ion exchange, and other techniques.

It is perhaps worth noting that the metallurgy of cadmium pinpoints the split-off point for processing cadmium residues at the dust collectors. This is the point of economic decision where additional costs must be balanced by expected revenues from the cadmium metal. However, this point is not entirely independent of what has gone before. The amount of cadmium in the zinc concentrates is determined not only by the cadmium grade in ores but also by prior metallurgical processes. Selective flotation practices will determine how much cadmium passes to the zinc concentrate during the milling of lead-zinc ores. For maximum transfer of cadmium into the zinc concentrate it is necessary that the depression of the zinc-cadmium minerals in the lead flotation cycle should be as effective as possible. It is then equally necessary that vigorous prolonged activation take place before zinc flotation. Likewise, the degree of saturation of individual particles in the flotation process affects the concentration of cadmium in the separate products of flotation. The metallurgy and the economics of cadmium, therefore, begins at the mill though the actual economic decision to produce cadmium is made at the smelter or refinery.

FIGURE 49: BY-PRODUCT CADMIUM SUBSYSTEM

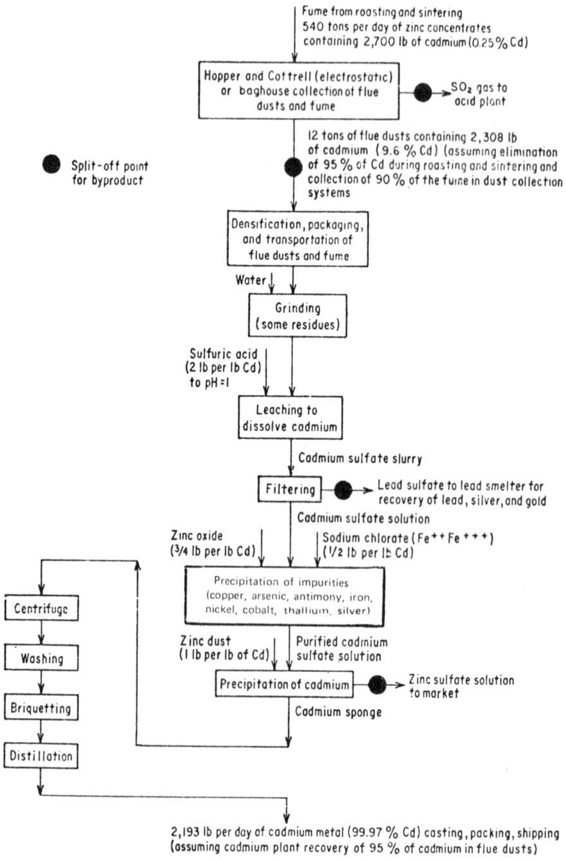

Source: Bureau of Mines I.C. 8570

The first step in the processing of cadmium-bearing materials shown in Figure 49 is the densification, packaging, and transportation of the cadmium fume. Inclusion of this step may be questioned because many cadmium plants are in close proximity to the source of raw material. Nevertheless, there is some transportation of flue dusts and concentration of cadmium production. Cadmium dusts produced at the New Jersey Zinc Smelter at Depue, Ill., for example, were shipped to Palmerton, Pa., for recovery of cadmium. Also, a considerable portion of the U.S. cadmium supply originates in flue dusts imported from smelters in Mexico.

The cadmium-bearing flue dusts are treated with sulfuric acid to dissolve the cadmium. The method shown is that of the acid process. Sulfuric acid is added in amounts to insure extraction of cadmium and other acid soluble components into solution. Lead remains as a solid residue as sulfate, which is subsequently filtered and shipped to a lead smelter for recovery of lead, gold, silver, and selenium, tellurium, and indium, if present.

The cadmium sulfate solution is purified by neutralization. Neutralization of the solution causes iron, arsenic, and various other elements to precipitate. The neutral agent in the hypothetical flowsheet shown is zinc oxide, which will precipitate iron hydroxide. The ferrous compounds are converted to ferric compounds by oxidation with sodium chlorate, thus promoting complete precipitation of the iron. The zinc oxide also precipitates arsenic, antimony, indium, gallium, germanium, and thallium, if they are present. Other impurities may require additional processing. Copper, for example, may be precipitated with zinc dust.

After the cadmium sulfate solution is purified, cadmium is precipitated by the addition of metallic zinc dust. The cadmium deposit is a spongy mass that can be separated from the zinc sulfate solution by centrifuging or by filtering. In some cases, the cadmium sponge may be redissolved in sulfuric acid and reprecipitated to obtain a cleaner product for distillation. The zinc remains in solution as zinc sulfate. In the hypothetical flowsheet shown in Figure 49 this step of the process is carried out by the addition of zinc dust. In some operations zinc plates are used, but this procedure is much less common.

In the cadmium subsystem, the cadmium sponge is purified by distillation rather than by electrolytic methods. The cadmium sponge is washed, compressed into briquets, and placed in oil-fired retorts. At the exit of the retort is a cast-iron cylinder in which the cadmium vapors condense. The distillation is a batch process carried out in a 20 hour cycle. The last 10% of the charge is redistilled.

The subsystem shown is assumed to handle 12 tons of flue dust per day containing 2,308 pounds of cadmium. These flue dusts are the result of roasting and sintering 540 tons per day of zinc concentrates containing 2,700 pounds of cadmium. The greatest losses in the system shown are in the elimination of cadmium from the concentrates (95% recovery of cadmium) and the capture of cadmium-bearing material in the dust collection system (90% recovery of cadmium). Thus overall recovery of cadmium at the first stage is 85%. A major factor in this recovery is the efficiency of the dust collection system.

The recovery in the cadmium-processing system istself is assumed to be 95%, which is believed representative in the industry, according to A. Patrick, Jr., H.J. Bennett, K.E. Starch and R.C. Weisner in Report I.C. 8570, Washington, D.C., U.S. Bureau of Mines (1973).

Recovery from Incendiary Alloys

The reader is referred to the same heading under "Magnesium" for a discussion of Bureau of Mines work on recovery of magnesium and cadmium from incendiary alloys.

Recovery from Scrap

During 1971, total U.S. cadmium production was 7.9 million pounds of which the secondary cadmium segment production was 400,000 pounds or about 5% of the total.

Cadmium-bearing raw material is available both as old and new scrap. Examples of old scrap are: cadmium plated parts from junked automobiles, nickel-cadmium batteries, cadmium coated electrical contact points, and used cadmium alloys. New scrap, which constitutes only a small percentage of the total, consists of factory rejects from manufacturing of bearings etc., Ni-Cd battery rejects, and scrap from plating (electrodeposition) cells, and scrap solder used to solder aluminum.

In the United States, cadmium is not recovered from junked automobile parts because of economics. Also, used nuclear reactor control rods (containing 80% cadmium and 20% silver and indium) are not processed by the secondary cadmium industry. Instead these rods are sent to fuel processors (e.g., Mallinckrodt Chemical Works, St. Louis, Missouri).

Products of the cadmium segment of the secondary materials industry are as follows: recast cadmium metal (sticks and balls) and recast cadmium alloys.

The three manufacturing operations employed in cadmium scrap processing are scrap pretreatment, smelting/refining, and casting. Of these, the major operation, smelting/refining, involves pyrometallurgy.

The only pretreatment process used with cadmium scrap is vapor degreasing. This process removes oil and grease adhering to the scrap alloys. The process steps are: heating solvent to generate vapors (perchloroethylene), circulating solvent vapors through the scrap in a pot, and condensing solvent for reuse. Heat energy is the main requirement of this process.

The major waste generated is the liquid waste, grease/oil residue from the solvent regeneration pot. The quantities of this waste are very small and are usually either incinerated at the municipal incinerator or dumped into sewer lines. Other waste generated is atmospheric emission composed of organic vapors. The potential for pollution from this process is not significant, and is regulated in many states. Three smelting-refining processes are used to recover cadmium from scrap.

Alloy Smelting/Refining Process: This process purifies the pretreated alloy scrap. The process steps involved are: heating the alloy scrap, adding the flux, skimming the slag, and pouring the molten metal into molds for casting. Heat energy is utilized in the process.

Wastes generated are atmospheric emissions containing metal fumes and solid wastes consisting of slag. The fumes are not significant to cause any pollution. The slag is recycled and ultimately disposed of in a landfill. Thus, the process has a low pollution potential.

Retort Smelting/Refining Process: This process recovers pure cadmium by distilling cadmium alloys. The process steps are: charging the retort, melting the charge, vaporizing the cadmium, condensing the cadmium, pouring the molten metal, casting the ingots, and discharging retort residue for shipping to residue processors.

Heat energy is utilized in the process. The potential pollutant generated by this process is atmospheric emissions containing cadmium fumes. If properly controlled, this process has little potential for producing serious pollution problems.

Melting Process: This process purifies the scrap cadmium metal collected from electrodeposition cells as sticks and rods. These rods and sticks are usually contaminated with acid salts or mixed with zinc occasionally.

The process steps are: charging scrap to pot, melting the charge, adding the flux, chlorinating when required for zinc removal, skimming the slag, pouring into molds, and casting the ingots. The process utilizes heat energy.

Wastes generated in the process are: atmospheric emissions containing cadmium fume, zinc chloride dust when dezincing operation is done, solid waste consisting of slag, and liquid waste composed of cooling water which may be recycled. The zinc chloride dust is collected with a baghouse. Slag generated is about 2,000 pounds per year in a plant producing 200,000 pounds of cadmium per year. The slag which contains some cadmium is disposed of in a landfill. Thus, this process has the potential for the production of pollution problems.

The casting operation is a simple operation and consists of one process step. The molten alloy and the molten metal from the retorting and melting processes are cast into required shapes by this process. The processing steps consist of pouring the molten material into molds and casting with water cooling. Electrical energy is used to circulate cooling water and auxiliary equipment. No wastes are generated in the process. The process has no pollution potential.

Recovery from Scrap Batteries

A process developed by D.A. Wilson and B.J. Wiegard, Jr., Report R.I. 7566, Washington, D.C., U.S. Bureau of Mines (1971) involves the recovery of cadmium and nickel from scrap batteries. Figure 50 is a flow diagram of the overall recovery process.

A cyclic leaching process was developed for selectively leaching and recovering cadmium from nickel-cadmium scrap battery waste. The six major steps in the process are the following: washing the plates to remove KOH electrolyte, roasting at 550° to 600°C to oxidize metallic cadmium and decompose cadmium and nickel salts, leaching with an ammonium nitrate solution, precipitation of the leached cadmium from solution as cadmium carbonate using CO_2 gas, removal of CO_2 from the lixiviator by adjusting to pH 4.5, heating, and evacuating the system, and removal of small amounts of leached nickel from the lixiviator using a solvent extraction technique. The regenerated lixiviator is then recycled to leach additional plates. Cadmium extraction was 94% and the cadmium carbonate product contained 0.14% nickel and 0.12% cobalt impurity. The leached plates were smelted to a high nickel-low iron alloy with no detectable amounts of cadmium.

FIGURE 50: PROCESS FLOWSHEET FOR CADMIUM AND NICKEL RECOVERY FROM SCRAP BATTERIES

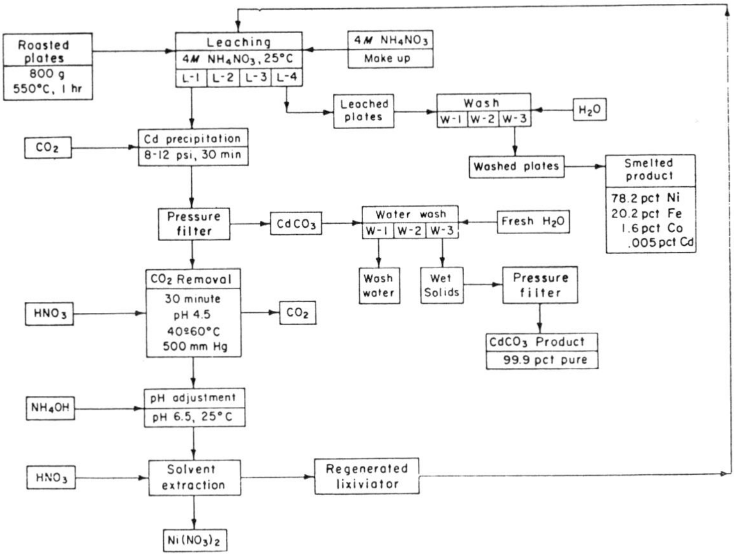

Source: Bureau of Mines R.I. 7566

CHROMIUM

Recovery from Liquid Wastes

Recovery from Cooling Tower Blowdown Water: A process developed by *A. Geinopolos, M.K. Gupta, and W.J. Katz; U.S. Patent 3,625,885; December 7, 1971; assigned to Rex Chainbelt Inc.* is one in which process water containing hexavalent chromium is pretreated to bring it to a state of dichromate dominance and the water is then subjected to reverse osmosis using a loose membrane. The membrane allows passage of most of the dichromate ions with the product water while concurrently effecting a significant reduction in the hardness and dissolved solids levels. The softened product water containing dichromate ions is recyclable with the recirculated process water and the wasted fraction contains substantially lower amounts of potentially harmful chromium.

Recovery from Pigment Manufacturing Wastewaters: A process developed by Mineral Pigments Corp. of Beltsville, Maryland and described by D.J. Robinson, H.E. Weisberg, G.I. Chase, K.R. Libby, Jr. and J.L. Capper in Report PB 233 641, Springfield, Va., Nat. Tech. Information Service (June 1974) can be used to remove chromate from wastewaters discharged from a zinc yellow pigment manufacturing plant. It was shown that the recovered chromate solution could be recycled into product manufacture without sacrificing product quality.

Strongly basic ion-exchange resins were shown to exhibit a preference for dichromate over many other anions in water solution in laboratory studies. From these laboratory studies, a full-scale ion-exchange treatment plant was designed, constructed, and demonstrated. The chromate composition of the plant effluent is being reduced from 2,700 ppm to one to two ppm.

The treatment system was designed to treat 60 gallons per minute of influent and to discharge an effluent which is within statutory limits for pH and for heavy metal content. The plant was designed to require minimal manual supervision. The steps in treatment and in resin regeneration are performed automatically and the control system is interlocked to make it fail-safe. Operators are required only to make up regeneration solutions, to clean pump strainers and filters, to answer to alarms and occasionally to differentiate between turbidity and color as seen by the colorimeter.

Recovery from Spent Chromic Acid Etching Solutions: A process developed by *R.R. Dougherty; U.S. Patent 3,531,262; assigned to Control Data Corp.* is one in which spent or used aqueous chromic acid etchants are treated to separate chromic acid from by-products of the etching process permitting reuse of the chromic acid and concentration or isolation of the by-products for waste disposal and/or recovery of the metal values contained therein.

In a typical process, chromic acid and water are separated from the by-products by acetone extraction. The extract is then stripped of the acetone to produce an aqueous solution of chromic acid which can be reused in the etching process. The raffinate is rich in the by-products of the etching process.

Recovery from Wash Water After Electrotinplating: A process developed by *R.B. Smith; U.S. Patent 3,847,757; November 12, 1974; assigned to National Steel Corporation* is one in which chromium values are recovered from wash water following a chemical treatment step in electrotinplating lines in a process comprising treating in a series of ion exchange zones. Figure 51 is a block flow diagram of this process.

As shown in the figure, steel strip having been preliminarily treated, for example in electrolytic cleaning and pickling steps, follows a travel path **10** and is subjected to a tinplating step **12**. This tinplating step can employ for example a halogen-type electrolyte such as, for example, an aqueous stannous fluoride solution containing minor amounts of additives. In such case, a plurality of tanks or plater cells are utilized, and the strip passes through these tanks in series. Each tank has a conductor roll which maintains the strip cathodic and the tanks contain grids which are anodic. The strip follows a horizontal travel path through each tank and is barely immersed in the electrolyte and is plated on the bottom side only in its passage through a tank. After the strip has passed through a number of the tanks, the travel path turns upward and the strip direction is reversed so that the original top of the strip becomes the bottom whereby both sides of the strip are plated.

Subsequent to the tinplating **12**, the strip is treated in a reclaim step **14**. In this reclaim step, rinse water as denoted by arrow **16** is added into a tank denoted a reclaim tank, and the strip is passed through the rinse water to recover plating solution residually wetting the strip. The resulting rinse solution is drained to plater storage, that is to the storage tank for plating electrolyte, as denoted by arrow **18**.

The strip is then subjected to a rinsing step **20**. In this step, plating solution still residual on the strip is rinsed therefrom. Rinse water provision is denoted by arrow **22**. Resulting rinse solution is diposed to waste as indicated by arrow **24**. The strip then progresses along travel path **10** through a drying step **26**.

Subsequent to the drying step **26**, the strip continues along travel path **10** and passes through a reflow tower where it is subjected to radio frequency energy to melt the electrodeposited tin.

Following passage through the reflow step, the strip passes through a quenching step **30**. In this step the strip is passed through a quench tank containing water sprays and also containing a body of liquid; the water line to the sprays is denoted by arrow **32**. Passage of the strip by the sprays and through the liquid controls the rate of freezing of the tin layer to produce the desired alloy coating and enhance brightness. The resulting solution is disposed to waste as denoted by arrow **34**.

Following the quenching step, the strip is subjected to a chemical treatment step **36** where a hexavalent chromium oxide coating is applied by electrolytic action to stabilize the surface of the tinplate.

FIGURE 51: PROCESS FOR CHROMIUM RECOVERY IN ELECTROTINPLATING

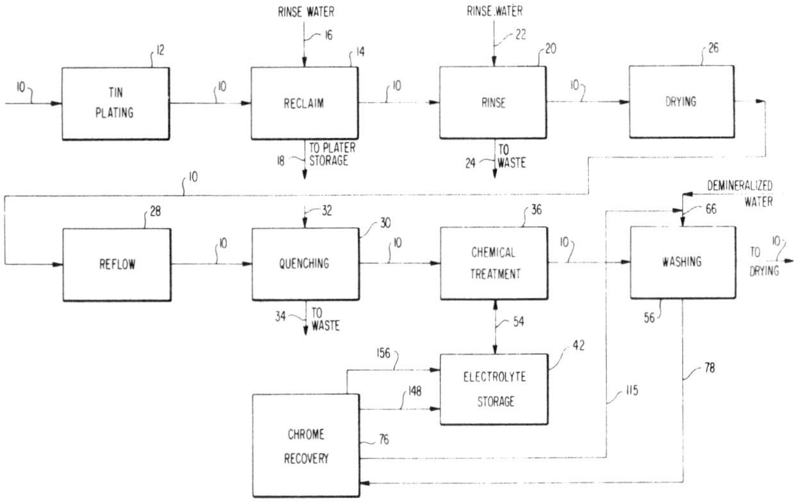

Source: R.B. Smith; U.S. Patent 3,847,757; November 12, 1974

The electrolyte is introduced into the chemical treatment tanks from an electrolyte storage tank **42**, and recirculation is carried out between tank **42** and the chemical treatment tanks via line **54**.

Subsequent to chemical treatment the strip follows travel path **10** through a spray washing step **56** whereby electrolyte from the chemical treatment step residual on the strip is washed therefrom and recovered. Demineralized water is added for washing purposes via line **66**.

The solution recovered from the washing step **56**, that is the solution accumulating in the bottom of tank **58** is passed to a chrome recovery step **76** as indicated by arrow **78**.

The chrome recovery step involves:

(a) Treating the aqueous solution with cation exchange resin in a cation exchange zone to replace cations with hydrogen ions to produce an aqueous effluent containing hydrogen ions and anions including anions containing hexavalent chromium;

(b) Treating the effluent with anion exchange resin in an anion exchange zone to replace anions with hydroxyl ions thereby recovering hexavalent chromium in the resin in the anion exchange zone and producing a demineralized water effluent;

(c) Regenerating the resin in the anion exchange zone to produce an effluent from the anion exchange zone comprising an aqueous solution containing sodium ions and anions containing hexavalent chromium;

(d) Treating effluent from step (c) with cation exchange resin in a cation exchange zone to replace cations with hydrogen ions and produce an effluent and passing the produced effluent into an electrolyte storage zone;

(e) Passing effluent from step (c) into an electrolyte storage zone without an intermediate cation exchange treatment.

Line **115** carries demineralized effluent from the anion exchanger for use in washing step **56**. Regeneration effluent (aqueous chromic acid) is carried by lines **148** and **156** to electrolyte storage **42**.

The apportionment between the first stream and the second stream, that is between lines **156** and **148** is carried out to provide a pH of 3.5 in the electrolyte storage zone. At such pH, the chromium containing anions are mostly in the form of dichromate. Thus, the apportionment is carried out to provide an aqueous solution of sodium dichromate at a concentration and a pH the same as that of the electrolyte for the chemical treatment step.

Recovery from Wastewaters: A process described by O.P. Case of Anaconda American Brass Co. in Report EPA-670/2-74-008, Washington, D.C., U.S. Environmental Protection Agency (Jan. 1974) involves chromium recovery from wastewaters by cementation.

Bench-scale experiments utilizing the "cementation" reaction (i.e., electrochemical reduction by contact with a metal of higher oxidation potential) for the precipitation of copper and the reduction of hexavalent chromium in industrial waste streams were performed. Reductants included: soft iron shot 4.37 mm diameter, particulate iron approximately -400 mesh and alkali earth metal-silicon alloys in granular form.

Soft iron shot was found to be an effective reductant. Below pH 3.0 chromium reduction is rapid and quantitative. Copper cementation is slower and is influenced more strongly by oxygen, speed of mixing, and available iron surface. Previously deposited copper does not significantly impede further copper deposition. Copper catalyzes the chromium reduction. Particulate iron was not as satisfactory as the iron shot due to clumping of the particles when they became coated with copper. Silicon alloys were effective for removal of copper, zinc and trivalent chromium, but not reliable for hexavalent chromium since the reaction is not primarily a reducing one.

Recovery from Solid Wastes

Recovery from Steel Furnace Flue Dusts: Over 5 million lb of chromium, 1 million lb of nickel, and 90,000 lb of molybdenum are lost annually in the dusts generated in the production of stainless steel. Additional losses of these metals in the form of mill scale and grinding swarf may be double the dust losses.

H.E. Powell, W.M. Dressel and R.L. Crosby in Report of Investigation No. 8039, Washington, D.C., U.S. Bureau of Mines (1975) describe a process for recovering these and other metals in the form of an alloy steel suitable for recycling. Zinc and lead, when present in the wastes, are recovered as by-products in the form of oxide fumes. The laboratory-scale process (10 to 25 lb charges) recovers up to 95% of the chromium, 100% of the nickel and 100% of the iron. The resultant alloy contains from 10 to 20% chromium, 6 to 12% nickel, and 55 to 65% iron, with lesser amounts of molybdenum and manganese. The alloy is essentially free of zinc, lead, and sulfur. The zinc-lead fume obtained from the treatment contains about 40% zinc and 20% lead and represents essentially complete recovery of these metals.

Recovery from Waste Sludges from Metal Finishing: A process developed by *C.O. Weiss; U.S. Patent 3,552,917; January 5, 1971; assigned to M & T Chemicals Inc.* involves recovering the chromium content from waste sludge produced from the separation of precipitated solids from the waste effluent streams of metal finishing processes, the chromium content of the sludge being in the form of barium chromate. The above is achieved by forming an aqueous slurry of the sludge, adding sulfuric acid thereto, agitating the resulting admixture for a period of time at ambient temperature, separating the insolubles therefrom, adding to the filtrate from the separation an amount of a carbonate selected from the group consisting of Ca^{++}, Sr^{++}, Ba^{++}, and Pb^{++} to precipitate out a substantial portion of the sulfate ion from the solution, again separating the insolubles from the reacting environment, adding to the filtrate of the second separation step strontium carbonate to precipitate out the remaining sulfate ion, and separating the insolubles from the reacting environment to obtain a solution containing all the CrO_3 content from the initial extraction.

Example: In this example, a representative sludge was prepared of a 55 kg sample of a mixture of $BaCrO_4$ (40%) and $BaCO_3$ (60%). This mixture was slurried with water to give a readily stirrable mixture. To the slurry was added 25.1 kilograms of H_2SO_4 (13.9 liters), and the admixture volume was adjusted to about 85 liters. The mixture was agitated with no external heating for 2½ hours.

The insolubles were then filtered off and the cake washed with several portions of water until the solution analyzed 5.5 g/100 ml of CrO_3 and 7.8 g/100 ml of H_2SO_4. The solution was then charged with 8.0 kilograms of $CaCO_3$ and agitated at room temperature for about 10 minutes. The insolubles were filtered off and the cake washed as described above to give a solution volume of about 115 liters. This solution still contained 2.0 kg of H_2SO_4, and this solution in turn, was charged with 4.5 kg of $SrCO_3$ and agitated at room temperature for about 15 minutes.

The insolubles were filtered off and the cake washed as described above to produce a final volume of 130 liters. The resulting solution contained no H_2SO_4 and virtually all the CrO_3 obtained in the initial extraction.

As will be seen from the above example, the process is effective for obtaining all of the chromium content from a slurried mixture containing barium chromate solids and with the solution obtained having substantially no sulfuric acid contained therein, thus making the final solution from the process particularly appropriate as a raw material for recirculation into a chromium electroplating process.

The recovery is performed in the absence of any high temperature operating conditions which reduces the need for expensive evaporative equipment, and in the presence of relatively dilute solutions for carrying out the various extraction steps so as to eliminate the requirement of expensive corrosion-resistant equipment.

COBALT

Recovery from Liquid Wastes

Recovery from Leached Nickel Ore Tailings: A process developed by *T.K. Wiewiorowski; U.S. Patent 3,848,054; November 12, 1974; assigned to Freeport Minerals Company* is one in which overall cobalt recoveries obtainable from a conventional ammoniacal leaching process for the recovery of nonferrous metals from reduced nickeliferous ores are significantly improved by the addition of a secondary cobalt recovery circuit in which the leached ore tailings of the process, which are normally discarded to waste with a substantial cobalt content, are treated with an ammoniacal solution in the presence of sulfite ions to recover a substantial portion of the cobalt values present in the tailings. Figure 52 is a flow diagram showing the application of the recovery process.

Referring to the drawing, stream **1** represents the leached ore tailings discarded from a conventional ammoniacal leaching process (not shown). The term leached ore tailings, as used herein, means the slurry that remains after the washing operation has been completed in a conventional ammoniacal leaching process. This slurry is normally the underflow from the last thickener of a conventional ammoniacal leaching process. The leached ore tailings usually comprise from about 5 to 60% by weight of a solid phase in the form of finely divided reduced ore, and from about 40 to 95% of a liquid phase in the form of an aqueous alkaline ammoniacal solution having a pH value higher than 7.

The nickel content of the liquid phase of this slurry is normally so low that additional nickel values cannot be economically recovered by additional washing steps. In fact, the liquid phase of the tailings is normally substantially free from both nickel and cobalt values. The liquid phase illustratively contains between about 0.05 and 0.5 grams of nickel per liter and between about 0.005 and 0.1 grams of cobalt per liter. These values are typically between about 0.10 and 0.30 for nickel and between about 0.01 and 0.05 for cobalt. This liquid phase normally also contains a considerable amount of ammonia, in the form of ammonium hydroxide and ammonium carbonate, and it is the common practice in the conventional ammoniacal leaching process to recover this ammonia, and reuse it to make up the wash liquor of the process.

The solid phase of the leached ore tailings is normally discharged as waste. Although a substantial amount of the nickel and cobalt values originally present in the solid phase have been removed, the solid phase still illustratively contains, on a dry basis, between about 0.1 to 0.6% by weight nickel and between about 0.02 to 0.15% by weight cobalt, on the same basis. In most cases the solid phase contains between about 0.2 to 0.4% nickel and 0.04 to 0.08% cobalt. The secondary cobalt recovery circuit significantly improves the overall cobalt recovery of a conventional process by recovering a substantial portion of the cobalt values which normally remain tenaciously associated with the solid phase of the leached ore tailings of the process. These cobalt values are completely lost in prior art ammoniacal leaching processes when this solid phase is discharged as waste.

Referring again to the drawing, the slurry of leached ore tailings **1** is subjected to a sulfite ion treatment in absorption tank **2**. The sulfite ions can be provided by introducing into the slurry **1** any compound **3** which will impart sulfite ions to the slurry. Thus the compound can be a salt which directly dissociates into sulfite ions upon introduction into the slurry or a material such as sulfur dioxide, which reacts with a component of the slurry (water) to generate sulfite ions in situ. Illustratively, any soluble sulfite salt can be employed in the method of this process.

The term soluble sulfite salt, as used herein, means any sulfite salt having a solubility in water in excess of about 0.1 g/l of water at 77°F. Exemplary of such sulfite salts are the ammonium sulfites and ammonium bisulfites such as ammonium sulfite $(NH_4)_2SO_3$ and ammonium bisulfite NH_4HSO_3, and the alkali metal sulfites and alkali metal bisulfites, such as sodium sulfite Na_2SO_3, sodium bisulfite $NaHSO_3$, and the like. Other sulfite ion source compounds such as sulfur dioxide and sulfurous acid can also be employed. Sulfur dioxide

FIGURE 52: FLOW DIAGRAM FOR COBALT RECOVERY FROM LEACHED ORE TAILINGS

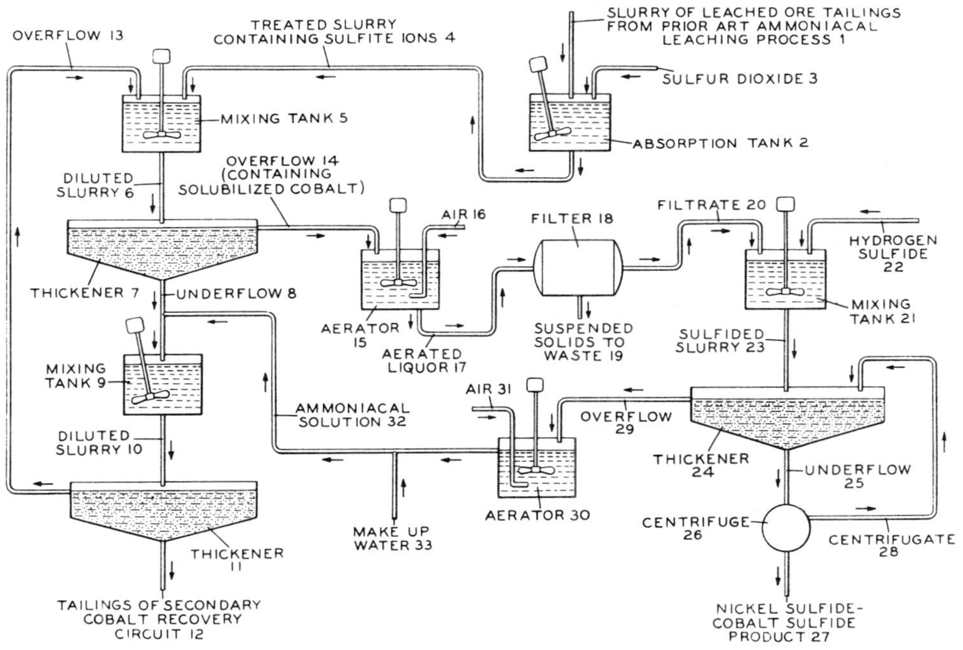

Source: T.K. Wiewiorowski; U.S. Patent 3,848,054; November 12, 1974

is a preferred source of sulfite ions as shown in the drawing. The amount of sulfite-providing compound used to treat slurry **1** can vary widely depending on a number of factors, such as, for example, the nature of the compound used. Illustratively, enough compound is introduced into the slurry to provide a sulfite ion content of between about 0.01 and 10% and preferably between about 0.02 and 1% by weight, expressed as SO_2 of the total weight of slurry treated.

The selection of temperature for the sulfite ion treatment is not critical. Illustrative temperatures range from about 50° to 200°F. The means for contacting the slurry with the sulfite ion-providing compound can likewise vary widely. It would be possible, for example, to inject the compound which is to provide the sulfite ions directly into stream **1**. A contact vessel with provisions for thorough mixing is desirable, however, and illustrative retention times of between about 30 seconds and 24 hours, and preferably between about 1 minute and 3 hours, in such a vessel have been found to be effective for the sulfite ion treatment. The treatment with sulfite ions solubilizes cobalt values associated with the solids of the leached ore tailings into the liquid phase of the slurry.

The treated slurry **4** containing sulfite ions is then fed to a countercurrent washing system where it is washed with an ammoniacal solution **32** which is a recycle stream in the secondary cobalt recovery circuit, as explained hereinbelow. Thus, treated slurry **4** is blended with overflow **13** from thickener **11** in mixing tank **5** and the resulting diluted slurry **6** is fed to thickener **7**. The solids enriched underflow **8** from thickener **7** is blended with ammoniacal solution **32** in mixing tank **9** and the resulting diluted slurry **10** is fed to thickener **11**.

Solids depleted overflow **13** from thickener **11** is used to dilute slurry **4** in tank **5** as explained above, and the solids enriched underflow **12** represents the tailings of the secondary cobalt recovery circuit. Solids depleted overflow **14** from thickener **7** represents the liquid phase separated by the countercurrent washing system, and contains the cobalt values solubilized from stream **1**. Thickener **7** could be replaced by two or more countercurrently operated thickeners as long as the ultimate result was to separate slurry **6** into overflow **14** and an underflow **8** for feeding to final thickener **11**. The separated solids depleted overflow **14** is aerated in aerator **15** with an oxygen-containing gas, such as air **16**. The aeration of the liquid phase **14** destroys any sulfite ions that might interfere with the sulfite operation which follows. Enough oxygen-containing gas is provided to oxidize sulfite ions present in the liquor **14** in accordance with the following reaction:

$$SO_3^{=} + \tfrac{1}{2}O_2 \rightarrow SO_4^{=}$$

The aerated liquor **17** is filtered **18** to remove suspended solids **19**, and the filtrate **20** then fed to mixing tank **21** where it is contacted with a sulfiding agent **22** to precipitate the cobalt values as cobalt sulfide. Any sulfide-providing compound may be used as the sulfiding agent for filtrate. Illustrative compounds include ammonium sulfide $(NH)_2S$, ammonium hydrosulfide NH_4HS, hydrogen sulfide H_2S, and the like. Hydrogen sulfide is a preferred sulfiding agent as shown in the drawing. Enough sulfiding agent is used to provide between 100% and 150%, and preferably about 150%, of the stoichiometric amount necessary to convert all of the nickel and cobalt present in filtrate **20** to nickel sulfide and cobalt sulfide, respectively, in accordance with the following reactions:

$$Ni^{++} + H_2S + 2OH^- \rightarrow NiS + 2H_2O$$

$$Co^{++} + H_2S + 2OH^- \rightarrow CoS + 2H_2O$$

The sulfided slurry **23** proceeds to thickener **24** to be separated into a solids enriched underflow **25** and a solids depleted overflow **29**. Underflow **25** is fed to centrifuge **26** to produce a mixed nickel-cobalt sulfide enriched underflow **27** representing the recovered cobalt values. Solids depleted centrifugate **28** is returned to thickener **24** for blending with sulfided slurry **23**. Overflow from thickener **24** is aerated with an oxygen-containing gas, such as air **31** in aerator **30** to destroy any sulfide ions present. Any suitable contact vessel may be used to aerate overflow **29**. Enough oxygen-containing gas is provided to oxidize the sulfide ions present, in accordance with the following reaction:

$$S^{=}+2O_2 \rightarrow SO_4^{=}$$

The resulting aerated stream **32** is an aqueous ammoniacal solution which is then reused as the wash medium for the countercurrent washing system of the secondary cobalt recovery circuit. Make-up water **33** can be added to the overflow **29** either before or after it is aerated to compensate for any water losses encountered in the circuit. Many modifications of the secondary cobalt recovery circuit will be apparent to those skilled in the art. For example, the positions in the circuit of aerator tank **15** and filter **18** can be interchanged, and the suspended solids removed from separated liquid phase **14** before it is aerated. Or filter **18** can be substituted with other suitable means for separating suspended solids, such as, for example, a settling tank. Or the combination of thickener **24** and centrifuge **26** can be replaced by a filter, or with a combination of a thickener and a filter, or with two thickeners, etc., as long as sulfided slurry **23** is separated into a nickel-cobalt sulfide precipitate **27** and recycle stream **29**. These and other modifications are all possible within the judicious choice of those skilled in the art.

Recovery from Solid Wastes

Recovery from Oxidation Catalyst Residues: A process developed by *Y. Ichikawa; U.S. Patent 3,525,762; August 25, 1970; assigned to Teijin Limited, Japan* involves recovering cobalt salt catalysts of lower aliphatic monocarboxylic acids of 2 to 4 carbon atoms, and comprises heating such cobalt salt catalysts at a temperature from 50° to 300°C in the presence of an aqueous solution of a lower aliphatic monocarboxylic acid of 2 to 4 carbon atoms in which

(a) not less than 2 mols of the aliphatic monocarboxylic acid, and not less than 4 mols of water, per gram-atom of the cobalt are present, and
(b) the concentration of the aliphatic monocarboxylic acid in the total aqueous solution present in the system being from not less than 0.2 mol percent to not more than 80 mol percent.

Such cobalt salt catalyst is one used for the oxidation of hydrocarbons or their oxidation derivatives with molecular oxygen, the used catalyst being in the form of a trivalent form of the cobalt salt.

Recovery from Scrap: The volume of scrap available to the cobalt segment of the secondary nonferrous metals industry is estimated at two million pounds annually. The majority of this is recovered as cobalt-base or nickel-base alloy in the U.S. Some of the refined high grade cobalt scrap is exported to Japan for recovery of pure metal. Approximately 5 to 10% of the scrap cobalt is used in the production of chemicals. Raw materials consist of cobalt bearing superalloy grindings and turnings and scrap superalloys (stellite, etc.) from jet engine components. The total scrap available can be classified into cobalt-base superalloy scrap containing 50 to 60% cobalt and nickel-base scrap consisting of 10 to 20% cobalt. Products from the secondary cobalt industry are as follows: (1) pig alloys (stellites, etc.), and (2) refinery grade cobalt scrap. Processing of scrap cobalt alloys involves three major operations, scrap pretreatment, smelting, and casting/refining.

A complex series of processes are involved in the scrap pretreatment operation. Some companies specialize in these processes and produce a clean scrap as their product for sale to smelting facilities. The four processes employed under this operation are described below.

Hand Sorting Process — This process identifies the scrap components and separates them into cobalt-base, nickel-base and nonprocessable components. The hand sorting operation is also aided by identifying the source of scrap and storing them in different lots to avoid mixing. The processing steps are: (a) spreading the dirty scrap on the floor, and (b) visually identifying and segregating the scrap. Human labor is the only requirement of this process. Waste produced is a minor amount of dust which is collected as sweepings and disposed of in trash cans and ultimately in a landfill. The process has no potential for pollution. The cobalt free scrap collected in the process is sent to other processing facilities.

Vapor Degreasing Process — This process removes the oil and grease from the scrap by using hot vapors of perchloroethylene. The processing steps are: (a) charging dirty scrap to degreasing unit, (b) heating to generate and circulate hot solvent vapor, and (c) discharging the degreased scrap to the blasting box.

Heat energy is the main energy requirement of the process. Waste emissions include minor amounts of solvent vapor to the atmosphere and grease and oil wastes to be incinerated or disposed of in sewers. Since the process is a closed loop operation, the only significant waste with a potential for pollution is the grease and oil from the solvent recovery unit.

Blasting Process — This process cleans the surface of the scrap material of all extraneous dirt, oxides, and rust. The process steps are: (a) blast the scrap either batchwise or continuously with grit, and (b) discharge the clean scrap. The process requires electrical energy to obtain compressed air used in the blasting equipment. Since the process is a closed loop operation, no significant wastes are generated. Hence the process has no potential for pollution.

Pickling and Chemical Treatment Process — Because the alloy produced as the final product of this segment has to be free of all impurities, the pickling process is utilized to clean the scrap of all residual oxide coatings. The chemical treatment operation is a trade secret and is known to remove lead from the scrap. The process steps are: (a) treating the scrap with a mixture of acids to remove rust and oxides, (b) treating with a chemical to remove lead, and (c) washing the scrap with water. The process requires electrical energy to drive pumps and other auxiliary equipment.

Waste discharges include small amounts of acid fumes and acidic wastewater containing metallic impurities. The acidic wastewater is neutralized before discharge to the waterways or sewer and hence has a reduced pollution potential. The acid fumes are usually not substantial to create serious ambient air pollution problems. The treated scrap is virtually free of all impurities except for some dissolved gases which are removed by smelting/refining. Usually this operation is carried out in an electric arc vacuum furnace or a vacuum induction furnace. This refines the scrap by removing volatile impurities under the influence of both heat and vacuum.

The process steps include: (a) charge scrap to the vacuum furnace, (b) evacuate the system, (c) melt the scrap by electric arc or induction, and (d) discharge the molten alloy into molds. Large amounts of electrical energy are used both to create vacuum and heat the charge. No significant wastes are produced. The vacuum pump discharge is contaminated with traces of gaseous wastes too insignificant to constitute any pollution. Vaporized heavy metals and other potential pollutants are trapped by the vacuum pump.

Casting is the final operation of the secondary cobalt industry. The castings produced by this operation are used as pig alloys or if the quality of the alloys produced is refinery grade cobalt alloy, it may be shipped to a cobalt metal recovery facility. This process produces castings of pig alloy from the molten charge obtained in vacuum melting. The process steps are: (a) pouring the molten metal into molds and (b) casting the ingots. Energy requirements of this process are not significant. No wastes are produced in the process.

COLUMBIUM

Separation from Titanium Chlorination Residues

The separation of columbium, tantalum, titanium and zirconium from titanium chlorination residues has been described by C.C. Merrill and D.E. Couch in Report of Investigation

No. 7671, Wash., D.C., U.S. Bureau of Mines (1972). Columbium, tantalum, titanium, and zirconium are present in solid-waste products generated by the chlorination of rutile used for production of titanium tetrachloride ($TiCl_4$). The columbium and tantalum were separated from titanium and zirconium by liquid-liquid extraction techniques using a secondary amine. Salts of these elements are in a leach liquor produced by water leaching of mineral waste from the chlorination of rutile.

This waste, produced at the rate of approximately 50,000 tons per year, is about 40 weight percent water soluble. Leach liquors, separated by filtration, were further separated into a solid fraction and a liquid fraction by pressure hydrolysis. The solid fraction containing Cb, Ta, Ti, and Zr was dissolved in HCl-HF and used as a feed solution. Columbium and tantalum were removed from the organic extractant with 1 N HCl. A miniature 10-stage reactor showed that the columbium and tantalum present in solution could be separated with a continuous flow.

COPPER

Recovery from Liquid Wastes

Recovery from Brass Wire Mill Effluents: A process for copper recovery from brass wire mill effluents has been developed by L.E. Lancy and C.A. Forbes of Lancy Laboratories, Inc., Zelienople, Pa. and applied at Volco Brass and Copper Co. of Kenilworth, New Jersey as described in Report 12010 DPF 11/71, Wash., D.C., U.S. Environmental Protection Agency (Nov. 1971).

An electrolytic system was installed to recover copper from the spent primary pickle solution and to regenerate the sulfuric acid for reuse. A hydrogen peroxide bright pickle replaced the chromate and fluoride bright pickles previously used. Copper from the bright pickle is also recovered in the electrolytic system. The electrolytic copper is reused on location in casting. An integrated copper treatment system was installed to treat bright pickle dragout. Sludge from the integrated system is recovered for sale. Rinse water consumption was reduced from 150 to 10 gpm. Former discharges of chromium, ammonium, and fluoride ions have been eliminated.

Figure 53 shows the operations prior to the EPA-sponsored study. Figure 54 shows the revamped operations giving less pollution and permitting copper recovery. Table 20 shows effluent quantity and composition for the old and new process configurations.

Recovery from Spent Copper Etching Solutions: A process developed by *E.H. Newton, J.M. Ketteringham, J.L. Sienczyk and C.M. Isaacson; U.S. Patent 3,783,113; January 1, 1974; assigned to The Shipley Company, Inc.* involves extracting copper as metal from a used etchant solution containing complexed cupric ions as an oxidant while simultaneously regenerating the etchant for further use. The etchant treated may have a pH from below 0 to 13, but preferably has a pH of at least 3 and more preferably has a pH between about 4 and 10.

The process comprises electrowinning a portion of the copper from solution under conditions effective for electrowinning but not etching. These conditions include a substantial freedom from oxygen in the vicinity of the cathode, high current density, a substantial freedom from solution agitation during the winning operation and preferably for good efficiency, low solution temperature at the interface of the solution and the cathode, typically below 140°F. The process is economical because in the preferred embodiment, a portion only of the copper in solution is removed, the remaining copper being left in solution and available as a source of cupric ion for reuse of the etchant. The process is an important contribution to pollution abatement efforts as it eliminates the need for dumping copper and other wastes resulting from an etching operation.

FIGURE 53: PICKLING OPERATIONS AND WASTE DISPOSAL PREVIOUSLY USED BY VOLCO BRASS & COPPER COMPANY

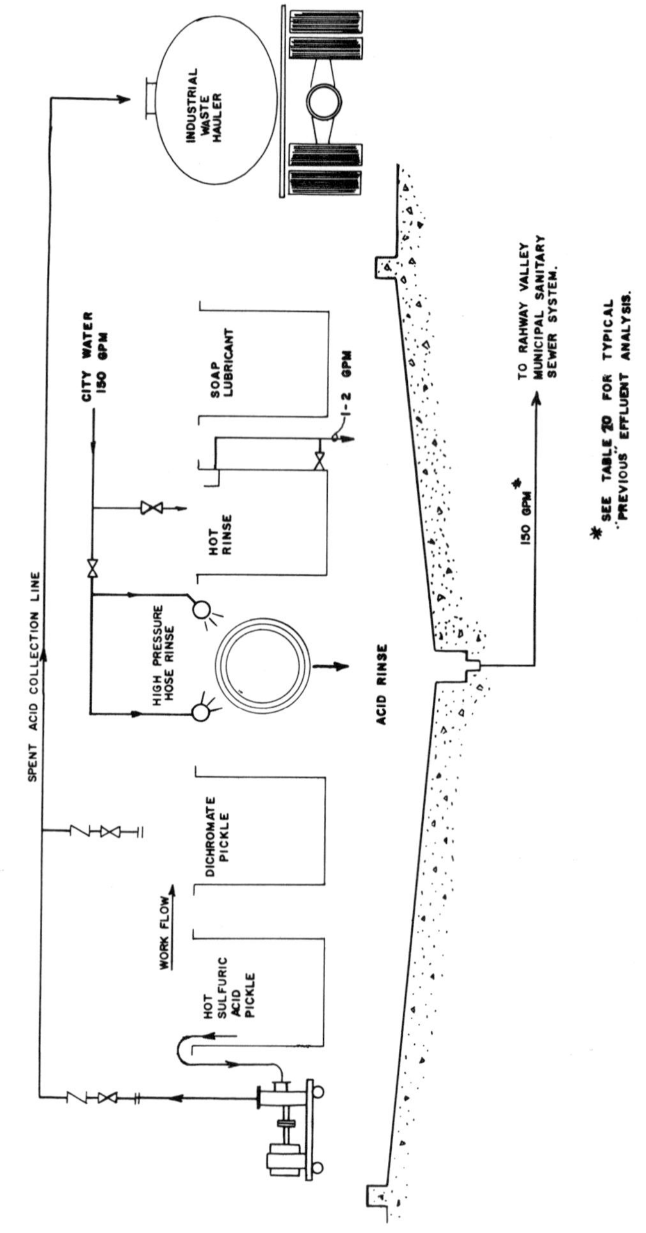

Source: EPA Report 12010 DPF 11/71

FIGURE 54: PICKLING OPERATIONS, COPPER RECOVERY, WASTE TREATMENT, AND WATER REUSE SYSTEM CURRENTLY IN USE AT VOLCO BRASS & COPPER CO.

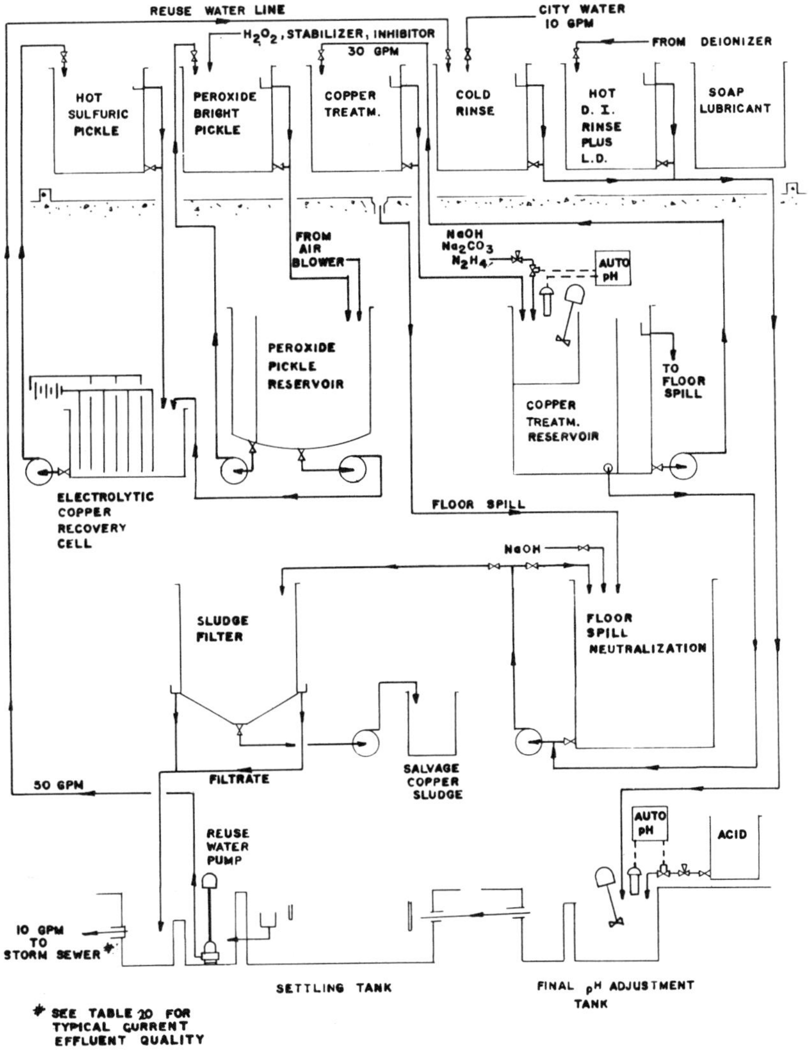

Source: EPA Report 12010 DPF 11/71

TABLE 20: PICKLING EFFLUENT QUALITY AND QUANTITY COMPARISON

Item (ppm, except pH)	Previous		Current				
	1/23/67[1]	7/7/67	1/25-2/5/71 Composite	2/8-2/17/71 Composite	2/17/71	5/12/71	5/13/71
pH	3.81	3.83	8.0	8.09	7.9	8.0	8.7
Cr^{+6} as CrO_3	70	96.9	No such process used	No such process used	No such process used	No such process used	No such process used
Cr^{+3}	27	---	No such process used	No such process used	No such process used	No such process used	No such process used
Cu^{+2}	124	75	.65	.50	.3	.25	.85
Zn^{+2}	374	36	.08	None	None	1.60	.30
Ni^{+2}	---	---	None	None	None	< .05	<.05
Sn^{+2}	---	---	---	---	None	None	None
Solids:							
Suspended, ppm	71.5	2.4	---	---	10	3.4	33.2
Dissolved, ppm	1551	---	---	---	---	760	814
Flow, GPM[2]	150	150	10	10	10	10	10

Notes:
1. United States Testing Company Report #90314 dated 2/10/67; all other results by Lancy Laboratories.
2. The 10 GPM, current effluent flow rate, corresponds to the freshening water input of 10 GPM into the Water Reuse System.

Source: EPA Report 12010 DPF 11/71

A process developed by *C.W. Ruff; U.S. Patent 3,855,141; December 17, 1974; assigned to Loewe-Opta GmbH, Germany* involves the electroless recovery of etched copper from a spent acidic copper chloride etching solution having a large concentration of a CuCl component. The spent solution is enriched by the addition of copper ions which are oxidized in the presence of an alkaline complex component to increase the concentration of $CuCl_2$ in the solution and to thereby decrease the concentration of CuCl. During a typical copper etching operation which employs copper chloride in the form of $CuCl_2$ in the starting solution along with the salt of an acid (e.g., HCl), at least a portion of the $CuCl_2$ is converted into CuCl as metallic copper accumulates in the solution during the etch.

In order to advantageously employ autocatalytic electroless techniques for recovery of the copper in the spent solution, a relatively low concentration of the saturated CuCl constituent should be present therein. Unfortunately, in the normal case the concentration of the saturated CuCl component is too high to produce stable long-term recovery of copper by the electroless method. As a result, it has been necessary as a practical matter to recover the copper directly from the spent etching solution by the use of relatively expensive electrolytic techniques, which in general require additional facilities for extracting impurities and other deposits from the electrolytic recovery bath. Figure 55 shows an arrangement of apparatus which is suitable for the conduct of the process.

FIGURE 55: SCHEME FOR COPPER RECOVERY FROM SPENT COPPER ETCHING SOLUTIONS

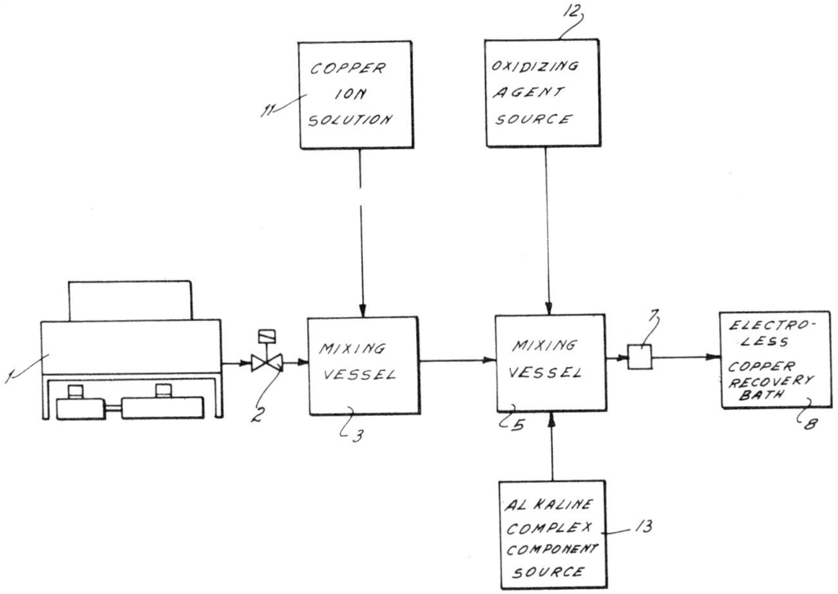

Source: C.W. Ruff; U.S. Patent 3,855,141; December 17, 1974

A copper etching operation carried out in a conventional etching tank 1 may typically employ a starting solution that includes copper chloride ($CuCl_2$) having a copper concentration in the range of 20 to 30 grams per liter, together with a salt of an acid such as HCl. During the etching of copper from the work piece (not shown) in the tank 1 etched copper enters the solution and serves to transform a portion of the starting $CuCl_2$ constituent in the solution to CuCl in accordance with the reversible equation

$$Cu + CuCl_2 \leftrightarrows 2\ CuCl$$

Since the presence of the CuCl in the solution serves to retard the etch, an oxidizing agent such as hydrogen peroxide (H_2O_2) may be added to the solution to oxidize a portion of the generated CuCl back into $CuCl_2$ in accordance with the reversible equation:

$$2\ CuCl + 2\ HCl + H_2O_2 \leftrightarrows 2\ CuCl_2 + 2\ H_2O$$

As the etch proceeds, however, the concentration of copper in the solution rises to a relatively high value (e.g., in the range of 120 to 140 g/l) and the solution becomes spent. At this point, the CuCl component is normally too large relative to the $CuCl_2$ component to permit efficient electroless separation of the etched copper. In accordance with the process, electroless recovery of the etched copper from the spent solution (which now contains constituents of CuCl, $CuCl_2$, HCl, H_2O_2 and H_2O) is facilitated by the procedure to be described. In particular, the spent solution from the etching tank **1** is conducted via a magnetic valve **2** and to a first mixing vessel **3** where it is enriched to a prescribed copper concentration by the addition of copper ions from a suitable source **11**. At this point, the concentration of CuCl in the enriched solution may be analytically determined.

Next, oxidizing agent (illustratively hydrogen peroxide) from a source **12** together with an alkaline complex component (illustratively EDTA) from a source **13** are then added to the enriched solution in a second mixing vessel **5**. The resulting oxidation of the added copper serves to increase the concentration of the unsaturated $CuCl_2$ component in the solution, in accordance with the following equation:

$$Cu + 2\ HCl + H_2O_2 \rightarrow CuCl_2 + 2\ H_2O$$

The relative amounts of the oxidizing agent and the alkaline complex component are determined in a conventional way in accordance with the previously measured concentration of CuCl in the enriched solution, although it has been found that when the oxidization agent is added in excess of such predetermined amount, improved performance is obtained in some cases. Also, the EDTA added during the same step preferably has a pH in the range of 11.5 to 12.5 to prevent hydroxide formation. Care must be taken to assure that the elevated temperatures generates in the vessel **5** generated during oxidation does not impair the efficiency of the EDTA.

Other combined forms of oxygen (e.g., a chlorate) as well as free oxygen may be used as the oxidizing agent in the mixing vessel **5**. Additionally, alkaline complex components other than EDTA (e.g., nitriloacetic acid or triethanolamine) may be used. From the vessel **5**, the regenerated etching solution is applied through a filter **7** into an electroless copper recovery tank **8** (together with suitable stabilizers, if necessary,) whereupon efficient recovery of the etched copper from the regenerated, alkaline solution may be accomplished without the requirement of extraction of impurities and other contaminants as in electrolytic processes.

Recovery from Spent Pickle Acids: A process developed by *F.C. Senior and D. MacGregor; U.S. Patent 3,787,306; January 22, 1974; assigned to Cationic Corporation* involves copper and beryllium metal recovery from the pickling of stainless, high alloy and super alloy steel. The acids most commonly used for these purposes are among the most corrosive and poisonous known, namely, sulfuric, nitric, nitric-hydrofluoric and hydrochloric. The steel and metal plant acid pickle solutions may be used until their acid content is decreased by the formation of salts of iron and other metals through the removal of mill scale and oxide. At this point, they must be discarded and fresh pickling acids used because of the acids' reduced effectiveness in removing mill scale and oxide from stainless steel and other alloy steels and metals. It has been found that such waste acid streams contain salts having approximately 3 to 10% iron, 0.4 to 1.2% nickel, 3.5 to 5% chromium, 0.04 to 0.06% cobalt and 0.1 to 0.2% zinc, 0.05 to 15.0% copper, 0.01 to 0.04% beryllium and other metals.

Depending on the pickle acid used, the oxides or scale enters solution in the acids as sulfates, chlorides, nitrates or fluorides. The iron oxide scale is at first an insoluble material. It is converted by the acid to a substantially soluble salt. In the case of iron, the ferric salt which is derived from ferric oxide is reduced to the ferrous state which is substantially soluble in the pickle acid. The extent of the use of pickling acids in steel manufacture or production may be appreciated by the fact that from 50 to 400 lb of strong acid solutions are used for each ton of finished steel produced in this country. The typical ranges of the percent acid in the initial solutions are 20 to 50% for nitric acid streams, 15 to 25% nitric plus 5 to 15% hydrofluoric in nitric-hydrofluoric streams and 30 to 50% for hydrochloric streams, parts of which are converted to salts and sludges.

The neutralization of waste acid streams of nitric, hydrochloric, hydrofluoric and nitric-hydrofluoric acids results in the production of large quantities of highly soluble salts. Discarding these salts in the earth, streams, rivers, etc. has resulted in a serious level of pollution and ecological disruption. In addition to the effect on the ecology of these highly corrosive obnoxious acids and the salts of these acids as they are dumped into the earth and into streams making them toxic to all life, there is a great economic loss of 4 to 5 million dollars which is required to replace up to 200,000 tons of various acid solutions which are annually consumed in pickling and electroplating operations. The spent waste acid streams may also be neutralized by the addition of a base such as calcium or sodium hydroxide forming large quantities of obnoxious sludge.

In addition, it has been estimated that in the production of a million tons of stainless steel there is a loss of 3 million pounds of nickel and cobalt in the waste acid streams. There is also a loss of some 8,000 tons of iron and several million pounds of zinc, copper, beryllium, tungsten and chromium. The total annual economic loss for the cost of new acids and the loss of metal values in the acids in the U.S. alone is conservatively 20 million dollars a year.

The process involves the controlled addition of sulfuric acid to convert the metallic salts to soluble sulfates and the evaporation and condensation of the acid to recover the acid in a pure state and in a concentration which may be higher than original and suitable for recycling in the acid pickling operations or for other uses. The method also provides for the concurrent recovery of metallic values from the thus treated waste acid streams and also from wet sludge and dry solid metal bearing wastes such as baghouse dust rolling mill scale and grinding dusts through the digestion of the sludge and dust with sulfuric acid to convert the metallic hydroxides and oxides to soluble sulfates, the oxidation and neutralization of the resulting solution to remove iron, chromium and calcium compounds by filtration, controlling the dilution and the pH as required by the addition of suitable acids and bases to redissolve and recover nickel, cobalt and copper values and the electrolytic deposition of the metal values from the clear filtrate and the recycling, if desired, of the demetallized sulfuric acid solution.

In addition to the recovery of acid and metal values from the waste acid streams, the process eliminates the problems associated with the disposal of highly soluble, toxic salts since the salts herein are ecologically harmless, substantially insoluble sulfate salts.

Recovery from Solid Waste

Recovery from Electrical Scrap: A process developed by *V.G. Leak and M.M. Fine; U.S. Patent 3,776,718; December 4, 1973; assigned to the U.S. Secretary of the Interior* is one in which scrap items containing copper and iron in physical admixtures, such as electrical and electronic scrap, are pretreated with an aqueous reagent, such as an alkali metal sulfate or silicate, prior to recovery of the copper by preferential melting in an inert, liquid medium. The chemical pretreatment reduces the tendency of copper to alloy or braze with the iron and results in a substantially higher copper recovery in the preferential melting step.

A process developed by *R.F. Stockman; U.S. Patent 3,807,321; April 30, 1974; assigned to The Air Preheater Company, Inc.* involves the recovery of metallic components from metal bearing scrap such as foil-backed paper, printed circuit boards and plastic coated wire. Incinerators for the combustion of organic components in scrap material containing a metallic base have long been utilized as a principle means for reducing the metal to a basic form whereby it may be used again. The salvaging of metallic components from a scrap presents problems not common for the complete incineration of organic waste because the high temperatures necessary for the elimination of the organic components melt, excessively oxidize, vaporize or otherwise damage the metal constituents of the scrap.

Various arrangements, an example of which is illustrated by U.S. Patent 3,636,873 have been developed wherein the scrap is initially heated by direct contact with a forced draft burner flame but is then quickly blown away to the cooler outer periphery of a combustion chamber that is maintained at a temperature well below the temperature at which the metals are harmed. Thus the scrap is not heated to an excessively high temperature that would damage its metallic content. Although a device of the type defined is theoretically effective, its operation is extremely critical and it has been found that satisfactory operation is highly dependent upon obtaining a specific workable relationship between the supply of oxygen for combustion, the type and amount of waste involved, the size and angle of the forced draft burner, and the size and disposition of the chamber in which the burner is located.

Accordingly, even though such apparatus may operate effectively, continuous satisfactory operation from a practical standpoint remains difficult to attain. The process is therefore directed to an incinerator and its method of operation whereby waste material having a significant content of metal scrap together with a quantity of organic waste may be incinerated by a controlled process that burns the organic waste without melting the metal, producing excess metallic vapors, or producing undesirable metallic oxides. Figure 56 shows a suitable form of apparatus for the conduct of the process.

FIGURE 56: APPARATUS FOR COPPER RECOVERY FROM ELECTRICAL SCRAP

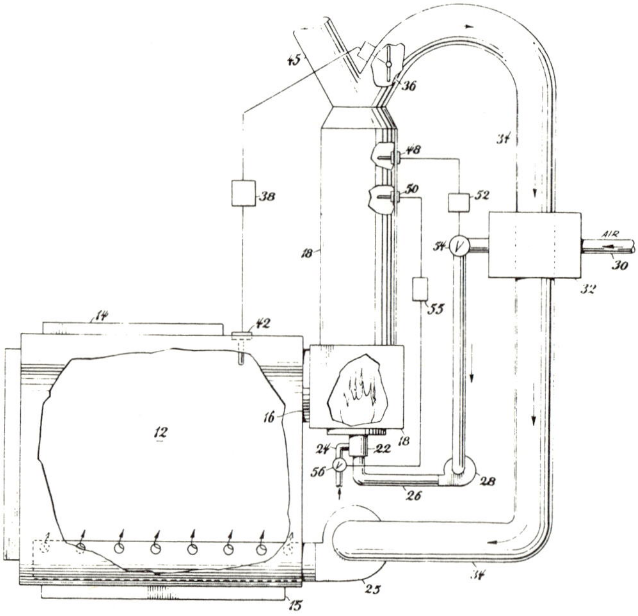

Source: R.F. Stockman; U.S. Patent 3,807,321; April 30, 1974

As shown in the drawing, the incinerator comprises a housing enclosing a chamber **12** having a loading opening **14**, an ash clean-out door **15** and a gas outlet **16** whereby gas produced therein may be exhausted to an afterburner chamber **18**. The chamber **18** contains an afterburner **22** having a supply for fuel **24** and a supply for air **26** that is provided by a fan **28**. The suction port of fan **28** is connected to passageway **30** containing air that is first circulated through a heat exchanger **32** over duct **34** that contains hot gas exhausting from afterburner chamber **18** whereby air for combustion in afterburner **22** is significantly heated before it is supplied thereto.

Inasmuch as controller **52** responding to thermocouple **48** is adapted to control valve **54**, a supply of air to afterburner **22** sufficient to maintain the temperature in chamber **18** at from 1600° to 1800°F is controlled by a proper setting of the controller. Thus, as the temperature in chamber **18** is maintained near its set-point, combustion of organic matter from the afterburner is so complete that the exhaust gas contains chiefly carbon dioxide, water vapor, metallic oxides, a small quantity of oxygen usually in the amount of 8 to 12%, and minute quantities of carbon monoxide. Inasmuch as the exhaust gas contains sufficient oxygen to pyrolyze all organic matter in chamber **12**, the exhaust gases from chamber **18** are recirculated to make use of the oxygen therein for the pyrolyzation of all raw waste; no other source of air to chamber **12** is provided.

Thus the exhaust gas from afterburner chamber **18** is directed back through duct **34** to the primary combustion chamber **12**. In order that a sufficient amount of exhaust gas including oxygen therein is returned to chamber **12**, a damper **36** directing the recirculation of exhaust gases is actuated by a controller **38** which is responsive to a thermocouple **42** in chamber **12**. The controller **38** is set to maintain the temperature in chamber **12** at from 700° to 1000°F, a temperature well below the melting point of most metals or the temperature at which the metals freely oxidize. Thus a temperature in chamber **12** below the set-point of 700° to 1000°F would call for an increased rate of combustion. To provide such an increase, damper **36** would be moved toward an open position to allow an increased flow of exhaust gas containing oxygen into chamber **12**.

An increase of temperature in chamber **12** would be similarly sensed by thermocouple **42**. Controller **38** responsive to thermocouple **42** would close damper **36** to accordingly reduce the flow of oxygen, the rate of combustion and the temperature in chamber **12** to meet the set-point temperature. Exhaust gas containing oxygen exhausted from the afterburner chamber **18**, not needed for combustion in chamber **12**, would be vented to the atmosphere through a duct **45**. A thermocouple **48** in afterburner chamber **18** acting through a controller **52** maintains the flow of oxygen to the afterburner **22** at sufficient levels to insure the burning of all combustibles in the exhaust gas being exhausted from chamber **12**. A valve **54** in air supply passageway **30** is therefore moved toward an open or closed position in accordance with the temperature in chamber **18** sensed by thermocouple **48** and controller **52**, usually set at from 1600° to 1800°F.

A thermocouple **50** in afterburner chamber **18** acting through a controller **55** maintains the flow of fuel traversing valve **56** in line **24** to the afterburner **22** in such a manner that the valve **56** gradually opens or closes until its set-point is reached. Thus at a predetermined setting of controller **55**, as for example 1600°F, the valve **56** closes completely so that no supplementary fuel is being added thereto and only gas exhausting from chamber **12** is being burned in the afterburner. However, if the temperature in afterburner chamber **18** should drop below the set-point, valve **56** would be opened to permit supplementary fuel to flow to the afterburner **22**.

In operation, chamber **12** is at first loaded with metal-bearing scrap through a charging door **14**. Air valve **54** is partially opened to permit air from inlet **30** to be supplied to afterburner **22** and afterburner chamber **18**. Inasmuch as there is at first no combustion in chamber **12**, and the gas exhausting therefrom is cool, damper valve **36** in duct **34** is wide-open as determined by controller **38** and thermocouple **42** to permit recirculated air from chamber **18** to be supplied by fan **25** to chamber **12**. When air sufficient for combustion is available, the charge is lighted manually or by an automatic means (not shown) and combustion in chamber **12** is thus effected, the temperature in chamber **12** rises, and damper **36** accordingly moves toward its closed position. As partial combustion or pyrolyzation of the waste material in chamber **12** is effected, exhaust gases given off therefrom pass to chamber **18** whereby they are mixed with oxygen from inlet **26** and burned in afterburner **22**.

When the temperature increases in afterburner chamber **18**, the airflow through valve **54** is also increased by controller **52** in accordance with thermocouple **48** commensurate with the flow capacity of the stack **45** so that excess or cooling air is flowing therethrough. As additional waste material in chamber **12** is subjected to pyrolytic action, more hot gases are given off through exhaust duct **16** to pass to the afterburner chamber **18** where they, along with fuel from source **24** and air being delivered through duct **26**, are subjected to the temperature of combustion so that the temperature continues to climb. Additional fuel is added to the afterburner **22** until controller **55** reaches a predetermined set-point of approximately 1800°F as determined by thermocouple **50** in chamber **18**.

At this point valve **56** moves to a closed position to shut off the quantity of fuel flowing to the afterburner through duct **24**. When the temperature in chamber **18** falls below its set-point, the valve **56** again is moved towards its open position to permit a flow of fuel to the afterburner **22** to compensate for the diminishing flow of pyrolyzed gas from chamber **12** whereby the temperature in the afterburner chamber may be maintained at approximately 1800°F.

Inasmuch as hot exhaust gas in afterburner chamber **18** contains a small amount of oxygen this exhaust gas alone is recirculated to chamber **12** to provide all the oxygen for pyrolysis of the waste in chamber **12** at a rate determined by thermocouple **42**. Thus a low temperature in chamber **12** below the set-point (approximately 900°F), damper **36** will be moved to open to allow more oxygen to be supplied via duct **34** to chamber **12**. As the temperature therein builds up to 900°F, the damper **36** closes and all excess gas is exhausted to the atmosphere through duct **45**.

Since the metal-bearing waste in chamber **12** is subjected to oxygen in an amount that supports only partial combustion of the waste, the oxidation of the metal is practically eliminated and the temperature is held below that at which the metal vaporizes or melts into a molten mass. However, the temperature within afterburner chamber **18** is at all times maintained sufficiently high to insure the complete incineration of gases given off from chamber **12**.

Recovery from Insulated Wire Scrap: A process developed by *B. O'Reilly; U.S. Patent 3,448,509; June 10, 1969; assigned to Nassau Smelting and Refining Company, Inc.* is one in which a quantity of plastic-insulated wire is placed in a chamber which is then sealed airtight both to enclose the chamber and to limit the amount of air therein. The wire is heated to a temperature sufficient to reduce the plastic to a crisp residue and to evolve gases from the plastic. The evolved gases are wet-scrubbed to remove obnoxious constituents therefrom. The heating step alone and in combination with the scrubbing step effects the reclamation of unoxidized wire and the control of the evolved gases. The crisp residue is ultimately removed.

A process developed by *W.D. Jones; U.S. Patent 3,719,471; March 6, 1973; assigned to American Metal Climax, Inc.* involves recovering copper from insulated copper scrap containing electrical insulation of the type comprising plastic, rubber and the like which, upon being thermally decomposed and partially combusted, forms a waste gas which may pollute the atmosphere. The scrap is thermally treated in a cupola to form blister copper and the thermally decomposed insulation is separated as a waste gas and passed through an afterburner where it is substantially completely combusted and the effluent gas therefrom then scrubbed and passed through an electrostatic precipitator following which it is vented to the atmosphere as relatively clean gas.

Figure 57 shows a suitable arrangement of apparatus for the conduct of the process. The figure depicts a pile of insulated scrap wire **10** which is sheared **11** and the sheared scrap fed to a vertically standing cupola **12** as shown. In starting the cupola, a bed of carbon is prepared at the bottom thereof and the carbon ignited and the combustion thereof maintained by injecting air **13** via tuyeres disposed around the cupola. The alternate layers of scrap and carbon are fed into the cupola to the top thereof.

FIGURE 57: SCHEME FOR COPPER RECOVERY FROM INSULATED WIRE SCRAP

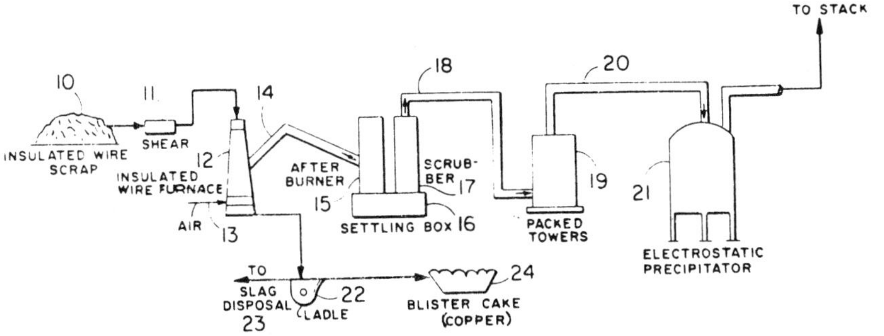

Source: W.D. Jones; U.S. Patent 3,719,471; March 6, 1973

As the temperature increases due to the combustion of the carbon, the insulation, which may comprise polyethylene, polyvinyl chloride, polytetrafluoroethylene, polypropylene, rubber, or the like, is destructively distilled or decomposed and the copper near the bottom where the temperature is the highest is melted and allowed to collect as a molten bath. The partially combusted organic vapors are drawn off intermediate the top and bottom of the cupola at a temperature of about 400° to 1200°F via conduit or pipe **14** and conducted to the afterburner **15** where the organic vapors are substantially completely combusted at a temperature ranging from 2000° to 2300°F or 2400°F, the resulting effluent gas being then passed through settling box **16** and up through scrubber **17** where the effluent gas is washed with an alkaline solution, e.g., an ammoniacal solution, to remove any acid in the gas, for example, HCl formed from the destructive distillation of polyvinyl chloride. The alkaline wash solution may be lime water, dilute sodium hydroxide or carbonate, or the like.

The effluent gas is removed from the scrubber through the top thereof via pipe **18**, passed through at least one packed tower **19** to complete the gas scrubbing and neutralize the saturated effluent gas and then fed via pipe **20** to electrostatic precipitator **21** where colloidal solids are removed and the relatively clean gas then vented to the atmosphere through the stack. The molten copper and slag are tapped at intervals from the bottom of the cupola into ladles **22**, the slag being poured off as shown at **23**, the blister copper being cast into cakes such as shown at **24**.

A process developed by *J.E. Perry; U.S. Patent 3,744,779; July 10, 1973; assigned to Horizons Research Incorporated* involves heating the contaminated scrap, in the absence of oxygen, to a temperature at which all of the volatile components are distilled off, using a continuous feed and removal of residue. At the temperatures involved, some volatile components are formed by destructive distillation. The volatile components, liquid and gas fractions, may be collected for reuse or they may be ignited and burned with a smokeless, odorless flame. Figure 58 shows a suitable form of apparatus for the conduct of the process.

As shown, the apparatus includes a furnace **10** which may be similar to a conventional powder metallurgy furnace or a rotary furnace. Furnace **10** has an inlet end **12** and an outlet end **14**, and is provided with a conveyor **16**, for example, a stainless steel belt for carrying scrap material through the furnace. Furnace **10** is provided with means to supply heat to the furnace for start-up and for maintaining the temperature of the furnace substantially constant at a controlled level during operation. A flame curtain **18** is provided at the inlet end of the furnace to keep air out of the furnace and to maintain a nonoxidizing atmosphere in the furnace **10**. Gases produced by the destructive distillation in furnace **10** burn at the flame curtain **18** at the inlet end. From the furnace, the product discharged at the outlet end falls through a chute **20**. Beneath the chute **20** is a hopper **24** with a bottom discharge through a star valve **26** and a side discharge through a side arm **28**.

As noted above, the scrap product emerging from the outlet end **14** of furnace **10** usually carries a coke or ash-like residue which adheres to the metal. This may be detached from the metal by means located in chute **20**, such as a rolling ring crusher **30**, or by other physical means. The scrap and loosened char fall onto a vibrating screen **34** which permits the coke-like residue to fall free of the metal to a collection bin. The metal, with any remaining residue, is conveyed from the screen **34** to a heated fused salt bath **40**, such as is described in U.S. Patent 3,000,766.

FIGURE 58: APPARATUS FOR COPPER RECOVERY BY DESTRUCTIVE DISTILLATION OF ORGANIC MATERIALS IN SCRAP

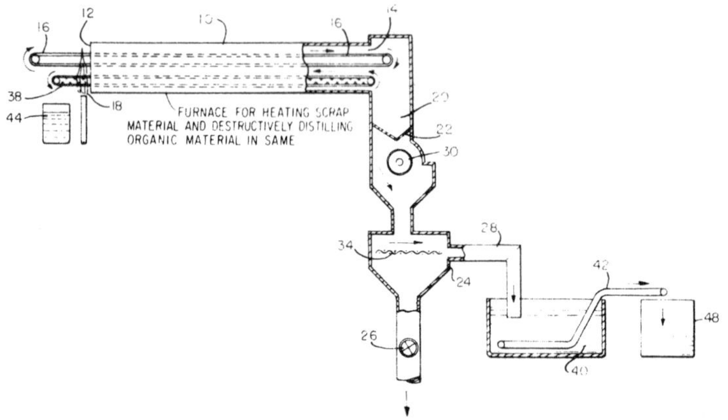

Source: J.E. Perry; U.S. Patent 3,744,779; July 10, 1973

It should be noted that the operating temperature is above the melting point of lead. Lead sheathed wire may be fed to the furnace and the lead removed by melting. The mechanism for removal of lead from the furnace would be the belt which would be designed with buckets to collect the molten lead and deposit the metal on the inlet end of the furnace. This additional feature is shown schematically in the drawing, at **44**. A second conveyor **38** positioned beneath conveyor **16** in furnace **10** catches nongaseous products formed when the insulation and other coating materials burn off the wire and discharges them at the inlet end of the furnace.

Crushing and screening should be done while the metal scrap, e.g., wire, is hot to conserve heat at the salt bath. These operations would be done in the atmosphere of the furnace. Coke is removed through a star (rotary) valve **26** and metal or wire is conducted into the salt bath **40** into which it passes below the surface of the salt. A continuous conveyor **42** removes the wire from the salt bath to a receiving bin **48**. It is also to be noted that the furnace **10** includes two parallel conveyors, one disposed above the other. These run in opposite directions so that scrap being processed is conveyed by the upper run of the uppermost conveyor away from the entry end and towards the discharge end of the furnace, while separated material is conveyed by the upper run of the lowermost conveyor towards the entry end and away from the discharge end of the furnace. Also the evolution of gas during the distillation process results in an excess of gas in the furnace and some gas passes down into hopper **20** and out valve **22** when it is opened, further protecting the furnace from contamination by air or oxygen. The metal product may be recovered without crushing or sieving or fused salt cleanup, if desired, or it may be melted or otherwise processed.

A process developed by *Z.J. Przewalski; U.S. Patent 3,777,679; December 11, 1973* provides an incinerating system for recovering copper wire following the burning of the insulation therefor as combustible waste while simultaneously insuring against wire overheating. The process includes the steps of: first burning the insulated wire in an ignition chamber, second preheating the secondary chamber, and third exploiting the fuel generated in the ignition chamber for maintaining the temperature of the secondary chamber while passing the smoke generated in the ignition chamber through the secondary chamber preliminary to passage to atmosphere.

A process developed by *E.H. Swartz; U.S. Patent 3,821,026; June 28, 1974* is one in which plastic coated scrap metal is heated within a vented combustion chamber to cause decomposition of the plastic without adversely affecting the metal core with respect to its subsequent separation from the decomposed plastic and recovery as a substantially pure metal product. A fuel gas mixture is fed at a controlled rate to the combustion chamber to initially elevate the temperature of the chamber to an optimum value that is then maintained constant. The products of combustion form a nonoxidizing atmosphere in the chamber within which the desired decomposition of the plastic occurs.

A process developed by *H.L. Wentworth and K.H. Seelandt; U.S. Patent 3,841,240; October 15, 1974; assigned to Sola Basic Industries, Inc.* is one in which material to be heated, such as insulated copper wire, is continuously introduced to a fluidized-bed furnace. Figure 59 shows such a fluidized-bed furnace system.

FIGURE 59: FLUIDIZED-BED APPARATUS FOR ORGANIC REMOVAL FROM COPPER SCRAP

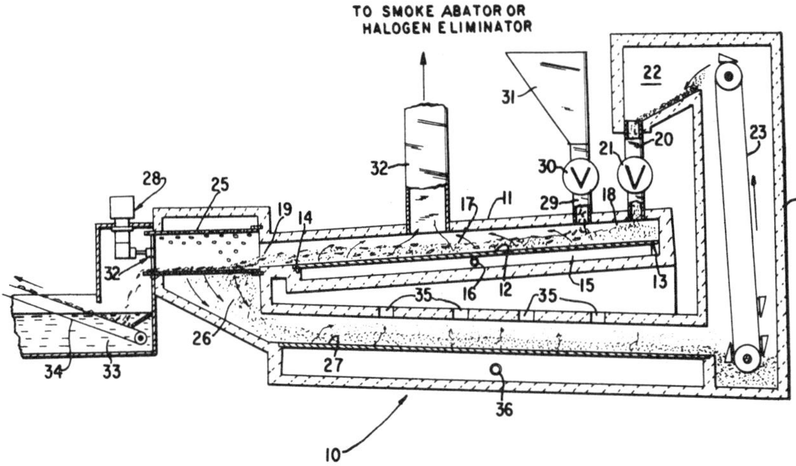

Source: H.L. Wentworth and K.H. Seelandt; U.S. Patent 3,841,240; October 15, 1974

The furnace **10** comprises a main frame in which is formed a chamber **11** fabricated of heat resistant material. The chamber **11** is inclined at a slight angle, and in the lower portion of the chamber, a bottom diffusion member **12** is positioned which may comprise a layer of porous refractory material. The diffusion member **12** may be supported upon a perforated plate (not shown but which is known in the art) which can have its ends supported on ledges **13** and **14** or in any suitable manner. A plenum chamber **15** is formed beneath the bottom member **12**. A fluid, which may be air or other suitable gas, is introduced through an air inlet **16** from a suitable supply.

Within the chamber **11** there is disposed upon the bottom plate **12** a fluid bed **17** which is a mass of nonagglomerative, granular material, through which a sufficient upward flow of fluid maintains a condition analogous to that of a fluid. The bed may comprise silica sand or granular particles of alumina, silicon carbide or other refractory materials. The bed **17** is about one foot in depth and the bottom member **12** can be inclined at a slope so that the fluidized material will move down the incline. Fluidizing the granular material constituting the bed cause the bed to move continuously downwardly along the sloping bottom member from the upper end of the chamber indicated at **18** to the lower end thereof indicated at **19**.

The granular material is introduced into the upper end **18** of the chamber through a conduit **20** in which is positioned a control valve **21**. The conduit **20** leads from the lower end of a hopper **22** which is supplied with the granular material by means of an elevating conveyor **23**. The granular material is dumped into the bottom of a vertical chamber **24** in which is mounted the conveyor **23**. As the granular material leaves the lower end **19** of the chamber, it enters one end of a rotating cylindrical perforated drum **25** through which the sand passes downwardly as shown at **26** to enter on a sloping surface **27** along which the granular material moves to the bottom of the vertical chamber **24**. The drum **25** is rotated by a suitable motor and a drive mechanism indicated at **28**.

The material which is to be heated or treated is introduced through conduit **29** having a control valve **30** and connected to the lower end of a hopper **31**. Where the furnace is to be used for burning insulation off of copper or conductors in electrical machinery so as to recover the scrap metal, the insulation covered work-pieces are dumped into the hopper **31** from which they are introduced onto the fluid bed **17**. As these pieces move along the fluid bed, the insulation is burned off and the resulting products of combustion are exhausted through exhaust conduit **32** where they are passed to a smoke abator or halogen eliminator.

As the granular material and pieces of copper from which the insulation has been burned off enter the rotating drum **25**, the heated material is separated from the granular material and the heated material is discharged from discharge end **32** of the drum into a water bath **33** for cooling purposes. A conveyor **34** may be provided for removing the stripped wire conductors from the cooling bath. As the granular material moves along the inclined surface **27**, it can be heated by burners at **35** or by electrical heaters which are known in the art and are not shown.

The inclined surface **27** may comprise a suitable conveyor for moving the granular material or it may comprise a porous diffusion member similar to the bottom member **12** in chamber **11**. In this event, a fluid such as air is introduced through inlet **36** and directed upwardly through the granular material to fluidize the granular material and to cause the fluidized material to move along the inclined surface **27** to the lower end of vertical chamber **24**. This bottom conveyor slope **27** may have an inclination of 2 to 7%. Describing the operation, as air under suitable pressure is introduced through inlets **16** and **36**, the air will be diffused upwardly through the diffusion member so as to cause the entire fluid bed to fluff slightly and then settle back to a state resembling a boiling liquid. The intensity of the boiling action can be varied by increasing or decreasing the air pressure.

Since the granular material constituting the fluid bed is continuously recirculated, it is apparent that this fluid bed furnace is continuous in operation. The material or wire to be heated are introduced at the upper end of the downwardly moving fluid bed and removed from the lower end thereof. The material being treated may become immersed in the granular material, but this is immaterial since the workpiece will be separated from the granular material by the rotary separator drum or other separating means. The bed may be heated to temperatures in the range of 200° to 1500°F or above depending upon the material of the bed, the heating unit, and the process to be carried out. Problems encountered in the burning of insulation from copper scrap to effect copper recovery are: (a) toxicity and corrosive nature of the gases due to the halogen, sulfur, and heavy metals content, (b) high concentration of unburned organic materials, and (c) the loss of metal values contained in the atmospheric emissions.

Efforts to control these emissions have had limited success. New or improved technology is still needed. One specific solution toward controlling atmospheric emissions being tried is that of the U.S. Metals Refining Company of Carteret, New Jersey, which is using a reactant-coated baghouse to collect HCl and other halides which are liberated when copper wire is burned to remove insulation, according to H. Nack, K. Murthy, E. Stambaugh, H. Carlton and G.R. Smithson, Jr. in PB 233 646, Springfield, Va., Nat. Tech. Information Service (May 1974).

A process developed by *M. Blackstone, B.K. Maddox, B.A. Rowland and S.L. Tate; U.S. Patent 3,858,776; January 7, 1975; assigned to Southwire Company* is one in which insulated wire, cable, and the like, is stripped of its insulation without damaging the wire. The insulated wire is preferably cooled to a low temperature, rendered brittle, and then continuously passed over roller means to effect fracturing and removal of the insulation. The bare wire is continuously recovered in an undamaged condition and the particles of insulation recycled where necessary. Wire straightening is optionally provided for straightening the bent wire after the insulation is removed. Figure 60 shows a suitable form of apparatus for the conduct of the process.

FIGURE 60: CRYOGENIC PROCESS FOR INSULATION REMOVAL FROM WIRE

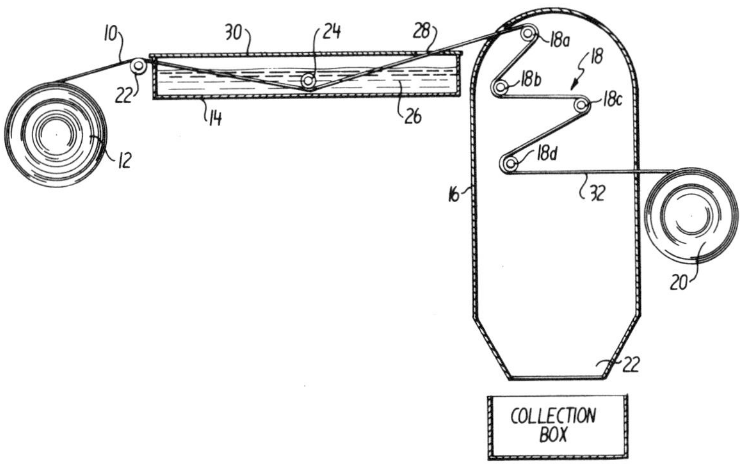

Source: M. Blackstone, B.K. Maddox, B.A. Rowland and S.L. Tate; U.S. Patent 3,858,776; January 7, 1975

Insulated wire, generally represented by reference numeral **10**, emerges from the supply reel **12** into and through the cooling chamber **14** and into the housing **16** which includes roller means **18**, comprising individual rollers **18a, 18b, 18c, 18d**. The insulated wire **10** passes over the roller means **18** in a tortuous path arranged to provide the necessary bending forces to cause fracturing of the chilled insulation. Upon leaving the roller means **18** the now bare wire **32** exits housing **16** and is rewound on a take-up reel **20**. Thus, in this particular instance the wire speed and the path length may be arranged to give the insulated wire a treatment time ranging from about 5 seconds to about 10 minutes or more and preferably from about 10 seconds to about 1 minute. The fractured insulation particles are collected at the bottom of the housing **16**, at **22**. Suitable valve means, or the like (not shown) is generally provided along the bottom portion of the housing **16** for recovering the particles collected therein.

Cooling chamber **14** comprises a trough-like insulated structure conventionally fabricated for handling liquefied cryogenic refrigerants, preferably liquefied nitrogen, in a manner designed to rapidly cool the insulated wire passing therethrough, as by immersion. The chamber **14**, as illustrated, includes two guide wheels **22, 24**, the former positioned in the vicinity of the entrance of the chamber, the latter positioned within the chamber **14** to direct the insulated wire **10** downwardly through the liquefied nitrogen **26**, where it emerges at **28** in an embrittled condition. A cover **30** is provided for protection to personnel.

The embrittled insulated wire **10** emerges from cooling chamber **14** and is directed into chamber **16**, over and around the four rollers **18a** to **18d**. It will be understood that the number of rollers used and their respective orientation within the chamber **16** can be modified to suit the type of insulated wire being processed. Under certain conditions even a single roller may be sufficient to effect the necessary fracturing, provided the insulation is rendered brittle and sufficient bending forces are imposed by the roller to fracture the insulation. Where the insulation is of a particularly brittle nature, e.g., polystyrene, a large, single roller can be used.

Alternatively, a single roller having means for receiving the insulated wire in more than one pass can also be employed. After the last roller, **18d**, has been traversed, the insulation is substantially totally removed, and the bare wire **32**, is recovered on the take-up reel **20**. Conventional means for driving take-up reel **20** is provided (not shown). In the event that some particles of insulation may cling to the bare wire **32**, a contacting pad, with or without lubricant, can be employed to remove these particles without damaging the wire. It should also be mentioned that under certain conditions the insulated wire being processed may be significantly deformed, requiring the additional processing step of straightening prior to being wound onto the take-up reel **20**.

Recovery from Mixed Metallic Scrap: Tonnage-wise, the copper segment (which produces refined copper) constitutes one of the major segments of the secondary nonferrous metals industry. Based on 1972 data, 1,210,070 tons of copper were recovered from scrap. Sources of raw materials to this industry include both new and old scrap. New scrap, which accounted for approximately 21% of the copper recovered from the scrap in 1972, refers to scrap produced in the manufacturing process by a fabricator of a finished product. Old scrap, comprising obsolete, worn out, or damaged articles, accounts for approximately 69% of the copper recovered by the copper segment in the U.S.

Segregating scrap according to classification standards is one of the most important steps in the recovery of secondary metals. Currently, copper-bearing scrap is classified into 54 types. Copper scrap as received generally contains metallic and nonmetallic impurities. Included among these are: lead, zinc, tin, antimony, iron, manganese, nickel, chromium, precious metals, and organic compounds such as plastics, oils, and greases. These impurities, which are removed during the recovery of copper, constitute major sources of emissions to the environment, if not collected and disposed of by an acceptable method or recovered as by-products. Products from the copper segment of the secondary nonferrous metals industry include:

(1) Babbit, lead, and solder recovered from the sweating process. This mixture of materials may be made into white-metal alloys, used for lead and tin additions to copper base alloys, or sold as produced to a refiner.
(2) Copper powder
(3) Copper shot

(4) Fire-refined copper
(5) Electrolytic-refined copper
(6) Oxygen-free high conductivity copper

The recovery of copper from scrap involves three manufacturing operations: scrap pretreatment, smelting, and refining/casting. Copper scrap is pretreated prior to smelting to remove some of the metallic and nonmetallic impurities or contaminants and to physically prepare the material for further processing. Three types of processing, mechanical, pyrometallurgical, and hydrometallurgical, are used. The pretreatment process varies according to the type of scrap. The specific pretreatment processes used are:

Stripping	Muffle/kiln burning/drying	Ammonium carbonate leaching
Briquetting	Sloping hearth sweating	Steam distillation
Shredding	Reverberatory sweating	Hydrothermal hydrogen reduction
Crushing	Pot sweating	Sulfuric acid leaching
Grinding	Tunnel sweating	Thermal void formation

Additional details of these processes follow.

Stripping Process — Insulation and lead sheathing are removed from electrical conductors by special stripping machines or by hand. Energy demand is that needed to drive the equipment. The process produces quantities of solid wastes. These wastes include lead and insulation. The lead is probably sold to a secondary lead smelter, whereas the insulation may be sent to a landfill or incinerator. The process has little potential for production of serious pollution.

Briquetting Process — Wire, thin plate, wire screen, and other bulky scrap are pressed into briquettes, bales, or bundles by hydraulic presses. In this form, the scrap is easier to handle, store, and load into the furnace. Energy required is that necessary to drive the presses. The process produces essentially no wastes or emissions. The process has no potential for production of pollution problems.

Shredding Process — Large items are reduced in size by pneumatic cutters, electric shearing machines, or manual sledging. The smaller pieces are easier to handle. Energy demand is limited to that needed to drive the equipment. The process may produce a small quantity of atmospheric emissions consisting of dusts with an approximate composition of the scrap. Collection of these dusts may be via a baghouse. Most likely, the emissions are not controlled. The process has essentially no potential for production of pollution problems.

Crushing Process — Brittle spongy turnings, borings, and long chips are densified by the crushing process. The process steps are: (a) crushing in hammermills or ballmills to reduce bulk, and (b) running over a magnetic separator to remove tramp iron. Energy demand is that required to drive the equipment. The process produces small quantities of atmospheric emissions consisting of dust particles composed of dirt, organic compounds from the scrap, heavy metals, and solid wastes. Most likely, the area may be hooded to carry the dust away from the vicinity of the crushing. The process has little or no potential for causing serious pollution problems.

Grinding Process — Prills or metallics are separated from the gangue in slags, drosses, skimmings, foundry ashes, spills, and sweepings by the grinding process. The process steps are: (a) grinding, (b) screening, and (c) gravity separation. Energy required is that to drive the equipment. The process produces atmospheric emissions, liquid wastes, and solid waste. The atmospheric emissions are dusts from the grinding and screening steps. These dusts may contain, in addition to fluxing materials and dirt, small quantities of heavy metals.

The liquid wastes containing both dissolved and suspended solids are generated in the gravity separation step. The solid wastes are generated during the screening step. As significant quantities of atmospheric emissions, liquid wastes, and solid wastes are generated by the grinding process, the process has a potential for the production of pollution problems.

Muffle/Kiln Burning/Drying Process — Much of the scrap contains oil, grease, cutting fluids, insulation, and moisture. These components may be removed prior to smelting by the muffle/kiln/burning/drying process. The process steps are: (a) charging of the scrap to the muffle or kiln, and (b) heating to remove the organic contaminants and moisture, afterwhich the treated scrap is removed for further processing. Energy required for this process is from gas or fuel oil to heat the furnace and electricity to drive the equipment.

The process produces significant quantities of atmospheric emissions. These emissions consist of both gases and particulate matter. The gaseous emissions are composed of the gases from combustion of the fuel and the organic contaminants. Depending on the organic contaminants, the gases may contain such materials as the chlorides, sulfur oxides, fluorides, and hydrocarbons. The particulate matter (fume and dust) is composed of a variety of heavy metals depending on the composition of scrap. Examples of heavy metals which may be found in the particulate matter are copper, zinc, tin, and lead.

Emissions data are not available for the burning/drying of all types of scrap. However, from burning of insulated wire, particulate emissions are estimated at 275 lb/ton of wire processed or an annual emissions total of 41,000 tons based on treating 300,000 tons of insulated wire. A variety of methods may be employed to control the atmospheric emissions. Included in the list are afterburners to destroy the hydrocarbons and other organic materials, wet scrubbers to remove the particulate matter and toxic gases, and baghouses to remove the particulates. In view of the toxic and hazardous nature of the atmospheric emissions, this process could be a source of serious pollution problems if the emissions are not controlled.

Sloping Hearth Sweating Process – Scrap such as journal bearings, radiators, and lead-sheathed cable contains lead, solder, or babbitt which would contaminate the melt. The contaminants are removed by the sloping hearth sweating process. The process steps are: (a) charging the preheated furnace with scrap, (b) melting the low-melting constituents, (c) raking the the sweated scrap over the hearth to remove all of the low-melting alloys, (d) collecting the solder, lead, or babbitt in a collecting pit, and (e) removing the sweated scrap for further processing.

Energy for this process is from gas to fire the furnace. In addition to the lead, solder, or babbitt which are recovered for subsequent use or sale, the process produces atmospheric emissions. These emissions are composed of gases and fumes. The gases contain the products from combustion of the fuel and any organic matter present in the scrap. Particulate matter is composed of such heavy metals as lead, tin, antimony, and copper, and organic materials such as soot.

Emissions data are not available for the sweating of the various types of scrap. However, the sweating of radiators results in the evolution of 15 lb of particulate matter per ton of scrap. Emissions may be controlled by baghouses, wet scrubbers, and electrostatic precipitators. Due to the hazardous nature of some of the pollutants and the quantity emitted, this process has a potential for the production of serious pollution problems.

A process developed by *B.F. Lambert; U.S. Patent 3,393,998; July 23, 1968* relates to the selective sweating of a definite portion of tin-lead solder from scrap vehicular radiators containing brass and/or copper components connected with solder, to directly produce a copper-base casting alloy having a particular specification. In the casting of valves, fittings and the like by brass foundries, it is required, in many instances, that the alloys employed for casting purposes correspond to a particular specification. In most instances, the alloy to be cast must correspond to an ASTM specification, such as a specification designated B-30-45T, which is classified as a leaded semired brass having numerical designation 5A. This alloy is known in the trade as valve metal and is one of the largest-selling copper-base casting alloys. Plumbing supplies and low pressure valves are cast from an alloy of this specification. The composition, based on a percentage by weight basis, consists essentially of copper 78 to 82%, tin 2.5 to 3.5%, lead 6.25 to 7.75% and zinc 7.50 to 10%. In making castings having this specification, the foundryman will generally melt ingots of one or more different compositions which have been provided by smelting companies who have used copper-bearing scrap materials including vehicular radiators, in preparing the compositions. These scrap materials are melted, refined and then cast into ingots by the smelting company. In preparing most copper-base casting alloys, only about half of the raw materials employed by the smelting company can be radiators due to the excessive lead content (e.g., 12%) arising from the solder in the radiators. The remainder of the raw materials must be more expensive, low-lead materials. Because the amount of radiators which can be used is limited, the cost of the ingots is increased.

Moreover, when a smelting company makes ingots, because of the melting and refining operations, there is generally a loss of raw materials either as slag or up the stack of the refining furnace; and these losses can amount to a substantial sum over a period of time, thereby affecting the over-all production cost in making ingots. The cost incurred by the foundry in purchasing the necessary supply of ingots is, of course, quite expensive and represents a substantial percentage of the over-all cost for casting the desired product. Scrap radiators or the like will, following processing, meet a particular ASTM specification without undergoing conventional melting, refining and ingot-casting operations. Because the solder is removed without subjecting the entire radiator to melting and refining operations, the loss of material arising in conventional procedures is eliminated.

In the process scrap radiators are briquetted after which the briquetted radiators are tumbled in a pitched, rotating tubular apparatus whereby a desired amount of solder is selectively sweated from the briquetted radiator. The melted solder rolls out of the tube machine into a slag buggy which collects the excess solder for resale or other purposes. The briquetted radiators upon removal from the tube can be agitated, if desired, on a conventional vibrating screen device where additional solder can be removed from the briquetted product. Subsequently, the briquetted product with the excess solder removed can then be used, without additional materials, for making castings calling for a particular ASTM specification.

Reverberatory Sweating Process – For removal of solder from scrap which is difficult to sweat, i.e., the solder remains in seams and folds when melted, the reverberatory sweating process may be employed. For this process a reverberatory furnace with a shaking grate is used. The molten solder removed from the items by a shaking motion falls to the floor and flows to the collecting sump. The process steps are: (a) charging the furnace, (b) heating the scrap to melt the low-melting components, (c) shaking to remove the low melting components, and (d) collecting the low-melting material. The sweated scrap is subsequently removed for further processing.

Energy demand is from the fuel oil or gas to heat the direct-fired reverberatory furnace and electricity to drive the equipment. Waste materials produced are atmospheric emissions consisting of gases and particulate matter. Composition of these emissions is similar to those from the sloping hearth smelting process. In fact, if the same type of scrap were sweated by both processes, the emission would be essentially identical with the possible exception that the emissions from the reverberatory sweating process may contain some dust resulting from shaking action. This dust is carried from the furnace by entrainment in the combustion gases. Particulate emission factor for the reverberatory sweating process is estimated at 15 lb/ton of radiators. The factor may be lower or higher for other types of scrap. Emissions are controlled with the same equipment as is used for other sweating processes. Atmospheric emissions produced by this process could cause pollution problems.

Pot Sweating Process – Removal of low-melting constituents from copper scrap is achieved by the pot sweating process. The process steps are: (a) dipping the scrap into a pot of molten alloy, (b) melting the low-melting constituents, and (c) removing the sweated scrap from the pot. Energy required is fuel, gas or oil, necessary to keep the pot of alloy molten. Potential pollutants generated by this process are atmospheric emissions containing fumes and possibly some gases if the scrap to be sweated contains any organic matter. Composition of emissions and emission factors should be comparable to those from other sweating processes. The process has the potential for producing pollution problems, as well as hazards to the operators of the equipment. A variant on the post sweating process uses molten salt baths in place of molten alloy baths as described by V.G. Leaf, M.M. Fine and H. Dolezal in Report of Investigation No. 7809, Wash., D.C., U.S. Bureau of Mines (1973). A technique for recovering copper and steel from metallic scrap was developed based on preferential melting in molten salt baths at 1150° to 1250°C. The molten salt, preferably barium chloride or calcium chloride, both minimizes oxidation of the metals and promotes heat transfer. Cleanest separations (0.01% copper remaining in iron) have been achieved by pretreatment (dipping or spraying) with sodium sulfate or sodium silicate, which alters and coats the surface of the iron to inhibit the alloying or brazing with copper. A flow diagram of the basic process is shown in Figure 61. Details of a salt bath furnace for copper reclamation by this technique are given by *E.L. Kemper; U.S. Patent 3,666,870; May 30, 1972; assigned to Upton Industries, Inc.*

FIGURE 61: RECOVERY OF COPPER FROM ARMATURES BY SALT BATH SWEATING

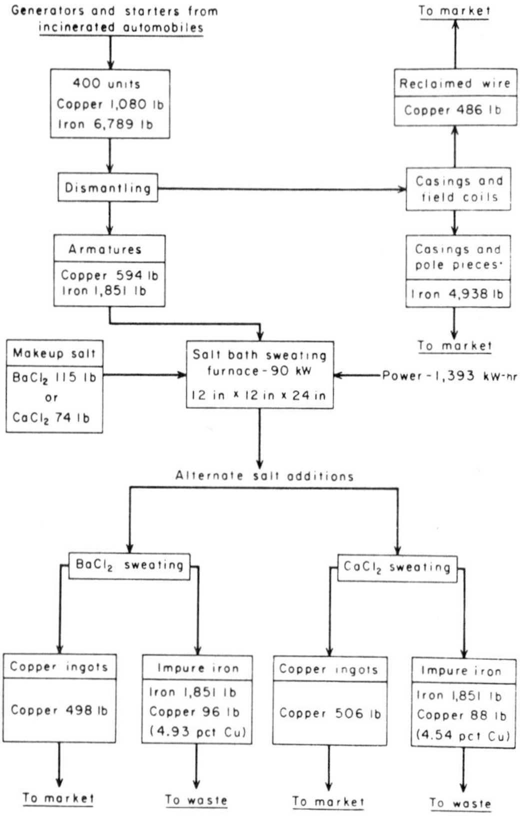

Source: Bureau of Mines R.I. No. 7809

Tunnel Sweating Process – This process is used by some smelters to sweat copper scrap. The process steps are: (a) load scrap placed in trays or on racks onto a continuous conveyor passing through the tunnel furnace, (b) remove the major portion of the low-melting constituents by passing the scrap through a heated tunnel furnace, and (c) dump hot partially sweated scrap onto tilted screen to remove remaining low-melting constituents. The molten low-melting constituents are collected in a floor sump. Energy required is from oil or gas to heat the scrap and electricity to drive the equipment. This process produces atmospheric emissions composed of gases (combustion gases from the fuel and gases from burning or volatilization of organic contaminants in the scrap), and particulate matter (fumes and possibly some dust from dropping of the partially sweated scrap onto the screen). The individual components in the emissions will be the same as those in emissions from other similar processes.

Particulate emission factor may be somewhat higher for this process (>15 lb/ton of scrap sweated) as the partially sweated scrap is exposed to the atmosphere while hot. This results in rapid oxidation of molten metal and subsequent volatilization of the oxides as fumes. The process produces atmospheric emissions which can cause pollution problems.

Ammonium Carbonate Leaching Process – The copper values are leached from copper scrap by the ammonium carbonate leaching process. The process steps are: (a) heating copper scrap in an ammonium carbonate solution while sparging the solution with air, (b) separating the crude copper ammonium carbonate solution from the leach residue, and (c) purifying the clarified copper solution. Copper values are then recovered from this intermediate product by the steam distillation process or by the hydrothermal hydrogen reduction process.

The energy required for this process is that necessary to heat the solution and to drive the equipment. The potential environmental pollutants generated by this process are atmospheric emissions, liquid wastes, and solid wastes. The atmospheric emissions are gases emitted from the leaching of the copper scrap with ammonium carbonate in the preparation of the crude copper ammonium carbonate solution. These emissions contain ammonia, water vapor, carbon dioxide, nitrogen, and excess air. Before being emitted to the atmosphere, the gases are cleaned via a wet scrubber which removes the ammonia and carbon dioxide.

Solid wastes generated by the process are the leach residues and residues from the various stages of solution purification. Essentially the only liquid wastes generated are those adsorbed on the surface of the discarded residue. Other liquid effluents generated are recycled to the process. Thus, this process has essentially no potential for the production of atmospheric pollution problems. The only atmospheric emission generated is nitrogen enriched air from the leaching operation. However, there is a potential for water pollution resulting from disposal of the leach residues from the ammonium carbonate leaching of the copper scrap and the disposal of purification residues. Pollution problems resulting from disposal of liquid wastes are negligible as essentially all of the liquid effluents are recycled.

A process developed by *W.L. Staker, K.C. Dean and C.J. Chindgren; U.S. Patent 3,660,022; May 2, 1972; assigned to the U.S. Secretary of the Interior* for recovery of copper from scrap comprising (1) leaching the scrap or ore with an aqueous cupric ammonium carbonate solution to dissolve the copper and (2) treating the resulting copper-containing leach solution with elemental sulfur to precipitate the copper as sulfide and to simultaneously regenerate the cupric ammonium carbonate leach solution.

The reader is also referred to Report of Investigation No. 7554, Wash., D.C., U.S. Bureau of Mines (1971) by W.L. Staker, C.J. Chindgren and K.C. Dean. Improved methods for the ammoniacal cupric carbonate leaching of copper were developed by this investigation. The improvements included: (1) The use of gas-bubble entrainment (air or nitrogen) to create a turbulent flow through the material being leached, (2) the use of elemental sulfur to oxidize the cuprous copper in solution and precipitate it as a sulfide, and (3) the combined use of bubble entrainment and elemental sulfur. Gas-bubble entrainment increases the leaching rate of copper by a factor of about 3 when either nitrogen or air is used.

Powdered elemental sulfur precipitates the dissolved copper as a sulfide as fast as it is leached, and it also regenerates the cupric ions. As a result there is no loss in the strength of the solution. The combined effect of nitrogen-bubble entrainment and the simultaneous addition of elemental sulfur accelerates the leaching rate by a factor of about 4. These improvements, which accelerate the leaching rate, potentially cut the cost of ammoniacal leaching by making possible the use of smaller, less costly equipment to produce a high-grade copper sulfide product.

A process developed by *T.W. Hummel and P.C. Schroy; U.S. Patent 3,865,580; February 11, 1975* is one in which copper is recovered from scrap sources, by accelerated solution thereof into ammonia plus ammonium carbonate solution with a catalyst which oxidizes copper from copper metal to the cuprous and from the cuprous to the cupric state, the catalyst being chelated iron, which is returned to its effective ferric catalytic state, and maintained in solution as a chelate by the oxidation of the medium in which the solution takes place. Following solution of the copper, recovery can occur in virtually any conventional manner.

Steam Distillation Process – Copper oxide is precipitated from the copper solution prepared from the ammonium carbonate leaching process by the steam distillation process. The process steps are: (a) heating the copper solution to precipitate the copper, (b) separating copper values from the slurry, (c) drying the precipitated copper oxide, and (d) recycling the aqueous effluent to the ammonium carbonate leaching process to prepare fresh copper. The copper oxide is an intermediate product for the production of copper by such processes as the blast furnace (cupola)/reverberatory smelting process of the electrolytic winning process. Step (a) is conducted at atmospheric pressure (boiling) or under pressure.

The energy requirement is that to produce the steam to heat the calciner. The process produces some atmospheric emissions from the copper ammonium carbonate decomposition step (Step a), and during drying of the copper oxide, and possibly some liquid wastes. The gaseous emissions from Step (a) are composed of essentially water vapor, ammonia, and carbon dioxide. These gases may contain a trace of copper carried from the system by entrainment in the gaseous emissions. Those emissions from drying of the copper oxide contain gaseous emissions composed of water vapor, carbon dioxide, and possibly some ammonia, and particulate emissions consisting of essentially copper along with trace quantities of other metals found in the copper oxide as impurities.

Control of the atmospheric emissions should be possible by using a wet scrubber for controlling those from Step (a), and by using a wet scrubber or baghouse to control those from the calciner. The liquid wastes are dilute ammonium carbonate solutions from washing the precipitated copper oxide. These contain, in addition to ammonium carbonate, copper compounds along with trace quantities of impurities associated with the copper. In most cases, these wastes can be recycled. Because of the small amount of wastes produced by this process, it is expected that pollution problems associated with the wastes are minor.

Hydrothermal Hydrogen Reduction Process – The copper value is precipitated as metallic copper from the ammonium carbonate leaching process by the hydrothermal hydrogen reduction process. The process includes the following process steps: (a) heating the copper solution in an autoclave with hydrogen at an elevated temperature and pressure to precipitate the copper, (b) removing the stripped liquor from the metallic copper, (c) washing the copper free of adsorbed liquor, (d) drying and sintering the copper powder under a hydrogen atmosphere, (e) pulverizing the sinter, and (f) screening the powder. The reduction of the copper solution is conducted at approximately 325°F and 500 psig of hydrogen. The stripped liquor is recycled to produce more copper solution by the ammonium carbonate leaching process.

The energy requirement is that needed to heat the copper solution in the reduction step, to dry and sinter the copper product, and to drive the pulverizing and screening equipment. Atmospheric emissions and possibly liquid emissions are generated by this process. Sources of atmospheric emissions are (1) drying and sintering of the copper powder, (2) pulverizing and screening of the copper sinter, and (3) reduction of the copper solution. The emissions from drying and sintering of copper powder contain water vapor, hydrogen, ammonia, and carbon dioxide as gases, and possibly particulates of copper powder or copper oxide. Those from the pulverizing and screening of the copper sinter contain particulates of copper. The gas in these emissions is entrained air. Hydrogen and water vapor are emitted from reduction of the copper solution. Liquid wastes are ammonium carbonate solutions containing such metal value as copper, nickel, lead, and zinc. Normally these are regenerated and recycled. If the ammonia and ammonium carbonate contents are low, the solutions may be discarded. No solid wastes are produced by this process. Thus, the process produces emissions and wastes which could cause pollution problems. However, the quantity of emissions is expected to be low; therefore, pollution problems associated with this process are expected to be minor.

Sulfuric Acid Leaching Process — Copper values are extracted from scrap copper as copper sulfate by this process. The process steps are as follows: (a) digesting copper scrap in hot aerated sulfuric acid solution to convert the copper values to copper sulfate, (b) separating the soluble copper sulfate from the undissolved portion of the scrap by filtration, and (c) purification of the copper sulfate solution. This intermediate product, aqueous copper sulfate solution, is a source of copper for the electrowinning process. The energy requirement is that to heat the digester solution and to operate the auxiliary equipment.

The wastes produced by this process are atmospheric emissions and solid waste. The source of atmospheric emissions is the digester. These consist of air containing fine droplets of the pregnant leach liquor. Most are equipped with demisters to reduce the emissions to the environment. The undissolved portion of the scrap and residues formed during the purification step are the sources of solid wastes. These are, most likely, disposed of via landfill. However, care must be taken to prevent contamination of the water by leachate from the landfill. A portion of the wastes may be reclaimed. This process produces essentially no atmospheric emissions and, thus, does not pose an atmospheric pollution problem. However, solid wastes disposal may result in a water pollution problem.

Thermal Void Formation — A process described by *J. Winter and A.J. Goldman; U.S. Patent 3,481,020; December 2, 1969; assigned to Olin Mathieson Chemical Corporation* is based on a discovery that in composites formed of dissimilar metals, the bond between the two metals can be weakened due to one or more of the following mechanisms: (1) the formation of voids which migrate to the dissimilar metal interface, (2) the formation of at least one brittle, intermetallic phase at the dissimilar metal interface, and (3) in hydrogen-containing atmospheres, the formation of H_2O vapor which causes grain boundary rupture in the area near the dissimilar metal interface.

One embodiment of the process relates to the U.S. Mint composite for dimes and quarters made of cupro-nickel (75% copper, 25% nickel) and copper (tough pitch copper). The tough pitch copper contains approximately 0.04% oxygen, mostly as copper oxide. The mint composite consists of layers of cupro-nickel (0.010") on either side of a tough pitch copper core (0.032") for quarter stock, and 0.007" layers on an 0.027" core for dime stock.

This composite can be separated by heating the composite to a temperature of from 1600° to 1800°F for a time between 20 minutes and three hours in an inert atmosphere, followed by mechanical separation of the cupro-nickel layer or layers from the copper by breaking the weakened cupro-nickel/copper bond. It is necessary to use an inert or slightly reducing atmosphere because if the atmosphere contains a significant amount of oxygen the cupro-nickel layer is oxidized at these temperatures.

The scrap pretreatment operation achieves a partial purification or densification of copper-bearing scrap. Further purification is achieved by the smelting operation. Via this operation, additional metallic and nonmetallic impurities are removed from the copper scrap to produce copper intermediate products (black copper and blister copper) containing 75 to 88% and 90 to 99% copper, respectively.

Blast Furnace (Cupola)/Reverberatory Smelting Process — Low-grade scrap (irony copper and brasses, motor armatures, foundry sweepings, slags, drosses, and skimmings) is treated by the blast furnace (cupola)/reverberatory smelting process to recover the copper as an intermediate product, black copper, for further processing. In this process, two major pieces of equipment are used, the blast furnace and the reverberatory furnace. The blast furnace is used to purify the scrap, whereas the reverberatory furnace is used to separate the slag from the molten metal. The process steps for this process are: (a) charging of the blast furnace (cupola), (b) melting the charge, (c) mixing of molten mass, (d) tapping of slag and metal into the reverberatory furnace, (e) allowing slag and metal to separate into layers, and (f) tapping of slag and black copper melt. Afterwards, the molten melt of black copper is either cast into ingots or transferred in the molten state to the converter for further processing.

Energy required for this process is from coke which serves both as a reducing agent and a source of fuel. Potential environmental pollutants produced by this process are atmospheric emissions, solid wastes, and liquid wastes. The atmospheric emissions are composed of gases and particulate matter. The gases are generated by the combustion of coke and burning of any organic constituents contained in the charge. Chemical composition will vary depending on the type of scrap but generally the gases will contain carbon dioxide, carbon monoxide, nitrogen, sulfur oxides, and possibly nitrogen oxides. The gases may also contain halogens and hydrocarbons, if present in the charge.

The particulate emissions are composed of a mixture of fumes, metal particles formed by the volatilization and condensation of metal values, and dusts removed from the furnace by entrainment in the gaseous emissions. These emissions normally contain such materials as zinc, lead, tin, copper, antimony, chlorine, and unburned carbon. Zinc is normally the major component. Data are limited on the composition of particulate emissions from the blast furnace. However, baghouse dusts from one installation had the following composition.

	Percent
Zinc	58 - 61
Lead	2 - 8
Tin	5 - 15
Copper	0.5
Antimony	0.1
Chlorine	0.1 - 0.5

At this installation, the cupola melt contained the following:

	Percent
Copper	75 - 88
Tin	1.5
Lead	1.5
Antimony	0.1 - 0.7
Iron	0.5 - 1.5
Zinc	4 - 10
Sulfur	0.5 - 1.25

Emission factors vary depending on the composition of the charge. However, for treating scrap and residues, the emission factor is estimated at 50 lb/ton of charge. Solid wastes generated are slags and baghouse dusts (discussed above). Composition of the slag is variable depending on such factors as type of scrap and fluxing agent. However, the slag will normally have a composition approximating the following:

	Percent
Iron oxide (FeO)	29
Calcium oxide	19
Silica	39
Zinc	10
Copper	$<$1.5
Tin	0.7

It will represent approximately 5% of the charge. Liquid wastes are generated as wastewater from cooling the blast furnace, in casting of the ingots, and quenching of the slag. Thus, the blast furnace/reverberatory smelting process has a high potential for the production of serious pollution problems. These problems could result from pollution of the atmosphere and the water systems, if the wastes and emissions are not collected and disposed of in a nonpolluting manner.

Converter Smelting Process — Black copper is purified, i.e., the heavy metals concentration is lowered to produce an intermediate product containing 90 to 99% copper, by the converter smelting process. The process steps are: (a) charging of the converter with molten black copper, (b) blowing of the molten charge (blow 1). (c) deslagging, (d) blowing of molten charge (blow 2), and (e) removing the slag, if any is formed. Afterwards, the copper, blister copper, is poured and cast into ingots for refining or transferred in the molten state directly to the refining furnaces. Some produce the copper shot product by quenching the blister copper to form small pellets.

Essentially no energy is required for this process. In fact, care must be taken to prevent overheating of the converter. The process produces atmospheric emissions, solid wastes, and liquid wastes. The atmospheric emissions contain both particulate matter and gas. The particulate matter is composed of fumes (metal particles resulting from volatization and condensation of metal values) and dusts entrained in the gaseous emissions.

The major sources of particulate emissions are the charging-melting and blowing steps. Emission (furnace) factors for these steps are estimated at 30 and 39 lb of particulate matter per ton of charge processed, respectively. Emission factor for pouring is estimated at only 0.6 lb of particulates per ton of charge processed. Emission factor for casting is comparable to that for the pouring step. The high emission factors for the charging-melting and blowing steps result from the heavy evolution of fumes when the charge is loaded into the furnace and melted down and when the melt is fluxed with a gaseous fluxing agent (air).

Quantitative analytical data are not available on the composition of these emissions. However, qualitatively, the particulate emissions contain such metal values as tin, lead, antimony, iron, zinc, sulfur, copper, silicon, and calcium. These emissions are in the form of dust and fumes. The dust is carried from the converter by entrainment in the blowing air, whereas the fumes result from volatization and condensation of the volatile metals. The fumes have a particle size in the submicron range. The dust particles are much larger, probably in the 5 to 20 micron range or larger. The particles have no definite shape, i.e., an irregular shape.

The atmospheric emissions can be controlled with baghouses, electrostatic precipitators, and scrubbers. Probably the most widely used device is the baghouse with control efficiency of approximately 95%. Other wastes include solid wastes and liquid wastes. The solid wastes generated are baghouse dusts and converter slags. Much of the solid wastes are recycled or treated for recovery of metal values. For example, the converter slags may be returned to reverberatory smelting for copper recovery. Liquid wastes are generated as wastewater from casting of the ingots, from wet scrubbers, if used, and from quenching of the slags. Thus, this process can be a source of serious pollution problems, if the pollutants are not collected and disposed of in a nonpolluting manner.

Electric Smelting Process — This process is very similar to the two previous processes in that blister copper and black copper, intermediate products for refining by the electric smelting process operation, can be produced by both processes. The advantage of this process over the combination of the preceding two processes is that the quantity of atmospheric emissions, both gases and particulates, is significantly reduced. Using electricity in place of coke as the the fuel and pure oxygen in place of air to oxidize the melt result in lower atmospheric emissions. Consequently, since much of the particulate emissions is carried from the smelting furnaces by entrainment in the gases, the reduced volume of gases results in a reduced amount of particulate matter. Thus, with the strong emphasis being placed on pollution abatement, this process is being viewed strongly as the preferred process for smelting of copper scrap. The process steps for the electric smelting process, which are essentially the same as those for the two preceding processes, are: (a) charging of the electric furnace, (b) melting of the charge, (c) removing the slag, (d) adding high iron scrap and virgin copper, (e) blowing the melt (blow 1), (f) skimming to remove the slag formed by blow 1, (g) adding more flux, (h) blowing the melt (blow 2), (i) skimming to remove the slag formed by blow 2, and (j) transferring the melt to the holding furnace for fire-refining or casting as blister copper. In some cases, the copper is given a third blow termed the finish blow.

The energy required for this process is the electricity necessary to melt the charge and drive the equipment. The heat evolved during the blowing is sufficient to keep the melt molten. In fact, in many cases, care must be taken to prevent overheating of the furnace. Atmospheric emissions, liquid wastes, and solid wastes are generated by this process. The atmospheric emissions consisting of gases, fumes, and dusts are generated throughout the process. The process steps which are the major contributors are (1) charging of the furnace, (2) melting of the charge, and (3) blowing of the melt.

The composition of the emissions will vary depending on the composition of the charge. Normally, the gases contain sulfur oxides, nitrogen, carbon dioxide, and carbon monoxide. The particulate matter includes fumes and dusts composed of such metal values as zinc, lead, tin, copper, antimony, chlorine, unburned carbon, precious metals (gold and silver), and constituents making up the fluxes. Particle size of the fumes and dusts ranges from 0.05 micron to approximately 1 micron for the fumes to approximately 5 microns or greater for the dusts. The dusts are readily collected using such devices as baghouses or electrostatic precipitators. Cyclones may be used in conjunction with these. However, because of the small particle size of the fumes, collection is difficult and much of the fumes pass on through the control devices into the atmosphere.

Emissions data are lacking and it is difficult to even estimate emission factors for this smelting process. However, it is estimated that the furnace emissions factors may range from about 80 to about 450 lb of particulates per ton of scrap processed. These estimates are based on the fact that the emission factor for smelting/refining in the brass and bronze segment was estimated at approximately 80 lb of particulates per ton of alloy produced, and emissions factor for similar processing in the primary copper industry was estimated at approximately 450 lb of particulates per ton of copper produced.

Solid wastes and liquid wastes generated by this process are the same wastes as generated by the two preceding processes. Data are lacking from which to calculate quantities. However, the slags contain such metal values as zinc, tin, lead, copper, iron, alumina, silica, antimony, and other nonvolatile metals present in the scrap and the fluxes. Many of the slags are recycled for copper recovery and the recovery of other metal values before being discarded. Liquid wastes from cooling of the equipment, quenching of the slags, and cooling of the ingots contain both soluble and suspended solids. These wastes are recycled, disposed of in settling ponds, or dumped into streams or sewers. Thus this process produces waste products which can cause serious pollution problems if the wastes are not collected and disposed of in a nonpolluting manner.

Blister copper, black copper, and high grade copper scrap are purified by the refining/casting operation. The products from this operation containing approximately 99.9% copper are: (1) fire-refined copper cast as ingots, billets, slabs, cakes, and bars for manufacturing plates, sheets, rods, and copper base alloys; (2) electrolytically refined copper cast as billets, wire bars, ingots, or cakes; (3) OFHC copper (oxygen-free high conductivity copper); and (4) copper powder. These are produced by the following processes.

Fire-Refining Process — The final purification of scrap copper may be achieved by the fire-refining process. The process steps are: (a) charging the furnace which may be either a reverberatory or cylindrical tilting type; (b) melting the charge in an oxidizing atmosphere until the melt begins to work, i.e., begins to bubble and liberate sulfur oxides; (c) skimming the melt; (d) blowing the melt with air; (e) covering the melt with a reducing agent such as coke, charcoal, or coal (anthracite); (f) poling with green wooden poles; (g) skimming, if necessary; (h) pouring; and (i) casting into ingots, slabs, wire bars, and billets. In some cases, molten metal from the smelting operation is charged directly to the reverberatory furnace. In regard to (f) above, the green poles used are largely maple and birch. They are approximately 10 inches in diameter at the base, at least 20 feet long, and weigh approximately 200 lb. Maple is preferred because it burns more slowly than other woods. The quantity of green poles consumed varies depending on the degree of reduction necessary. In the manufacture of wire bars from scrap copper, approximately 100 lb of poles per ton of copper are required. As an alternative natural gas is being developed and in some cases may be used in place of green poles.

Energy required for this process is that necessary to keep the melt molten. Atmospheric emissions, solid wastes, and liquid wastes are produced by this process. Essentially each process step is a source of atmospheric emissions which are composed of gases and particulate matter made up of fumes and dusts. Emissions data are lacking for the fire-refining process. However, based on the chemistry involved and the composition of the melt, the gases are known to contain sulfur oxides, carbon dioxide, nitrogen, carbon monoxide, and most likely, hydrocarbons if green poles are used, or methane if natural gas is used in the reduction phase of the refining. The particulate emissions volatilized as fume or carried from the furnace by entrainment in the gases contain such metal values as zinc, tin, copper, lead, and others contained in the melt. In addition, the emissions may also contain unburned carbon and materials from the fluxing agents.

The atmospheric emissions are controlled with baghouses, electrostatic precipitators, or wet scrubbers. A cyclone may be used in conjunction with these devices. Other wastes generated by the process are solid wastes and liquid wastes. The solid wastes are slags containing the fluxing materials and metal impurities from the black, blister, and scrap copper and the impurities from the green poles. The liquid wastes are cooling wastewater from cooling or quenching of the castings and cooling of the furnace. The water contains soluble phosphate used to condition the molds and the inhibitor from the cooling water. Although emission factor data are not available, this process does generate large quantities of materials which can cause serious atmospheric and water pollution problems.

Fire (Partial) Refining Process – When the copper is to be cast into anodes, an intermediate product for electrolytic refining, the melt is not completely fire-refined by oxidizing and poling. Instead, the copper is partially fire-refined and cast into anodes. The anodes contain about 99% copper and small amounts of silver, gold, lead, selenium, tellurium, and other metals. The process steps, energy required, pollutants generated, emission factors, and control of pollutants are essentially the same for this process as for the full fire-refining process. This process does generate significant quantities of potential environmental pollutants and is, therefore, a source of potential pollution problems.

Electrolytic Refining Process – Partial fire refined copper, anodes, is converted to high purity copper cathodes by the electrolytic refining process. The impurities in the anode copper either dissolve in the electrolyte or fall to the bottom of the cells as slime. The slimes may contain such impurities as sulfur, arsenic, antimony, lead, nickel, selenium, tellurium, gold, and silver, if present in the anode copper. The cathode copper assaying about 99.9+% copper contains trace quantities of the above impurities and is the intermediate product for the production of electrolytic refined copper product and for the production of copper powder.

The process steps for the electrolytic refining process are: (a) making up the electrotype containing water, sulfuric acid, copper sulfate, and addition agent; (b) placing the copper anode in the electrolyte bath as one of the electrodes, the anode; (c) dissolving the anode; (d) depositing the copper on the starting sheets, the cathode; (e) removing the cathode for melting and casting; (f) regenerating the spent electrolyte; and (g) removing slimes for recovery of metal values. Energy required for this process is electricity to produce the cathodes and to drive the equipment. This process produces essentially no atmospheric emissions if the anode copper does not contain arsenic. If arsenic is present, arsine (AsH_3)

is formed and evolved during electrolytic refining of copper. This has not presented a serious problem for the operators as the tank houses are generally well ventilated and the arsine is ejected outside the house. However, arsine formation can be a problem in electrolyte regeneration.

The major wastes produced are solid wastes (aqueous slurries composed of the slime) and liquid wastes (spent electrolyte). The slimes are sent to precious metal recovery where such metal values as gold, silver, nickel, and lead are recovered. A black liquor, liquid waste, is formed during electrolyte purification. This material is either discarded, generally after neutralization, or used in leaching operations in associated plants. Additional liquid wastes are generated during recovery of nickel, iron, zinc, and other metal values from the spent electrolyte. Thus, the electrolytic refining process can present serious pollution problems. Arsine, an atmospheric pollutant which is volatile and poisonous, is generated, if arsenic is present in the cathodes. Furthermore, large quantities of liquid wastes and solid wastes are generated which can cause water pollution problems.

Electric Melting Process – Cathode copper is melted and cast into the desired shape by the electric melting process. Normally the process steps are: (a) feeding the cathodes into the furnace through a charge slot to maintain a molten bath; (b) removing molten copper from the tap hole; and (c) casting the molten copper as billets, wire bars, ingots, or cakes. Periodically, flux is added to the melt to refine the copper and slag is removed as the waste product. Energy required is the electricity to melt the copper and drive the auxiliary equipment.

Wastes generated by this process are small amounts of atmospheric emissions, liquid wastes and solid wastes. The atmospheric emissions result primarily from the evolution of gases from the melting operation and from pouring and casting of the pure copper. The solid wastes are the slags or skimmings from the melting furnace, whereas liquid wastes are formed during ingot casting and cooling of the equipment. Emissions data are not available for this process. It is expected that this process would constitute a minor source of pollution problems in this segment of the secondary nonferrous metals industry.

Reverberatory Melting Process – This process is an alternative process for converting cathode copper intermediate product to the desired shapes of electrolytic refined copper. The process steps are: (a) charging the reverberatory furnace, (b) melting the cathodes, (c) blowing (flapping) of melt, (d) skimming of melt, (e) poling of melt, (f) skimming of melt, (g) pouring, and (h) casting. Energy required is the fuel to melt and keep the charge molten and that required to drive the auxiliary equipment. Atmospheric emissions, liquid wastes, and solid wastes are produced by this process. The atmospheric emissions, gases, fumes, and dusts, are formed during each processing step with the majority of the emissions being evolved during blowing (flapping) and poling of the melt.

Emissions data are not available on the quantity and composition of these emissions; however, it is expected that the emissions contain gases such as carbon dioxide, carbon monoxide, sulfur oxides, and hydrocarbons. The fumes and dusts are composed of, primarily, copper, and small quantities of such metal values as lead, nickel, selenium, and tellurium. Liquid wastes result from casting of the melt into the desired shapes and from cooling of the equipment. These liquid wastes, primarily wastewater, contain soluble inhibitors and insoluble solids such as the bone ash to condition the molds. In some cases, phosphates may be used.

Solid wastes are the slags or skimmings formed and removed from the melt during the melting (refining) of the cathodes. These wastes contain the impurities from the cathode copper and the fluxes, if added. In view of the waste products and emissions generated by this process, pollution problems can result if these pollutants are not collected and disposed of in a nonpolluting manner. However, it is expected that this process is a minor source of pollution problems when compared to some of the other processes.

Induction Melting Process – Oxygen-free-high-conductivity (OFHC) copper is produced by the induction melting process using cathodes as the source of copper. The OFHC copper is oxygen-free with a reported purity of 99.99%. The process steps are: (a) clean the cathodes, (b) preheat the cathodes in an atmosphere of reducing gas which is low in hydrogen, (c) melt the clean, preheated cathodes, in a low-frequency induction heater, (d) cover the melt with graphite granules, (e) discharge molten copper to a low-frequency induction-heated pour hearth which distributes the copper to the casting machine, and (f) cast the OFHC copper. OFHC is a trade name of the U.S. Metals Refining Company and also means oxygen free high purity copper. A nonoxidizing atmosphere is maintained throughout melting and casting.

Energy required is the electricity to melt the cathodes and keep the melt molten and to operate the auxiliary equipment. Waste products produced by this process are atmospheric emissions, solid wastes, and liquid wastes. The atmospheric emissions are produced during preheating of the cleaned cathodes, melting of the cathodes, and casting of the copper melt into ingots and other shapes. Emissions data are not available; however, the emissions no doubt consist of reducing gases and carbon monoxide from melting of the cathodes. The gases contain a small quantity of particulate matter which is composed of copper along with other metals found in the melt and graphite particles. The solid waste is basically graphite used to cover the melt. The liquid wastes are wastewater used in the casting of the melt and cooling of the equipment. This process produces minor quantities of potential environmental pollutants. Therefore, the process has little or no potential for producing serious pollution problems.

Electrolytic Winning Process — Copper in the form of impure copper cathode, an intermediate product to electrolytic refining process, is recovered from copper sulfate solutions by the electrolytic winning process. The deposit is not as pure as the cathodes from the electrolytic refining process and, therefore, must be purified. The process steps are: (a) prepare electrolyte, (b) electrolytically deposit the copper on thin copper starter sheets, (c) remove the copper cathodes from the electrolytic cells, (d) clean cathodes by soaking in water, (e) replace copper starter strips, (f) regenerate electrolyte, and (g) repeat series of above steps. Spent electrolyte is continuously bled from the system, regenerated, and returned.

Energy required is the electricity to operate the process. The waste products generated by the process are primarily liquid waste consisting of spent electrolyte and solid wastes consisting of slimes formed during the electrolysis and resulting from regeneration of the electrolyte. If the electrolyte contains arsenic, gaseous arsine (AsH_3) is liberated. This material is generally vented to outside of the building. Emissions data are lacking. However, from data which are available on electrolytic refining, most likely, the waste products produced by the electrolytic refining process can cause water pollution problems.

Electrolytic Powder Production Process — Copper powder is prepared from copper sulfate solutions by the electrolytic powder production process using pure copper cathodes from the electrolytic refining process as the source of copper. The process steps are: (a) prepare electrolytic cell for powder production using cathode copper as the anode, (b) electrolytically deposit copper powder, (c) remove copper powder from cell, (d) return electrolyte to cell, (e) wash electrolyte from copper powder, (f) heat powder in hydrogen-carbon monoxide atmosphere to remove water, (g) classify dried powder, (h) blend powders of different particle size to obtain correct particle size distribution, and (i) package product.

Energy required is the electricity to produce the copper powder and drive the equipment and the fuel to heat the dryer. Waste products produced by this process are atmospheric emissions and liquid wastes. The atmospheric emissions contain gases composed of water vapor, carbon monoxide, hydrogen, possibly arsine, and unburned fuel. The particulate portion of the atmospheric emissions is copper powder from the dryer, blender, and classifier. The liquid waste is the spent electrolyte from the electrolytic cell and wash liquor from washing the copper powder. The spent liquor is normally regenerated and recycled to the electrolytic cell. The wash liquor containing sulfuric acid, water, and some soluble copper is probably neutralized and disposed of in a holding pond. Thus, waste products produced by this process can cause pollution problems if not properly controlled.

Recovery from Oil-Contaminated Scrap: A process developed by *R.K. Rawlings; U.S. Patent 3,839,016; October 1, 1974; assigned to U.S. Reduction Co.* is one in which metallic scrap contaminated with combustibles is melted in a well-type furnace to recover the metallic component with the aid of preheated air and heat generated by burning the fumes generated by the combustible contaminants. The scrap is fed into a depository well which communicates with a main furnace well under a combustion chamber in which burners are operated. Melting heat is transmitted through the molten material within the wells to the incoming scrap within the depository well to maintain its temperature at a level sufficient to melt the scrap and both significantly, heat the air above the depository well and vaporize the combustibles.

A negative pressure is maintained above the depository well by a hood covering the depository well thus preventing the escape of fumes but admitting air. The hood communicates with a duct system which incorporates a blower and leads to the burners. The fumes and air from the depository well are thereby directed into the burners, which ignite and burn the same, sometimes with conventional fuel such as natural gas, the latter being needed in only a reduced amount.

Recovery from Reverberatory Furnace Slags: A process developed by *H.W. Wilson; U.S. Patent 3,744,990; July 10, 1973; assigned to Golden Cycle Corporation* involves the beneficiation of waste copper slag to be used for the entrapment of sulfur dioxide from waste gases by contacting the waste slag with an aqueous solution of ammonia gas to remove the reactive copper content, and zinc content, if any present in the waste slag. The removal of the copper by the aqueous ammonia solution is accomplished without the presence of any dissolved ammonium salts or the addition of air or other ozidizing agent to the reaction. After dissolving the copper values and zinc values, if desired, they may be recovered from the ammonia gas solution by displacement.

A process developed by *K.O. Lindblad and R.E. Dufresne; U.S. Patent 3,868,440; February 25, 1975; assigned to The Anaconda Company* is a hydrometallurgical process for treating slag material, especially copper smelter slag material, comprising vigorously mixing the slag material with about one part by weight of sulfuric acid and about one part by weight of water for each part by weight of the slag material, and allowing the resulting mixture to react to produce a dry, solid material, containing the metal values of the slag in water-soluble form. These solubilized metal values may be separated by leaching from the insoluble silicious residue of the dry solid material.

A study of the recovery of copper from converter slags by flotation has been presented by V.E. Edlund and S.J. Hussey in Report of Investigation No. 7562 (Revised), Wash., D.C., U.S. Bureau of Mines (1972). Laboratory batch flotation tests were conducted on copper converter slags to evaluate the relative merits of recovering copper from slow-cooled versus water-quenched slags. Three slags containing 1.6, 5.0, and 6.6% copper were used. More than 90% of the copper was recovered in a rougher concentrate leaving a 0.2 to 0.3% copper tailings when treating slow-cooled slag. Lower recoveries and higher copper tailings ranging from 0.5 to 0.6% copper were obtained from quenched slag.

Grindability studies were made on the respective heat-treated slags. Quenched slags proved more difficult to grind than slow-cooled slags. Cost studies showed that quenched slags can be treated at slightly lower cost than slow-cooled slags. However, the cost advantage of processing quenched slags is more than offset by the higher copper recoveries obtained from slow-cooled slags.

Recovery from Strip Mine Waste: A process developed by B.W. Madsen, R.D. Groves, L.G. Evans and G.M. Potter is described in *Prompt Copper Recovery from Strip Mine Waste,* Report of Investigation No. 8012, Wash., D.C., U.S. Bureau of Mines (1975). Research by the Bureau of Mines has shown that treatment of open pit waste by crushing and sizing followed by flotation of the fines and leaching the coarse fraction results in greater metal recoveries than obtained by leaching only. Tests with a chalcocite-bearing quartz monzonite mine waste sample that had been crushed to minus 6 inches and screened on ½ inch yielded copper and silver recoveries of 71 and 22%, respectively, by the combined flotation-leaching process, compared with 48% of the copper and no silver recovered leaching the run-of-mine material. Copper was more readily leached from the coarse material with the minus ½ inch fines removed than from material with the fines included. The improved leaching was attributed to improved permeability and the associated improved circulation of air to the interior of the ore column.

Cost estimates showed that the proposed intensive treatment of mine waste will require a greater capital investment and higher annual operating costs than conventional milling and dump leaching process. However, the income from the additional metal production offsets the increase in capital investment and operating costs, resulting in additional profit. The process for the prompt recovery of copper and by-products from porphyry strip material consists of a crushing step (if advantageous), screening at an optimum size to obtain enriched fines, froth flotation of the fines for copper, molybdenum, gold and silver recovery and dump leaching of the coarse fraction with an acid ferric sulfate solution as shown in Figure 62.

FIGURE 62: PROCESS FLOWSHEET FOR PROMPT RECOVERY OF COPPER

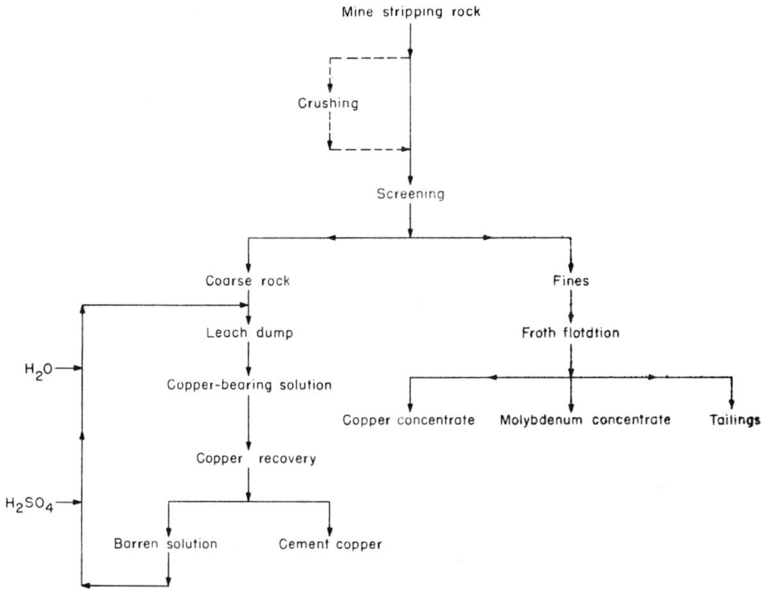

Source: Bureau of Mines R.I. 8012

GALLIUM

Recovery from Liquid Wastes

Production from Alumina Plant Liquors: Gallium, a by-product of the aluminum system, is produced by tapping a portion of the green liquor generated in the production of alumina (Al_2O_3). Gallia (Ga_2O_3), present in bauxite, dissolves along with the alumina during the digestion of the bauxite ore. The soluble gallia builds up in recycled plant liquors by dissolving to saturation (about 0.2 g/l) and partial precipitation. After the solution is saturated with gallia, its concentration in the liquor remains relatively constant with the excess exiting the process via the red mud waste. Gallia is produced by withdrawing a small portion of the plant liquor and subjecting it to a separate precipitation. In this process, a small amount of alumina is lost equal to about 100 parts alumina for each part of gallia produced. Once the alumina-gallia precipitate is obtained, the gallium is separated from it electrolytically and is refined by subjecting it to as many as 16 steps of processing on a laboratory scale.

The sand removed in classification prior to filtration of the green liquor is of interest because it is an example of the variety of minerals contained in the plant liquor. The black sand is presently considered waste, though it might some day be considered worthy of by-product processing. If the sand removed by the classifier is subjected to magnetic separation, a black (magnetic) portion can be separated from the total. A sample of this material at Alcoa's Bauxite, Ark., plant assayed 40% iron oxide and 3.8% titania with smaller amounts of zircon, manganese, magnesium, vanadium, niobium, gallium, chromium, molybdenum, nickel, lead, and copper oxides. The system depicted in Figure 63 is based on the alumina plant of the Aluminum Company of America at Bauxite, Ark.

FIGURE 63: ALUMINA SYSTEM SHOWING GALLIUM BY-PRODUCT

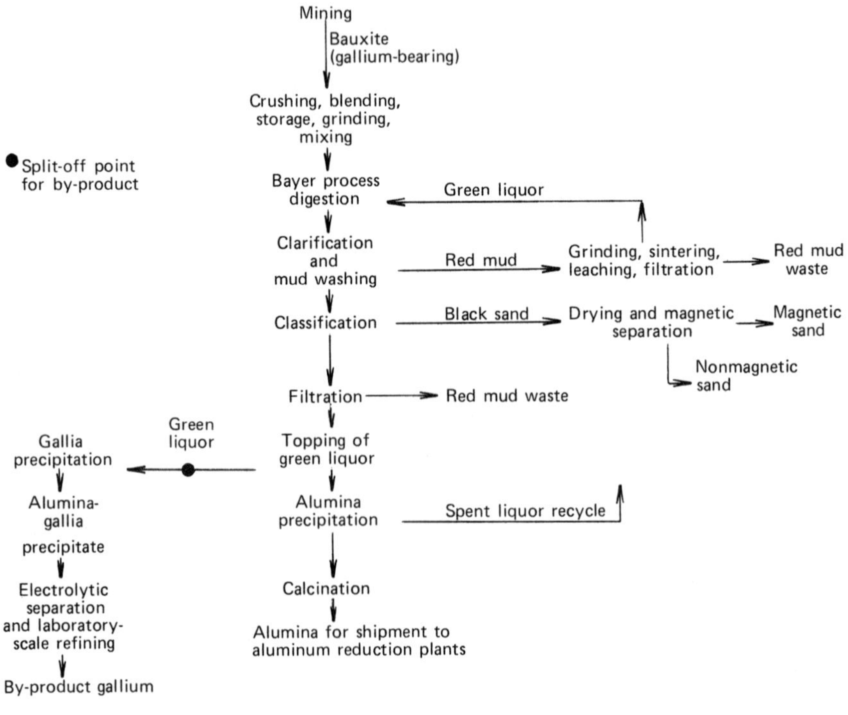

Source: Bureau of Mines I.C. 8570

Gallia is produced by extracting a very small portion of plant liquors which flow through the plant at 300,000 gal/min. Labor is by far the major item of cost in the production of gallium.

Even the methods of packing and shipping have significant impact on incremental costs. There is no fulltime labor devoted to gallium production. The labor force that extracts gallia is also concerned with the production of several other specialty products. Because there is a market for only small quantities of gallium, large quantities are discarded in the red mud wastes. Bauxites are known to contain up to 0.01% gallium or 0.22 pound per long ton. 50% recovery of contained gallium would yield 0.11 lb/ton, and 90% recovery would yield 0.20 lb/ton. The U.S. production of bauxite in 1970, principally by Alcoa and Reynolds Mining Co. in Arkansas, totaled 2,082,000 long tons. On this basis, a potential production of 229,000 to 416,000 lb of gallium is indicated. The recovery of gallium from aluminum production is far below this figure.

Recovery from Solid Wastes

Production from Fly Ash and Flue Dusts: The extraction of gallium and germanium from coal fly ash and from phosphorus furnace flue dust have been described by R.F. Waters and H. Kenworthy in Report of Investigation No. 6940, Wash., D.C., U.S. Bureau of Mines (April 1967).

Laboratory-scale selective volatilization methods developed by the Bureau of Mines to recover germanium and gallium concentrates from coal fly ash and phosphorus furnace flue dust are summarized. In the better experiments between 85 and 98% of the germanium was recovered as the lower oxide and as sulfides, and between 75 and 97% of the gallium was recovered as the lower oxide and as the trichloride. Grades of the condensates were inconsistent and varied between a few tenths of a percent and 8%, depending on the method used. In the primary extractions of germanium alone the degree of enrichment surpassed 100 to 1, and for gallium the best ratio was 30 to 1. Upgrading of condensates by vaporization retreatment and by leaching and precipitation is discussed. An appreciable degree of mechanical preconcentration by attrition scrubbing and elutriation was possible on two fly ash samples.

GERMANIUM

Recovery from Cd-Ge Fume from Zinc Smelting

The production of by-product germanium has been described by A. Petrick, Jr., H.J. Bennett, K.E. Starch and R.C. Weisner in Information Circular No. 8570, Wash., D.C., U.S. Bureau of Mines (1973). Particular reference to germanium has been drawn from A.P. Thompson and J.R. Musgrave in *J. Metals*, 4, 1132-37 (Nov. 1952). The most relevant by-product subsystem for germanium includes all processing of germanium-bearing residues after their precipitation with zinc dust and separation from cadmium-bearing solution. The production of germanium can be viewed as a by-product cadmium system if the cadmium system is profitable in itself.

The flow diagram shown in Figure 64 is based primarily on the article by Thompson and Musgrave printed in 1952. The processing is essentially unchanged today. Some additional technical material was reviewed to assure consistency with modern practice. The process shown in Figure 64 begins with the collection of the cadmium-germanium-bearing fume from the roasting and sintering operation in processing zinc sulfide concentrates. The split-off point for germanium separates the collection, leaching, and filtration of the fume from the first partial precipitation which was done expressly to separate the germanium-bearing precipitate from the cadmium-bearing solution. This partial precipitation is achieved by adding zinc dust sufficient to precipitate copper and germanium without precipitating the cadmium.

Some other impurities, principally arsenic, are also thrown down with the copper-germanium precipitate. From this point on, the process is similar whether the 1% germanium product comes from zinc smelter residues, zinc flue dusts, or coal ashes. The germanium-bearing sludge is redissolved in sulfuric acid, and the germanium is concentrated by a second precipitation. The germanium concentrate is dried and roasted and is then dissolved in strong hydrochloric acid. The resulting germanium tetrachloride and arsenic chloride are distilled from the solution and are then ready for purification. Purification involves multiple distillation of the tetrachloride, hydrolyzing with water, and calcination to produce a pure germanium dioxide. The metal is produced by hydrogen reduction of the germanium dioxide in a tube furnace.

FIGURE 64: BY-PRODUCT GERMANIUM SUBSYSTEM

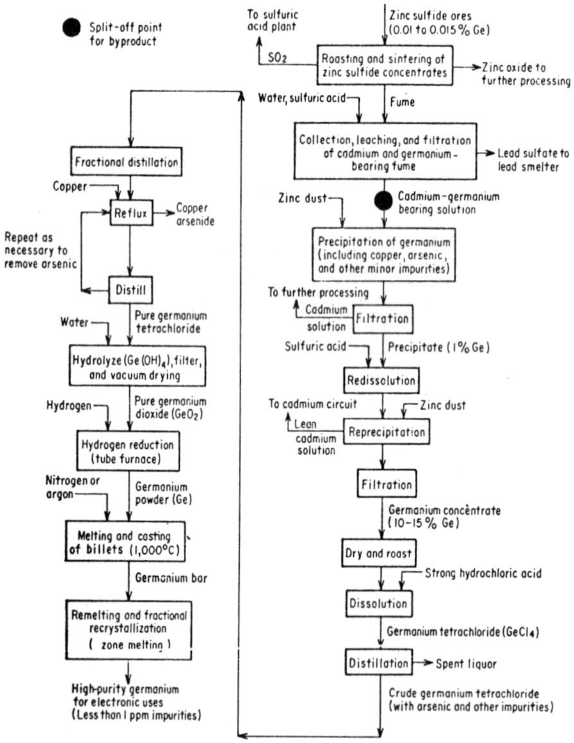

Source: Bureau of Mines I.C. 8570

The resulting germanium metal powder is melted in an inert atmosphere to prevent oxidation and is cast into bars. The product is not pure enough for electronic applications and must be further purified by such techniques as zone refining.

Recovery from Fly Ash and Flue Dusts

The extraction of gallium and germanium from coal fly ash and from phosphorus furnace flue dust have been described by R.F. Waters and H. Kenworthy in Report of Investigation No. 6940, Wash., D.C., U.S. Bureau of Mines, April 1967, (page).

Recovery from Scrap

According to H. Nack, K. Murthy, E. Stambaugh, H. Carlton and G. R. Smithson, Jr. in

PB 233 646, Springfield, Va., Nat. Tech. Information Service (May 1974), annual production volume of secondary germanium is about 8,000 lb. Because of the high cost of germanium (~ $50/lb), process losses through air, solid waste, and water emissions are reduced to a minimum. Also, the emphasis placed on the highest possible purity of the final product insures that all processing operations are conducted in closed systems to prevent atmospheric contamination of the product. Germanium scrap sources are as follows:

(1) Solid scrap (100% solid metal) obtained as scrap ends from zone refining.
(2) Slicing compounds from single crystals obtained as sludge containing 50 to 80% germanium.
(3) Acid etching scrap.
(4) Grindings and polishing wastes from the infrared industry.
(5) Scrap from the electronics industry.

High purity germanium is the only product of secondary germanium processing. The process used in handling germanium scrap is hydrometallurgical processing. This operation involves three processes, chlorination, hydrolysis, and hydrogen reduction. These are discussed in detail below.

Chlorination Process — In this process, germanium is converted to germanium tetrachloride. The process steps are (1) charge scrap to vessel, (2) heat and add sodium chloride. The volatilized tetrachloride with any impurities is distilled to get germanium tetrachloride. The process requires minor amounts of heat energy. Small quantities of gaseous wastes are generated. The emissions are water scrubbed and the solids recycled. Scrubber water is also recycled. The process has no significant pollution potential.

Hydrolysis Process — This process converts the tetrachloride to germanium dioxide by ice water hydrolysis. Details of the process steps are not known. Minor amounts of water waste (with insignificant pollution potential) are generated.

Hydrogen Reduction Process — This process converts the germanium dioxide to the pure metal. The process steps are (1) pass hydrogen, (2) heat to 650°C. Heat energy is utilized in this process. The process produces no pollutants.

GOLD

Recovery from Gold Mud in Silver Refining

The block flow diagram for by-product gold recovery is shown in Figure 65. The Wohlwill electrolytic process is used to recover gold from the residue of the silver cells after the excess of silver is removed by leaching with hot sulfuric acid. The Wohlwill process consists of electrolyzing gold anodes in a hot solution of silver chloride. The cathode consists of a thin ribbon of gold. The deposit which adheres to the cathode is exceedingly pure, averaging 99.975% purity with minute impurities of platinum and palladium.

In the copper and silver refining preceding gold refining, the more noble metals are insoluble and form mud and slime residues. The process is repeated again in gold electrolysis so that the platinum and palladium appear in the slime and become the by-products of the by-products of the by-products.

The scale of operations in the gold subsystem is very small compared to the silver circuit. In the hypothetical circuit shown, only 1 ounce of gold is handled for each 50 ounces of silver. Equipment and labor required are correspondingly small. Rates of operation are kept high to keep the high value in processed metal at a minimum level.

FIGURE 65: BY-PRODUCT GOLD SUBSYSTEM

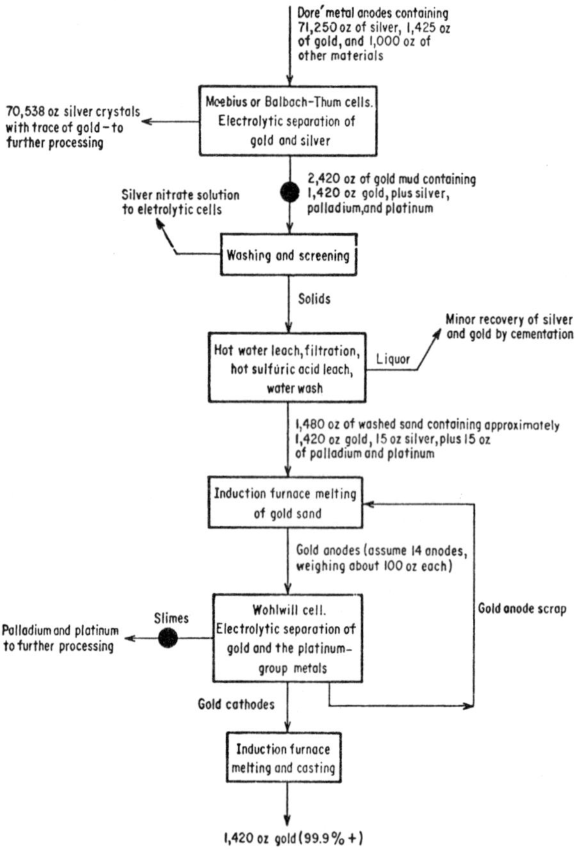

Source: Bureau of Mines I.C. 8569

Recovery from Mine Waste

The experimental leaching of gold from mine wastes has been described by H.J. Heinen and B. Porter in Report of Investigation No. 7250, Wash., D.C., U.S. Bureau of Mines (April 1969). Excellent gold recovery was demonstrated in a laboratory-scale leaching test on waste material from an open-pit mine in northeastern Nevada. An injection system for leaching was applied in which the sodium cyanide solution was introduced into the interior of the dump at selected points. Dissolution of gold from the waste material, which assayed 0.06 ounce gold per ton, was readily achieved because the gold occurred as submicroscopic particles in a porous host rock. Gold recovery of 80% was obtained in 11 days from material crushed to 1 inch in size, and recovery was 90% after 21 days.

Recovery from Miscellaneous Scrap

The reader is referred to the subsequent section of this volume on Precious Metals for techniques which are applicable to gold.

Recovery from Scrap Electronic Solders

The recovery of gold from scrap electronic solders by fused salt electrolysis has been described by E.K. Kleespies, J.P. Bennetts and T.A. Henrie in Bureau of Mines Technical Progress Report No. 9, available as PB 184 309, Springfield, Va., Nat. Tech. Information Service (March 1969).

The Bureau of Mines developed this process recovering gold and refining tin-lead solders from scrap solder discarded by the electronics industry. During assembly of electronic systems, gold and other contaminant metals are introduced into solder baths which are periodically discarded. Contaminated solder can be refined by electrotransport in a molten-salt chloride electrolyte. Refined solder is recovered at the cathode, and the gold and other metal impurities are concentrated at the anode, increasing gold concentration from 60 to 15,000 oz/ton without any significant loss. The gold is then reclaimed by conventional fire-refining methods.

Recovery from Scrap Printed Circuit Boards

The reader is again referred to the section on Precious Metals.

HEAVY METALS

Recovery from Mine Drainage by Precipitation

Such a process has been described by L.W. Ross in EPA-670/2-73-080, Wash., D.C., U.S. Environmental Protection Agency (Sept. 1973). Heavy metals in mine drainage waters of the Rocky Mountains can be removed by a two-stage process consisting of (1) neutralization followed by (2) sulfide treatment. The first stage removes ferric and aluminum hydroxides, and the second (sulfide) stage precipitates the heavy metals that are most objectionable as pollutants, and that are of possible interest for economic recovery. The two-state process has been demonstrated in the laboratory and in a field experiment.

In the field, powdered lime was employed for neutralization, and barium sulfide was employed as sulfide source in a two-stage treatment tank. The ferric and aluminum hydroxides failed to settle when even the slightest winds disturbed the surface of the settling pond, but the sulfides settled within a few feet downstream. The measured pH of treated streams has proven entirely satisfactory for control of chemical additions.

A computer program based on published values of equilibrium constants and solubility-product constants is capable of adequate prediction of the required chemical quantities for treatment, and of the resulting metal concentrations in solution. The economics of chemical addition indicates that recovery of sulfides (CuS, ZnS) for sale to smelters can partially offset the cost of treatment, but never completely.

INDIUM

Recovery from Alloys

A process developed by *Y.E. Lebedeff; U.S. Patent 2,433,770; December 30, 1947; assigned to American Smelting and Refining Company* involves recovering indium from metals or alloys such as those with lead, zinc and the like. Methods have been developed heretofore for treating indium-conaining metals to recover indium which comprise heating the metal with caustic soda and niter to selectively oxidize the indium whereby it can be removed and recovered. When appreciable amounts of zinc or other impurities such as Sn, As, Sb, Te, etc. are present, however, the caustic-niter fusion process is not effective in bringing about a clean-cut selective separation of indium from these impurities. According to the process, a method is provided for removing indium from impure metals or alloys particularly zincky lead metal containing indium by converting the indium into indium chloride. A melt of the metal containing indium is treated with a suitable chlorine supplying agent or gaseous chlorine to selectively remove the zinc and indium as chlorides using a low melting cover slag and temperatures sufficiently low to avoid loss of indium by volatilization.

Recovery from Scrap

The Indium Corporation of America is the only U.S. producer of secondary indium and information obtained from this company is summarized below by H. Nack, K. Murthy, E. Stambaugh, H. Carlton and G.R. Smithson, Jr. in PB 233 646, Springfield, Va., National Technical Information Service (May 1974).

(1) The total annual U.S. production (1972) of both primary and secondary indium is less than 2 million ounces ($<$62 tons). The market value of indium is about $4.50 per ounce. This high cost of indium makes the recovery process very profitable and necessary.
(2) The process for recovery of indium consists of using a small blast furnace followed by a series of typical chemical and electrical recovery processes, details of which are proprietary.
(3) Indium is not toxic. There are no environmental problems or emissions.

Recovery from Zinc Processing Fumes

It will be recalled that, like cadmium, indium commonly occurs as a minor constituent of certain zinciferous ores and, upon roasting such ores, the indium passes into the fume together with cadmium, lead, arsenic and other components. In the usual practice of treating the fume for the recovery of the bulk of its cadmium content, there remains a so-called lead sulfate residue comprising the indium and such ingredients as arsenic, cadmium, lead, and iron.

While the bulk of the indium will be found in the aforementioned sulfate residues, nevertheless the actual quantity of indium in such a residue is small so that one is confronted with the problem of separating a small quantity of indium from large quantities of other residue components. Thus, while the indium may be extracted readily from the residue by leaching with a suitable acid, such as sulfuric acid, or a saturated aqueous solution of sulfur dioxide, large quantities of other constituents of the residue, such as cadmium and arsenic, are also dissolved by the leach so that one still faces the problem of separating the small amount of indium from the other components of the leach solution. Further, if it is sought to precipitate the indium by simple neutralization of the leach solution, it will be found that large quantities of cadmium and arsenic will precipitate.

A process developed by *C. Zischkau and H.B. Linford; U.S. Patent 2,241,438; May 13, 1941; assigned to American Smelting and Refining Co.* involves the precipitation of indium from the indium-containing leach solution as indium phosphate, the conversion of the indium phosphate into indium oxide, and the reduction of the oxide to indium metal. The indium precipitates selectively as phosphate. In carrying out of the process, the initial step ordinarily is that of extracting the indium from the indium-bearing material by leaching with dilute acid, sulfuric acid being the preferred solvent.

The presence of iron in the leach solution may or may not interfere with the subsequent selective precipitation of the indium depending upon whether such iron is present in the ferric or ferrous state. Accordingly, as a precautionary measure, the leach solution should be treated with a reducing agent in order to assure that iron, if present, is maintained in the reduced condition. This can be accomplished readily by the addition of sulfur dioxide to the leach solution either simultaneously with, or subsequent to, the leaching. Following the reducing treatment, the indium is precipitated selectively from the leach solution to the substantial exclusion of such elements as arsenic, cadmium, and iron, by adding a soluble phosphate to the leach solution and adjusting, if necessary, the hydrogen ion concentration of the latter to a proper value. The soluble phosphate used may be any of those containing the PO_4 radical. As to hydrogen ion concentration, it has been found that a pH value within the approximate range of 2.5 to 4.0 should be observed. This treatment precipitates the indium substantially completely. This indium phosphate now is converted into indium oxide, it being important that this conversion be complete in order that reduction of the indium may be made smoothly and to avoid too much indium run-around in the residue from the reduction.

The conversion of the indium phosphate into metallic indium is accomplished by way of indium hydroxide. To convert into the hydroxide the indium phosphate is treated with caustic soda. Various procedures for so doing are open, but it is found that the simplest procedure is to boil the indium phosphate in a solution of caustic soda. The resulting indium hydroxide is filtered, washed, and dehydrated or roasted, to indium oxide, which then may be reduced to metal, preferably by passing hydrogen thereover at temperatures below such as will vaporize the indium substantially. A reduction temperature of from 800° to 850°C is found to be satisfactory. The resulting metallic indium may be refined electrolytically, if desired.

A process developed by *H.M. Doran, M.A. Jackson and A.I. Alf; U.S. Patent 2,384,610; September 11, 1945; assigned to Anaconda Copper Mining Company* also deals with indium recovery from zinc processing. The usual method employed for the recovery of indium from acid sulfate solutions comprises first precipitating the indium from solution as indium hydroxide. This is accomplished by neutralizing the acid solution with a base such as zinc oxide. The indium hydroxide precipitate is filtered away from the solution and dissolved in sulfuric acid.

This dissolution, filtration and precipitation may be repeated a number of times until a sufficient concentration of indium is obtained. The indium hydroxide precipitate is then finally dissolved in sulfuric acid, the undissolved portion of the material filtered away from the solution, and the indium-bearing acid sulfate solution treated with hydrogen sulfide gas. By suitable adjustment of the acidity of the solution, any arsenic present in the solution is removed as arsenic sulfide. By a second adjustment of the acidity of the solution, the indium is precipitated as indium sulfide. The indium sulfide precipitate is leached with concentrated hydrochloric acid, and the acid chloride solution, after suitable purification, forms the electrolyte from which indium metal is recovered electrolytically.

The process provides an improved method for the recovery of indium from indium-bearing material such as zinc calcines or fumes which avoids the necessity of treating large volumes of leach solution to recover indium present therein in very low concentrations. The method is described below with particular reference to the treatment of indium-bearing zinc calcine or zinc oxide fume containing impurities including iron and arsenic, but it is understood that the process can be applied equally well to the treatment of any other indium-bearing material in which the indium present is soluble in sulfuric acid.

A complete process for the recovery of indium from an indium-bearing material containing zinc in acid soluble form and further containing impurities including iron and arsenic involves first leaching the material with a very dilute acid solution in such manner that the bulk of the zinc is dissolved and the indium remains in the undissolved residue. In this manner the bulk of the zinc is separated from the indium in the initial leaching operation, and further treatment of the indium-bearing material is very much facilitated by the reduction of the volume of the material to be treated.

The indium-bearing residue from the initial very dilute acid leach is then dissolved in a relatively strong acid, advantageously sulfuric acid, to obtain an acid solution containing indium. This solution is treated with a reagent of the group consisting of zinc oxide, sodium bisulfite, and sodium sulfite in such manner as to form an insoluble precipitate containing indium free from the bulk of the iron and arsenic originally present. In this manner the indium-bearing material is freed, at an early stage in the process and without substantial loss of indium, from the greater part of the mineral impurities with which it is associated in the original material. In consequence the recovery of metallic indium of high purity is greatly facilitated. The indium-bearing precipitate thus obtained may be treated with an aqueous solution of a strong alkali to dissolve aluminum and a further amount of arsenic. After this treatment, the precipitate may be washed with water and decanted. The precipitate may then be treated with dilute sulfuric acid in such manner as to dissolve at least a portion of the remaining zinc without acidifying the precipitate. It is important that the precipitate be not acidified, or the indium will dissolve and the benefit of the alkali treatment and the dilute sulfuric acid wash will, to a large extent, be lost.

After completion of the dilute sulfuric acid treatment or wash, the indium-bearing precipitate is dissolved in a relatively strong acid solution, and this strongly acid indium-bearing solution is treated with a sulfide such as hydrogen sulfide gas so as to precipitate substantially all of the remaining arsenic. The precipitated sulfides are separated from the residual solution and the resulting arsenic-free solution is treated with metallic zinc, whereby metallic sponge indium is precipitated. It is important that all of the arsenic be eliminated from the solution prior to treatment with metallic zinc so as to avoid danger of the formation of extremely poisonous arsine gas. The sponge indium obtained by treatment of the solution with metallic zinc is then treated for the recovery of metallic indium in commercial form.

LEAD

Recovery from Lead Scrap

The principal sources of lead scrap are: soft lead, hard lead, cable lead, battery-lead plates, mixed common babbits, solder and tinny lead, type metals and drosses and residues. New scrap in the form of purchased drosses and residues make up 18% (143,400 tons) of the total of some 785,000 tons of lead scrap consumed by the secondary lead industry. The remainder, old scrap, is predominantly battery scrap with small amounts of the other scrap. The scrap is essentially all from domestic sources. In addition to containing metallic lead or compounds and alloys of lead, the scrap contains a variety of organic materials such as insulation, grease, and oil. Common alloying agents found in the scrap are tin, antimony, arsenic, cadmium, copper, indium, silver, zinc, tellurium, and bismuth.

Products from the lead segment of the secondary nonferrous metals industry are: (1) lead ingots (pure/soft), (2) lead ingots (hard/semisoft), (3) lead ingots (alloys), (4) lead oxide (battery lead oxide), and (5) lead pigments (Pb_3O_4 and PbO). Lead is recovered from lead scrap by three manufacturing operations: (1) scrap pretreatment, (2) smelting, and (3) refining/casting. In the scrap pretreatment operation, the lead scrap is treated to make it more amenable to further processing. This treatment involves densification and/or partial removal of metallic and nonmetallic contaminants by one of the following processes: (1) battery breaker process, (2) crushing process, (3) rotary/tube sweating process and (4) reverberatory sweating process. These processes are described in somewhat more detail in the paragraphs which follow.

Battery Breaker Process — The major source of lead is obsolete automobile batteries. By this process, the lead is separated from the nonmetallic portion of the battery. The process steps are: (a) draining the battery if not received dry, (b) breaking (crushing) the battery to break the lead portion from the nonlead portion, and (c) separating the lead from the battery case. This lead scrap in this form can be fed into the reverberatory or blast furnace for further processing and recovery of the lead. Energy required for this process is that to drive the equipment.

Waste products generated are primarily liquid wastes and solid wastes. A small quantity of dusts may be generated if the batteries are treated dry. These dusts are composed of dirt, plastic, or battery case materials and lead compounds such as lead oxide and lead sulfate. The liquid wastes are composed of sulfur acid, water, lead compounds, and the alloying agents from the lead plates. The solid wastes, which make up the bulk of the waste products from this process are composed of the organic materials used in battery case fabrication. In addition, these solid wastes contain some sulfuric acid and lead compounds. In view of the waste products generated by this process, serious pollution problems can arise if the wastes are not disposed of in a nonpolluting manner.

Crushing Process – Much of the drosses, residues, and possibly slags are received in large pieces, i.e., larger than desired for further processing. These scrap materials are crushed with jaw crushers to a suitable size. The process steps are: (a) load the crusher and (b) crush the scrap. Afterwards, the scrap is removed and further processed to recover the lead content. Energy required is limited to that necessary to drive the equipment. Waste products from this process are dusts resulting from the handling and crushing of the scrap. Emissions data are not available. However, significant quantity of emissions could be generated. Therefore, the process can present pollution problems, if the dusts are not collected..

Rotary/Tube Sweating Process – Lead sheathed cable and wire, aircraft tooling dies, type metal drosses, and lead dross and skimmings are treated by the rotary/tube sweating process to recover the lead value. The process steps are (a) charge the furnace with the scrap, (b) melt the lead value, (c) collect the molten lead, (d) cast the molten lead, and (e) remove the residue from the furnace. Energy required for this process is the fuel to heat the rotary furnace or sweating tube and electricity to drive the equipment. Atmospheric emissions and solid wastes are generated by this process. The emissions contain gases composed of sulfur oxides, entrained air, nitrogen oxides, and the fuel combustion products.

The particulate portion of the atmospheric emissions is composed of fumes, dusts, unburned fuel, soot, and fly ash. Furnace atmospheric emissions data are not available. However, based on data from similar processes, emissions factors are estimated to range from 70 lb of particulates per ton of metal processed (raw emissions factor for rotary reverberatory furnace smelting of lead) to 32 lb of particulates per ton of scrap processed (raw emissions factor for sweating of residual zinc).

Particle size of the emissions varies. The larger particles (dusts) are probably in the 5 to 20 micron or larger range, whereas the unagglomerated lead fumes vary in diameter from 0.07 to 0.4 micron with a mean particle size of 0.3 micron. Solid wastes are the metallic and nonmetallic portion of the scrap after removal of the lead. As significant quantities of atmospheric emissions and solid wastes are generated by this process and since lead and lead compounds are toxic or hazardous materials, the waste products from this process can cause serious pollution problems if the waste products are not controlled. Failure to control the atmospheric emissions can result in pollution of the atmosphere, whereas disposal of the solid wastes in an unapproved manner can result in water pollution.

Reverberatory Sweating Process – This process is normally used to sweat scrap containing high lead content such as lead battery plates. The process steps are the same as for the Rotary/Tube Sweating Process described above. Energy required is the fuel, gas or oil, to heat the furnace. Waste products from this process are atmospheric emissions and solid wastes. The atmospheric emissions are composed of gases and particulate matter containing both fumes and dusts. The gases discharged from the furnace contain the combustion products, sulfur oxides, unburned fuel, and hydrocarbons from pyrolysis of the organic compounds present in the scrap. The fumes and dusts contain metallic materials such as lead, antimony, and other heavy metals used to prepare the lead alloys, and nonmetallic materials such as the fluxes and carbon or other organic contaminants from the scrap.

Although emissions data are lacking, it is estimated based on data from similar processes that raw emissions factor for the reverberatory sweating process will range from approximately 30 lb per ton of charge to 70 lb per ton of metal processed. Unagglomerated lead oxide fume particles vary in diameter from 0.07 to 0.4 micron with a mean diameter of about 0.3 micron. The particle size of the dust is expected to be much larger, normally in the 5 to 20 micron or larger size range. The atmospheric emissions are controlled by the same system as is used for the smelting operation and discussed under that process. Solid waste materials are the residues remaining after removal of the lead. Composition of these wastes may vary somewhat, but generally will contain lead and other heavy metals such as antimony. This process has a high potential for the production of pollution problems if emissions are not controlled. Pretreated scrap is partially refined, i.e., some of the metallic and nonmetallic contaminants are removed by the smelting operation, which may take the form of reverberatory smelting or blast furnace smelting as described in the following paragraphs.

Reverberatory Smelting Process — Lead scrap (pretreated or untreated, mixed) is partially purified and densified by the reverberatory smelting process. This is achieved by the following process steps: (a) charge the furnace with the lead scrap and flue dusts, (b) melt the scrap, (c) allow the antimonical slag to rise to the surface of the melt, (d) tap the antimonical slag to the blast furnace, and (e) tap the molten lead melt to the refining/casting operation. Alternately, the molten lead may be poured and cast into ingots as a semisoft/hard lead product. The energy required for this step is from the fuel oil or gas to smelt the scrap.

Atmospheric emissions are produced by this process. No solid or liquid wastes are produced. The atmospheric emissions are composed of gases and particulate matter. The gases contain the products from combustion of the fuel and sulfur oxides along with any other gases from the scrap. The particulate matter (fume) is composed of oxides, sulfides, and sulfates of lead, tin, arsenic, copper, and antimony. Recoverable lead from the charge is about 47%. Approximately 46% of the charge is slag and 7% is fumes and dusts. The unagglomerated particulate matter was found to have a particle size range of 0.07 to 0.4 micron with a mean value of 0.3 micron. The particle shape was essentially spherical.

The reverberatory smelting process accounts for approximately 75% of the emissions from the lead segment of the secondary nonferrous metals industry. Particulate emission factor for reverberatory smelting is estimated at 130 lb (uncontrolled) and 1.6 lb (controlled) per ton of metal processed. Raw sulfur oxide emission factor is estimated at 85.0 to 190 lb/ton of metal processed. Particulate emissions from this process are controlled with a baghouse, scrubber, or both. The baghouse is arranged as the final collector. These systems also include auxiliary items such as gas cooling devices and settling chambers. Sulfur oxide emissions are controlled in some plants with a wet scrubber. The process produces significant quantities of atmospheric emissions and is, therefore, a potential source of pollution problems.

Blast Furnace Smelting Process — Semisoft/hard lead which is a product and an intermediate product, is produced from pretreated scrap, antimonical slag (reverberatory slag), and rerun slag by the blast furnace smelting process. In this process the molten lead flows almost continuously and the slag is tapped at intervals. A typical charge which is added as the material melts down is: 4.5% scrap iron; 3% limestone; 5.5% coke; and 82.5% drosses, oxides, and reverberatory slag. The drosses are miscellaneous drosses consisting of copper drosses, caustic dross, and dry drosses from the refining process in the pot furnace. Thus, the process steps are: (a) add the charge with a composition as noted above, (b) continuously remove the molten lead, (c) remove the slag at intervals, and (d) cast the lead ingots or transfer molten lead to the refining kettles. The energy utilized in this process is from coke, which also serves as a reducing agent.

Waste products produced by this process are atmospheric emissions and solid wastes. The atmospheric emissions consist of gases containing carbon monoxide, hydrocarbons, sulfur oxides, nitrogen oxides, and probably nitrogen. The particulate matter in the emissions contains oil vapor, smoke, fume, and dust. The fume and dust are composed of lead, tin, zinc, coke dust, sulfur, and other metals contained in the charge. The fumes are volatized and condensed from the charge, whereas the dusts are carried from the furnace by entrainment in the furnace gases.

Particle morphology of the particulate emissions are similar to those from the reverberatory furnace. Particle size ranges from 0 to 16 microns with approximately 15% between 0 to 1 micron, 45% between 1 to 2 microns, 20% between 2 to 3 microns, 15% between 3 to 4 microns, and 10% between 4 to 16 microns. No data are available on particle shape;

however, the dust is most likely irregular and the fume (smaller size) is spherical. Emissions rate is high from the blast furnace smelting process. Raw particulate emissions factor (uncontrolled) is estimated at 190 lb/ton of metal processed. Sulfur oxide emissions factor is estimated at 90 lb/ton of metal processed. Controlled emissions factors are 2.3 and 0.8 to 46 lb/ton of metal processed, respectively.

The emissions are subject to control. The baghouse is the most acceptable device for controlling the particulate emissions. The gas stream containing the emissions is first passed through a series of water or air cooled tubes and then to the baghouse. To prevent tar volatiles from blinding the bags, the temperature of the baghouse is kept high. Lime is also added to the gas stream to prevent the blinding action. The gas stream from the bag house is discharged to a stack or possibly to an electrostatic precipitator in an attempt to further clean the gases before emitting them to the atmosphere.

Wet scrubbers may be employed to clean the gaseous portion of the stream. The caustic (NaOH) scrubber is more effective than the water spray chamber for removing sulfur oxides as evidenced by the fact that caustic scrubbing reduced the emissions factor from 90 lb/ton of metal processed to 0.8 lb/ton of metal processed, whereas water scrubbing reduced the emission factor to only 46 lb/ton of metal processed. Solid wastes from this process are the baghouse and flue dusts and fumes, and the slag. For the most part, these are recycled to the process because of the high lead content. Thus, uncontrolled, the blast furnacing process produces large quantities of atmospheric emissions and solid wastes which can cause serious pollution problems, both from the atmospheric and water pollution aspects.

The intermediate products from the reverberatory and blast furnace processes, soft lead and antimonical lead, required additional purification or refining to produce the desired products. In addition to the metal products, some of the lead is converted to lead oxide products. The processes employed to achieve conversion of the impure lead intermediate products to the final products are: (1) kettle (softening) refining, (2) kettle (alloy) refining, (3) kettle oxidation, and (4) reverberatory oxidation. These processes are described in somewhat more detail in the paragraphs which follow.

Kettle (Softening) Refining Process – The intermediate products from the smelting operation may contain copper (generally not present) and antimony which makes the lead hard. Removal of these two contaminants by the kettle (softening) refining process produces soft lead, one of the products from the lead segment of secondary nonferrous metals industry. The process steps are: (a) charge the preheated kettle, (b) melt the charge, (c) stir the molten charge to mix in the flux, (d) remove the skimmings, (e) pour the molten metal, and (f) cast into ingots (bullion). In some cases, the molten lead is poured directly from the smelting operation to the refining kettle. Several fluxes or purifiers are used. Included in this list are sodium hydroxide, sodium nitrate, aluminum chloride, or aluminum. Air is occasionally used to blow the melt.

Source of energy for this process is generally gas to heat the kettle indirectly and electricity to drive the equipment. Waste products produced are atmospheric emissions and solid wastes (skimmings). The atmospheric emissions containing mainly fumes and possibly some dust are composed of such metals as lead, antimony, sodium, and trace quantities of copper, zinc, arsenic, bismuth, and tin. The gaseous portion of the emissions contains primarily the fuel combustion gases. Particle morphology should be similar to the fumes from the reverberatory smelting process, i.e., the particle size is in the submicron range and the shape is generally spherical.

Although the quantity of emissions from the refining process is much less than from the other lead processes, for example, 0.8 lb of particulates per ton of metal processed as compared to 190 lb/ton of metal processed for reverberatory smelting, the toxic or hazardous nature of lead requires that the emissions be controlled. The most widely used device is the baghouse. The fumes are collected via a hood over the kettle. The solid wastes are skimmings containing the flux and other metal values removed from the melt during the refining operation. These wastes along with the particulate matter from the baghouse are generally recycled to the blast furnace for recovery of the lead value. Thus, the kettle (softening) refining process produces waste products which can cause pollution problems, if wastes are not collected and disposed of in a nonpolluting manner.

Kettle (Alloy) Refining Process – Kettle (alloy) refining process involves treatment and adjustment of the metals content of lead to produce the desired lead alloy. This involves the following process steps: (a) charging the preheated kettle, (b) melting the charge, if not added molten, (c) stirring to mix the flux into the charge, (d) skimming to remove collected impurities, (e) pouring, and (f) casting. Alloying agents commonly used are antimony, copper, silver, and tin. Source of energy for this process is natural gas and electricity to drive the equipment. Emissions data—waste products produced, emissions factor, control of emissions, and particle morphology—are essentially the same for this process as for the kettle (softening) process above. Thus, the process can be a source of environmental pollution.

Kettle Oxidation Process – Battery lead oxide (PbO containing approximately 20% lead metal) is produced by the kettle oxidation process. The process steps are: (a) charge kettle with molten lead from a melting pot, (b) rapidly agitate (mechanically) the lead, (c) draw air over the surface of the melt through the duct leading to the baghouse, and (d) collect the lead oxide (fumed from the surface of the melt) in the baghouse. Source of energy is natural gas to heat the melt furnace and the oxidation furnace, and electricity to drive the equipment.

Atmospheric emissions are produced by this process. These include (1) the combustion gases which are separate from those from the melting chamber, oxidation chamber, and the baghouse, and (2) fumes and dust from the melting and oxidation chambers and the baghouse. The gases are composed of carbon dioxide, carbon monoxide, sulfur oxides, nitrogen oxides, and unburned fuel, whereas the particulate matter contains lead oxide, lead metal, and other metal values from the melting and oxidation chambers and the baghouse.

The major source of emissions is the baghouse since the air is being swept over the oxidation chamber into the baghouse. Emission factor for the baghouse will vary somewhat, but in most oxide collection systems, collection efficiency averages greater than 98%. On this basis, emissions factor is estimated at less than 40 lb of lead oxide per ton of oxide produced. Particle size of the oxide is in the submicron range (approximately 0.2 to 0.5 micron) and particle shape is spherical. Thus, the kettle oxidation process is a source of pollution problems.

Reverberatory Oxidation Process – The most widely used lead oxide pigments are lead monoxide (PbO) and red lead (Pb_3O_4). These are produced by the reverberatory oxidation process. The process steps are: (a) charge the hot reverberatory furnace with molten lead, (b) agitate the melt while oxidizing the lead to lead oxide, and (c) remove the lead oxide from the furnace and cool rapidly. Lead oxide (PbO) is produced by fully oxidizing the lead, whereas red lead (Pb_3O_4) is produced by overoxidizing the PbO. Energy required for this process is from the fuel (natural gas or fuel oil) to heat the furnace.

Waste products produced by this process are atmospheric emissions. These emissions are gases containing the combustion gases, excess air, and nitrogen and particulate matter containing lead oxide along with the impurities found in the lead charge. These emissions are carried from the furnace by entrainment in the furnace gases or from volatilization and condensation of the lead oxide fumes. Particle size is in the submicron range and the particles are spherical in shape. Emissions data are not available for the reverberatory oxidation process. However, it is estimated that the emission factor will range from approximately 40 to 150 lb/ton of lead oxide produced. These emissions are most likely controlled by a baghouse. Thus, the reverberatory oxidation process for producing lead oxides can be a source of serious pollution problems.

Recovery from Steel Mill Dusts

A process developed by *J.E. Allen; U.S. Patent 3,850,613; November 26, 1974; assigned to Ferro-Carb Agglomeration Ltd., Inc.* is one in which steel mill and foundry waste dusts containing zinc and/or lead oxides in addition to iron oxides are briquetted by means of a hydrocarbonaceous binder, together with sufficient carbon to reduce at least the lead and zinc oxides to the metal.

The briquettes are preferably preferentially treated with an oxygen-containing gas at 350° to 600°F to selectively react with the binder to obtain strong briquettes; and the briquettes are then heated to a temperature of between 1800° to 2500°F to cause any zinc or lead present to first be reduced to metal and be volatilized as metal, and then be oxidized to ZnO and PbO respectively in the gaseous phase. The zinc and lead oxides are recovered from the exhaust gases in an appropriate dust collector while the iron-bearing briquettes are cooled in a nonoxidizing atmosphere, and are then used as ferrous charge in the iron or steelmaking processes.

Recovery from Used Storage Batteries

A process developed by *M.E. Elmore; U.S. Patent 3,393,876; July 23, 1968; assigned to Bunker Hill Company* involves recovery of metallic lead and/or lead alloys from waste storage batteries having lead compound dust in the cells thereof. The process is carried out in a system having sink flotation and smeltering stages therein, and comprises subjecting a first portion of a stream of the batteries to the sink flotation stage by comminuting at least the battery cells therein, and charging the mass of comminuted particles into a liquid suspension of a solid densification medium consisting essentially of lead compound dust, the quantity of which is adjusted to cause the valued and nonvalued particles in the mass to separate into diverse regions of the suspension.

The valued particles are removed from the suspension and charged into the smeltering stage, together with the lead compound densification medium adhering thereon. Meanwhile or subsequently thereto, a second portion of the stream of batteries is subjected to the sink flotation stage by comminuting at least the battery cells therein and liberating the lead compound dust therefrom, and charging the mass of comminuted particles from such second portion into the aforementioned suspension, together with at least enough of the lead compound dust liberated from the second portion to compensate for the loss of lead compound dust densification medium adhering to the removed particles of the first portion.

A process developed by *M. Shoeld; U.S. Patent 3,395,010; July 30, 1968* is one for reclaiming lead-acid storage batteries, which are constituted by a laminar construction of cellulosic paper, lead foil and lead oxide and/or lead sulfate with sulfuric acid, substantially free from antimony. The carbon of the paper is present in approximately stoichiometric amounts to reduce the lead compounds, and the materials are heated out of contact with the atmosphere to about 800°F, which is above the melting point of lead and the dissociation temperature of sulfuric acid. Molten metallic lead is tapped off and the exit gases are water vapor and sulfur trioxide and carbon dioxide, from which useful values may be recovered. Figure 66 shows a suitable form of apparatus for the conduct of the process.

There is shown a furnace **1** of the upright shaft type, whose upper end **3** is reduced in diameter for the reception of the lead-acid battery materials to be reclaimed. A pair of vertically spaced horizontally sliding gates **5** define between them a hopper portion **7** of the furnace. With the upper gate **5** open and the lower gate **5** closed, material is introduced into the hopper portion **7**. The upper gate **5** is opened to transfer the material by gravity to the top of the furnace. In operation, the furnace is kept substantially full. Atmospheric air or other oxidizing gases are thus largely excluded from the furnace.

The reduced upper end **3** of the furnace then tapers downwardly outwardly in the form of a substantially conical hood **9** which at its lower end is of the same diameter as a vertical cylindrical shell **11**. The portions of the furnace thus far described are metallic and

FIGURE 66: PROCESS AND DETAIL OF FURNACE FOR RECOVERY OF LEAD FROM SCRAP LEAD-ACID STORAGE BATTERIES

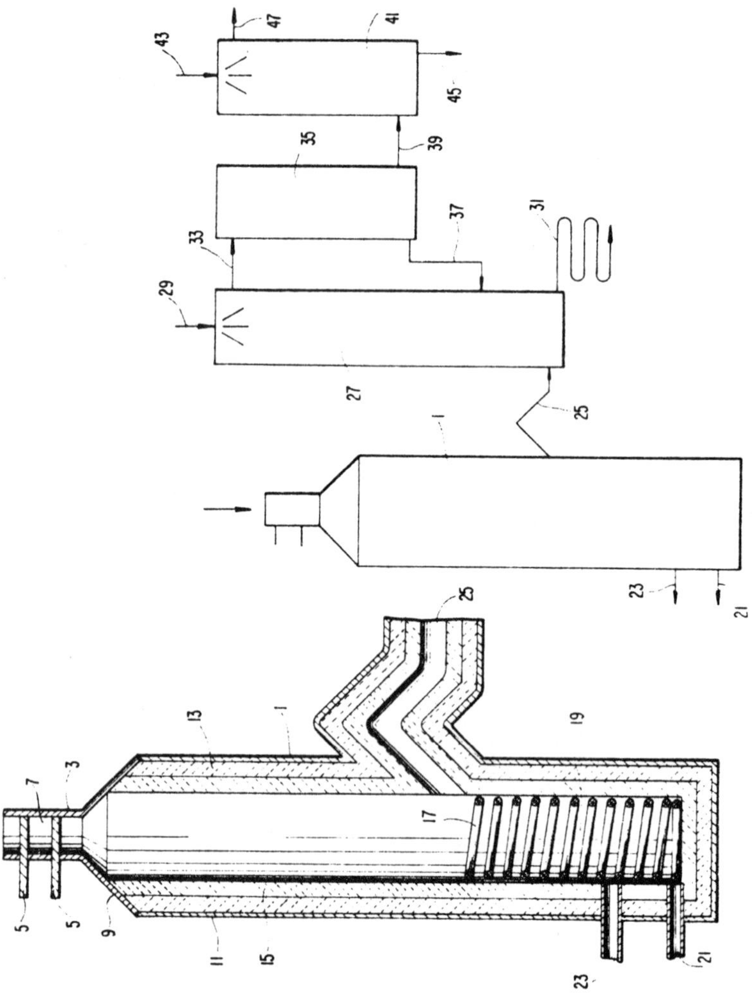

Source: M. Shoeld; U.S. Patent 3,395,010; July 30, 1968

preferably of stainless steel. Alternatively, they may be lead-lined steel, or of any other material which presents an inner surface that is inert to hot sulfuric acid vapor. Shell **11** forms a support for refractory linings in the form of an outer lining of insulating brick **13** and an inner lining **15** of acid-proof brick. The linings **13** and **15** cover not only the side walls but also the bottom of furnace **1**.

Disposed in a lower portion of furnace **1** are electric resistance heating elements **17** of nichrome or the like, heated by a source of electric current (not shown), which define within their confines a reaction zone **19** which is the highest temperature portion of furnace **1**. Material introduced into the furnace thus moves by gravity downwardly through the length of the furnace, rising in temperature as it goes, until it reaches the highest temperature in the reaction zone, which is preferably a temperature about 800°F, and in any event is above the melting point of lead and the dissociation temperature of sulfuric acid, which under the contemplated circumstances is each about 560°F. The top of the furnace, by contrast, is at a temperature of about 150° to 200°F. The furnace is operated under a slight negative pressure, to prevent the escape of potentially corrosive gases.

In the reaction zone **19**, the lead oxide or lead sulfate is reduced by the carbon of the cellulosic paper. The molten lead thus produced is tapped through conduit **21**, while any dross or slag may be tapped through a conduit **23**. The carbon is present in about a stoichiometric amount for this reduction, and the reactions in zone **19** proceed as follows:

$$PbO_2 + C \longrightarrow CO_2 + Pb$$

$$2PbSO_4 + C \longrightarrow 2Pb + 2SO_3 + CO_2$$

$$H_2SO_4 \longrightarrow H_2O + SO_3$$

$$H_2O \text{ liquid} \longrightarrow H_2O \text{ vapor}$$

The reduction of lead oxide is slightly exothermic. The reduction of lead sulfate is somewhat endothermic. The dissociation of sulfuric acid is endothermic, as is also the evaporation of water. As a result, the heat required to be supplied by the heating elements **17** is not great and the cost of electric resistance heating is not high. For example, let it be assumed that the composition of a spent battery is approximately as follows: 12 lb of lead metal foil; 24 lb of lead in active masses; 3 lb acid-proof porous paper; and 11 lb sulfuric acid, as such, and in lead sulfate plus oxygen in lead oxides plus water in the acid.

The carbon in the cellulosic paper is accordingly about stoichiometric with regard to the lead compounds, and the products are essentially molten metallic lead and a gaseous mixture of water vapor and carbon dioxide and sulfur trioxide, the gaseous mixture leaving by the conduit **25**. From conduit **25**, the gaseous affluent passes to a brick-lined acid tower with Gay Lussac packing, shown at **27**. Conduit **25** is insulated and removes the gas from a point contiguous to reaction zone **19**, so that the gas is at a temperature substantially above the dissociation temperature of sulfuric acid, preferably about 800°F.

There is accordingly no opportunity for reassociation to sulfuric acid, and the gas enters tower **27** in an essentially dry and noncorrosive condition. Water is sprayed in by conduit **29**, the temperature of the tower being about 250°F, and hot sulfuric acid at a specific gravity of about 1.28 is removed and cooled through conduit **31** for reuse in batteries. Acid at this specific gravity is in fact preferable for use in lead-acid storage batteries.

The gaseous overhead from tower **27**, which is mostly carbon dioxide and water vapor, leaves through conduit **33** and passes through Cottrell precipitator **35**. Water practically free from acid is returned to tower **27** through conduit **37**. A packed tower may be used instead of a Cottrell precipitator. The gas leaving precipitator **35** passes through conduit **39** to a condenser cooler **41** into which water is sprayed through conduit **43**. The condensate of water leaves through a conduit **45**; and the gaseous overhead leaving through conduit **47** is moist carbon dioxide, which may be dried and used as relatively pure carbon dioxide.

A process developed by *F.C. Smyers and J.H. Garrison; U.S. Patent 3,679,395; July 25, 1972; assigned to United States Smelting, Refining & Mining Co.* for recovering lead and other metallic components from whole storage batteries and/or battery scrap involves the preparation of a unique feedstock which is useful as a charge for a blast furnace or other reducing apparatus.

This unique feedstock is prepared, for example, by treating batteries or battery scrap to form a comminuted mass of battery particles, preparing a porous, pelletized bed of battery particles, and sintering the pelletized bed of battery particles for a time and under conditions sufficient to form an agglomerated sinter cake and to reduce or eliminate the sulfur content of the pelletized bed. The sinter cake is used alone or in combination with various other materials as a blast furnace charge. Figure 67 is a flow diagram showing the various steps in the process.

FIGURE 67: PROCESS FOR RECOVERING LEAD FROM BATTERIES BY SINTERING COMMINUTED SCRAP AS FURNACE FEED

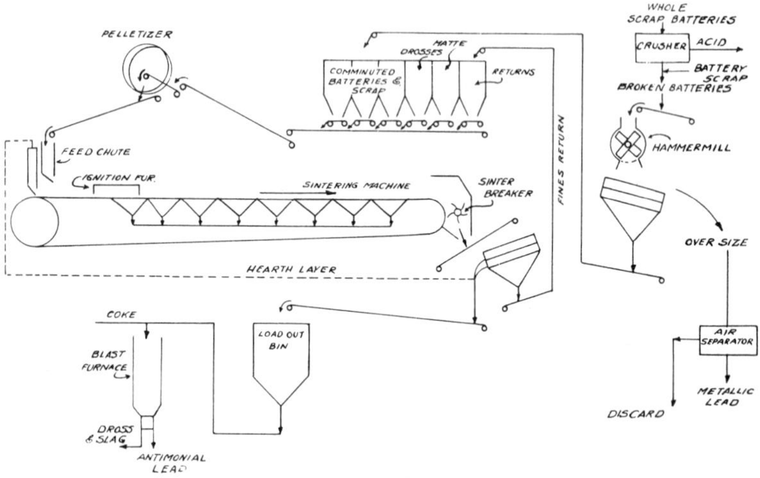

Source: F.C. Smyers and J.H. Garrison; U.S. Patent 3,679,395; July 25, 1972

A process developed by *R. Fischer; U.S. Patent 3,777,994; December 11, 1973; assigned to Klockner-Humboldt-Deutz A.G., Germany* is one in which lead accumulators or used storage batteries are coarsely ground into pieces in an impact crusher and the fine material containing metallic lead is separated and ground again and then classified.

The remaining part of the coarsely ground material is subjected to a sink-flotation-separation and the fine material recovered as a sludge is introduced into a tubular vibrating mill that is filled with grinding bodies and ground therein, whereupon the ground material is screened and classified to recover the lead therein. Figure 68 is a flow diagram of the process and the operation of the process will now be described in a typical example with reference to that diagram.

Example: For the recovery of the component of metallic lead contained in used lead accumulators, the accumulators are first broken up in an impact crusher **1** and preliminarily ground. The material leaving the impact crusher **1** consists substantially of broken casing pieces, the grid lead with the filler mass still adhering thereto, already ground filler mass, connecting straps consisting of lead, the separators consisting of synthetic material, wood or paper and the like, as well as a certain portion of iron parts, such as handles, fastening elements or the like.

FIGURE 68: PROCESS FOR LEAD RECOVERY FROM BATTERY SCRAP BY SINK-FLOAT SEPARATION

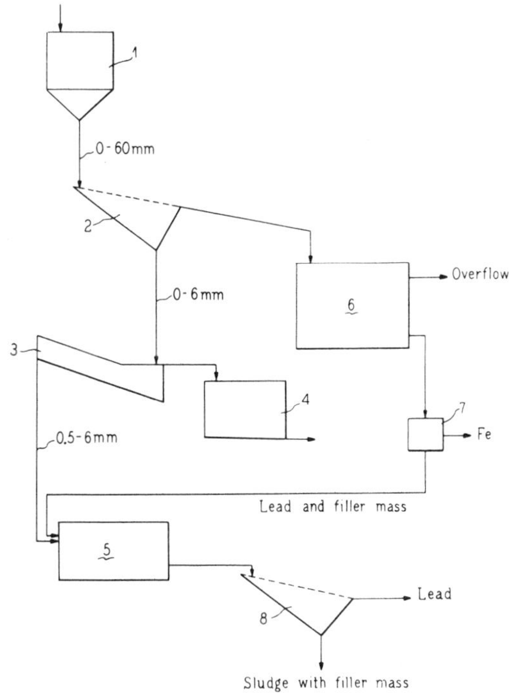

Source: R. Fischer; U.S. Patent 3,777,994; December 11, 1973

During the coarse grinding operation, a fine material is produced having grain sizes of 0 to 6 mm, which constitutes between 30 and 40% of the total charge and still contains 7 to 10% metallic lead with reference to the entire charge. In order to obtain a smooth or faultless operation of the sink-floatation device, in which one uses either a separate dense liquid or a heavy medium assisted by the produced sludge, the fine material produced by the impact grinding operation has to be first separated from the other material.

Such a separation of the fine material may be performed, for example, by employing a wet sieve **2**, whereby the fine material as screenings having a grain size from 0 to 6 mm is desludged and to certain degree is dehydrated. This may be accomplished to advantage in a spiral classifier in which grain sizes of 0.5 to 6 mm are withdrawn as solid material, while the finest components are withdrawn together with the liquid and subsequently are thickened to form a sludge in a slurry thickener **4**.

The fine material which has been desludged in this manner is now introduced in a tube mill **5** filled with grinding bodies, and is ground. The grinding bodies preferably consist of rods. After passing through such a tube mill, the material preferably is delivered to a wet sieve **8** and classified.

The wet sieve **8** depending upon the degree of grinding and the size of the lead particles has a sieve layer with a mesh size up to approximately 1 mm. The sieve residue contains, apart from low quantities of impurities resulting from accumulator or storage battery casings made of synthetic material, solely the metallic lead component, while the screenings consist of the filler mass which is withdrawn in the form of a sludge. The metallic lead thus recovered may then be melted.

Since during the coarse grinding operation according to the type of machine used, not the entire filler mass is knocked from the grid-lead, a certain amount of the filler mass reaches the sink-floatation device **6** and is here, on account of its high specific weight, withdrawn together with the lead and with any iron components with the heavier phase.

In fact it may happen that the portion of lead-sulfate-containing filler mass is so great that in the subsequent melting process, the limits of tolerance permissible for the sulfur dioxide content of the exhaust gases in the subsequent melting process is exceeded.

Therefore, after separation of the iron components, for example with the aid of magnets **7**, which are arranged over corresponding delivery members, the entire heavier phase is introduced into the tube mill **5**, in which due to the action of the grinding bodies, the filler mass still adhering to the lead particles is knocked off and is completely pulverized. In a subsequent classification by means of the sieve **8**, the metallic lead is separated from the finely ground filler mass.

A process developed by *F.A. Acoveno and T.W. Freudiger; U.S. Patent 3,883,348; May 13, 1975; assigned to R.S.R. Corp* is a process for the conversion of waste battery scrap materials containing lead sulfate comprising an initial leaching operation to convert such lead sulfate to lead carbonate while at the same time converting the aqueous ammonium carbon solution to ammonium sulfate.

Such is followed by filtration to thus separate soluble ammonium sulfate from the insoluble lead carbonate. The ammonium sulfate is then suitably crystallized resulting in a commercial product namely, a useful fertilizer compound.

The lead carbonate referred to in the foregoing is decomposed by application of heat to produce lead oxide and carbon dioxide, the latter then being reacted with aqueous ammonia to produce more ammonium carbonate solution for another beginning leach cycle. The advantages of the system reside in the reduction of emission of noxious sulfur oxides (SO_x) in an amount from at least about 80 to 90% while at the same time producing a useful commercial product, ammonium sulfate.

A process developed by *A.E. La Point; U.S. Patent 3,892,563; July 1, 1975* involves separating the constituents of a lead-acid storage battery of the type having at least grid plates of antimonial lead, active material consisting of lead, lead oxide and lead sulfate, a case and separators of organic material and acid, comprising the steps of:

(a) crushing whole batteries into fragments in a ball mill in the presence of water and sodium carbonate, to form with the active material a heavy medium suspension of density in which the organic material floats and the fragmented grid plates sink; and to react the sodium carbonate chemically with the lead sulfate to transpose it into lead carbonate and to neutralize the acid;

(b) removing the organic material and the suspension continually through an opening at one end of the ball mill onto a first screen adapted to retain the organic material and to pass the suspension therethrough;

(c) advancing the fragmented grid plates toward the opposite end of the ball mill; and

(d) discharging the fragmented grid plates mechanically through an opening in the other end of the ball mill onto a second screen adapted to retain the fragmented grid plates and pass therethrough any active material adhering thereto, thereby separating the battery constituents into three distinct streams consisting respectively of antimonial lead, active material and organic material.

Battery plate scrap is contaminated with residual sulfuric acid. Failure to remove the acid prior to melting results in evolution of sulfur oxides. The SO_x emissions vary, being high during initial melt-down and low at the end of the processing stage. Solutions being tried for the abatement of SO_x emission from lead recovery are:

(1) Lime-sulfur dioxide scrubbing system at the General Battery Corporation, Reading, Pennsylvania. This system uses wet scrubbing with hydrated lime. The underflow sludge, the flyash resulting from burning pulverized coal and the crushed plastic battery cases are being tried to produce a composition that is sanitary, nonreverting, nonleaching and dimensionally stable. The process was developed by G. & W.H. Corson Inc., Pymouth Meeting, Pennsylvania.

(2) The U.S. Bureau of Mines in College Park, Maryland, is developing a process for reducing the sulfur oxide emissions from recovery of lead from scrap battery plates. This process entails:

   (a) heating a mixture of the lead scrap and calcium hydroxide in a calciner to convert the sulfur content to calcium sulfate;

   (b) adding solid reducing agent and flux and reducing the charge at 650°C which is below the decomposition temperature of calcium sulfate, and

   (c) removing molten lead for further processing and the slag for disposal. Consideration is being devoted to developing a process for the regeneration of the calcium sulfate to lime for recycle and generation of elemental sulfur for use in other processes.

This problem has been discussed by H. Nack, K. Murthy, E. Stambaugh, H. Carlton and G. R. Smithson, Jr. in PB 233 646, Springfield, Va., National Technical Information Service (May 1974).

MAGNESIUM

Recovery from Incendiary Alloys

The recovery of magnesium and cadmium from incendiary alloys by vacuum distillation has been described by H.S. Caldwell, Jr. and M.J. Spendlove in Report of Investigation No. 5476, Washington, D.C., U.S. Bureau of Mines (1959).

A magnesium-base incendiary alloy, containing about 20% cadmium, 4% aluminum, and 0.5% zinc, was investigated in a series of vacuum-distillation experiments to determine the feasibility of recovering magnesium and cadmium as refined metals. Alloys prepared in the laboratory and actual bomb-casing alloy were studied at temperatures of 550° to 750°C and pressures of 100 to 150 microns.

Simple distillation and fractional distillation methods were only partly effective in separating the cadmium and magnesium. They were separated satisfactorily, however, with a special vapor baffle designed to condense the metals selectively. Magnesium and cadmium products were 99% pure and together comprised 82% of the total condensate. About 17%

of the condensate was intermediate-grade alloy that could be recycled with new charge and contained 70% magnesium and 30% cadmium.

Recovery from Scrap

Magnesium recovered from scrap constitutes approximately 10% of the United States production of magnesium metal. In 1972, this amounted to 15,662 short tons, most of which was recovered from aluminum alloys. Emissions from the magnesium segment include minor amounts of particulates and gases emitted to the atmosphere and solid waste to the land.

Major sources of raw material to this segment include new scrap (79%) and old scrap (21%). Most of the new scrap comprises magnesium-aluminum alloys generated in-house in the fabrication and machining of finished components. The old scrap, also comprised of magnesium-aluminum alloys, is generated from scrap automobile and aircraft parts and therefore, contains hand sortable ferrous and nonferrous metallic parts. Other sources of scrap include drosses and sludges.

Products from the magnesium segment of the secondary nonferrous metals industry include: (1) magnesium alloy ingots, (2) magnesium alloy castings and shapes, (3) aluminum-magnesium alloys (4) zinc and other alloys and (5) cathodic protection rods. Items (1), (3) and (5) constitute more than 90% of the products. The recovery of magnesium from scrap involves three manufacturing operations: scrap pretreatment, smelting and casting. Magnesium scrap is hand sorted to separate the large and easily identifiable pieces of other metals from the scrap magnesium parts.

The hand-sorting process consists of separating the magnesium and magnesium-alloy parts from all the other metals present in the scrap. The processing steps involve spreading the scrap on the floor and hand picking the lightest pieces which are magnesium and magnesium-alloys. This hand picking of light metal becomes second nature to the person experienced in the process. The process does not require any energy except human labor. Fine particles left on the floor constitute the solid waste generated by this process. It is usually insignificant as a source of pollution. Thus, the hand sorting process does not offer any potential for pollution.

The smelting of sorted scrap magnesium and the smelting of drosses and sludges are two types of smelting processes used for magnesium scrap. However, the same company does not necessarily do both these operations. The scrap melting facility finds it convenient to ship its sludge (slag) to a sludge processing facility, thus avoiding a potential solid waste disposal problem at the scrap melting plant.

Open Pot Melting Process: The sorted solid magnesium and alloy scrap is transferred to a melt crucible of about 1,000 lb capacity, where the scrap is melted and the magnesium separated from the contaminants. The process steps are: heat the crucible to melt the scrap; add flux which is a mixture of calcium, sodium, and potassium chlorides; pour the molten metal into molds; cast the ingots; and remove the slag or sludge. Thermal energy is the main energy requirement of this process.

The wastes generated are: gaseous effluents consisting of chlorine, hydrogen chloride, smoke from oil and grease, and particles of dust, including small amounts of magnesium oxide (MgO); and solid waste as sludge containing slag which is sent for reprocessing at a sludge smelter. The estimated uncontrolled particulate emission factor for a pot furnace is 4 pounds per ton of metal processed.

This process has a moderate air pollution problem arising from the gases generated. In fact, the in-house pollution is a greater threat to the plant personnel than it is to the public by way of ambient air pollution. Some installations use hoods to vent the gases to ambient air. The potential pollution from this process is significant by way of air pollution if not controlled properly.

Sludge Smelting Process: This process recovers magnesium from sludges produced in primary and secondary smelters. The process steps are: heating, flux addition, slag removal, and casting. Energy required for this process is thermal energy.

Potential pollutants from this process are solid wastes and atmospheric emissions. The solid waste—slag—contains the fluxes and contaminants removed from the scrap. The atmospheric emissions are composed of chlorine and hydrogen chloride gases from the melting pot, combustion gases, and dust. The atmospheric emissions, if uncontrolled, present an atmospheric pollution problem, while the slags are a source of water pollution if disposed of in an unapproved landfill.

The molten magnesium is cast into ingots in the ingot casting process. The process steps are: covering the melt with a protective flux, pouring the melt into casting dies, and casting the melt. Energy requirements of this process are not significant. Minor quantities of atmospheric emissions are generated. Thus, the quantity of emission is too small to represent any pollution potential.

MANGANESE

Recovery from Open Hearth Slags

A process developed by *J.C. Vignos; U.S. Patent 2,546,936; March 27, 1951; assigned to Ohio Ferro-Alloys Corp.* involves fractionally separating the phosphorus from the slag as an iron-phosphorus alloy under certain conditions and then separating the manganese, instead of reducing the combined oxides of phosphorus, sulfur, iron, manganese, etc. and then attempting to separate the phosphorus from the resulting alloy.

A method of fractionally separating phosphorus from molten slag outside of a furnace by the addition of a reducing agent involves reaction with the slag to maintain it in the molten condition and at the same time cause substantially continuous agitation of the slag during its reaction. The agent will, if added in the proper amounts, reduce the phosphorus oxide and sufficient iron oxide to combine with the phosphorus in the form of an iron-phosphorus alloy which may be readily separated from the remaining manganese oxide containing material.

Most efficient separation may be had if the composition of the iron-phosphorus alloy has a melting point at or below the melting point of the slag. Preferably the composition of any reducing agent is adjusted so that it not only maintains the slag in the molten condition but also substantially maintains the thermodynamic balance of the slag. The remaining manganese oxide containing material may then be treated by any of the known reduction methods to produce a satisfactory manganese alloy.

A reducing agent which has the aforementioned required properties is a mixture of carbonaceous and noncarbonaceous materials, preferably a mixture of carbonaceous material such as coke breeze and ferro-silicon. Ordinarily a mixture of 40% carbonaceous material and 60% noncarbonaceous material will form a satisfactory reducing agent although these proportions may vary depending upon the composition of the slag and the nature of both the carbonaceous and noncarbonaceous material.

Under certain conditions as little as 5% carbonaceous reducing agent may be used in combination with 95% noncarbonaceous reducing agent but generally the proportion of carbonaceous material should be larger. While ferro-silicon is preferred as the noncarbonaceous reducing agent, other noncarbonaceous reducing agents including aluminum, magnesium, calcium, and titanium and their alloys may be used. A starter or initiator, such as sodium nitrate or other oxidant, may be added to the mixture to accelerate the reaction.

A process developed by *R. Perrin; U.S. Patent 2,732,293; January 24, 1956; assigned to Societe d'Electro-Chimie d'Electro-Metallurgie et des Acieries Electriques d'Ugine, France* involves recovering manganese from basic open-hearth slags. These slags contain substantial proportions of manganese, for example 5 to 10%. They also contain large quantities of iron, and as impurities, the sulfur and phosphorus they have absorbed in the course of the operation of refining steel. It is essentially the presence of these two elements which makes these slags unsuitable for being reused directly in an open-hearth furnace.

One of the proposed processes for recovering manganese from basic open-hearth slags consists in reducing them in a blast-furnace so as to obtain a ferro-manganese or a spiegel; but the metal thus obtained contains prohibitive quantities of phosphorus and must be subjected to a second treatment, for example a blowing in a converter, to obtain a new slag rich in manganese and poor in phosphorus, which, in its turn, is subjected to a new reducing operation. Such a process is expensive and requires the use of expensive equipment.

The process enables one to avoid these disadvantages and conduces to an easy and inexpensive recovery of manganese. The process consists essentially, first in converting the basic open-hearth slag into an acid slag by incorporation of acid elements, in effecting a partial reduction of the acid slag so as to remove the greatest part of the phosphorus and sulfur, while keeping most of the manganese in the slag, and in reusing the slag thus refined as a raw material for producing manganese.

The recovery of manganese from open-hearth slags and low-grade ores by smelting and selective oxidation has been described by R.C. Buehl, M.B. Royer and J.P. Morris in Report of Investigation No. 6596, Washington, D.C., U.S. Bureau of Mines (1965).

A pyrometallurgical process for recovering manganese from open-hearth furnace slags or low-grade manganiferous iron ores is described. The slag or ore is smelted in a blast furnace to produce spiegeleisen; the spiegeleisen is then blown with air in a basic converter to give an oxide slag. This slag product contains 60 to 75% manganous oxide and may be considered a synthetic manganese ore.

In the smelting operations coke rates for five test periods ranged from 3,255 to 4,130 lb per ton of spiegeleisen. The recovery of manganese in the spiegeleisen varied from 60 to 76%. Recovery increased to 83% when the open-hearth slag feed to the blast furnace was diluted with a high-grade iron ore sinter to facilitate smelting. In blowing the spiegeleisen a cyclic procedure was developed that gave nearly 90% recovery of manganese in the slag product (synthetic ore) yet prevented excessive contamination of the slag by phosphorus and iron. The overall recovery of manganese was about 75% when iron ore sinter was used with the open-hearth slag in the smelting step. The metal remaining after oxidizing the manganese from the spiegeleisen contained approximately 3.5% phosphorus. It was refined in a basic converter by blowing with air in the presence of lime. A high-grade iron and a basic phosphate slag were produced.

A one-step operation for the recovery of manganese as the chloride from open-hearth slags and low-grade ores has been described by A.A. Cochran and W.L. Falke in Report of Investigation No. 6859, Washington, D.C., U.S. Bureau of Mines (June 1967).

A one-step operation was developed on a laboratory scale in which manganese and iron compounds are treated with chlorine, volatilized, and separated by selective condensation. Manganese recoveries of 90 to nearly 100% were obtained from 9 of the most important domestic ores. Analyses of the manganese chloride product indicated that it was suitable for aqueous electrolysis with little additional purification. A complete, integrated process that provides for the recovery and recycling of chlorine is proposed. This process appears to have distinct advantages over previously proposed manganese recovery processes, and, in sharp contrast to leaching procedures, it is applicable to practically any manganese-bearing material.

Recovery from Mill Rejects

As part of a Bureau of Mines program to investigate recovery of manganese from low-grade materials, laboratory studies were made of four samples of mill rejects: three from Montana and one from Peru. These studies demonstrated that certain mill-waste products can be a potential source of manganese according to W.A. Stickney and C.W. Sanders in Report of Investigation No. 5692, Washington, D.C., U.S. Bureau of Mines (1960).

Both manganese oxide and carbonate minerals were recovered by flotation in an acid circuit with a fuel oil-tall oil emulsion as collector. Sintering, required to prepare a physically acceptable product, also was used as a method of beneficiation. In addition to being low-grade, some of the samples presented problems of intimate mineral association and complex textures.

A manganese carbonate middling from the Taylor-Knapp Co. magnetic-separation plant near Philipsburg, Montana, was upgraded to only 35.5% Mn with, however, more than 95% recovery. Results of testing a manganese oxide gravity tailing sample from the Trout Mining Division, American Machine & Metals, Inc., Philipsburg, Montana, showed that 80% of its manganese could be recovered in a nodule containing 43% Mn. From a composite of these two samples, more than 94% of the manganese was recovered at a grade of about 41% Mn.

The flotation tailing sample from the Domestic Manganese & Development Co., Butte, Montana, was extremely fine grained, and, although about 90% of the manganese could be recovered by flotation, manganese content of the concentrate was only 34%. Less than 53% of the manganese was recovered when the concentrate grade was increased to 40% Mn by repeated cleaning. About 94% of the manganese in the Condoroma (Peru) mill-tailing sample was recovered by flotation. Sintering the concentrate increased the grade from 28 to 41% Mn.

MERCURY

Recovery from Gaseous Wastes

A process developed by *L.L. Fornoff, J.J. Collins and C.M. Madigan; U.S. Patent 3,755,989; September 4, 1973; assigned to Union Carbide Corp.* is one in which mercury vapor is removed from gas streams containing more than 60% relative humidity by raising the gas stream temperature to lower the relative humidity to less than 50% and thereafter selectively adsorbing the mercury vapor by passage through an activated carbon adsorbent bed under conditions such that water does not condense on the carbon adsorbent. Figure 69 is a flow diagram showing the essential features of the process.

With reference to the drawing, a by-product hydrogen stream at 50°F at essentially atmospheric pressure and saturated with mercury vapor from a chlor-alkali plant sodium amalgam decomposer is introduced by line **10** to cooler **12** and separator **14** which condenses some water vapor and some mercury from the gas stream. Liquid water and any entrapped mercury are removed from the system through line **16**. The gaseous constituents, comprising hydrogen, water vapor at a relative humidity of 100% and saturated with mercury vapor and containing some mercury mist are passed through line **18** and valve **20** to compressor **22** which raises the gas stream pressure to 9 psig and forces the gas stream through line **23** into cooler **24** which lowers the gas stream temperature to the range of 50° to 65°F before passing it to separator **25** wherein additional water vapor and mercury is condensed and removed from the system through line **26**.

The gas stream remaining is passed by line **27** into heater **28** which raises the temperature 30°F. Thereafter, the gas stream, which now has a relative humidity of about 40% is fed via line **29** and valve **30** into adsorbent bed **31** containing activated carbon. Mercury is adsorbed on the carbon adsorbent as is a small percentage of the water vapor of the gas stream, but the major proportion of water vapor leaves the adsorbent bed through line **32** along with the hydrogen. Valve **33** is controlled such that a major proportion of the purified hydrogen stream is removed from the system through line **34** for ultimate consumption and a minor proportion, 15 to 20% is directed through line **35**, heater **36** where it is heated to 500°F to 650°F, then through line **37**, valve **38**, line **39** and adsorbent bed **40** having the same configuration and adsorbent as bed **31**. Bed **40**, at the beginning of the adsorption stroke in bed **31**, has just terminated a similar adsorption step and is hence ready to undergo desorption of the mercury it contains.

The hot purified hydrogen gas stream entering bed **40** through line **39** heats the bed and desorbs the mercury from bed **40** and the entire effluent is passed through line **41**, valve **42** and line **43** to cooler **44** and separator **45** which serve to reduce the mercury and water content before passage through valve **20** in combination with the untreated gas stream newly entering the system through line **18**. After desorption of the mercury from bed **40**, heating of the gas stream from line **35** is terminated and the resulting cool gas stream entering bed **40** returns the bed to temperature for a subsequent adsorption stroke, which is commenced prior to breakthrough of the mercury from adsorbent bed **31**.

FIGURE 69: PROCESS FOR RECOVERY OF MERCURY FROM GAS STREAMS BY ADSORPTION

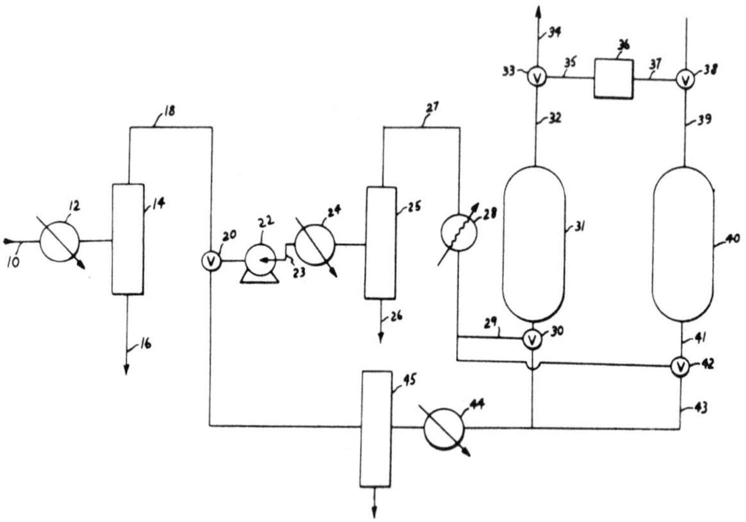

Source: L.L. Fornoff, J.J. Collins and C.M. Madigan; U.S. Patent 3,755,989; September 4, 1973

A process developed by *L.L. Fornoff; U.S. Patent 3,793,806; February 26, 1974; assigned to Union Carbide Corp.* is one in which the partial pressure of mercury as vapor in a gas stream is lowered to a maximum of less than 40 parts per billion by volume of mercury in a gas at one atmosphere. The product gas is made substantially free of condensed mercury as mist by cooling a mercury containing flowing gas stream to 0°F or below, providing that the gas stream also contains water vapor at a minimum partial pressure of 6×10^{-3} atmosphere and the cooling is effected by indirect cooling means wherein condensed water is collected as crystalline ice. Mercury can be recovered from the ice-mercury agglomerate produced by this process.

A process developed by *L.G. Birke, A. Jebens, I. Eystein and F. Dyvik; U.S. Patent 3,838,190; September 24, 1974; assigned to Det Norske Zinkkompani A/S, Norway* is one in which removal of mercury from gases, in particular gases from combustion or roasting is achieved by washing the gases with sulfuric acid of at least 50% having a temperature of 60° to 180°C. The washing acid and the gases contain chlorine or chlorine compounds or both chlorine and chlorine compounds in such an amount that the atomic ratio Cl/Hg is at least 1:1. Practically all mercury and other mercury compounds are converted to mercury chlorides which are separated by condensation in a succeeding cooling of the gas.

A process developed by *H. Hilgen and C.J.N. Rekers; U.S. Patent 3,849,267; November 19, 1974; assigned to Akzo N.V., Netherlands* involves mixing chlorine with the mercury-containing gas, then passing the mixture through a gas-permeable bed of a nonporous solid material which has a large surface area in relation to the bed volume to collect necessary chloride thereon and thereafter recovering the mercury by either washing the bed with chlorine containing brine and passing the resulting mercury-containing brine to an electrolysis cell with a mercury cathode or dissolving the bed material in an aqueous process stream and passing the stream to an electrolysis cell with a mercury cathode.

Recovery from Liquid Wastes

Recovery from Aqueous Solutions: A process developed by *J.H. Entiwisle and R.W. Griffiths; U.S. Patent 3,718,457; February 27, 1973; assigned to Imperial Chemical Industries, Ltd., England* is a process for recovering mercury from waste brine from mercury cells comprising adding sulfide or hydrosulfide ions to the brine to precipitate mercury sulfide, trapping the mercury sulfide on a mixed media filter in which the average size of particles in successive superimposed layers decreases and the specific gravity of the particles increases from the top of the filter downwards, removing the mercury sulfide from the filter by backwashing with an aqueous liquor, separating the mercury sulfide from the backwash liquor and treating the sulfide to produce metallic mercury.

A process developed by *A.C. Waiss, Jr., J.A. Kuhule and M.E. Wiley; U.S. Patent 3,719,473; March 6, 1973; assigned to the U.S. Secretary of Agriculture* is one in which mercury is effectively removed from water by contact with a nut waste such as peanut skins, walnut expeller meal, peanut hulls, and the like. Figure 70 shows a suitable form of apparatus for the conduct of the process.

Chamber **1**, lined with a nonamalgamatable material such as polyethylene, is filled with peanut skins ground to 1 mm in diameter. To prevent excessive packing of the peanut skins, they may be mixed with a dispersing agent such as charcoal, sand, gravel, and the like. Water containing dissolved mercury is introduced into filter **1** by inlet conduit **2** fitted with screen **3**. This water is impelled by pump **4** through filter **1**. The purified water leaves the system via outlet **5**, fitted with screen **6**. When the peanut skins have taken up so much mercury that they lose their effectiveness, the flow of water is cut off, and residual water drained out of the system via valve **7** and drain pipe **8**.

The peanut skins are removed through star valve **9** to combustion chamber **10** wherein they are heated to first evaporate water and eventually heated (in the absence of air) to a temperature on the order of 375°C to carbonize the skins and expel mercury in vapor form. Finally, air is admitted so that the carbonized residue is burned to CO_2 and ash.

The vapors from combustion chamber **10** are conducted through condenser **11** maintained at a temperature below the boiling point of mercury (356.7°C) but high enough to prevent condensation of water and empyreumatic substances. The liquid mercury which collects in condenser **11** is collected and may be reused for various industrial purposes. To start up the water treatment cycle again, a fresh supply of peanut skins is introduced into chamber **1** from reservoir **12** via star valve **13**.

The process may be applied on a continuous basis. In operating under such conditions, the water to be purified is continuously fed into column **1** through inlet **2** and purified water is discharged continuously through outlet **5**. Concomitantly, fresh peanut skins are continuously fed at a predetermined rate into the system via star valve **13**, and spent peanut skins are removed from the system at the same rate via star valve **9**.

In an alternative embodiment, gravity flow is used. Water contaminated with dissolved mercury is passed downwardly through a column of peanut skins. The gravity flow-rate through peanut skins ground to 1 mm in diameter is approximately 2 gallons per square foot per minute. This simplified filtration procedure constitutes but a minor alteration to either a municipal or an industrial water purification system.

FIGURE 70: DEVICE FOR RECOVERING MERCURY FROM AQUEOUS STREAMS

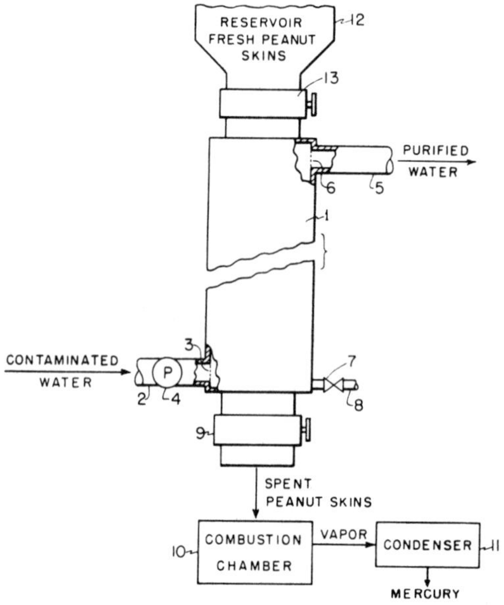

Source: A.C. Waiss, Jr., J.A. Kuhule and M.E. Wiley; U.S. Patent 3,719,473; March 6, 1973

A process developed by *F.L. Moore; U.S. Patent 3,785,803; January 15, 1974; assigned to U.S. Atomic Energy Commission* involves a method of quantitatively removing mercury from aqueous waste streams with certain selected liquid organic amines. The method in its preferred form is particularly appropriate for removing mercury from aqueous brine solutions. Efficient and quantitative recovery of the extracted mercury is also provided.

Several methods are available for removing mercury from aqueous solutions. The precipitation of mercury by a more active metal such as aluminum, copper, iron, or zinc is effective, but it substitutes another metal in solution, usually in considerably more than an equivalent amount. There are methods which are based on the precipitation of mercuric sulfide, but these suffer from high toxicity and offensive odor of sulfide reagent. Sorption of mercury on preformed metal sulfides such as iron, cadmium, or zinc sulfides is available, but, as in the precipitation methods, it results in replacing the mercury with another metal ion.

Ion exchange methods which may be available are expensive and suffer from problems arising from resin attrition and fouling arising from suspended solids present in industrial solutions. A limited class of organic analytical reagents such as dithizone are known to form organic complexes with mercury, but such reagents are expensive and chemically unstable.

Any promising pollution control process for removing mercury from aqueous solutions is one characterized by the use of a cheap, relatively available selective reagent for mercury which can quantitatively remove mercury from solution. The process entails the use of selected liquid aqueous-insoluble high molecular weight primary, secondary, tertiary, and quaternary amines which extract at least 99% of the mercury from a mercury containing aqueous solution.

A process developed by *K. Motani, T. Kiyotoshi and A. Nakahara; U.S. Patent 3,790,535; February 5, 1974; assigned to Tokuyama Soda KK, Japan* involves contacting waste liquor with a chelate resin to thereby cause the resin to adsorb mercury. This method is superior to the method using an anion exchange resin in that mercury contained in the waste liquor can be substantially completely removed and that the method is applicable irrespective of the form of the mercury present in the waste liquor.

While the anion exchange resin used for mercury recovery can be regenerated merely by washing it with hydrochloric acid, nitric acid, etc., to remove the ion exchange mercury, most chelate resins having mercury adsorbed thereto cannot usually be regenerated. Therefore, in spite of excellent mercury adsorbing properties, the chelate resin having mercury adsorbed thereto is disposed of as such to great economical disadvantage. It is therefore an object of the process to provide an economical method of recovering mercury substantially from a mercury-containing waste liquor and of easily regenerating a chelate resin having mercury adsorbed thereto.

The above objects can be achieved by a method which comprises contacting a chelate resin having mercury adsorbed thereto with a solution with a pH of at least 11.0, of at least one sulfur compound of the general formula Me_nS_x wherein Me is an alkali metal, barium or ammonium, n is an integer of 1 or 2, n is 2 when Me is an alkali metal or ammonium and n is 1 when Me is barium, and x is an integer of 1 to 5, to thereby liberate the mercury from the chelate resin.

A process developed by *R.F. Gerow and S.B. Soule; U.S. Patent 3,802,910; April 9, 1974; assigned to GAF Corp.* is based on the discovery that by bringing an aqueous effluent having a pH between 2 and 11 and containing from 1 to 500 ppm of dissolved or metallic mercury into intimate contact in a reaction zone with a substantially water-stable solid metallic reducing agent, having a greater solution potential than mercury, elemental metallic mercury is removed therefrom. The liberated mercury amalgamates on the surfaces of the reducing agent and also coalesces into droplets on these surfaces. Impure mercury so recovered may be, if desired, purified by standard methods, such as acid washing or retorting, or by a combination of methods. If desired, mercury may be recovered by removing the reducing agent from the reaction zone periodically, along with accumulated reaction products, and retorting the entire mass.

A process developed by *L.H. Dunsmoor; U.S. Patent 3,835,217; September 10, 1974; assigned to PPG Industries, Inc.* for removing soluble mercury from a brine containing more than about 5 parts per million of mercury involves the following steps:

adding a 100 to 500% excess of a sulfur compound chosen from the group consisting of alkali metal sulfides, alkaline earth metal sulfides, hydrosulfides, and hydrogen sulfide, to the brine while maintaining the pH of the brine between 4 and 8;

adding a flocculant chosen from the group consisting of compounds of aluminum, magnesium, and nickel in a quantity sufficient to provide a flocculant concentration of between 5 and 20 parts per million by weight in the brine;

adjusting the pH of the solution to convert the flocculant to an insoluble hydroxide;

maintaining the temperature of the solution between 130° and 150°F; and

thereafter recovering a brine solution containing less than one part per million mercury.

Recovery from Solid Wastes

Recovery from Brine Filter Sludge in Chloralkali Plants: Previous methods for removal of mercury from cathode sludge involved ion exchange chemical treatment and roasting. U.S. Patent 3,536,597 shows an ion exchange process, which in practice is too expensive and complex for extensive commercial use. Chemical processes involving sodium hypochlorite leaching, or a modification thereof involving the generation of chlorine in situ by electrolysis of sodium chloride, have been employed. Although the mercury content of the sludge is considerably reduced by such treatment, the amount of mercury remaining is still too large from an ecological standpoint. Roasting the sludge, a process analogous to that employed since ancient times for separating mercury from other materials, likewise reduces the mercury content of the sludge. However, the mercury remaining in the residue is too large.

A process developed by *E. Wygasch and G. Weiss; U.S. Patent 3,755,110; August 28, 1973; assigned to Badische Anilin- & Soda-Fabrik AG, Germany* is one in which the brine filter sludge is treated with solutions containing active chlorine for the extraction of mercury or its compounds. Insoluble impurities are filtered off, and the resulting mercuric filtrate is passed to the electrolytic cell where the dissolved mercury is reduced to metal at the cathode under the operating conditions of the cell. Metallic mercury may also be recovered from the filtrate with the aid of sodium amalgam.

A process developed by *G.A. Carlson; U.S. Patent; 3,785,942; January 15, 1974; assigned to PPG Industries, Inc.* is one in which mercury is recovered from electrolytic cell caustic filter solids by treatment with a brine containing elemental chlorine in an amount sufficient to solubilize the mercury. The mercury containing brine can then be returned to the electrolytic cell.

A process developed by *D.G. Rachor and R.A. Perry; U.S. Patent 3,814,685; June 4, 1974; assigned to U.S. Environmental Protection Agency* is one in which mercury is recovered from mercury cathode sludge by treating the sludge with acid and then roasting. The mercury content of the residue is reduced significantly lower than when the acid treatment step is omitted.

A process developed by *W.E. Dean and W.E. Makris; U.S. Patent 3,836,442; September 17, 1974; assigned to PPG Industries, Inc.* is one in which mercury is removed from electrolytic cell solids, comprising magnesium hydroxide, ferric hydroxide, and calcium carbonate, by contacting the solids with an aqueous sulfide solution to form soluble mercury polysulfide. In a preferred embodiment, the mercury containing solids are first treated with a hypochlorite oxidizing agent.

Recovery from Scrap Mercury: According to H. Nack, K. Murthy, E. Stambaugh, H. Carlton and G.R. Smithson, Jr. in PB 233 646, Springfield, Va., National Technical Information Service (May 1974), the United States production of recovered mercury from secondary sources represents some three-fourths of all mercury produced. Reported high volume sources of scrap for reprocessing include the following: salvage from instrument and electrical manufacturers, research laboratories (dirty liquid mercury from research organizations, educational institutions, etc.); mercury battery scrap; industrial scrap and dental amalgam.

The most common single source mentioned is dental amalgam, though the absolute volume of this source is relatively small. Obviously, in this case, silver, rather than mercury, provides the primary incentive for recovery. The recovery of mercury from scrap involves three manufacturing operations: scrap pretreatment, refining, and bottling.

There is no pretreatment of the scrap when vacuum distillation is employed as the refining method. The dirty liquid mercury scrap is fed directly to the distillation vessel. However, during transfer of the scrap mercury from the storage to the distillation flask, a small quantity of mercury may escape as vapor. This does not constitute any potential for widespread pollution. Building ventilation at the plant is sufficient to reduce the mercury levels to below the OSHA's limit of 100 $\mu g/m^3$ (*Federal Register,* Volume 27, No. 202, October 18, 1972).

Prefiltration is a pretreatment operation used only with the oxygenation process and when the mercury scrap contains insoluble impurities that need to be removed before oxygenation starts. In fact, this is the only scrap pretreatment operation in the entire secondary mercury industry. The process removes insoluble impurities by filtering which is a single process step.

Energy, mostly electrical, is required for the filtration process. The solid waste (dirt) generated is sent to the retorting process for recovery of trace amounts of mercury that may be present. The prefiltration process presents no potential serious pollution problem providing the area is vented and gases are cleaned prior to emitting to the atmosphere.

There are four different processes used for refining scrap mercury. The process employed depends on the type of scrap. The processes are as follows: vacuum distillation, solution purification, oxygenation and retorting. Some details of each of these processes are given in the paragraphs which follow.

Vacuum Distillation Process – When the vapor pressure of the impurities is substantially lower than that of mercury, purification by vacuum distillation which separates a high purity mercury from the impurities is employed. The distillation unit consists of a still pot for heating the impure mercury, a water cooled condenser, and a pure mercury receiving vessel. The processing steps are: heating the still pot by external heat to vaporize the mercury at an absolute pressure of 0.01 torr (0.01 mm Hg pressure), condensing the mercury in water-cooled condensers, and collecting the clean mercury in a receiver. Afterwards, the still residue is removed from further processing.

Energy requirements for this operation are: heat energy for heating the distillation still and electrical energy to operate the vacuum pump and associated equipment. The main source of potential pollution is the exhaust gas from the vacuum pump. Estimated maximum mercury emissions by distillation are 0.02 lb per 760 lb of mercury produced. Solid waste residue from the still pot is normally recycled to other refining processes. No liquid effluent is produced in this process. The process has no serious pollution potential if good housekeeping practices are employed and precautions are taken to control emissions.

Solution Purification Process – This process removes metallic and/or organic impurities contained in the dirty liquid mercury by washing with a dilute acid. The processing steps are: leaching metallic impurities with dilute nitric acid using compressed air to mix the acid and mercury; separating the mercury layer from the aqueous nitric acid slurry by decantation; water washing to remove residual acid, and filtering, probably through a bed of activated charcoal and silica gel, to remove last traces of water. If the mercury contains organic impurities, the scrap mercury is treated in the same manner with organic solvents.

Electrical energy is required to operate the compressor and associated equipment. Pumping the mercury is avoided by use of gravity flow scheme from overhead tanks. Potential pollutants generated by this process are atmospheric emissions, liquid wastes, and solid wastes. The atmospheric emissions containing mercury vapor resultsfrom sparging of the liquid mercury during the leaching step. The spent leach liquor is the source of liquid waste, while the solid waste consists of the contaminated filter media. In view of the hazardous nature of mercury, pollution problems can result.

Oxygenation Process – This process, also called oxification, removes the metallic impurities contained in dirty liquid mercury by oxidation with sparging air. The process steps are: sparging the dirty mercury with air in a closed, agitated vessel for several hours and filtration to remove solid metal oxides. Finally, the filtered mercury is bottled which is discussed in the bottling process. Electrical energy is required for sparging and water scrubbing process steps.

Sparging air which contains saturation levels of mercury at 77°F is a potential atmospheric pollutant. However, before venting to the atmosphere, it is generally water scrubbed and filtered through a bed of charcoal to remove the mercury. The water is recycled many times before discharge because mercury settles out of the water. In most cases, the solid waste as filter cake is sent to the retorting process. Thus, the potential pollutants from this process are atmospheric emissions, liquid wastes, and solid wastes. The potential for pollution from this process is not significant, providing pollutants are treated before discharging to the environment.

Retorting Process – This process produces pure mercury by volatilizaiton of the mercury contained in solid scrap. Examples of scrap are dental amalgams, mercury battery scrap, and distillation sludges. The process steps are: external heating of solid scrap in a closed still pot or stack of trays to volatize the mercury, cooling mercury vapor in water-cooled condensers to condense the vapors, and bottling which is discussed in the paragraphs which follow.

Heat energy is required for volatilization of mercury and electrical energy for cooling water circulation. Potential pollutants generated by the process are atmospheric emissions from the retort, liquid wastes from the condenser, and cleaning of the atmospheric emissions and solid wastes as the retort residue, each containing mercury. The degree of pollution can be reduced by scrubbing of the process gases prior to emission to the atmosphere, packaging of the retort residue and disposing of it in a hazardous landfill, and treating the liquid wastes prior to disposal. Because of the hazardous nature of mercury, this process can be a source of serious pollution problems.

Bottling of the purified mercury is done to enable marketing of the product in 76 lb flasks. The steps are: pumping or manual pouring of the mercury into the 76 lb bottles and sealing the bottling. Electrical energy is used for pumping the mercury to the bottles. The only waste emission is mercury vapor (at 77°F), produced when pumping or pouring mercury into the bottle. This is not a significant source of air pollution but can cause high levels of mercury vapor inplant. Building ventilation air is used to reduce the mercury levels to below the TLV. The bottling process does not offer any potential for serious air pollution.

Recovery from Spent Catalysts: A process developed by *A. Jacobowsky; U.S. Patent 3,537,843; November 3, 1970; assigned to Knapsack A.G., Germany* involves the recovery of mercury from a contaminated inactive mercuric chloride/active carbon-catalyst by burning the active carbon with a deficiency of oxygen or air, expelling or subliming off, together with the combustion gases, a mixture of gaseous mercury, mercurous chloride and mercuric chloride, condensing the gaseous mixture by cooling it, and, by adding a suitable reducing agent, reducing to mercury the mercurous chloride and mercuric chloride contained in the condensate.

Recovery from Waste Mercury Batteries: As noted in *Environmental Science and Technology* 9, No. 5, 416 (May 1975), the Mercury Refining Co. of Albany, N.Y. has been in operation since 1954 and has two retorts adapted to receive waste mercury containing batteries (old or new rejects) as well as various types of sludges containing 12 to 80% of mercury for recovery.

MOLYBDENUM

Recovery from Milling of Copper Sulfide Ores

The split-off point for by-product molydenum occurs at the flotation stage in the milling of copper sulfide ores according to A. Petrick, Jr., H.J. Bennett, K.E. Starch and R.C. Weisner in Information Circular 8569, Washington, D.C., U.S. Bureau of Mines (1973). From a combination of a literature review and plant visits, the above authors prepared the hypothetical flow sheet shown in Figure 71.

FIGURE 71: BY-PRODUCT MOLYBDENUM SUBSYSTEM

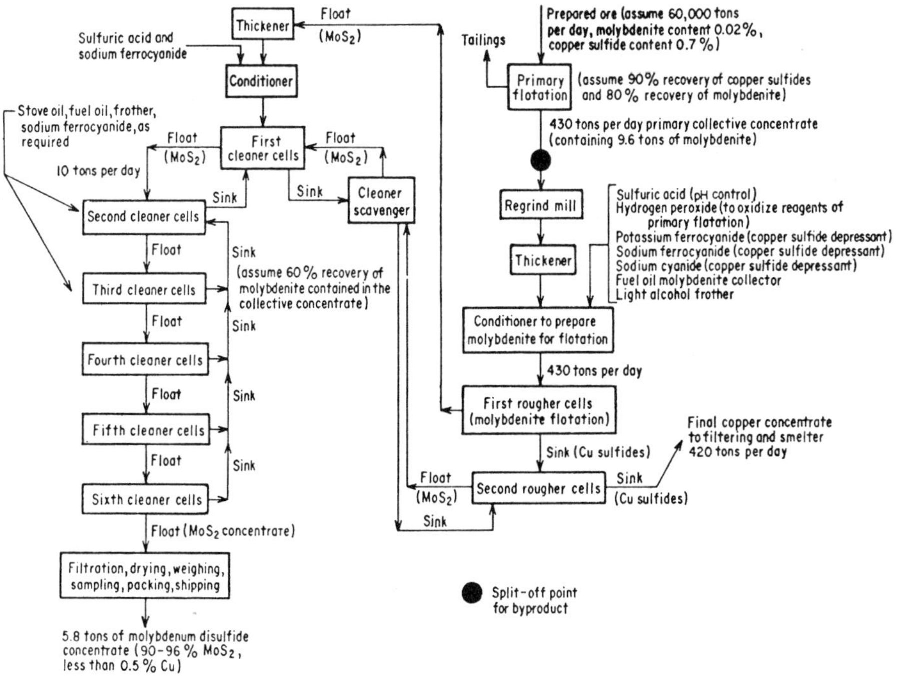

Source: Bureau of Mines I.C. 8569

The plant has a 6 ton-per-day throughput of by-product molybdenum. The tonnage of copper concentrates handled per ton of molybdenite production varies widely. The ratio of copper sulfide production to molybdenum sulfide production varies from less than 10 to 1 to thousands to 1.

There are several approches used in producing separate molybdenum and copper concentrates. The one chosen for the hypothetical subsystem is "differential" flotation. First a collective float of both copper and molybdenum sulfides is made, then a separation is made by selective flotation in a second circuit. This is the approach used when the molybdenum content of the ores is so low that selective flotation in the primary circuit is not justified. This occurs in the great majority of cases.

The collective flotation of molybdenite along with copper sulfide minerals presents no major problems. The process used for copper-molybdenum ores preparatory to flotation is similar to that used for copper only ores. The ores are crushed and ground to a particle size sufficient to free the copper minerals for flotation. They are then treated with reagents that promote flotation of both molybdenum and copper sulfides. The primary collective concentrate in the hypothetical flow sheet is assumed to be at a rate of 430 tons per day. The collective concentrate is thickened, reground, and conditioned for molybdenum disulfide flotation.

The major separation of copper and molybdenum sulfides is made in the first and second rougher cells. The nonfloatable material from the second rougher cells becomes the final copper concentrate, and the float material which is molybdenum disulfide continues through several cleaning operations. It was decided to include all operations beyond primary flotation in the by-product subsystems.

The by-product molybdenum subsystem consists of all processing of molybdenum disulfide subsequent to the production of the collective concentrate. There is considerable variation in the approach. The hypothetical flow sheet is confined to the operations shown in Figure 71.

This consists of thickening, grinding, and conditioning prior to the first flotation as well as six-stage cleaning of the molybdenite concentrate. Recovery of molybdenum disulfide in primary flotation is assumed to be 80% and recovery in the molybdenum subsystem beyond that point is assumed to be 60%. The result is a total of 5.8 tons of 11,600 pounds of molybdenum disulfide daily derived from the original processing of 60,000 tons per day of copper ore averaging 0.02% molybdenum disulfide.

Recovery from Particulate Silicate Slags

A process developed by *P.R. Ammann; U.S. Patent 3,896,210; July 22, 1975; assigned to Kennecott Copper Corp.* involves recovering molybdenum from ferrosilicate slags containing iron, silica, molybdenum and other nonferrous metals including a member from the group consisting of: the alkaline earth metals, calcium and magnesium; the alkali metal, sodium; and, mixtures thereof. The process involves the steps of:

(a) comminuting the slag to produce particulate slag;

(b) oxidizing the particulate slag at a temperature between the range of 800° and 1200°C to convert the particulate ferrosilicate slag into particles containing an iron oxide phase and a silicate phase, the oxidation step also oxidizing the molybdenum in the slag to its highest oxidation state so that molybdenum complexes with the alkali and alkaline earth metals in the slag to form stable molybdates that remain with the silicate phase;

(c) selectively leaching the molybdenum from the oxidized slag of step (b) to form a molybdenum bearing leach solution; and

(d) recovering molybdenum from the leach solution of step (c).

NICKEL

Recovery from Liquid Waste

Recovery from Waste Phosphate Solutions: The solvent extraction of nickel and zinc from waste phosphate solutions has been outlined by H.E. Powell, L.L. Smith and A.A. Cochran in Report of Investigation No. 7336, Washington, D.C., U.S. Bureau of Mines (Jan. 1970).

The selective recovery of nickel and zinc from a waste phosphate solution by solvent extraction was investigated by the Bureau of Mines as part of a broad program to recover valuable metals and compounds from industrial wastes, and to reduce the associated land and water pollution caused by dumping these wastes.

The waste, which resulted from the phosphation of automobile bodies, yielded 99% of both the nickel and the zinc by extraction with dinonyl naphthalene sulfonic acid (DNSA) and di-2-ethylhexyl phosphoric acid (EHPA) respectively. Dilute sulfuric acid was used to recover the metals and regenerate the solvents. A flow diagram of this process is shown in Figure 72.

FIGURE 72: FLOW SHEET OF ZINC AND NICKEL CIRCUITS IN RECOVERY FROM WASTE PHOSPHATE SOLUTIONS

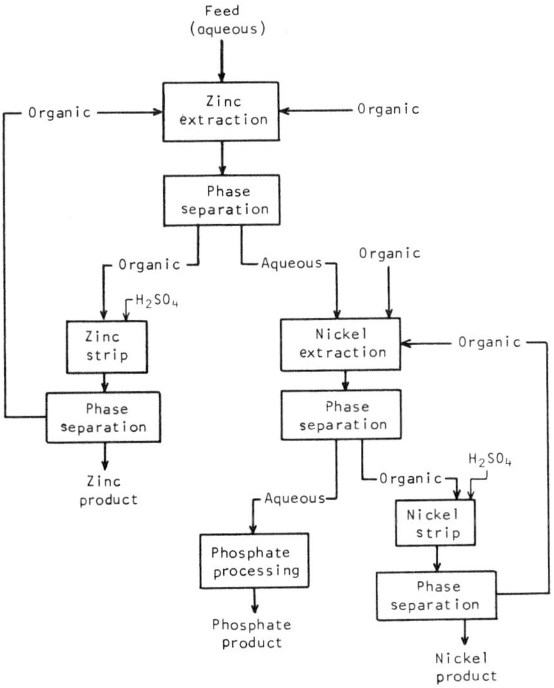

Source: Bureau of Mines R.I. 7336

Recovery from Solid Waste

Recovery by Dissolution of Superalloy Scrap: A process developed by *P.T. Brooks and D.A. Martin; U.S. Patent 3,607,236; September 21, 1971; assigned to the U.S. Secretary of the Interior* is one in which valuable metals are recovered from superalloy scrap by a series of process steps comprising dissolution of the scrap in chlorinated acidic aqueous solution, removal of tungsten and silica by adsorption on activated carbon, extraction of molybdenum with a water-immiscible organic phosphate or phosphonate, extraction of iron with a water-immisible secondary amine, extraction of cobalt with a water-immiscible tertiary amine, and precipitation of chromium as a hydrous basic sulfate, and nickel recovery. The process is shown in block flow sheet form in Figure 73. The following is an example of the operation of the process.

Referring to the flow sheet of the figure, debris from dry grinding alloy billets was screened to remove oversize material, chiefly pieces of grinding wheel, that would plug the equipment. The grindings were fed continuously into the dissolver, and the metallic components were dissolved by the boiling, chlorinated leach liquor containing 1.2 to 1.8 grams of hydrogen ion per liter.

Solution was bled from the dissolver at a specific gravity of 1.35 and an EMF of -450 mv. Water and hydrochloric acid were added to make up for the volume of solution bled off and to maintain the desired acidity. Operation in this manner dissolved, in percent by weight of the metal charged, 99.6 Ni, 90 Mo, 99.5 Co, 91 Cr, 97 Fe, and 10 W. Much of the tungsten was suspended in the leach solution as insoluble tungstic acid. Material not dissolved was mainly abrasive materials and metallic carbides. The leach liquor assayed, typically, in grams per liter, 105 Ni, 36 Cr, 28 Fe, 11 Mo, 13 Co, 1.3 W, 310 Cl and 1.3 H^+.

Tungsten and traces of silica were removed from the leach liquor by agitation for 30 minutes at room temperature with 5 grams of activated carbon per liter. Filtration gave a spent carbon cake containing 21% by weight tungsten and a filtrate in which the remaining tungsten was 0.01 g/l. Continuous countercurrent solvent extraction of molybdenum was done next in seven actual extraction stages using 0.5 molar trioctylphosphate in kerosene at a solvent-to-feed ratio of 0.65 by volume. Stripping the extract phase was done in four stages using 0.3 molar HCl at an organic-to-aqueous phase ratio of 4.9 by volume.

Molybdenum remaining in the raffinate was 0.01 g/l and the molybdenum-rich strip liquor contained, in grams per liter, 92 Mo, 15 Fe, 110 Cl, and 3.3 H^+. The barren organic phase was continuously recycled. Molybdic acid was precipitated from the strip solution by adjusting the acidity to pH 0.5 with sodium carbonate followed by heating to boiling. Better than 99% of the molybdenum was precipitated to give a product that, when dried at 105°C contained 60% by weight Mo and 2% by weight Fe.

FIGURE 73: PROCESS FOR NICKEL RECOVERY BY DISSOLUTION OF SUPERALLOY SCRAP

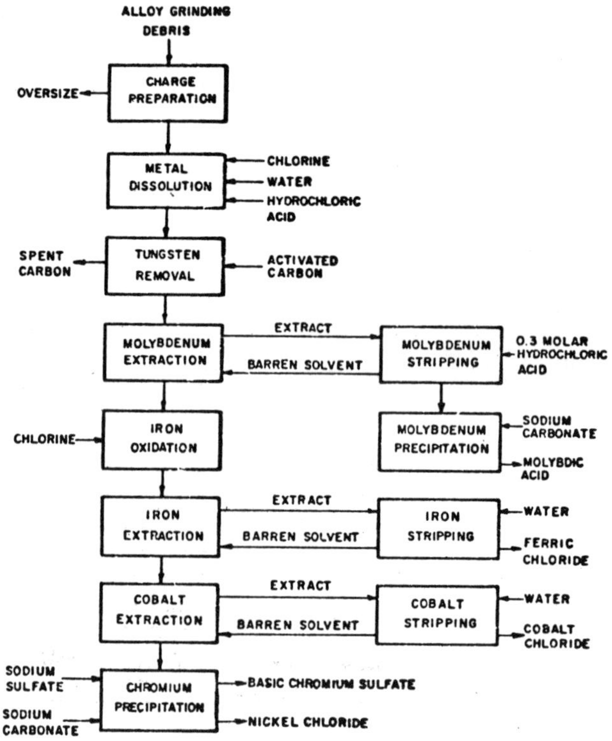

Source: P.T. Brooks and D.A. Martin; U.S. Patent 3,607,236; September 21, 1971

Molybdenum-free leach liquor was next chlorinated to a solution EMF of -900 mv to oxidize the iron to the ferric state. Iron was extracted on a continuous countercurrent basis using 0.25 molar dodecenyl-trialkyl-methyl amine hydrochloride dissolved in aromatic naphtha. Five actual extraction stages were employed at an organic-to-aqueous phase ratio of 2.2 by volume. The extract phase was stripped of iron with water in four stages.

Iron remaining in the solvent extraction raffinate was less than 0.02 g/l, and the iron-rich strip liquor contained typically, in grams per liter, 39 Fe, 0.18 Co, 0.04 Ni, 0.04 Cr, 88 Cl, and 0.6 H^+. The barren organic phase was recycled. The iron-free leach liquor was next extracted in a continuous countercurrent manner to remove cobalt using 0.5 molar triisooctylamine hydrochloride dissolved in an aromatic naphtha. A solvent-to-feed phase ratio of 1.33 by volume was used in seven actual extraction stages, and cobalt was stripped with water from the extract in five stages at an organic-to-aqueous phase ratio of 4.4 by volume. Cobalt remaining in the solvent extraction raffinate was 0.02 g/l, 40 Co, 0.3 Ni, 57 Cl, and 0.2 H^+. The barren organic phase was recycled.

The final solvent extraction raffinate containing mainly chromium and nickel chlorides was next treated batchwise to separate the chromium from the nickel. The solution was diluted to 25 grams of chromium per liter, 37.5 grams of sodium sulfate per liter were dissolved, sodium carbonate was added to a pH of 2.5, and the solution was aged overnight. The aged solution was vigorously agitated and cooled with ice to 8°C and the necessary saturated sodium carbonate, determined by titrating a sample with sodium carbonate, was added. The solution was heated to 95°C for 30 minutes and the precipitated basic chromium sulfate was filtered, water washed, and dried at 105°C. The precipitate assayed, typically, in percent by weight 31 Cr, 3.4 Ni and 5.1 S.

Nickel contained in the filtrate and washings from chromium precipitation was next precipitated as a basic carbonate by adding sodium carbonate solution to pH 8. The precipitate was filtered, water washed, and calcined to give nickel oxide assaying, in percent by weight, 69 Ni, 0.025 Cr, 0.05 S, $<$0.001 P, and 0.02 Co.

The chemical reclaiming of superalloy scrap has been discussed in some detail by P.T. Brooks, G.M. Potter and D.A. Martin in Report of Investigation No. 7316, Washington, D.C., U.S. Bureau of Mines (November 1969).

Recovery from Scrap Batteries: See this same heading under "Cadmium" in this volume for details of a process developed by D.A. Wilson and B.J. Weigard, Jr., R.I. 7566, Washington, D.C., U.S. Bureau of Mines (1971) involving recovery of cadmium and nickel from scrap batteries.

Recovery by Smelting of Scrap: According to H. Nack, K. Murthy, E. Stambaugh, H. Carlton and G. R. Smithson, Jr. in PB 233 646, Springfield, Va., National Technical Information Service (May 1974), recovered scrap accounts for about 30% of total United States nickel consumption. It is estimated that only 40% of the available scrap is recycled. The remaining 60%, such as used nickel-cadmium batteries and spent nickel-base catalysts, are disposed of in landfills. Emissions to the environment from this segment are atmospheric emissions and solid wastes.

Raw materials include old scrap (74%) and new scrap (26%) both being comprised of nickel-, copper-, and aluminum-base alloys. The scrap originates as used burners (inconel), airplane parts, sheet scrap, electrical scrap, and so on. An exact percent breakdown of the contribution of these sources is not available. Products from the nickel segment are slag and the following alloys: stainless steels (18/8 and other grades); nickel-base alloys (cupronickel, nickel silver, etc.); base alloys (low alloy steels) and nickel metal (quantitatively not significant). The recovery of nickel as nickel alloys from scrap involves three manufacturing operations: scrap pretreatment, smelting, and refining/casting.

Scrap pretreatment by hand sorting is done usually at the scrap merchants facilities rather than at the scrap smelting facilities. The process segregates the nickel bearing alloys and metals from the nickel-free components in the raw scrap. The process steps are: spread the scrap on a platform and separate the nickel scrap from the nonmetallics and nonnickel components. Manual labor is the only energy requirement for the process. Wastes produced are limited to a small quantity of solid wastes which are composed of dirt and nonnickel scrap. The pollution potential of the waste is insignificant.

In the smelting operation, partial purification of the nickel scrap is achieved or the nickel scrap is melted and mixed with selected alloying agents in one of the following steps: electric arc process and rotary reverberatory furnace smelting. Some details of these processes are given in the paragraphs which follow.

Electric Arc Process – This process produces a partially purified product which is purified further in the refining/casting operation or nickel alloys which are cast into alloy ingots. The process steps are: charging the scrap to the electric arc furnace, adding a reductant (lime), melting the charge, pouring into molds or transferring the molten metal to a reactor for refining and removing the slag.

Electric energy is required in substantial quantities for this process to operate the furnace. Additional energy is required to operate the auxiliary equipment. Wastes generated are: atmospheric emissions consisting of large quantities of dust particles and limited quantities of gases and solid wastes consisting of furnace slag.

The off-gases are invariably treated in a baghouse which removes approximately 99% of the dust particles. Hence, the gases are clean when vented and do not constitute any serious air pollution problem. However, the collected dust (about 4 tons per day at a medium size facility) is sent to a landfill and may be considered a potential source of water pollution by leaching to surface and ground waters. The metal value of this dust is unknown and the industry would like to investigate possibilities of recovery of any metals in the dust. The slag may or may not be a waste as, in many cases, it is ground at another facility for use in sandblasting or road surfacing operations.

Rotary Reverberatory Furnace Smelting Process – This process is the same as the electric arc process except a reverberatory furnace is used in place of an electric arc furnace. The process steps are: charging the scrap to the furnace, adding a reductant (lime), melting the charge, pouring the melt in the molds or transferring the melt to a reactor for refining, and removing the slag.

Energy required for this process is from the fuel to operate the furnace and other forms of energy to operate the auxiliary equipment. Potential environmental pollutants produced by this process are atmospheric emissions and solid wastes. The atmospheric emissions are composed of dust particles and gases. These are, in most cases, treated in baghouses which collect approximately 99% of the particulate matter. The solid waste is generated as slag and, in most cases, is ground at another facility for use in sandblasting or road surfacing operations.

Thus, the smelting process is a potential source of pollution problems. These problems may result from disposal of the baghouse dusts via landfill which could result in contamination of the water system and possibly contamination of the atmosphere with gases and particulate matter from pollution abatement equipment. However, atmospheric emissions do not probably constitute a serious pollution hazard.

Then the reactor finishing process, very similar to a foundry operation, purifies the metal further and adjusts the composition to the required values. The process steps are: charging the molten metal to the furnace; adding cold metals such as base scrap and pig nickel; adding lime, silica, and trim (manganese, columbium, titanium, etc., in trace quantities as required by alloy composition); melting the charge; pouring into molds, and removing the slag. Large amounts of heat energy are required by the process.

Potential pollutants generated by this process are atmospheric emissions containing gases and particulate matter and solid wastes in the form of slag. The atmospheric emissions are treated via a baghouse which removes approximately 99% of the particles. The slag is sold as a product for use as a sandblasting material or in road surfacing operations. Thus, the reactor finishing process has a low potential for the production of serious pollution problems.

Finally, in the ingot casting process, the molten alloy from the reactor or electric furnace is cast into required shapes. The process steps are: pouring into molds and removing the ingots from the mold after they are cast by air cooling. Energy requirements of the process are minimal. Wastes produced are minor amounts of metallic vapor which stays inplant. The potential pollution of the wastes is insignificant.

ORE TAILINGS

Use as Mine Backfill

A process developed by *R.B. Booth; U.S. Patent 3,508,407; April 28, 1970; assigned to American Cyanamid Co.* involves filling subterranean cavities by charging into a cavity a high density water suspension of particulate mineral matter such as ore flotation tailings, sand, slag, which contains a water-reactive setting agent and a high molecular weight water-soluble polymer as a fines retaining agent. The water drains from the mixture after placement without simultaneous exudation of finely divided solids such as the setting agent. The remaining solid material sets to form a cohesive, stabilized load supporting structure.

Use in Building Bricks

The preparation of dry-pressed building blocks from copper mill tailings has been described by P.G. Pigott, E.G. Valdez and K.C. Dean in Report of Investigation No. 7537, Washington, D.C., U.S. Bureau of Mines (July 1971).

This Bureau of Mines laboratory research demonstrated that building bricks meeting ASTM specification C62-66 for grades SW and MW of FBS-type bricks can be produced from waste copper mill tailings. Waste tailings from accumulations at Magna, Utah, Anaconda, Montana, and Hayden, Miami, and San Manuel, Arizona, were dry-pressed into bricks, and the products were tested in accordance with ASTM Designation C67-66 on sampling and testing of brick. Utilization of such mill tailings on a commercial scale would eliminate the need for digging new pits for brick raw material production, conserve natural resources, and simultaneously remove potential sources of air pollution.

PLATINUM

Recovery from Scrap Printed Circuit Boards

See Section on Precious Metals.

Recovery from Used Auto Pollution Converters

As described by J. McDermott in *Catalytic Conversion of Automobile Exhausts,* Park Ridge, N.J., Noyes Data Corp. (1971), platinum catalysts play an important part in this aspect of pollution control.

D.J. Kusler has more recently reported in Information Circular No. 8565, Washington, D.C., U.S. Bureau of Mines (1973) on the demand for platinum to reduce pollution from automobile exhaust and Figure 74 is taken from that report showing the phase of resource recovery in the use of platinum in catalytic converters.

FIGURE 74: PLATINUM SUPPLY-DEMAND RELATIONSHIP ASSUMING USE OF A PLATINUM CATALYST IN AUTOMOBILE EXHAUST CONVERTERS

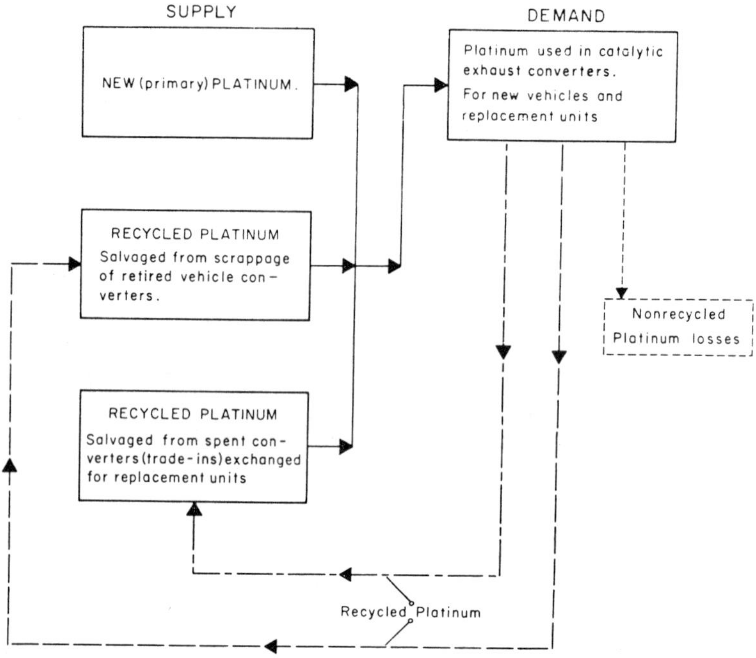

Source: Bureau of Mines I.C. 8565

PRECIOUS METALS

Recovery from Miscellaneous Scrap

This section includes the secondary recovery methods for gold, silver, and platinum group of metals. These are grouped together because they derive from the same scrap and the same general recovery operations are used to obtain the individual metals. Exact statistics on scrap volume are not available. However, estimates are that about 10,000 tons are processed per year, excluding low-grade scrap which is disposed of in landfills at present.

Electronic components such as gold-plated contacts and printed circuit boards from military and civilian scrap equipment are the chief sources of scrap for this segment. The second important source is scrap from the dental industry. Scrap jewelry usually goes to special smelters as these do not require complicated processing. Thus, raw materials to this segment include: electronic scrap consisting of relay contacts, switch contacts, wires, solders, etc; dental amalgams, inlays, dentures, and crowns; and new scrap generated in manufacturing Products from this segment include: refined gold metal, refined silver metal, refined platinum metal, and refined palladium and trace quantities of iridium and other metals. The recovery of precious metals from scrap involves three manufacturing operations; scrap pretreatment, smelting, and refining/casting. Although a variety of processes are possible in scrap pretreatment, only two processes are notably important for detailed consideration—hand sorting and crushing, and incineration. Details are given below.

Hand Sorting and Crushing Process – Bulk precious metals scrap consisting of identifiable aluminum contaminants (aluminum chassis) is hand sorted to separate the aluminum from the scrap. The aluminum-free scrap is then crushed in a hammer mill. The process steps are: spread bulk scrap on floor, pick out aluminum parts, and shred aluminum-free scrap in a hammer mill shredder. The process requires electrical energy for crushing. Wastes generated consist of minor in-house dust emissions during charging and discharging of the shredder. The potential for pollution of wastes in this process is insignificant.

Incineration Process – The crushed scrap is incinerated to burn off plastics and organic liquids and prepare a smelter feed. The processing steps are: charge to incinerator, burnoff combustible constituents, and discharge smelter feed. The process requires heat energy. Wastes generated are: atmospheric emissions consisting of organic vapors and dust particles and liquid waste if wet scrubbing is used to control dust particles in the atmospheric emissions.

Untreated wastes have considerable ambient air pollution potential. However, the atmospheric emissions are treated in an afterburner to complete combustion of organic matter. Thus, the effluents from the afterburner contain particles of dust and metal. Usually a wet scrubber or baghouse is used to collect these particles. The amount of dust collected by either of these control devices is not sufficient to constitute serious solid waste problems. Similarly, any liquid waste generated is insignificant. Therefore, the wastes from this process have no serious pollution potential.

The pretreated scrap is then treated by blast furnace smelting which purifies the pretreated scrap to produce black copper. The process steps are: charging the segregated scrap with coke and flux, melting the charge, slagging the melt, discharging black copper, and removing the slag. Heat energy is the main energy requirement of the process. Copious amounts of atmospheric emissions and moderate quantities of solid waste—hard slag—are generated in the process. The atmospheric emissions are treated in a baghouse. The collected dust is landfilled. The hard slag is crushed and used in sand blasting operations. Therefore, the wastes generated in this process do not have significant pollution potential.

The black copper is further purified by oxidizing the contaminating metals with air blown through the heated copper. The end product is bullion (copper rich in precious metals). The process steps are: charging the converter, blowing the melt, removing the slag, pouring the melt into molds, and casting the ingots. Heat energy is the main energy requirement. Wastes generated are: minor amounts of atmospheric emissions and slag metal oxides. The slag metal oxides are recycled to the blast furnace. Therefore, the process has no significant pollution potential if the atmospheric emissions are controlled. The final refining/casting step may take one of two forms.

Electrolytic Refining Process – The bullion from the converter (as the anode) is electrolytically separated into pure copper (cathode) and precious metal slimes. The process steps are: charging the electrolyte (copper sulfate—$CuSO_4$), electrolysis, and slime recovery. Electrical energy is required in this process. Waste products are limited to minor amounts of gaseous hydrogen and arsine. Quantitatively, these wastes do not represent significant potential for pollution. Arsine is very much below TLV levels.

Chemical Refining Process – The precious metals slime from the electrolytic process is refined to isolate each of the precious metals by complex chemical processes. The process steps are: treating the slime with aqua regia to dissolve the precious metals; precipitate the gold, silver, and platinum with suitable chemical solutions; and melt or ignite to collect gold and silver as grains and platinum and palladium as sponge.

The process requires electrical and thermal energy in moderate quantities. Wastes include waste liquor discharged to waterways and atmospheric emissions generated during melting and ignition. The process has a low potential for the production of atmospheric emission problems if emissions are controlled. Discharge of the spent liquor to the waterways could result in pollution problems.

Recovery from Mixed Electronic Scrap

Most of the precious metals presently being recovered from scrap are derived from selected items high in precious metal value and free from complicating contaminations such as aluminum chassis. Items must be plastic-free or incinerated to remove plastics. Thin exposed platings of gold or silver are successfully recovered by cyanidation, sometimes accelerated with electrolysis. Proprietary cyanide-based, organic compounds are available that dissolve gold and silver much faster than ordinary cyanide solutions as described by *R. Springer in U.S. Patent 2,649,361; August 10, 1953.*

In at least one installation, gold plating is removed from copper or brass objects by reverse electrolysis in strong sulfuric acid. In this process, the gold dissolves at the anode but immediately reprecipitates as it leaves the anode and falls to the bottom of the cell as a sludge. Nitric acid and/or aqua regia leaching followed by treatment with sodium chloride or ferrous sulfate to precipitate the silver or gold, respectively, is used in some cases where the ratio of precious metals to acid-soluble base metals in high.

Hand stripping to separate aluminum and components high in value as electronic items, followed by smelting to form assayable products for sale to a custom smelter, is the most widely used procedure to treat complex scrap. The cost of hand stripping is high and is usually justified only if the scrap contains items which are resalable for their electronic value.

Chemical treatment is generally applicable only to gold, silver, and palladium-bearing scrap as noted by R.O. Dannenberg, J.M. Maurice and G.M. Potter in Report of Investigation No. 7683, Washington, D.C., U.S. Bureau of Mines (1972). Other platinum-group metals dissolve too slowly in chemical reagents unless they are very finely divided. All presently used commercial processes for the recovery of platinum-group metals depend upon concentration by hand-sorting or smelting-electrolysis, usually followed by dissolution of the precious-metal-rich fraction in molten lead. Dissolution of the lead in dilute nitric acid dissolves some of the precious metals but leaves most of the less soluble ones in the form of a finely divided residue from which they can be selectively dissolved.

Recovery from Scrap Printed Circuit Boards

A process developed by *B.F. McLouth; U.S. Patent 3,448,702; June 10, 1969* is one in which scrap material containing precious metal, in particular printed circuit boards having strips of metal contained in epoxy-fiber glass laminates, is reduced and the metal reclaimed in an incinerator having a forced combustion burner from which a high velocity stream of pressurized combustion air impinges directly on the scrap material, and scatters the metallic residue and glass fibers to an area away from the burner flame where their temperatures will not exceed their melting points. The volatile plastics in the scrap material are vaporized and then completely burned in a secondary combustion chamber having two burner nozzles at different levels projecting radially inwardly towards the rising stream of combustion gases.

The feeder chute for the incinerator has an inner fire door, and an outer, swingably mounted access door which blocks the feeder chute when the fire door is open, but which may be conveniently swung to an open position when introducing scrap material into the incinerator.

RHENIUM

Recovery from Gaseous Effluents

Rhenium is a relatively scarce material and, in recent years, it has achieved attention as a technological material in a variety of fields. It is employed in electrical contacts, thermocouples, filaments for electronic devices, as a ductilizing addition agent in molybdenum and tungsten for electronic uses and has also been proposed for use as a catalyst in the processing of petroleum.

The presence of rhenium in ores and minerals does not generally exceed a fraction of 1% (e.g., up to 0.5%) and usually rhenium is found in the greatest quantities in certain sulfide ores. During the oxidation roasting of sulfide ores, such as molybdenum and copper sulfides, the rhenium is converted to a large extent to rhenium heptoxide (Re_2O_7) which is volatile and passes off with the gaseous effluents resulting from roasting. The roasting gases may include N_2, O_2, H_2O, SO_2, SO_3, and the like. The roaster gas stream also contains solid particles in suspension, such as molybdenum sulfide, molybdenum oxide, silica and other extraneous materials which are generally removed by precipitators, scrubbers, etc.

A method most often used to recover rhenium oxide from the gaseous effluent is wet-scrubbing with a solvent, such as water. In this connection, reference is made to U.S. Patents 2,414,965 and 2,809,092. Rhenium heptoxide (Re_2O_7) is soluble in water and forms perrhenic acid when it goes into solution. The scrubbing is usually performed in well known scrubbing and/or absorption equipment, such as venturi scrubbers, cyclonic scrubbers, packed towers and other gas-cleaning equipment. Thus, a solution is obtained containing rhenium values and other soluble and suspension products, such as molybdenum, iron, sulfuric acid and other substances. The rhenium is then recovered from solution by known methods consisting of, for example, the selective precipitation of a rhenium compound, or by using known ion-exchange techniques that enable the recovery of rhenium salts, such as potassium perrhenate, ammonium perrhenate and other compounds.

However, the foregoing techniques have their limitations in that the collection efficiency is generally limited at best to 65% by weight using high energy venturi or very large packed tower scrubbers which are expensive in installation cost or power requirements in the case of venturi scrubbers. It is generally more economical to use low or medium energy scrubbers, however, the collection efficiency tends to be lower, e.g., 50 to 60%. The unrecovered rhenium oxide particles are generally in the submicron range and leave the scrubbers with the exit gases which may or may not be cleaned of SO_2 and then sent to the stack.

It would be desirable to provide a method of improving efficiency of rhenium recovery in view of the fact that the rhenium is a very valuable material. As far as is known, no methods are presently available for achieving high recovery of rhenium from roaster gases.

Recovery from Molybdenum Sulfide Roaster Gases: A process developed by *R.J. Hevia and L. Soto-Krebs; U.S. Patent 3,739,549; June 19, 1973* is one in which the recovery of rhenium from roaster gases is improved in extraction efficiency in that, following the removal of solids from the gas, the dust-cleaned gaseous effluent remaining which contains particles of rhenium oxide ranging down to submicron size is then subjected to wet-filtering, whereby to effect a substantial improvement in the recovery of rhenium in the gas.

A process developed by *R.N. Platzke and H.R. Spedden; U.S. Patent 3,783,158; January 1, 1974; assigned to Kennecott Copper Corp.* is one in which certain organic compounds in solution other than aqueous are used to directly recover oxides of metals, such as rhenium, that form anionic species therewith, from gas streams carrying such metal oxides in volatilized form. The organic compounds are selected from the group consisting of organic phosphates and those alcohols and water-insoluble amines which have carbon chains made up of a minimum of four carbon stoms. The tertiary amines and tributyl phosphate are especially effective.

Although the process is particularly applicable to rhenium, other typical metals are selenium, lead, zinc, and gallium. One or more of such organic compounds is dissolved in a suitable organic vehicle, such as kerosene, and the resulting organic solution is brought into intimate contact with the gas stream. The resulting solution pregnant with the metal oxide values is treated, as by solvent extraction procedures, for the recovery of such metal values.

The essential features of the Kennecott and Chase Brass processes are summarized in Figure 75. The flow sheet with the exception of the hydrogen reduction step appears in the 1967 edition of the Minerals Yearbook and is based on the new Kennecott Copper Corp. molybdic oxide plant near Salt Lake City, Utah.

FIGURE 75: BY-PRODUCT RHENIUM SUBSYSTEM

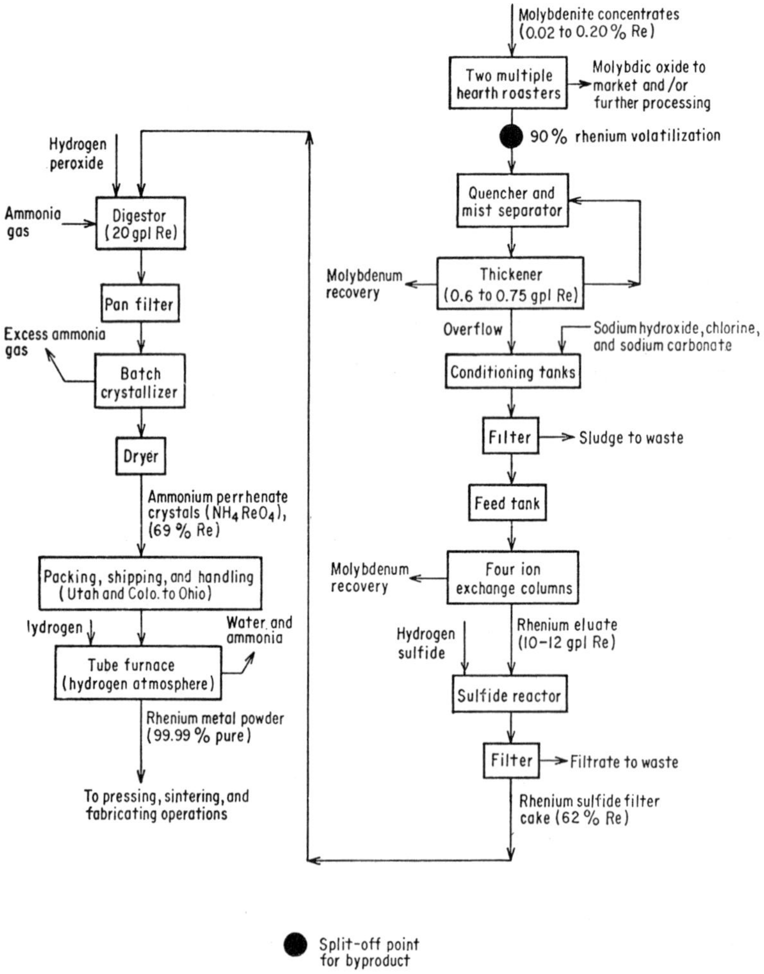

Source: Bureau of Mines Report I.C. 8659

The process involves roasting of molybdenite concentrates to molybdic oxide in multiple hearth roasters. A portion of the contained rhenium is volatilized and appears in the effluent gas stream along with entrained molybdenite particles. If sulfur in the molybdenite is reduced to 0.2 to 0.4% somewhat better than 50% of the rhenium is released. The rhenium oxide is then dissolved in a liquid solvent, separated from the solvent by an ion exchange technique and subsequently recovered in the form of rhenium salts from ion exchange resins.

Recovery from Liquid Wastes

A process developed by *K.J. Richards and C.N. Wright; U.S. Patent 3,894,866; July 15, 1975; assigned to Kennecott Copper Corp.* involves recovering rhenium values from dilute solutions containing from 0.00001 to 1.0 g/l of rhenium. The process consists of adding particles of a base metal which is more electropositive than copper to such a solution to react with and to precipitate the dissolved rhenium values therein; introducing into the solution, simultaneously with or prior to the addition of the base metal thereto, particles of a collector metal selected from the group consisting of copper and silver to collect the precipitated rhenium values, the particles of the collector metal having sufficient surface area to retain and carry the precipitated rhenium values; separating the solids and the liquids; and treating the solids to recover the rhenium values therefrom.

Recovery from Solid Wastes

Overall, the demand for rhenium has been increasing with its increased use as an addition to tungsten and molybdenum to form alloys for high-temperature application. This rising demand, coupled with the limited supply, has led the Bureau of Mines to conduct a series of studies of the conservation and more effective utilization of rhenium.

Recovery from Scrap Tungsten-Rhenium Alloy: A report by M.J. Ferrante, F.E. Block, A.D. Fugate and F.A. Skirvin, Report of Investigation No. 7254, Washington, D.C., U.S. Bureau of Mines (April 1969) describes an attempt to develop a process for converting tungsten-rhenium scrap to high-purity rhenium powder having an average particle size of 2μ as measured by the Fisher Sub-Sieve method. Volatile rhenium heptoxide was separated from less volatile tungsten trioxide by oxidation of tungsten-rhenium scrap with dry oxygen. The condensed rhenium heptoxide was dissolved in water, and minor metallic impurities were removed by selective precipitation from a rhenium solution. Ammonium perrhenate was then obtained by crystallization; before it was reduced with hydrogen to form rhenium powder, experiments were conducted to determine a perrhenate crystal size and reduction temperature that would yield 2μ rhenium powder, the size conventionally used for consolidation by powder metallurgy techniques.

SCANDIUM

Recovery from Uranium Plant Sludge

As described by A Petrick, Jr., H.J. Bennett, K.E. Starch and R.C. Weisner in Information Circular 8570, Washington, D.C., U.S. Bureau of Mines (1973), scandium can be recovered from uranium process sludges. The relationship of the scandium by-product subsystem to the uranium production system is illustrated in Figure 76, based on the Vitro process.

Two split-off points for by-product scandium are shown. In the Vitro Chemical Co. process, thorium and titanium and scandium building up in the organic stream of the uranium solvent extraction system are selectively tapped and subjected to a special stripping with hydrofluoric acid. The "white cake" precipitated contains 10% scandium oxide and is further refined to produce high-purity scandium oxide. The system is based on processing of an iron sludge containing scandium, titanium, thorium, uranium, and iron. This sludge is filtered from the loaded strip liquor and is either recycled to recover uranium, discarded, or is used as a raw material for scandium recovery as shown.

Recovery from Wolframite Concentrates

J.R. Ross and C.H. Schack have described process details for scandium recovery from uranium plant iron sludges and from wolframite concentrates in Report of Investigation No. 6580, Washington, D.C., U.S. Bureau of Mines (1965).

FIGURE 76: BY-PRODUCT SCANDIUM SUBSYSTEM

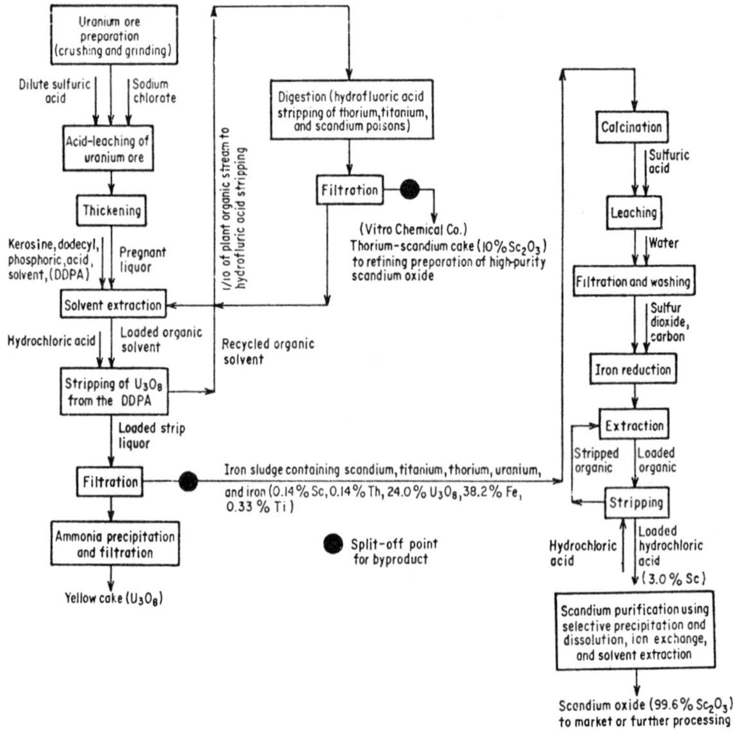

Source: Bureau of Mines I.C. 8570

SELENIUM

Recovery from Electrolytic Copper Muds and Slimes

As described by A Petrick, Jr., H.J. Bennett, K.E. Starch and R.C. Weisner in Information Circular 8569, Washington, D.C., U.S. Bureau of Mines (1973), there are three main methods of recovering selenium from copper muds and slimes. The first is roasting with soda, the second roasting with an acid, and the third smelting with soda and niter; the last is illustrated in Figure 77. The metallurgy and the economics of selenium and tellurium, are closely interlocked. At a precious metals refinery the volatile selenium in gases from the Cupel furnace is collected by water scrubbing to yield an acid solution. The addition of alkaline fluxes to the Cupel yields a soda slag containing both selenium and tellurium. When the high-acid selenium solution is contacted with the soda slag in a neutralizing tank tellurium dioxide is precipitated and the neutralized solution can then be treated to precipitate the selenium.

The metallurgy and economics of selenium alone can best be described by following the processing of the volatile gases and soda slag from the Cupel furnace. Figure 77 indicates a production of 1,900 lb of selenium from the 2,000 lb of selenium in the charge. This is based on an assumed recovery of 95% overall in selenium processing. Such a recovery assumes recirculation of part of the selenium content via precipitation of a low-grade selenium product, which is returned to the Cupel furnace. Actual recovery is likely to be higher than 95% but the 100 lb loss is assumed to account for a portion of the recirculated selenium.

FIGURE 77: BY-PRODUCT SELENIUM SUBSYSTEM

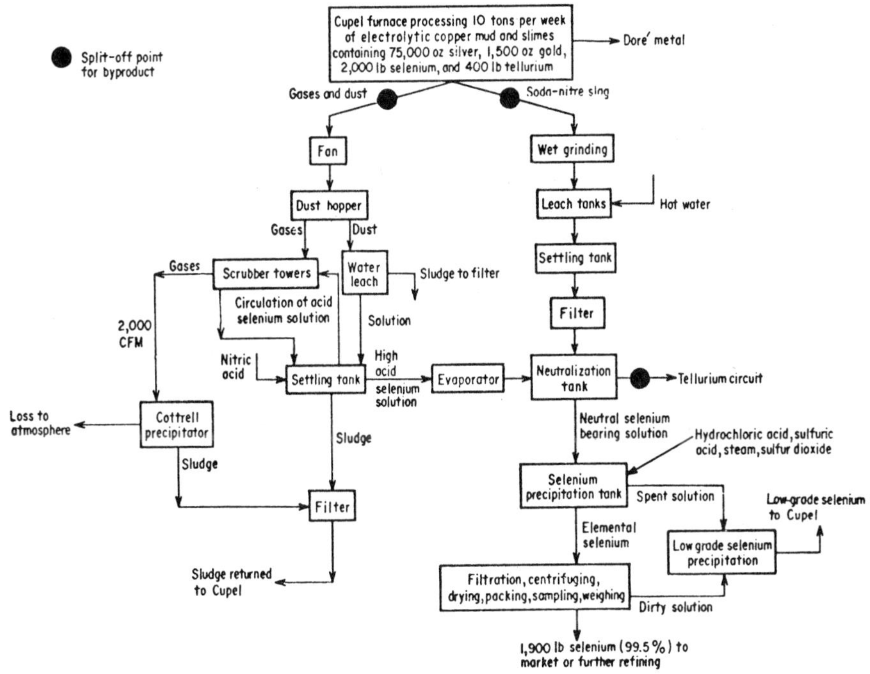

Source: Bureau of Mines I.C. 8569

Recovery from Scrap

Tonnage-wise, the selenium segment constitutes one of the minor segments of the secondary nonferrous metals industry. as described by H. Nack, K. Murthy, E. Stambaugh, H. Carlton and G.R. Smithson, Jr. in PB 233 646, Springfield, Va., National Technical Information Service (May 1974). Based on 1971 data, 30,000 lb or 3% of the total United States consumption was recovered from scrap. Wastes from this segment are atmospheric emissions, liquid wastes, and solid wastes.

Sources of raw materials for this segment, including both new and old scrap, are: selenium rectifiers (burned out and factory rejects), spent catalysts, used xerographic plates. The product from this segment is metallic selenium in the form of shot, refined powder, and selenium ingots. The recovery of selenium from selenium scrap involves three manufacturing operations: scrap pretreatment, smelting/refining, and casting (product formation).

Selenium is separated from the majority of the other scrap components in the scrap pretreatment operation by a mechanical process. The selenium is separated from the other scrap components by hammer milling, shot blasting, and similar methods. The recovered selenium is further processed or sold as a selenium metal. Energy demand is that needed to drive the equipment.

The process produces atmospheric emissions and solid wastes. The atmospheric emissions contain particulates of selenium metal, compounds, alloys, and oxide, whereas the solid wastes are composed of metallic and nonmetallic scrap. In the smelting/refining operation, the pretreated scrap is purified to produce selenium metal for subsequent processing. The processes by which the purification is achieved are: retort smelting, hydrometallurigical refining and distillation. Some details of these processes are given in the paragraphs which follow.

Retort Smelting Process – In retort smelting, partial purification of the pretreated scrap is achieved and the scrap is densified. The process steps are: charging the retort, melting the scrap, separating the selenium from the impurities by distillation, and discharging the residue. The energy required is that needed to heat the retort. Potential environmental pollutants are atmospheric emissions composed of gases and particulate matter containing selenium, solid wastes containing the impurities from scrap selenium, and liquid wastes.

These wastes may go to landfill. The liquid waste is cooling water which may be recycled in most cases. Thus, retort smelting has the potential for producing atmospheric pollution problems as well as water pollution problems from leaching in the landfill.

Hydrometallurgical Refining – Hydrometallurgical refining also achieves a partial purification of the scrap selenium. The process steps are: dissolve the scrap selenium or the selenium from retort smelting in a suitable solvent such as aqueous sodium sulfite, remove insoluble impurities by filtration, and precipitate the selenium for further purification. The energy required is that needed to heat the solutions and operate the auxiliary equipment. Potential environmental pollutants are atmospheric emissions if gases are used to precipitate the selenium, liquid wastes, and solid wastes. The process has the potential for the generation of pollution problems.

Distillation Process – This process yields high-purity selenium. The process steps are: charge the distillation unit, melt and distill the selenium, condense the selenium vapors, transfer molten selenium to the product formation operation, and remove residue from the still. The energy required is that needed to operate the distillation unit. Potential pollutants are atmospheric emissions composed of distillation wastes and particulates of selenium, solid wastes, and liquid wastes This process also has the potential for the production of pollution problems.

In product formation, the molten selenium is converted to the desired shape. The processes used to achieve this are a quenching process or a casting process.

Quenching Process – The quenching process produces selenium shot and selenium powder. The process steps for the production of selenium shot are: quench the selenium melt in a quenching liquid, separate the shot from the cooling liquid, and dry the shot. Powder is produced by quenching the selenium vapors from the distillation unit. Energy required is that needed to cool the quenching medium and operate the auxiliary equipment. Potential pollutants are liquid wastes such as cooling water which is recycled in most cases. The process has a low potential for the production of pollution problems.

Casting Process – By this process, the molten selenium is cast into ingots or other shapes. The process steps are: pour the molten selenium into the molds and solidify the melt. Energy is needed to operate the equipment. Essentially no pollutants are emitted. Thus, the process has no potential for the production of pollution problems.

SILVER

Recovery from Liquid Wastes

Recovery from Spent Photo Processing Solutions: A process developed by *J.S. Bentley; U.S. Patent 3,715,291; February 6, 1973; assigned to Photographic Silver Recovery Ltd.,*

England is one in which the recovery of silver from waste silver solutions, especially spent photographic processing liquids, is effected with a silver recovery machine, such as a rotating cathode machine, and an auxiliary tank installed immediately upstream of the silver recovery machine to receive the silver waste solution from film processing equipment or other source. A metering pump delivers the waste solution from the auxiliary tank to the silver recovery machine, the pump being started and stopped in dependence upon changes in the level of the waste solution in the auxiliary tank. The silver recovery machine is arranged to operate only while the metering pump is running, this operation being at a constant current setting.

A process developed by *K.R. Larson, J. Drew and H. Ott; U.S. Patent 3,792,845; February 19, 1974* provides an improved unit for the recovery of silver from spent photographic processing solutions by a metallic replacement technique. The unit is designed to pass silver-bearing photographic processing solution through a roll of steel wool, and includes a high-density polyethylene cartridge to hold the steel wool, fitted with suitable plumbing for passing the solution through the roll in the cartridge.

The polyethylene cartridge is essentially a drum with a generally flat lid, and includes a rim clamp for locking the lid tightly on the drum. The rim clamp fits around the top of the drum and has a lever lock operable from one side of the drum. The plumbing attached to a drum includes fittings for routing the solution into and out of the drum through a pair of openings in the cartridge lid. When the unit is in operation, feed solution passes into the top of the cartridge, contacts the steel wool roll in the cartridge drum, and then passes out of the drum for recycle use or discard to a sewer. The unit operates efficiently until the steel wool in the cartridge is exhausted, at which point it becomes necessary to replace the cartridge. This requires removal of the cartridge lid, but the removal can be easily accomplished without disconnecting the plumbing from the lid since the lever lock is operable from the side of the drum to permit loosening of the clamp for each movement of the lid with the plumbing attached.

A process developed by *R. Anderson and R.B. Beyer; U.S. Patent 3,829,549; August 13, 1974; assigned to Future Systems, Inc.* involves treating waste photographic processing solutions to destroy the pollutant nature of the solution and recover residual silver from the solution. Waste photographic processing solutions containing silver thiosulfate complexes and/or sulfite, such as exhausted fixing solutions, are treated with a chemical formulation including an oxidant of sodium chlorate in the presence of an acidic buffering agent. The formulation oxidizes the thiosulfate and/or sulfite ions of the solution to the sulfate state and precipitates the silver ion as an insoluble halide.

A process developed by *I. Slovonsky, T.J. Dagon and T.W. Bober; U.S. Patent 3,832,453; August 27, 1974; assigned to Eastman Kodak Co.* is one in which silver is recovered from an exhausted photographic fix or bleach-fix solution by bringing together just at the entrance of a static mixer a stream of a sodium sulfide solution and a stream of fix or bleach-fix solution and passing the mixture at high velocity through the static mixer, wherein it is repeatedly divided and changed in flow pattern, to precipitate silver sulfide particles.

The flowing mixture containing silver sulfide particles is then passed into a separation chamber of enlarged cross section, and its flow is slowed and reversed by baffles to cause the silver sulfide particles to drop out and collect in the bottom. A fix solution is regenerated for reuse by the above procedure. When a bleach-fix is treated, it can be regenerated by introducing air, either upstream or downstream of the separation chamber, to reoxidize to ferric ions any ferrous ions present as a result of the bleaching operation, or those ferrous ions produced by reaction of sodium sulfate with residual ferric ions in the exhausted bleach-fix. A small amount of acid also should be added to adjust the pH to below 7.

A process developed by *M. Korosi; U.S. Patent Application (Published) B 282,819; January 28, 1975; assigned to Eastman Kodak Co.* is one in which silver is recovered from photographic waste liquids containing emulsified silver, silver compounds, and gelatin by reacting therewith a proteolytic enzyme while the reactants are maintained in an alkaline condition. Then the reactants are acidified to a pH of 4.2 or less by introducing an acid such as HCl or H_2SO_4.

Precipitation of gelatin-bound silver and silver compounds occurs upon acidification, and the precipitate is allowed to settle out in a setting tank. Supernatant liquid is removed from the top, neutralized and sent to sewer. Settled sludge is removed from the bottom and silver recovered, as by incineration. The operation can be conducted either batchwise or continuously.

A process developed by *D.T. Thompson; U.S. Patent 3,875,032; April 1, 1975; assigned to Foresight Enterprises, Inc.* is one in which the plating current in a system for recovering silver from photographic fixing solution is controlled in response to the threshold voltage required to pass plating current in the solution, and to the variation of current flow at voltages near the threshold as the silver ion concentration in the solution varies.

Silver recovery from waste photographic solutions by metallic replacement has been described by R.O. Dannenberg and G.M. Potter in Report of Investigation No. 7117, Washington, D.C., U.S. Bureau of Mines (April 1968).

A metallic displacement process, utilizing easily available steel wool or steel window screen, was devised and successfully used for the recovery of silver from waste photographic fixing solutions. Laboratory and prolonged commercial-scale testing established that the iron filament type unit will efficiently precipitate silver, as a high-grade sludge, from acidic waste photographic fixing solutions ranging in grade from less than 1 to over 10 grams of silver per liter. Data and design information are presented to facilitate the construction and operation of a practical unit of any desired size. A smelting process for recovering pure silver from precipitated sludge containing 27 to 80% silver was devised.

Recovery from Solid Waste

According to R.L. Rawls, *Chemical and Engineering News,* July 21, 1975, pp 9-10, silver recycling has grown appreciably in importance recently. According to information obtained by Rawls silver recycling also has grown in importance recently. According to data from the Silver Institute, refining of scrap silver is up 15% for the first five months of 1975 over the same period in 1974 and 30% over the first five months of 1973. Much of this increase comes from more efficient methods of silver recovery that becomes economical when silver's price is high.

Recovery from Copper Refinery Anode Muds and Slimes: Silver production involves processing of anode mud and slime from electrolytic copper refining. As decribed by A. Petrick, Jr., H.J. Bennett, K.E. Starch and R.C. Weisner in Information Circular 8569, Washington, D.C., U.S. Bureau of Mines (1973), the silver production scheme starts from the point where the silver-bearing material is stripped of copper to the point of production of silver crystals or bricks as shown in Figure 78.

Where silver production is obviously justified economically by comparative revenue and incremental costs, as is usually the case, one may view the production of the other by-products, selenium, tellurium, gold, and platinum metals as by-products rather than as co-products of the silver subsystem. This further highlights the economic dependence of one by-product upon another.

The hypothetical system shown in Figure 78 assumes as its raw material 10 tons of leached slimes per week from copper electrolytic refining. The assumed mud-slimes analysis is 7,500 ounces of silver and 150 ounces of gold per ton of slimes, 10% selenium, and 2% tellurium. The slimes are first washed, dried, mixed, packaged, and transported from an electrolytic copper refinery to the facility for recovering precious metals. The metal content of the mud and slime are hypothetical based on review of available data on treated slime analyses from several electrolytic copper refineries.

When the mud-slimes are received at the precious metal refinery, they are loaded into a 15-ton Cupel furnace, and fluxes are added. The Cupel is a reverberatory-type furnace with wall construction of a steel shell, lined with basic refractories and having a roof of alumina brick.

FIGURE 78: BY-PRODUCT SILVER SUBSYSTEM

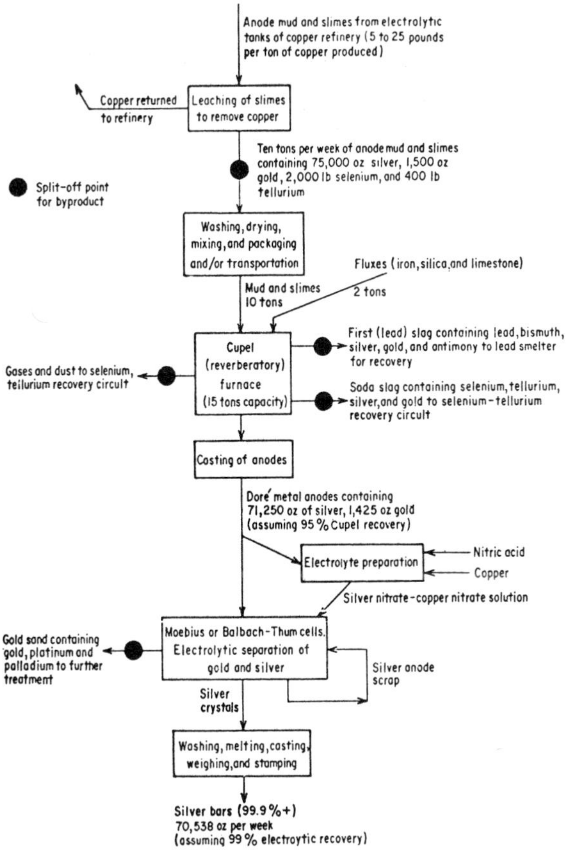

Source: Bureau of Mines I.C. 8569

The Cupel furnace is fired with oil or gas. The operation of the Cupel is a batch process, consisting of 5 to 8 days smelting and refining in the same furnace. The length of time in the furnace depends upon the impurities to be removed. Selenium and tellurium recovered in gases, dust, and slag can be regarded as separate subsystems involving economic decisions, essentially separate from the production of the silver.

The product of the Cupel furnace is cast into Doré (gold-silver alloy) anodes and moved to the Balbach-Thum electrolytic cells for separation of the silver and gold. Recovery of silver in the Cupel furnace is estimated at 95% of that in the anode-mud slimes.

The Balbach-Thum cells look like washtubs made of fiber glass and plastic, 36 by 24 inches square and 9 inches deep. The flat Doré metal anodes, about 20 inches square and 1 inch thick, weighing about 1,000 ounces each, are suspended 3 to 4 inches above and parallel to the bottom of the cell. The bottom of the cell is the cathode. The electrolyte is a mixture of copper nitrate and silver nitrate. There is no circulation or external heating of the electrolyte.

The anode basket is a filter cloth bag resting on glass rods that supports the weight of the anodes. Silver crystals precipitate on the cathode and are scraped out every 3 to 4 hours. Silver crystals recovered from the cells are then washed, melted in an induction furnace, and cast into silver bricks.

The gold mud collecting in the filter bag during the silver electrolysis contains gold and platinum metal. This is removed and subjected to further treatment. Recovery in the Balbach-Thum electrolytic cells is estimated at 99%.

Recovery from Metallic Electrical Scrap: A process developed by *G.A. Miller and K.N. Johnson; U.S. Patent 3,616,332; October 26, 1971; assigned to Texas Instruments Inc.* is one in which substantially pure silver is economically and effectively recovered from scrap material, such as scrap material containing silver, silver oxide and other metals such as cadmium, copper, nickel and the like in elemental or combined form, by electrolysis of an aqueous electrolyte solution containing silver sulfamate, a glutamic acid compound such as monosodium glutamate monohydrate and an excess of sulfamic acid. The scrap material itself serves as the anode and fine grained silver is deposited at a cathode of corrosion-resistant material without substantial codeposition of any other metals from the scrap. During electrolysis, silver and any other metals present in the anode scrap material are dissolved electrolytically and the metals in combined form are dissolved chemically in the electrolyte.

FIGURE 79: PROCESS FOR SILVER RECOVERY FROM ELECTRICAL SCRAP

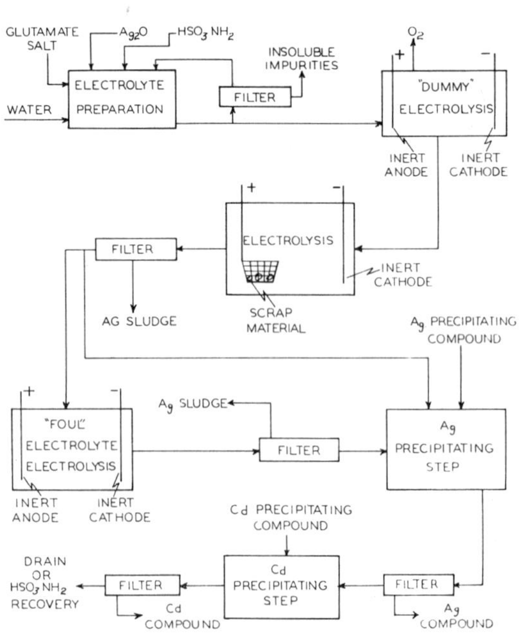

Source: G.A. Miller and K.N. Johnson; U.S. Patent 3,616,332; October 26, 1971

Because of the high solubility of silver sulfamate in the electrolyte, high current densities on the order of 30 to 300 asf may be employed. When substantial quantities of cadmium and metal ions other than silver build up in the electrolyte, substantially all of the silver therein may be recovered by subjecting the electrolyte to further electrolysis using inert electrodes as the anode and cathode followed by precipitation of the remaining silver as silver chloride or other insoluble silver compound. Cadmium may then be recovered from the resulting electrolyte by precipitation as cadmium hydroxide or other insoluble cadmium compound. Figure 79 is a flow diagram showing the essential elements of the process.

Recovery from Scrap Photographic Film: A process developed by *E.B. Johnson; U.S. Patent 3,754,743; August 28, 1973* is a process for the recovery of silver from photographic waste in which the film or other waste is ignited and brought up to a satisfactory burning temperature in a primary chamber and maintained at this temperature until the film or waste is burned by controlling the amount of oxygen supplied to the primary chamber. During the ignition and burning of the photographic waste the smoke and particulate matter is directed to a secondary chamber where it is subjected to burning at a higher temperature than in the primary chamber. After the photographic waste is burned, the silver bearing ash is cooled and removed from the primary chamber. The ash is then processed by conventional techniques to refine the silver. This apparatus permits the economical rendering of silver from photographic waste by burning with the elimination or reduction of smoke and other air pollutants.

A process developed by *R.B. Lilly; U.S. Patent 3,793,168; February 19, 1974* is a process for recovering metallic silver from scrap photographic film in which the film is first immersed in a solution of hydrochloric or hydrobromic acid and a chromate; thereafter immersed in a hypo bath; the free silver thereafter being removed from the hypo bath by an electrolytic process.

Recovery from Scrap Printed Circuit Boards

See Precious Metals, page 278.

SODIUM METAL

Recovery from Insulated Alkali Metal Conductors

A process developed by *G.I. Addis; U.S. Patent 3,428,447; February 18, 1969; assigned to Union Carbide Corporation* is one in which alkali metal is recovered from insulated alkali metal conductors by reducing the conductor to segments of a suitable size, heating the segments in an inert media to a temperature above the melting temperature of the alkali metal but below the decomposition temperature of the insulation. The segments are preferably heated in an oil bath.

TANTALUM

Recovery from Scrap Capacitors

A process developed by *W.M. Baldwin, E.O. Fuchs, D.J. Sharp and J.H. Swisher; U.S. Patent 3,697,255; October 10, 1972; Baldwin and Sharp assigned to Western Electric Co. and Fuchs and Swisher assigned to Bell Telephone Laboratories, Inc.* is one in which scrap tantalum bodies having impurities therein are reclaimed as substantially pure tantalum metal by a process which includes immersing the bodies in a molten metal such as calcium or magnesium at elevated temperatures, and under nonoxidizing conditions for a prolonged period of time sufficient to cause entrapped impurities within the body to diffuse out and react with the molten metal. Figure 80 is a block flow diagram showing the essential steps in this process.

FIGURE 80: TECHNIQUE FOR TANTALUM RECOVERY FROM SCRAP CAPACITORS

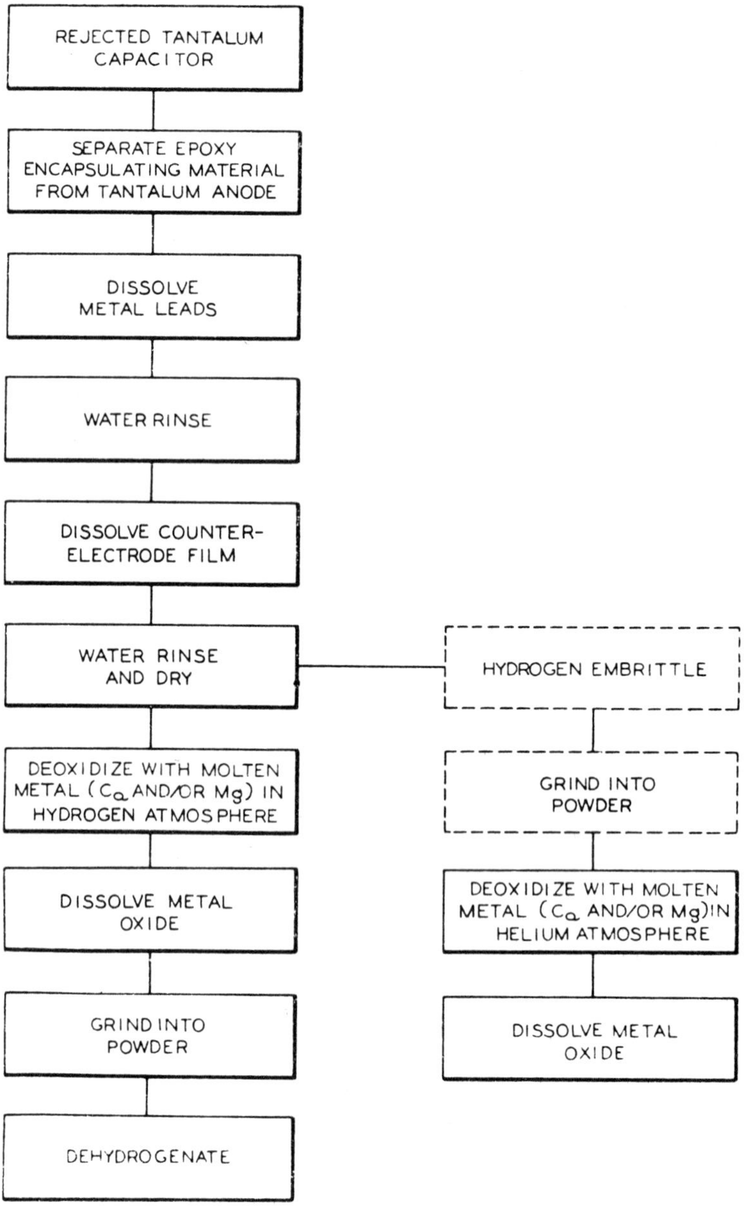

Source: W.M. Baldwin, E.O. Fuchs, D.J. Sharp and J.H. Swisher; U.S. Patent 3,697,255; October 10, 1972

TELLURIUM

Recovery from Electrolytic Copper Anode Muds and Slimes

As described by A. Petrick, Jr., H.J. Bennett, K.E. Starch and R.C. Weisner in *Information Circular 8569,* Washington, D.C., U.S. Bureau of Mines (1973), the by-product tellurium subsystem duplicates a part of the selenium subsystem during the early stages of processing. This is due to the fact that the soda-nitre slag from the Cupel furnace carries all of the recoverable tellurium and about one-half of the recoverable selenium. The two are separated at the neutralization tank where tellurium dioxide is precipitated and the neutralized selenium solution goes on to further processing. Therefore, the tellurium subsystem can be confined to the metallurgical flowsheet concerned with further processing of the tellurium dioxide pulp precipitated in the neutralization tank. Such a subsystem is illustrated in Figure 81.

Assuming 90% recovery of tellurium in the Cupel furnace and 90% recovery in the balance of the subsystem shown, the 400 lb of tellurium in the anode mud and slimes results in 324 lb of tellurium metal.

FIGURE 81: BY-PRODUCT TELLURIUM SUBSYSTEM

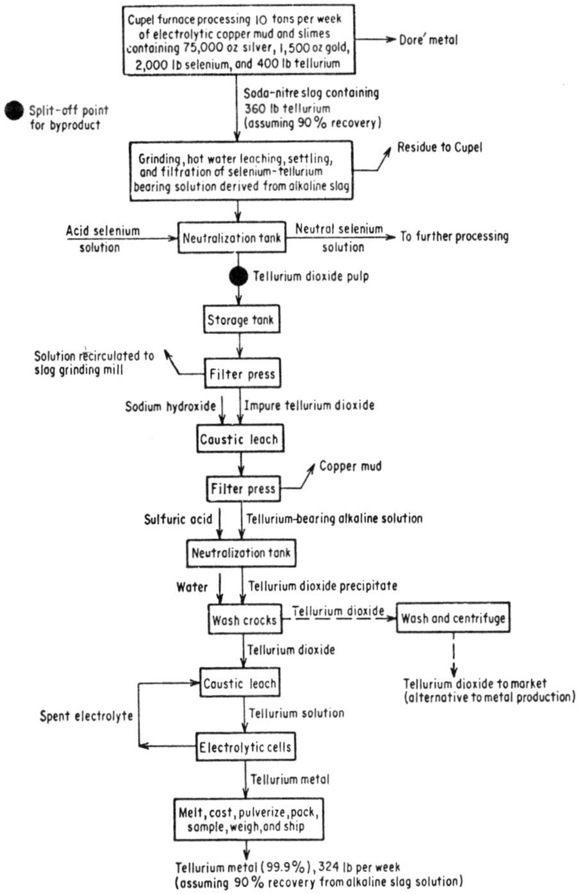

Source: U.S. Bureau of Mines I.C. 8569

THALLIUM

Recovery from Zinc Smelter Flue Dusts

The U.S. supply of thallium is derived principally from the processing of zinc concentrates. The American Smelting and Refining Company at Denver, Colorado, has been the sole domestic producer of thallium for many years. Commercial thallium is derived generally from flue dusts generated in the roasting of sulfides or from cadmium-rich material derived from zinc ores. It is commonly associated with the cadmium by-product system. The first steps of thallium recovery are part of the cadmium subsystem. The final steps of thallium recovery involve precipitation with zinc, purification by reprecipitation and subsequent electrolytic recovery of the metal.

TIN

In recent years the worldwide consumption of tin has steadily increased to a point where annual consumption exceeds annual output. This growing demand for tin has been attributed largely to the increased utilization of tinplate throughout the world by the food container industry, as well as the increased utilization of this metal in electronics applications. There has therefore been an increasing need for more efficient processes for the recovery of tin from tin-containing substances if the growing demand for tin throughout the world is to be met. For example, tin mining methods employed in some countries of the world produce waste by-products containing up to about 50% of the available tin, with waste materials containing about 5 to 30% tin being considered common.

Also, detinning operations in which metallic tin is recovered from tin scrap may leave up to 10% of the available tin unrecovered in the discarded refuse material. These and other readily available materials of low tin content are, therefore, recognized as holding great potential as sources for additional quantities of needed tin. Standard techniques, such as roasting, reduction, and physical-metallurgical separation methods, have largely proven to be commercially unsuccessful, however, when attempts have been made to recover tin from most tin-containing waste substances, i.e., from solids containing less than about 10% tin.

Recovery from Hardhead

The recovery of tin from hardhead by amalgam electrolysis has been described by P.M. Sullivan and D.H. Chambers in *Report of Investigations No. 5827,* Washington, D.C., U.S. Bureau of Mines (1961). Hard tin or hardhead is a secondary product from tin smelting which contains crystals of $FeSn_2$ and a small quantity of free tin. Relatively pure tin metal formed in the first stage of smelting tin ore is filtered at a temperature near the melting point of tin, leaving a hardhead residue. The largest source of hardhead, however, is the second (slag) smelting stage where the tin in the first smelting slag is entirely reduced to the approximate composition $FeSn_2$. The general practice in tin smelting is to return hardhead to the first smelting stage where the iron from $FeSn_2$ acts as a reducing agent for the tin minerals in the ore. This method is effective, but it imposes a heavy recirculating load on the smelter.

The principles and application of amalgam electrolysis to the recovery of zinc and tin from secondary products were studied in the report cited above. This Bureau of Mines project was one phase of investigating the development of more efficient methods for processing iron-bearing zinc and tin to produce both base metals of high purity. Zinc of 99.99% purity was recovered from galvanizer's dross with recovery efficiency of 98%. Tin of 99.9 to 99.99% purity was recovered from hardhead with recovery efficiency of 81%. Both materials were treated by the same method in separate systems. Mercury was used to leach zinc from dross and tin from hardhead to form amalgams of each metal, both of which were subsequently refined electrolytically to produce high-purity metal. Impurities that were contained in both secondary products behaved as theoretically predicted; impurities that were insoluble in mercury remained as sludge in the leaching column. Except for lead, the readily amalgamated impurities were prevented from fouling the electrolyte by the blocking action of the amalgam electrode. Lead impurity in the tin-refining system dissolved and immediately precipitated as the sulfate. The primary advantage of the amalgam refining method lies in direct recovery of the extremely pure base metal from highly contaminated secondary material.

Recovery from Mine Wastes and Slimes

A process developed by *E.M. DeForest, W.C. Bradbury and T.W. Royer; U.S. Patent 3,394,061; July 23, 1968; assigned to Vulcan Detinning Division of Vulcan Materials Company* is a process particularly suited to the recovery of tin from a wide variety of substances of low tin content, such as refuse by-products from detinning operations, tin mine wastes, mining slimes, etc. Figure 82 is a flow diagram of this process.

FIGURE 82: PROCESS FOR TIN RECOVERY FROM MINE WASTES AND SLIMES

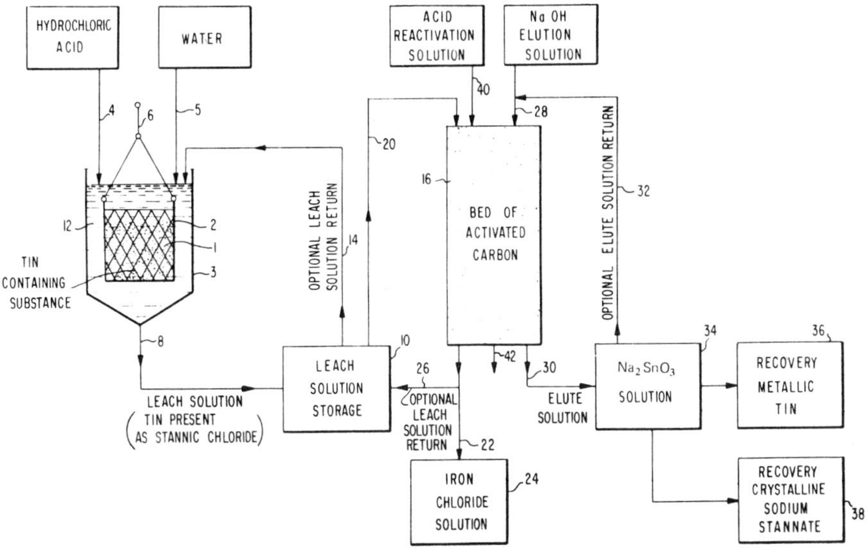

Source: E.M. DeForest, W.C. Bradbury and T.W. Royer; U.S. Patent 3,394,061; July 23, 1968

A satisfactory batch application of this process may be conducted, as shown in the figure, by placing the refuse from a detinning process or other tin-containing solid **1** within a permeable acid-resistant plastic basket **2** which may be lined with a filter cloth made of polyethylene terephthalate or similar acid resistant material. Hydrochloric acid and water of the proper concentration and temperature are added to the leaching vessel **3** by means of conduits **4** and **5** respectively to form leach solution **12**. If desired the leaching vessel **3** may be provided with suitable heating means to accurately maintain the preferred acid leach temperature. The basket **2** may be suspended within leaching vessel **3** by means of a suitable suspension linkage **6**.

Following an appropriate leach period the leach solution containing stannic chloride dissolved therein is withdrawn from leaching vessel **3** by means of conduit **8** and conveyed to to a suitable storage vessel **10**. If desired, however, the leach solution **12** may be continuously recycled by means of conduit **14** during the leach period. The resulting leach solution is next conveyed from storage vessel **10** to column **16** containing a bed of activated carbon by means of conduit **20** where stannic chloride present in the solution is preferentially adsorbed. At the lower portion of column **16** the solution is withdrawn by means of conduit **22**. Iron chloride present in the solution may be recovered at **24**. If desired, however, the solution may be recycled through the bed of activated carbon to obtain greater tin adsorption by use of conduits **26** and **20** prior to the recovery of iron chloride at **24**. The adsorbed stannic chloride is next eluted from the bed of activated carbon by the adding of an aqueous solution of sodium hydroxide to the column **16** by means of conduit **28**.

This solution may be withdrawn by use of conduit **30** and if desired recycled through conduits **32** and **28**. The sodium stannate solution formed is collected at **34** and may be further treated for the recovery of metallic tin at **36**, or crystalline sodium stannate at **38** as heretofore discussed. The column **16** may be reactivated for further stannic chloride adsorption by the addition of a hydrochloric acid solution through conduit **40** with its removal through conduit **42**.

Example: A 1,000 gram charge of an essentially solid tin-containing mud-like residue obtained as a by-product from a detinning operation was selected for treatment according to the process. An initial analysis of this detinning waste revealed the following composition:

Components	Weight %
Sn as SnO_2	3.6
Fe as Fe_2O_3	9.4
Organics (combustibles at 600°C)	23.6
Water (volatile at 110°C)	37.0
SiO_2	26.4
Also, traces of Al and Pb, etc.	--
Total	100.0

The tin-containing waste was subjected to a dilute solution of 7% hydrochloric acid by weight for a period of 30 minutes at a temperature of 85° to 90°C. The mol ratio of hydrochloric acid/tin was 25 and the liquid/solids ratio was 5.4. During this period 84.2% of the tin present in the tin-containing waste was converted to stannic chloride. This resulting solution containing stannic chloride was filtered, and then passed across a carbon bed at a flow rate of approximately 0.5 gal/min/sq ft of cross-sectional area at ambient temperature. The bed was formed in a column three feet tall and two inches in diameter with 12 x 40 mesh activated carbon, type CAL, available from the Pittsburgh Chemical Company.

Prior to use the carbon bed was packed by flotation with water in the usual manner. The adsorbed tin was eluted from the carbon bed by dilute sodium hydroxide at 70°C having a concentration of 160 grams sodium hydroxide per liter of water at a flow rate of approximately 0.5 to 1 gal/min/sq ft of cross-sectional area. 100% of the adsorbed tin was recovered by employing a single pass of the solution having a mol ratio of sodium hydroxide to tin or 72. Metallic tin was recovered from the resulting solution of sodium stannate by electrodeposition.

Recovery from Scrap

The importance of the tin segment of the secondary nonferrous metals industry is highlighted by the fact that virtually all U.S. primary tin requirements (69,029 tons in 1972) are imported and that the secondary production constitutes about one-third of total U.S. production. In 1972, this amounted to 20,180 long tons. The forecast demand for tin for the year 2000 is about 120,000 long tons. Minor sources of air pollution are the dross smelting and gas fired refining processes. The water pollution potential from the chemical detinning operation is the major pollution problem of the tin scrap recovery segment.

Major raw material sources for scrap recovery are the trimmings and rejects from the can companies and rejected plating coils from the steel industry. Although the used food cans and containers constitute a vast potential source of raw material, their present contribution to the recoverable scrap is only 1%. Even this 1% is collected by the ecology minded groups like the Isaac Walton group, etc. Other than these groups, there is no organized method of used can collection for scrap recovery. Almost 99% of the used cans find their way to municipal incinerators.

The recovery of scrap from used cans is not expected to exceed the current level because of the aluminum and plastic used in production of cans. The high cost of segregating the tin and iron metal portion from the garbage and the difficulties in processing the metal cans to recover the tin in the presence of aluminum are negative incentives to increased usage of used cans in scrap recovery. Thus, the raw materials are mostly new scrap consisting of:

(1) Tin drosses and sludges
(2) Solder drosses and sludges
(3) Can rejects from can companies
(4) Rejected plating cells
(5) Bronze used parts and rejects
(6) Type metal scrap

Products from the tin segment of the secondary nonferrous metals industry include:

(1) Pig tin
(2) Solder and type metal (tin-lead alloy with varying amounts of antimony)
(3) Tin oxide
(4) Fire-refined tin
(5) Sodium stannate
(6) Stannous sulfate and other tin chemicals
(7) Babbit metal

Three manufacturing operations are employed to recover tin from scrap: (1) scrap pretreatment, (2) detinning/smelting, and (3) refining/casting. The pretreatment process varies according to the type of tin-containing scrap. It may involve dealuminization or simply a batch mixing process.

Dealuminization Process – This process is conducted only if the scrap tin cans are suspected to contain substantial quantities of aluminum ends. About 90% of the time dealuminization is not done. This process removes aluminum as sodium aluminate in solution. The steps are: (a) leach the scrap in hot sodium hydroxide, (b) remove the sodium aluminate solution, (c) pump it to the evapocentrifugation refining operation for recovery of soluble tin values, and (d) recover dealuminized tin scrap for the chemical detinning process. Heat energy for alkali solution and electrical energy for pumping are required. The process produces no wastes; hence, there is no potential pollution from this process.

Batch Mixing Process – This mechanical operation mixes the drosses with limestone and coke to prepare a feed suitable for smelting. Mixing is the only process step involved. Electrical energy for the rotary drum mixer is required. Dust emissions are produced. However, these are collected by baghouses and recycled. The process, therefore, has no pollution potential. By the detinning-smelting operation, the tin values in scrap are separated from metallic and nonmetallic impurities or contaminants. Two types of operation (hydrometallurgical and pyrometallurgical) are used. The hydrometallurgical process is chemical detinning. The pyrometallurgical process is the dross smelting process. A dust leaching and filtration process may be used to prepare the feed for the smelter. These subprocesses are described briefly in the paragraphs which follow.

Chemical Detinning Process – The process extracts the tin value in the scrap as sodium stannate solution and also generates detinned steel scrap for use in secondary ferrous metallurgy. The process steps are: (a) addition of hot alkali (NaOH) and sodium nitrite or nitrate ($NaNO_2$ or $NaNO_3$) solution to the scrap, (b) draining and pumping the solution to the refining/casting operation after the reaction is complete, and (c) washing the detinned scrap with water sprays (as many as four) to obtain clean scrap for the steel industry. Three of the sprays are pumped to the crude sodium stannate solution tank and the fourth spray is used for recycle. Heat energy is required to heat the caustic solution and electrical energy for pumping stannate solution and spray water. No significant quantities of wastes are generated in the process and hence there are no serious pollution problems.

Dross Smelting Process – This process partially purifies the drosses and produces crude furnace metal. The process steps are: (a) charging the furnace, (b) melting the charge, (c) tapping the crude furnace metal, and (d) tapping matte and slag. Heat energy is required in smelting. It is reported that 90 gallons of fuel oil per ton of charge are used in the process. The matte produced is shipped to the copper smelter. The slag is discarded if low in metal values. Cooling water is recycled in a closed loop operation. Hence, the only wastes are minor atmospheric wastes and major amounts of slag. Almost 99.9% of dust in the furnace gases is collected in an electrical precipitator and leached in the dust leaching and filtration process whose description follows. The process has reasonable potential for solid waste pollution.

Dust Leaching and Filtration Process – The tin and chlorine values in the Cottrell dust collected in the dross smelting process are separated from the dust by sulfuric acid leaching. The process steps are: (a) leaching the Cottrell dust with dilute sulfuric acid to remove zinc and chlorine, (b) filtering to remove acid and dissolved zinc and chlorine, (c) drying the leached dust, and (d) conveying the dried leached dust to the batch mixing process. Energy required for this process is electrical energy and that derived from fuel oil. Potential environmental pollutants generated by this process are atmospheric emissions from handling of the dust and drying of the leached dust, liquid wastes from the leaching and filtering steps, and solid wastes collected via a baghouse from the atmospheric emissions. Thus, the process has a potential for the production of pollution problems if the pollutants are not collected and disposed of in an acceptable manner.

In the refining/casting operation, the tin is recovered as sodium stannate of commerce, as stannic oxide, as pure metallic tin, and as tin alloys. The individual processes to produce these products are:

(1) Settling and leaf filtration
(2) Evapocentrifugation
(3) Electrolytic refining
(4) Acidification and filtration
(5) Fire refining
(6) Smelting
(7) Calcining
(8) Kettle refining

These processes are each described briefly in the following paragraphs.

Settling and Leaf Filtration Process – This process produces a purified stannate solution from which tin can be recovered by a number of processes. The impurities–silver, mercury, copper, cadmium, some iron, cobalt, and nickel–are precipitated as metal sulfides. The process steps are: (a) addition of sodium sulfide with agitation, (b) settling, (c) decantation, and (d) filtration. Electrical energy is required for agitation and filtration steps. The process has marginal potential for solid waste disposal problems. These solid wastes are generated as metal sulfides.

Evapocentrifugation – This process produces a sodium stannate of commerce (Na_2SnO_3) product. The process steps are: (a) heating the solution to concentrate the sodium stannate solution, (b) crystallization of sodium stannate, and (c) recovery of the stannate by centrifugation. Thermal energy for evaporation and electrical energy for centrifugation are required. The wastes produced are atmospheric emissions consisting of water vapor and liquid wastes. The process has no substantial pollution potential.

Electrolytic Refining Process – This process produces cathodic pure tin metal from the purified sodium stannate solution. The process steps are: (a) passing the alkaline electrolyte (sodium stannate solution) through a cascade of cells at about 190°F, a characteristic condition with an initial concentration of 110 to 120 g/l of tin; (b) removal of cathode as tin builds up; and (c) stripping the tin from the cathode. Electrical energy is required for electrolysis and to operate auxiliary equipment. Spent electrolyte, an alkaline solution, is generated as a liquid waste. This waste has moderate water pollution potential.

Acidification and Filtration Process – The process produces tin hydrate which is subsequently processed to tin and stannic oxide. The process steps are: (a) acid neutralization with sulfuric acid to form tin hydrate and (b) filtration to separate the hydrate as filter cake. The filtrate is recycled to the settling and leaf filtration process described above. Electrical energy is used for stirring and filtration. No wastes per se are generated in the process and hence the process has no pollution potential.

Fire Refining – This process produces purified tin from cathodic tin. Natural gas is the heat source. The process steps are: (a) charging the furnace, (b) melting of the charge, (c) removing the impurities as slag and dross, (d) pouring of the molten metal, and (e) casting of the metallic tin. Energy required for this process is that derived from natural gas. Potential pollutants are atmospheric emissions–gases and metallic fume–and solid wastes. The solid wastes are recycled for metals recovery. Thus, the process is not a source of serious pollution problems.

Smelting Process – This is an alternate route to producing pure tin when electrolysis is not feasible. The process converts the tin hydrate to tin metal by reducing the stannic oxide (SnO_2) with a reducing agent. The process steps are: (a) heating the hydrate, (b) adding the reducing agent, (c) reducing the tin hydrate, (d) melting the tin, (e) skimming the dross, (f) pouring the molten metal, and (g) casting the metal. Heat energy derived from the fuel is the main energy requirement of the process. Potential pollutants are solid wastes, liquid wastes, and atmospheric emissions. The solid wastes produced as dross are recycled. The water used to cool castings is recycled although eventually some of it is rejected to the waterways. Atmospheric emissions are collected via baghouses or wet scrubbers. The pollution potential of the process does not appear to be significant.

Calcining Process – This process converts tin hydrate to anhydrous stannic oxide (SnO_2). The process steps consist of: (a) charging the calciner, (b) calcining the tin hydrate, and (c) removing and packaging the stannic oxide. Energy required is thermal energy. The primary pollutant is atmospheric emissions consisting of water vapor, dusts, and fume. Solid wastes generated are those collected in the baghouse. Thus, this process has a potential for the production of pollution problems.

Kettle Refining Process – This process purifies the crude furnace metal. The process steps are: (a) dry drossing to remove the impurities as slag and matte, (b) decopperizing using sulfur, (c) removing antimony using aluminum, and (d) casting into required shapes. The process uses large amounts of heat energy as derived from fuel oil. The waste products generated are atmospheric emissions and solid wastes. The solid wastes are slag and matte from the purification of the crude furnace metal. Atmospheric emissions consist of gases from combustion of the fuel and fume and dust entrained in the gases. These wastes constitute reasonable potential for pollution if not collected and disposed of by an approved method.

Recovery from Smelter Wastes

Techniques have been described by A.G. Starliper and H. Kenworthy in *Report of Investigation No. 6107,* Washington, D.C., U.S. Bureau of Mines (1962) for recovery of tin, tungsten and other metals from tin smelter wastes. One waste product was a spent acid solution that contained tin, tungsten, iron, lead, copper, zinc, and silver and that came from presmelting treatment of impure tin concentrates. The waste acid was used in other experiments to dissolve an iron-tin-tungsten alloy (hardhead) obtained by direct reduction of the slag. An insoluble fraction containing tungsten, silver, silicon, columbium, and tantalum was removed, and copper was cemented on metallic iron. Tin was recovered by cementation on zinc plates, and lead was precipitated by adding sulfate ion. Iron was precipitated by increasing the pH which left a solution containing principally zinc.

TITANIUM

Recovery from Scrap

The titanium segment of the secondary nonferrous metals industry constitutes one of the minor segments. Annual recovery of titanium is only approximately 8,000 tons. This, however, is some 20% of total ingot output. Much of the scrap is in-house material generated by primary producers of ingots and mill products. Because of the nature of the operations, this segment has a low potential for the production of serious atmospheric emission problems. However, problems resulting from water pollution could be significant.

Raw materials to this segment include the following:

(a) Bulk scrap (trim sheets, plate sheets, cuttings, etc.)
(b) Noncarbide machine turnings and borings
(c) Carbide-containing machine turnings and borings. This scrap is generated by the use of carbide tools in machining.

Products include titanium alloy ingots and electrodes. The recovery of titanium alloys from scrap titanium involves three operations: (1) scrap pretreatment, (2) smelting, and (3) casting. Some statistics and processing details concerning generation and disposal of titanium scrap are covered by D.J. Maykuth of Battelle Memorial Institute in AD 648 565, National Technical Information Service, Springfield, Va., (March 1, 1969).

Two processes used in scrap pretreatment are: vapor degreasing and acid pickling. Normally, if the scrap is not contaminated by oxides, acid pickling is not necessary and degreasing will suffice to produce a clean scrap.

Vapor Degreasing: This process removes grease and cutting oils from machine turnings and borings. The process steps are: (1) vaporizing a solvent such as trichloroethylene, (2) mixing the vapors with the scrap, (3) condensing the oil-bearing vapor, (4) filtering the condensed solvent, and (5) removing the scrap for subsequent processing. As the solvent goes through this cycle, the oil dissolves in the solvent and gradually lowers the boiling point of the solvent mixture. Eventually, the oil-solvent mixture is discarded due to loss of deoiling ability of the solvent or the solvent is regenerated.

Heat energy is required by the process for vaporizing the oil and electrical energy for water circulation through condenser. Wastes generated are the oil-solvent mixture, minor amounts of solvent vapor and insignificant quantities of solid dust particles. The oil-solvent mixture has a considerable water pollution potential. Atmospheric emissions and solid wastes generated are negligible and thus, present no serious pollution problems.

Acid Pickling Process: This process is done only when the scrap is known to contain oxide-scale formed during machining and storage. The process follows vapor degreasing and consists of treating the scrap with a mixture of hydrochloric and hydrofluoric acids. The process steps are: (1) leach the scrap with the mixed acid to remove the oxide, (2) remove the acid from the scrap with a water wash, and (3) dry the scrap for further processing. Energy requirements are not significant. The major waste generated is a liquid waste consisting of the spent acid mixture. This spent acid is normally disposed of by neutralization. The process has little potential for the production of serious atmospheric problems. However, the potential for water pollution by spent acids could be significant.

The scrap pretreatment operation removes primarily the surface impurities such as the oils and the oxide. Via the smelting refining operation the scrap is densified and the internal impurities are removed by vaporization in the vacuum electric arc melting process. In this process, the clean scrap is melted to obtain molten alloys and the volatile impurities are removed. The process steps are: (1) charging clean scrap to the electric furnace, (2) melting the scrap, (3) vaporizing the impurities, and (4) pouring the molten metal. Energy required is electricity to operate the process. The process is conducted in a completely closed system and wastes generated are well controlled. Therefore, the process has little potential for the production of pollution problems.

In the casting process, the molten titanium metal is cast into ingots. The process consists of: (1) pouring the molten alloy into molds and (2) casting. Energy requirement is necessary to operate the equipment. Wastes generated are insignificant and do not present any pollution problem.

Recovery from Titanium Chlorination Residues

C.C. Merrill, M.M. Wong and D.D. Blue have reported on the beneficiation of titanium chlorination residues in *Report of Investigations No. 7221,* Washington, D.C., U.S. Bureau of Mines (February, 1969). The Bureau of Mines beneficiated the wastes generated by commercial titanium chlorination operations to reclaim metal values and reduce the problem of disposing of solid wastes. A water leach solubilized about 40% of the dry weight and over 75% of the metallic impurities. The residue from the water leach was rich in titanium dioxide and carbon; it was combined with rutile and recycled to a laboratory chlorinator. The contained carbon was used in lieu of part of the coke requirement.

TUNGSTEN

Recovery from Tin Smelter Wastes

Techniques have been described by A.G. Starliper and H. Kenworthy in *Report of Investigations No. 6107,* Washington, D.C., U.S. Bureau of Mines (1962) for tungsten recovery from tin smelter wastes. One waste product was a slag that contained tin, tungsten, iron, tantalum, and columbium. Molten slag was treated with sulfur or sulfur compounds and with coke, calcium carbide, or gaseous reductants to form tin sulfide that volatilized from the charge and to reduce tungsten along with iron and silicon to form a carbon-bearing metallic layer. The metallic layer was treated by an oxidizing alkali roast and then was water-leached to extract the tungsten.

URANIUM

Recovery from Scrap

The electrolytic recovery of scrap uranium metal has been described by W.S. Fleming, R.D. Piper and J.A. Kennelley of Mallinckrodt Chemical Works in MCW-1494, National Technical Information Service, Springfield, Va., (August 1966).

As reported by these authors, an electrolytic process for recovery of scrap uranium metal was developed on a laboratory scale. In this process, uranium oxides on the scrap react with a consumable graphite anode to form uranium metal and carbon monoxide.

Operation at 1150°C in a fused BaF_2-LiF-UF_4 electrolyte produces molten metal at the cathode by a combination of electrolytic reduction of the oxides, electrolytic transfer of the metal, and thermal melting of the metal. Scrap type, cell design, electrolyte composition, and cathode current density were found to be major factors affecting current efficiency and metal quality. Figure 83 shows the overall outline of this scrap recovery process.

FIGURE 83: FLOWSHEET FOR ELECTROLYTIC RECOVERY OF SCRAP URANIUM

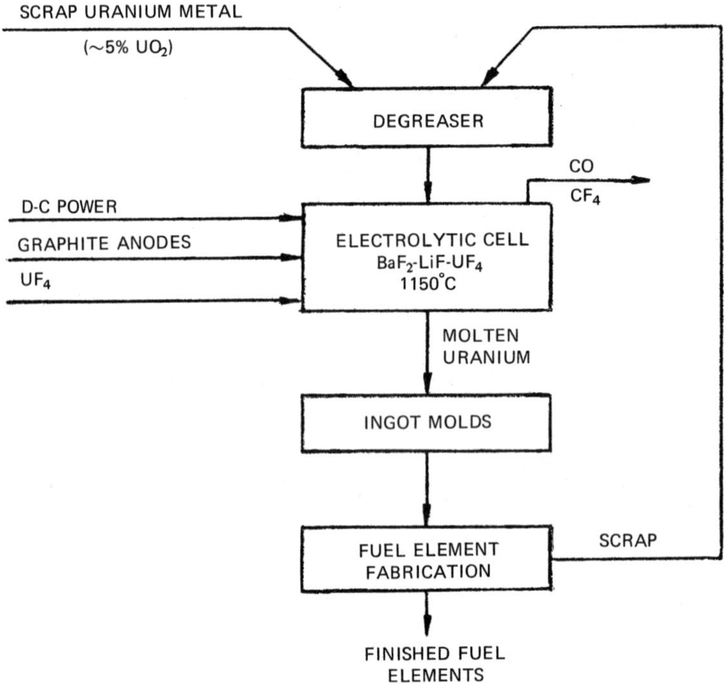

Source: MCW-1494

A process developed by *R.K. Welty; U.S. Patent 3,578,419; May 11, 1971; assigned to General Electric Company* involves recovering hard scrap nuclear reactor fuel material. In the preparation of fuel pellets, a significant quantity of scrap nuclear fuel material which has been pressed and sintered is produced. In the process, the scrap is first oxidized in a fluidized bed method which simultaneously oxidizes and comminutes the scrap, then, if desired, the oxidized material is reduced back to the original chemical form with improved sinterability.

VANADIUM

Recovery from Liquid Materials

Recovery from Crude Oil: A process developed by *W. Whigham; U.S. Patent 3,416,882; December 17, 1968* involves obtaining a vanadium concentrate by first desalting crude oil. The desalted crude is then fractionally distilled to obtain a residual oil component. From the residual oil a coke product is obtained which is ground to a fine mesh. The coke product is subsequently subjected to combustion and from the flue gas of the combustion, the solid particles are trapped and recovered to obtain a vanadium concentrate.

A process developed by *K.D. Vesely; U.S. Patent 3,522,001; July 28, 1970; assigned to Universal Oil Products Company* involves recovering metals from coal and/or hydrocarbonaceous coke. Particular metals are nickel and vanadium, being found in the greatest concentration in coal and petroleum crude oils; these are recovered in a readily marketable form as nickel oxide and vanadium pentoxide. Salt, preferably sodium chloride, is admixed with coke and the coke is burned at temperatures which allow the salt and metals to fuse to nickel chloride and sodium vanadate. Subsequent treatments result in substantially pure nickel oxide and vanadium pentoxide products. Where the source of the metals in a petroleum crude oil, or heavy hydrocarbonaceous fraction derived therefrom the salt resulting from an initial desalting operation is utilized within the metals recovery process.

Recovery from Solid Wastes

Recovery from Scrap: The molten salt electrorefining of vanadium scrap has been discussed by K.P.V. Lei and T.A. Sullivan in *Report of Investigations No. 7036,* Washington, D.C., U.S. Bureau of Mines (October 1967). The Bureau of Mines investigated molten-salt electrorefining as a practical method for reclaiming vanadium from vanadium scrap. An electrolytic process utilizing a $KCl\text{-}LiCl\text{-}VCl_2$ electrolyte at 615°C was developed for producing ductile vanadium with a purity over 99.9%. The recovery of the vanadium in the cell feed ranged from 88 to 93% with a current efficiency from 83 to 95%.

ZINC

Recovery from Liquid Wastes

Recovery from Waste Phosphatizing Solutions: A process developed by *H.E. Powell and L.L. Smith; U.S. Patent 3,649,220; March 14, 1972; assigned to the U.S. Secretary of the Interior* is one in which zinc and nickel are recovered from waste phosphate liquor by solvent extraction. Zinc is first extracted with di-2-ethylhexyl phosphoric acid in an organic diluent. Nickel is then extracted from the raffinate of the zinc extraction with dinonyl naphthalene sulfonic acid in an organic diluent. The reader is also referred to the earlier section on the recovery of nickel for additional information on this topic.

Recovery from Solid Wastes

Zinc die-cast scrap and galvanizers' dross represent two of the largest sources of secondary zinc in the U.S. Die-cast scrap is composed of discarded, broken, or obsolete manufactured items which have been partially segregated according to composition into a specific type of commercial zinc scrap. Much of the die-cast scrap originates from grills, fuel pumps, carburetors, trim, and hardware from scrapped automobiles and other machinery, or from punchings or factory rejects of cast articles. A typical mixture of such die-cast scrap would contain, in addition to zinc, from 3 to 4% aluminum and approximately 1% copper. In recent years, however, there has been a trend toward replacement of the copper with considerably smaller amounts of magnesium.

Recovery from Brass Smelter Flue Dusts: The recovery of zinc, copper and lead-tin mixtures from brass smelter flue dust has been described by H.E. Powell, H. Fukubayashi, L.W. Higley and L.L. Smith in *Report of Investigations No. 7637,* Washington, D.C., U.S. Bureau of Mines (1972). The smelting of brass and bronze generates a dust composed primarily of zinc oxide, with lesser amounts of other metal oxides, such as copper, lead, and tin. The small quantity of this waste generated by individual smelters and the relatively low unit value of the contained metals, suggests the desirability of reclaiming the metals onsite or at some nearby point. The Bureau of Mines has developed a process that fits these requirements.

The technique consists of (1) dissolving zinc and copper from the dust with sulfuric acid, (2) recovery of copper by cementation, (3) recovery of zinc as zinc sulfate crystals, and (4) recovery of lead and tin as a salable residue. Alternatively, zinc may be recovered by electrodeposition. Free acid is recycled and no major pollution problems are generated by the process.

A process developed by *E.G. Sunday and P.P. Taubenblat; U.S. Patent 3,123,465; March 3, 1964; assigned to American Metal Climax, Inc.* involves the pyrometallurgical smelting of finely divided oxidic materials containing principally lead, zinc and tin as, for example, bag house dusts that are evolved and collected in various metallurgical operations. More particularly, the process relates to an improved process for carrying out the smelting of such dusts whereby enhanced yields of the metal values contained in the original dust may be obtained.

The practice of recovering lead, zinc and tin from metallurgical dusts has been carried out for a considerable time by various means. One process consists of smelting such dusts in admixture with carbon and soda ash and other fluxing materials in stationary, indirectly or electro-thermally heated shafts or retorts as described, for example, in U.S. Patent 2,139,065. This mode of processing is generally satisfactory in that it enables recovery of the zinc in the dust directly as metallic zinc by the appropriate condensation of vapors educted from the smelting chambers.

The lead and tin values being essentially nonvolatile at the smelting temperatures utilized are recovered as a molten metallic alloy mixture that collects in a pool at the bottom of the smelting chamber from which it is withdrawn by tapping at appropriate intervals. The residue resulting from the smelting process at the same time is likewise but separately removed from the bottom of the retort and normally discarded.

One drawback of this process has been the difficulties encountered in obtaining satisfactory recovery of lead and tin in the form of the molten metal product. The residue has the tendency to entrain a large amount of the molten metal product. The incomplete separation of the molten metal from the residue may be remedied by subsequent treatment of the residue after its removal from the furnace to recover the entrained metals and particularly the tin which is the more valuable metal. Any additional treatment is however, time consuming and increases processing costs. When the smelting operation is being carried out continuously, the need for reworking the residue to separate the mechanically entrained molten metal is particularly objectionable.

It is an object of this process to provide an improved technique for smelting metallurgical dusts of the nature hereinbefore indicated which provides for the substantially complete separation of any physically entrained molten metal from the otherwise barren residue during the course of smelting in a stationary smelting chamber without resorting to extraneous treatment of the residue after smelting for completing the recovery of lead and tin.

It is another object to provide an improved process for smelting metallurgical dusts which is readily adaptable for continuous operation while minimizing the loss of valuable metals to the residue produced in the process. These and other objects and advantages are realized according to this process by feeding agglomerates consisting of the oxidic dust in admixture with suitable amounts of carbonaceous material and soda ash through an inclined smelting chamber to provide a columnar charge which upon appropriate smelting results in the production of zinc vapors, a molten metallic product and a residue.

The zinc vapors educted from the top of the columnar charge are condensed in accordance with conventional practice. The molten metal product formed during the smelting operation filters advantageously through only a relatively small portion of the columnar charge due to the tilted arrangement of the smelting chamber and channels along the lower sidewall eventually accumulating at the bottom of the furnace. In this manner a greatly improved separation of the molten metal product is effected leaving an essentially barren residue which can be discarded without incurring any significant metal losses.

Recovery from Electrolytic Zinc Plant Residues: A process developed by *B. Orlandini and W. Schmittroth; U.S. Patent 3,510,414; May 5, 1970; assigned to The Bunker Hill Company* is a three-stage process for recovering metals particularly zinc, from zinc plant residues. The zinc plant residue, in first stage, is first leached in strong sulfuric acid solution for 6 to 8 hours at a temperature between 90° and 94°C to dissolve the zinc ferrites and other zinc compounds, and then settled or filtered to produce a first-stage residue and solution. The first stage residue may be returned to a smelter for recovery of contained metal values. In the second stage, the clarified solution from the first stage is neutralized by a suitable neutralizing agent such as zinc oxide to reduce the acid concentration to a pH of 5.0 to 5.3 to precipitate iron compounds and other impurities.

After settling or filtering, a second stage residue and clarified solution are produced. The clarified solution is returned to the normal zinc plant leach tank for recovery of zinc values. The second stage residue is then dried by heating the residue to a temperature between 110° and 140°C. In the third stage, the dried residue is leached in a weak sulfuric acid solution to dissolve the contained zinc. The resulting leach pulp is filtered to produce a final low zinc-containing residue, and a third stage filtrate which is returned to the normal zinc plant leach circuit.

Recovery from Ferrous Smelter Dusts: A process developed by *A.L. Fitch; U.S. Patent 3,754,890; August 28, 1973; assigned to Harsco Corporation* is a process for producing metalized iron agglomerates and for recovery of zinc values from waste materials, in dust form, including comminuting and drying the dust. The process steps include (1) collecting and separating the dust in accordance with the content of the zinc values and proportionately blending the dust to provide a product having a substantially uniform zinc value content for pelletizing into green balls, (2) drying and indurating the green balls and reducing the indurated balls with a reductant to provide highly metalized balls and volatilization of the zinc values, (3) recovering the volatilized zinc values, as exhaust gases, to provide metalized balls having a high iron content substantially free of zinc values.

A process developed by *C.L. Stevenson; U.S. Patent 3,756,804; September 4, 1973; assigned to Deltech, Inc.* involves the treatment of flue dust from iron-making furnaces to recover minor constituents therefrom which are volatile at high temperature and to reduce the iron components in the dust to a stage where they can be recycled to the furnace or otherwise used. The method consists essentially in preheating the flue dust to a temperature approaching 1700°F. Thereafter, the flue dust particles are passed into a reducing, essentially oxygen-free, atmosphere where the temperature is increased to as high as 2200°F. A portion of this reducing atmosphere is continuously diverted and cooled to condense the metallic volatiles, primarily zinc. Thereafter, the remaining components of the dust are cooled and released, at a temperature less than 700°F, to recover the iron without reoxidation when it is exposed to the atmosphere.

A process developed by E.G. Valdez and K.C. Dean and described in *Report of Investigation No. 8000,* Washington, D.C., U.S. Bureau of Mines (1975) describes experiments in treating iron foundry flue dusts to recover zinc. Iron foundries that melt automobile scrap produce flue dusts averaging about 32% zinc and 6% lead. Some is sold to smelters, but because of the meager return and difficulty in handling, most of the dust is discarded.

As part of the Bureau of Mines program to find uses for mineral wastes, research was initiated to develop an economical process for recovering zinc (and lead) from flue dust. Controlled vacuum fuming of zinc was found most effective and produced prime zinc and a lead-rich residue. This process involved two distillations or fumings with carbon or methane as the reductants and resulted in 91% recovery of zinc and 98% recovery of the lead.

Recovery from Galvanizers' Dross: Galvanizers' dross is a semisolid pasty mixture that forms in the bottom of the galvanizers' kettle. It is created by the attack of molten zinc on iron salts that coat the surface of the piece to be galvanized. These iron salts are formed by the customary pickling and pregalvanizing treatments. Lesser quantities of dross are formed by the solvent action of zinc on the galvanizing pot and the steel or iron piece being coated. Accumulation of this dross eventually interferes with the operation of the galvanizing bath so that periodic removal is necessary. Dross, as it is first formed, is believed to correspond to the formula $FeZn_7$ (10.9% Fe).

However, the average iron content of the final dross is considerably below this value because the $FeZn_7$ gradually absorbs zinc without loss of its crystalline character until it exists as a solid solution. Dross, as it is removed from the galvanizing bath, usually contains from 2 to 4% iron and approximately 1% lead.

The recovery of zinc from galvanizers' dross has been described by J.A. Ruppert and P.M. Sullivan in *Report of Investigation No. 6417,* Washington, D.C., U.S. Bureau of Mines (1964). A process was developed on a laboratory scale in which iron contained in galvanizers' zinc dross and aluminum contained in scrap zinc die-cast alloy were removed simultaneously. Melting both types of scrap together in the proper proportions resulted in formation of solid intermetallic Fe_2Al_5, which was readily removed from the melt by filtration and centrifugation. Recovery was 87 to 92% of the available zinc. Refined filtrate contained 0.02% iron and 0.20% aluminum. Aluminum content was further reduced to less than 0.01% by fluxing with zinc chloride. Low temperatures employed and inexpensive equipment requirements make the process commercially attractive.

Recovery from Galvanizing Sal Skimmings: The generation of preflux solutions from galvanizers' sal skimmings has been described by P.M. Sullivan, D.H. Chambers and P.J. Berney in *Report of Investigations No. 5421,* Washington, D.C., U.S. Bureau of Mines (1958). Sal skimmings are essentially spent zinc ammonium chloride flux. In its original form, this flux is a crystalline double salt containing 3 molecules of ammonium chloride per molecule of zinc chloride. It is used on the surface of the zinc bath to provide a final cleaning of materials to be galvanized.

The composition of the flux gradually changes owing to chemical reactions occurring during cleaning and to losses by volatilization. The blanket then becomes viscous and ineffective and must be skimmed off and replaced. These skimmings, containing part of the original flux mixed with oxides, free zinc, and some foreign material, are known as sal skimmings. Generally, they contain about 50% zinc and 10 to 20% chlorine, with minor amounts of ammonia. Galvanizers' sal skimmings have been treated in the Bureau of Mines study cited by two methods of leaching in attempts to produce preflux solutions suitable for use in galvanizing operations. Both water-leaching and acid-leaching procedures are effective, except in severely exhausted skimmings when water-leaching becomes impractical.

Recovery from Scrap: The zinc segment of the secondary nonferrous metals industry consists of those plants that process discarded and scrap materials for the purpose of recovering zinc. In the recovery of the zinc, waste products—atmospheric emissions, liquid wastes, and solid wastes—are generated which can cause serious pollution problems. The principal sources of zinc scrap as classified by the Bureau of Mines are:

- New clippings
- Old zinc
- Engravers' plates
- Skimmings and ashes
- Sal skimmings
- Die-cast skimmings
- Galvanizers' dross
- Diecastings
- Rod and die scrap
- Flue dust
- Chemical residues

In addition to containing metallic zinc or compounds and alloys, thereof, the scrap contains a wide variety of organic materials such as oil, grease, lubricants, and electric insulation. Alloying elements commonly found in the scrap are aluminum, copper, and lead. Products from the zinc segment of the secondary nonferrous metals industry are: (1) specification zinc alloys, (2) zinc ingots (slab) containing essentially 100% zinc, (3) zinc dust, (4) zinc oxide approaching 100% purity, and (5) crude zinc oxide for reduction to zinc metal by the primary smelters.

The products from the zinc segment are recovered from the various types of scrap zinc employing two manufacturing operations—scrap pretreatment and refining. Zinc scrap is pretreated prior to refining to remove a portion of the metallic and nonmetallic impurities and to physically prepare the material for further processing. Three types of preprocessing (hydrometallurgical, pyrometallurgical, and mechanical) are used. The pretreatment varies depending on the type of scrap. The pretreatment processes used are:

(1) Crushing/screening (mechanical)
(2) Kettle sweating (pyrometallurgical)
(3) Reverberatory sweating (pyrometallurgical)
(4) Rotary sweating (pyrometallurgical)
(5) Electric resistance sweating (pyrometallurgical)
(6) Muffle sweating (pyrometallurgical)
(7) Sodium carbonate leaching (hydrometallurgical)

Some details of these various pretreatment processes are given in the paragraphs which follow.

Crushing/Screening Process – The concentration of metallic zinc in skimmings/residues is increased by the crushing/screening process. The process steps involved are: (a) pulverizing or crushing the skimmings/residues to separate the metallic zinc from the flux, and (b) screening or pneumatically treating the crushed material to separate zinc from nonmetallic constituents. Energy is required to drive the equipment.

The process generates atmospheric emissions and solid wastes. The atmospheric emissions are composed of dusts from both the pulverizing (crushing) process step and the screening step. These emissions contain, in addition to zinc, those materials–nonmetallic constituents, metallic values such as zinc, aluminum, copper, iron, lead, cadmium, tin, and chromium and fluxes–separated from the scrap zinc during the sweating process. Emissions data are not available; however, based on studies conducted of other industries, the raw emissions factor is estimated to range from 0.9 to 7.5 lb per ton of skimmings/residues processed. The solid wastes generated by this process are nonmetallic constituents from the screening operation and baghouse dusts, if emissions from crushing and screening process steps are controlled. This process produces a small quantity of emissions and is, therefore, a minor source of pollution.

Kettle Sweating Process – The kettle sweating process separates the metallic zinc from metal attachments, having higher melting points, and from nonmetallic residues. This is achieved by the following process steps: (a) charging the scrap into the furnace containing generally a molten heel, (b) melting of zinc fraction of the scrap, (c) working of charge to effect separation of the zinc, (d) fluxing to aid in the separation, and (3) skimming to remove impurities removed by the flux. Afterwards the molten metal may be removed and cast in blocks for further processing, fed to a distillation furnace or alloyed to obtain specification composition and then cast into ingots. Operating temperatures of the kettle-sweating baths range from 800° to 1000°F. Production is batchwise, with one batch requiring 6 to 8 hours to complete one heat.

Energy required is the electricity to drive the equipment and the fuel, generally, natural gas, to melt and keep the zinc molten. Fuel oil is used as fuel in some cases. The process produces significant quantities of potential environmental pollutants. Atmospheric emissions consist of: (1) combustion gases (which are separate from the emissions from the molten metal), and (2) emissions consisting of gases and particulates from the molten metal. Solid wastes generated are metal attachments and skimmings or slag from the flux present in the scrap or that added to purify the zinc metal and baghouse dusts.

Composition of the particulate atmospheric emissions varies somewhat depending on the charge. However, in most cases, the emissions contain ammonium chloride, zinc, aluminum, tin, nickel, copper, iron, lead, cadmium, magnesium, manganese, and chromium. The gases are composed of, in addition to the combustion products, carbonaceous materials from the rubber, plastic, and other organic impurities or contaminants in the scrap. Some carbonaceous materials are also mixed with the particulate emissions. Emissions factors also vary depending on the composition of the charge. When sweating metallic scrap with zinc chloride as the flux, raw emissions factor was reported to be 10.8 lb/ton of scrap processed. These particulate emissions contained 4, 77, 4, 4, and 10%, respectively, of zinc chloride, zinc oxide, water, other metal chlorides and oxides, and carbonaceous materials. In other cases, the raw emissions factor for sweating residual scrap (the flux was residual zinc chloride) was reported at 24.5 lb/ton of scrap processed. Composition of particulate emissions was similar to that from the sweating of the metallic scrap. Morphology of the particles is dependent on the type of scrap. For example, the particle size of the emissions from sweating of residual scrap ranged from less than 1 micron to greater than 20 microns, whereas those from sweating of metallic scrap were less than 2 microns. Particle shape is acicular and/or irregular. Those emissions that are predominately zinc oxide are composed of primarily acicular particles which is the characteristic shape of zinc oxide. Low zinc emissions have an irregular shape or are composed of a mixture of particle shapes.

Generally, the atmospheric emissions are controlled or collected using baghouses with orlon filter medium. Collection efficiencies of approximately 96% are reported. No liquid wastes are generated by the kettle sweating process unless possibly some cooling water. However, significant quantities of solid wastes are generated. These wastes are: (1) slag (skimmings or drosses), (2) high melting attachments, and (3) baghouse dusts. Thus, the kettle sweating process is a source of potential pollution problems if the wastes are not collected and disposed of properly.

Reverberatory Sweating Process – This process separates or recovers zinc metal from mixed zinc scrap. The recovered impure zinc is processed further to obtain specification alloys or fed to a distillation furnace for refining. The process steps are: (a) charging of furnace, (b) melting of zinc, and (c) removal of unmeltables. In this process the zinc flows from the furnace as it melts and is collected in vessels or allowed to flow to the next process in operation. Energy required for this process is the electricity to drive the equipment and the gas or fuel oil to fire the furnace. The reverberatory furnace is direct-fired whereas the kettle furnace is heated indirectly.

Both atmospheric emissions and solid wastes are generated. The atmospheric emissions are composed of the combustion gases, gases or vapors from burning of the organic content of the scrap, and volatile metallic materials. The particulate emission is the fume volatilized from the furnace and dust carried from the furnace by the entrainment of particles in the furnace exit gases. Composition of the atmospheric emissions varies depending on the source and type of scrap. However, in general, the particulate emission will contain such metals as zinc, aluminum, copper, iron, lead, cadmium, manganese, and chromium, in addition to carbonaceous materials and components from the residual fluxes. The gases contain primarily carbon dioxide, nitrogen, oxygen, and water, and possible sulfur oxides, chlorides and fluorides.

Particulate emission factors vary depending on the type of scrap. For metallic scrap, raw emissions factor was reported to be approximately 13 lb of particulate matter per ton of charge processed, whereas raw emissions factor for residual scrap was reported to be approximately 32 lb per ton of feed process. Baghouses with orlon filter media are used to control these emissions. Particle morphology of atmospheric emissions is the same or comparable to the morphology of emissions from the kettle furnace. No liquid wastes are generated. Solid wastes from this process are: (1) baghouse dusts; (2) skimmings, drosses or slag; and (3) unmeltable attachments. The process does produce significant quantities of atmospheric emissions and solid wastes which could cause pollution problems.

Rotary Sweating Process – Zinc metal is separated from zinc scrap such as die-castings by the rotary sweating process. During the sweating, the furnace is rotated on its axis. The zinc metal is collected in a kettle (outside the furnace) and the residue is skimmed off the molten metal. The process steps are: (a) charging of furnace, (b) melting of zinc values, (c) removing (skimming) of residue, and (d) removing the unmeltables from the furnace. Afterwards, the impure zinc melt is transferred to a distillation furnace or alloyed and cast into ingots. Energy required is the electrical energy to rotate the furnace and fuel (gas or oil) to heat the furnace which is direct fired. Potential environmental pollutants generated by this process are: (1) atmospheric emissions, and (2) solid wastes. The atmospheric emissions include both gases and particulate emissions.

The gases are composed of those produced by combustion of the fuel and burning of organic matter contained in the scrap. The particulate emissions generally are composed of such metal values as are found in the emissions from the reverberatory sweating process. Included in this list are zinc, aluminum, copper, iron, lead, cadmium, manganese, and chromium. Particulate emission factor data are not available. However, based on reported values from other sweating processes, it is estimated the raw emission factors will range from 11 to 25 lb of particulate matter per ton of feed.

Particulate morphology of the particles should be similar to the morphology of the particles from other sweating processes. The solids generated by this process are skimmings, unmeltable attachments, and baghouse dusts.

The rotary sweating process produces significant quantities of atmospheric emissions and solid wastes and, therefore, is a potential source of atmospheric and water pollution problems.

Electric-Resistance Sweating – This process is employed in small plants to recover zinc from clean scrap. The process steps are: (a) charging of furnace, (b) sweating of zinc, (c) pouring of zinc melt for subsequent processing, and (d) removing residue from furnace. Energy required is the electricity to sweat the zinc. Potential environmental pollutants are: atmospheric emissions and solid wastes. Atmospheric emissions are composed of primarily fume and dust since only clean scrap is used as the source of zinc and no combustion products are formed. The fume and dust contain primarily zinc oxide from the oxidation of zinc vapor along with trace quantities of other heavy metals and possible a small quantity of chloride and fluxing materials. The gaseous portion of the emissions is primarily nitrogen. Emissions are controlled with baghouses, if pollution controls are used.

Particle morphology is the same as that of emissions from other processes. Emissions factor is probably low, <10 lb of fume and dust per ton of feed processed. Solid wastes from this process are drosses, if fluxes are used, unmeltable components of the scrap, and baghouse dusts. Although this process produces small quantities of atmospheric emission, there is potential atmospheric and water pollution problems associated with disposal of these materials and with the process.

Muffle Sweating Process – Zinc metal is recovered from zinc scrap materials by the muffle sweating process. The process steps are: (a) charging of furnace, (b) sweating of zinc values, (c) tapping of furnace to remove molten zinc for subsequent refining or alloying, and (d) removing residue with a shaker screen to separate the dross from the unmeltable attachments. Energy required is the fuel–gas or oil–to heat the muffle furnace which is indirectly fired.

Potential pollutants from this process are: (1) atmospheric emissions composed of gases from the combustion of the fuel and from the sweating chamber and particulate emissions from the sweating chamber, and (2) solid wastes–drosses and unmeltable attachments. The gaseous emissions include those from the combustion chamber which are emitted separately to the atmosphere and those from the sweating chamber. The combustion gases contain primarily carbon dioxide, nitrogen, unburned fuel, and possibly trace quantities of heavy metals and sulfur oxides if oil is used as the source of fuel. Those from the sweating chamber are composed of a small quantity of gases from air leaking into the chamber and from combustion or decomposition of organic compounds added with the scrap.

Composition of the particulate emissions from the sweating chamber will vary depending on the type of scrap but, in general, contains the metals found in particulate emissions from other sweating processes, with the major component being zinc. Particle morphology of the particulates is the same as that of emissions from other processes. Emission factors for the muffle furnace process are not available. However, they are estimated to range from approximately 10.8 to 32 lb per ton of feed processed. Baghouses are used to control the emissions. A cyclone may be used in series with the baghouse. The solid wastes generated are drosses, baghouse dusts, and unmeltable attachments. The process generates a considerable quantity of atmospheric emissions and solid wastes, and thus is a source of pollution problems.

Sodium Carbonate Leaching Process – The sodium carbonate leaching process removes the nonmetallic contaminants from skimmings and residues and converts the zinc values to zinc oxide which can be reduced to zinc metal. The process steps are: (a) crushing to make the scrap more amenable to leaching, (b) washing with water to separate nonmetallic contaminants and recover metallic zinc, (c) treating the aqueous stream with sodium carbonate to precipitate the zinc values, (d) drying the precipitate to remove the water, and (e) calcining to convert zinc hydroxide to zinc oxide. The product (zinc oxide) is shipped to a primary smelter for reduction to zinc metal. Energy required for this process is electrical energy to drive the equipment and fuel, either gas or oil, to dry and calcine the zinc hydroxide.

The process produces waste products–atmospheric emissions, liquid wastes and solid wastes–in modest quantities. The atmospheric emissions consisting of both gases and particulate matter are produced during the crushing, drying, and calcining process steps. The gases are from the crushing step and present no problem since they primarily consist of air. Those from the drying and calcining steps contain primarily the combustion gases and water vapor. Since any residual zinc chloride is vaporized during the calcination, the gases may also contain hydrogen chloride and zinc chloride. The particulate matter consists primarily of zinc oxide fume contaminated with other metal values either vaporized with the zinc oxide or entrained in the gaseous stream. The atmospheric emissions are controlled with baghouses.

Liquid wastes are generated in steps (b) and (c). These wastes contain sodium chloride, sodium carbonate, and other water-soluble compounds not precipitated by the sodium carbonate. Because of the corrosive nature of the atmospheric emissions, allowing them to escape to the environment could cause pollution problems. Liquid wastes may cause water pollution problems if not treated before discharge.

The pretreated zinc scrap still contains a wide variety of metallic and nonmetallic contaminants which are removed in the refining operation to produce pure zinc ingots, zinc dust, zinc oxide, and specification zinc alloys. These products are produced by the following refining operations:

(1) Retort distillation
(2) Muffle distillation
(3) Retort distillation/oxidation
(4) Muffle distillation/oxidation
(5) Alloying
(6) Graphite rod resistor distillation

Some details of each of these processes will be given in the paragraphs which follow.

Retort Distillation Process – This process produces pure zinc ingots and pure zinc dust from pretreated zinc scrap–molten zinc-rich metal from the sweat furnace or cast zinc-rich metal ingots from a sweating process–and/or zinc dross. The process steps are: (a) charging the furnace with the pretreated zinc scrap or dross, (b) melting the charge, (c) distilling the zinc, (d) condensing the zinc vapor rapidly to produce zinc dust or slowly to produce pure liquid zinc. Afterwards, the zinc metal is removed and cast into ingots (slabs). Zinc dust, if the final product, is removed and packaged. The energy required for this process is the fuel–gas or oil–to heat the furnace which is indirectly fired.

This process produces significant quantities of atmospheric emissions and solid wastes. Sources of atmospheric emissions are the retort opening, combustion chamber, the condenser, and casting of ingots. Emissions as zinc oxide fumes are evolved from the retort opening (charging port) during removal of the hot residue from the retort after a distillation heat and recharging of the retort for the next distillation.

These emissions contain primarily zinc oxide along with aluminum, copper, and other metal values found in the residues. Raw emissions factor from this source and the condenser as discussed below are estimated to be 45 lb of particulates per ton of zinc distilled. Emissions factor may be much higher due to leaks in the grout seal around the condenser neck. Particle size of the fumes is estimated to range from approximately 0.05 to 1.0 micron. Since the fumes are primarily zinc oxide resulting from air oxidation of zinc metal, particle shape is acicular, the characteristic shape of zinc oxide.

Combustion gases which are emitted separately from the other emissions contain carbon dioxide, unburned fuel and nitrogen, and possibly trace quantities of heavy metals if oil is used as the fuel. During distillation of zinc, the condenser must be kept at postive pressure to prevent air from entering the condenser and contaminating the zinc metal with oxygen. To ensure that there is a positive pressure, a small hole, called a speise hole, is provided through which a small amount of zinc vapor is allowed to escape continuously into the atmosphere where it reacts with oxygen. These emissions (fumes) are essentially 100% zinc oxide. The particles, acicular in shape, are extremely fine, i.e., approximately 0.05 to 1 micron.

Only a small amount of atmospheric emissions is formed during the pouring and casting processing steps. Emission factors are estimated to range from 0.4 to 0.8 and 0.2 and 0.4 lb of particulates, respectively, per ton of zinc produced. These emissions are emitted as fine particulates with the same particle size and shape as those from the condenser. Solid wastes generated are the distillation residues and the particulate matter in the atmospheric emissions. The atmospheric emissions are primarily zinc oxide, whereas the distillation residues are composed of heavy metals such as residual zinc (10 to 50%) not removed by the distillation process and contaminants such as aluminum, copper, lead, and fluxes not removed during the pretreatment process. Serious pollution problems may arise if the atmospheric emissions are not collected and the wastes are not disposed of by a nonpolluting method.

Muffle Distillation Process – This process produces pure zinc ingots, generally from pretreated zinc scrap. However, in some cases, untreated scrap such as zinc die-castings may be substituted for pretreated scrap. This process may be operated continuously whereas the retort distillation process is batch. The process steps involve: (a) continuously adding molten zinc from a melting pot or a sweating furnace to the muffle section, (b) continuously distilling and condensing the pure zinc, (c) periodically tapping the zinc from the condenser into the molds, (d) casting the zinc ingots, and (e) periodically removal of residue from muffle.

Energy utilized by this process is fuel oil or gas to supply the heat for vaporization of the zinc. The process produces significant quantities of atmospheric emissions and solid wastes. Sources of atmospheric emissions are the combustion gases, distillation residues, pouring and casting of the ingots, and the orifice in the condenser. Emission factors, composition of emissions, and morphology of particulates are the same for both distillation processes. Atmospheric emissions may be controlled with baghouses alone or in series with cyclones. Solid wastes generated are distillation residues and baghouse dusts which are essentially of the same composition as the solid wastes from the retort distillation process. Thus, the process is a source of pollution problems.

Retort Distillation/Oxidation Process – Zinc oxide produced by the retort distillation/oxidation process involves vapor oxidation of zinc metal. The process steps in this process are the same as for the retort distillation process except zinc vapor from the retort is discharged through an orifice into a stream of air where zinc oxide is formed inside a refractory-lined chamber. The combustion gases from oxidation of the metallic zinc vapors and excess air carry the zinc oxide to a baghouse collector where the zinc oxide is collected. The combustion gases and excess air pass through the baghouse into the surrounding atmosphere.

Energy required for this process is the fuel—gas or oil—to vaporize the zinc metal. Waste products from this process are atmospheric emissions and solid wastes. The atmospheric emissions, which consist of gases and fumes, are emitted to the atmosphere from the retort opening (charging port), baghouse, and combustion (furnace) chamber. Those from the retort opening result from oxidation of the hot residue removed after a distillation heat and upon charging of the retort with molten zinc for the next distillation heat. See the discussion on the retort distillation process for pertinent data on composition of emission, emission factors, and morphology of the particles. Combustion gases from the retort furnace consist of primarily carbon dioxide, nitrogen, and unburned fuel. If oil is used as the fuel, the gases may also contain sulfur oxides, chlorides, and trace amounts of heavy metals.

Atmospheric emissions from the baghouse contain pure zinc oxide fume and the gases from burning of the zinc with air. The zinc oxide may contain trace quantities of heavy metals such as cadmium, depending on the purity of the feed. The particles of zinc oxide are very small, ranging in size from approximately 0.05 to about 1.0 micron. The particle shape is predominantly acicular, i.e., needle-like. Emission factors for the baghouse are estimated at 20 to 40 lb of zinc oxide per ton of oxide produced. These data are calculated based on a collection efficiency of 98 to 99%.

Solid waste generated is the distillation residue remaining in the retort after removal of the zinc. This residue contains 10 to 50% zinc and contaminants–aluminum, copper, lead, and residual flux–not removed from the zinc scrap during the pretreatment operation. Thus, the retort distillation/oxidation process is a source of potential environmental pollutants which can cause problems.

Muffle Distillation/Oxidation Process – This process produces zinc oxide from pretreated zinc scrap. The process steps are the same as for the muffle distillation process except that in the production of zinc oxide, zinc vapors are allowed to escape through an orifice at the top of the first chamber of the condenser and are burned or oxidized in the air. The resulting zinc oxide is transported by the combustion gases and excess air through conduits and collected in a baghouse. Energy required for this process is gas or oil to heat the muffle furnace.

The potential environmental pollutants are: (1) atmospheric emissions, and (2) solid wastes. Sources of atmospheric emissions are: (1) the furnace combustion chamber which is separate from the other sources as the furnace is indirectly fired, (2) muffle opening (charging port), and (3) baghouse. See discussion on the muffle distillation process for details concerning composition of emissions, emission factors and morphology on particulates from sources (1) and (2), and the above discussion on the retort distillation/oxidation process for pertinent data on composition of emissions, emission factors, and morphology of particles for conversion of zinc to zinc oxide and subsequent collection of the zinc oxide.

Solid wastes generated by the muffle distillation/oxidation process is the distillation residue from the muffle. This residue has the same composition as the residue from the retort distillation/oxidation process. Based on these data, this process is a source of pollution problems.

Alloying Process – The alloying process produces zinc alloys from pretreated scrap. The process steps are: (a) melting the pretreated scrap, (b) adding the alloying agent to molten metal, (c) homogenizing the mixture, (d) pouring the melt into ingot molds, and (e) casting the ingot. In many cases, the alloying process is combined with the sweating process, whereby the molten pretreated scrap is used directly to produce the alloy. Energy requirement for the alloying process is fuel–oil or gas–to keep the melt molten or to melt pretreated scrap ingots.

The process produces atmospheric emissions and solid wastes. If alloying is carried out using molten zinc from a sweating furnace, the atmospheric emissions contain zinc oxide, small amounts of the alloying agent, and the flux cover. If pretreated scrap ingots are used as the source of zinc, the atmospheric emissions will contain, in addition to the particulate matter, gaseous emissions from combustion of the fuel. Particulate matter may be controlled using a baghouse; however, in many cases, no pollution control is employed. Solid waste is the flux cover used to protect the melt. Flux covers commonly used are carbon or zinc chloride. The process produces minor quantities of emissions and solid wastes and, therefore, is not a major source of pollution.

Graphite Rod Resistor Distillation Process – This process produces zinc dust from pretreated scrap ingots. The process steps are: (1) charge the ingot or melt to furnace, (2) heat with electric power, and (3) collect zinc dust as distillate using closed-loop water cooling system. Electric energy is used in significant quantities to vaporize the zinc. The process generates vapors containing traces of zinc, zinc oxide, lead, and lead oxide. The pollution potential of these vapors is not significant because the metal content is recovered in baghouses and returned to the process. The process has no pollution potential.

Recovery from Scrap Die-Cast Materials: Zinc-rich materials are in relative abundance since numerous types of short useful life articles manufactured in large volume contain zinc in commercially recoverable quantities. The largest source of such material is in the form of discarded zinc die-cast articles. Such articles include automobile carburetors and fuel pumps, certain types of household applicance and office equipments, and aircraft forming dies. Unfortunately, because of low efficiencies, high cost and atmospheric pollution problems inherent in presently known zinc recovery processes, much zinc-rich scrap is being lost to land fill dumps rather than being collected for recovery because of the low price being paid for such scrap.

Of the numerous methods for recovering and refining zinc which are disclosed in the prior art, the oldest and most common method is the batch-type retort method which typically uses horizontal retorts for melting the scrap materials. In practicing this method retorts, typically constructed of graphite, silicon carbide or other refractory materials, are charged with the zinc-bearing material and are placed in a furnace wherein the material is melted by conduction heating. When sufficiently high temperatures are reached in the retort the zinc in the scrap will vaporize and the zinc vapors will ascend into a space provided within the retort above the molten metal. The zinc vapors are continuously evacuated from the retort and are diverted into a condenser wherein they are condensed to zinc metal or zinc dust, depending on the type of condenser used.

Batch-type methods using horizontal retorts or muffle furnaces are quite inefficient and present numerous processing difficulties. In a small retort such as is commonly used in the prior art, the percentage of wall surface in contact with the liquid material is very limited and because heating is by conduction through the walls of the retort, the efficiency of vaporization of the zinc is extremely low. If the size of the retort is increased, numerous structural difficulties are encountered because of the very large hydrostatic pressures exerted by the heavy molten material.

Additional problems encountered in typical retort-type processes are attributable to the fact that the bath of molten metal within the retort is of necessity quite deep. Carrying out vaporization in a deep bath of molten metal results in increased boiling points at the bottom of the retort, and also results in violent ebullition, which is objectionable because it contaminates the vapor above the liquid with a spray of impure metal. Retorts will quite often break while in operation. This results in the molten zinc material flowing into the combustion zone of the furnace and voluminous quantities of zinc oxide are then either released to the atmosphere or must be collected by very costly air pollution equipment.

In methods such as the muffle furnaces all the heat for distillation is obtained by radiation from above the bath which is relatively deep, and irregular vaporization at the end of the cycle is experienced. Very low fuel efficiencies are characteristic of this method. Keeping an atmospheric seal between the combustion zone and the distillation zone of the furnace prevents problems because the seal is usually made of relatively small tiles of silicon carbide with the resultant numerous joints between the tiles, any of which may open. Leaks or breakage of the seal result in: (1) evolution of zinc oxide with the same problems as related above in the description of retorts; and (2) oxidation of zinc vapors which can affect the quality of zinc dust or result in lower recoveries of zinc if zinc is being produced.

A process developed by *C.R. Brown and M. Templer; U.S. Patent 3,778,044; December 11, 1973* is claimed to overcome the problems of many prior art devices and provides a method and apparatus for recovering and refining zinc from scrap metals having a high zinc content. The scrap metals are melted in a conventional furnace and introduced in a molten state into a distillation apparatus which comprises an atmospherically sealed vertical column having a distillation zone and a communicating superposed fractionation zone. Arranged within the vertical column is a plurality of vertically spaced horizontally staggered trays constructed of a suitable refractory material.

Molten material is continuously introduced into the apparatus through a metal seal and flows downwardly in a circuitous path over the trays disposed in the distillation zone. The metal in each tray is heated to temperatures sufficient to vaporize zinc by means of precisely controllable radiant heating elements located immediately above and below each tray. The zinc vapors rise through the distilling zone in countercurrent with the descending molten metal, pass through the fractionating column where various impurities are removed, and then are directed into a condenser wherein they are condensed to metallic zinc of high purity.

The refining of zinc-base, die-cast scrap using low cost fluxes has been described by D. Montagna and J.A. Ruppert in *Report of Investigation No. 7315,* Washington, D.C., U.S. Bureau of Mines (Ocotber 1969). Large quantities of zinc-base die-cast scrap containing 3 to 4% aluminum are presently refined by mixing with other secondary zinc products such as galvanizer dross and distilling in a retort. The Bureau of Mines investigated alternate aluminum removal processes utilizing low-cost fluxes. Basing the investigation on the metallurgical principle that a reactive impurity can be removed from a less reactive matrix metal by melting under a flux consisting of a salt or oxide of a metal less reactive than the impurity, three separate flux mixtures were studied.

In the first flux system, $\underline{FeCl_3}$ + NaCl (the active component is underlined), part of the aluminum is eliminated by the formation of volatile aluminum chloride and the rest by the creation of a high melting point iron-aluminum intermetallic compound that can be separated from the remaining molten zinc by filtration. Aluminum removal in the second system, $\underline{Fe_2O_3}$ + $CaCl_2$ + NaCl, is effected by the formation of the iron-aluminum compound and possibly aluminum oxide. The third system, spent sal skimmings—essentially a mixture of $\underline{ZnCl_2}$, NH_4Cl, and ZnO, depends solely on aluminum chloride formation. Of the three flux systems investigated, the best results obtained were with the third system. Aluminum was reduced in the molten die-cast alloy from 4.10% to 0.01% when depleted sal skimmings were used.

Recovery from Zinc Ash from Galvanizing Plants: As described by W.P. Ruemmler of Pacific Smelting Company in *Chemical Engineering,* June 12, 1972, p 70–71, zinc ash is formed at galvanizing plants when metallic zinc is oxidized and floats on the surface of molten-metal zinc baths. The ash contains entrained particles of metallic zinc, and it is frequently contaminated with large quantities of pernicious chlorides—these come from ammonium chloride and other chloride fluxes used by galvanizers. The ash contains 50 to 75% zinc.

A three-stage recovery process is described involving (1) milling and metallic separation of zinc and zinc oxide, including trash removal; (2) crude zinc oxide treatment involving soda ash treatment followed by drying and calcination; and (3) zinc metal recovery involving furnace melting of zinc from the first process step.

Recovery of Zinc from Slime: The recovery of lead and zinc from slimes and mill tailings has been described by J.G. Donaldson in *Report of Investigation No. 6263,* Washington, D.C.,

U.S. Bureau of Mines (1963). A bench-scale method for recovering lead and zinc from slimes and mill tailings from the Tri-State District of Missouri, Oklahoma, and Kansas was developed. Dilute hydrochloric acid, containing sodium chloride, and chlorine gas were used to decompose the mineral constituents. Lead was electrolyzed from the pregnant acid solution, and zinc was precipitated from the depleted electrolyte by neutralization with lime. Up to 98% of the lead was recovered in an impure lead electrodeposit assaying approximately 85% lead. Zinc recoveries up to 90% of the total were attained in a precipitate assaying approximately 40% zinc.

ORGANIC MATERIALS

AROMATIC AMINES

Recovery from Used Color Photography Solutions

A process developed by *B.A. Hutchins and R.S. Walsh; U.S. Patent 3,502,577; March 24, 1970; assigned to Eastman Kodak Company* is one in which potential water pollutants including primary aromatic amine color developing agents are recovered, alone, or simultaneously with color forming couplers or benzyl alcohol from developer solutions in a separable, second phase by admixture with a water-soluble salt under limited pH conditions.

Large quantities of color developing solutions are used in the processing of exposed photographic elements in order to produce color photographs. It is the color developing solutions that produce the dye images that form the color picture in the processed element. In the processing of color subtractive type photographic elements, the final picture is usually made by the formation of a cyan dye image, a yellow dye image and a magenta dye image in superposed relationship.

The magenta dye image development, for example, is accomplished with an alkaline developer solution which contains a primary aromatic amine color developing agent, such as N-(2-amino-5-diethylaminophenethyl)methane sulfonamide hydrochloride, and a magenta forming coupler which is usually a 2-pyrazolin-5-one or a coumarone type coupler. Where exposed silver halide in the magenta image layer is developed to silver, color developing agent is oxidized to a compound that couples with the magenta forming coupler present to form a nondiffusing dye image. The cyan and yellow dye images are formed in a similar manner with cyan dye forming and yellow dye forming couplers, respectively.

In commercial color processing of color film, a continuous flow of color film is conducted through various processing solutions including the color developers in a processing machine. The concentrations of the components of the developer solutions must be maintained within certain specified ranges in order to give good sensitometric results. Thus, it is usual practice to replenish the various developer solutions by a controlled flow of the appropriate replenisher solution at the necessary rate.

The flow of color developer replenisher into a processing machine color development tank, for example, displaces used color developer solution which still contains substantial amounts of color developing agent, color forming coupler and other such materials which are relatively expensive to replace. Not only is it undesirable to lose these materials, but it is highly undesirable to sewer such solutions, since some of them contain materials such as benzyl alcohol in addition to the developing agents which have a substantial biochemical oxygen demand and an adverse effect on fish and wildlife that eventually come into contact with the waters into which these materials are deposited.

It has been proposed to recover the various components of the color developing solution for reuse by means of certain techniques including solvent extraction, evaporative or freezing out procedures, ion exchange and the like. However, none of these processes has been completely satisfactory for commercial purposes. Certain of the proposed processes are complicated and difficult to operate. Furthermore, while some of the proposed processes permit the recovery of certain of the components of the color developing solution, e.g., the color forming coupler, the color developing agent and benzyl alcohol (when used in the developer) still remain in solution and must be discarded.

These problems are overcome in this process which involves admixing a color developer solution having a pH in the range of between about 9 and 14 with a water-soluble salt with the solution in an amount sufficient to result in the formation of a developing agent containing second phase that is separable from the main body of color developer solution phase. The newly formed second phase can be either in the form of a liquid or a solid phase depending upon the materials being recovered.

The developing agent phase is easily recovered or separated from the main body of developer solution by any suitable means including filtration, centrifuging, decanting, and the like, thus substantially reducing the biochemical oxygen demand of the main body of the developer solution which is then discarded. The recovered developing agent or developing agent with color forming coupler or developing agent with benzyl alcohol is then advantageously used to make fresh color developer solution for processing.

CARBON MONOXIDE

Recovery by Cuprous Chloride Scrubbing

A process developed by *G.C. Blytas and E.R. Bell; U.S. Patent 3,415,615; December 10, 1968; assigned to Shell Oil Company* is one in which carbon monoxide, useful in various carbonylations, is recovered from admixture with other gases, such as hydrogen and lower aliphatic hydrocarbons, including olefins such as propylene, by selective complexation with a cuprous salt, e.g., cuprous chloride, in propionitrile at 20° to 100°C, after which the carbon monoxide is recovered from the complex by heating.

Figure 84 shows a suitable form of apparatus for the conduct of this process. Referring to this figure, a feed stream containing about 20 mol percent CO and about 3 mol percent propylene, the remainder being inert gases which will not react with the salt, including carbon dioxide, hydrogen, and lower saturated hydrocarbons of from 1 to 3 carbon atoms, is fed via line **11** into the upper portion of an extractive distillation column **10**, operated at a pressure of about 220 psig, and a bottoms temperature of about 20 to 60°C. The solvent, about 50% by weight cuprous chloride in propionitrile, is introduced through line **12** near the top of the column, and flows down the column (provided with suitable gas-liquid contacting trays) cocurrent with the feed.

The carbon dioxide, hydrogen, saturated hydrocarbons and other inerts, which are not complexed by the solvent mixture, are taken off from the top of the column through line **14**, and are subsequently passed into and through a gas scrubber **44** to remove any entrained propionitrile from the gaseous effluent, the light gases being removed from the top of the column **44** through line **46**, and the solvent being tapped off for recycle through lines **48** and **49**.

The complex comprising carbon monoxide in the solvent mixture is taken from the bottom of column **10** via line **17** and passed into a solvent flasher **20**. The flasher is operated at a pressure of about 50 psia, with cooler **30** being provided to lower the temperature from about 115° to about 45° to 50°C. A reboiler **18** may be provided, fed by take-off line **19** from line **17**. In the solvent flasher **20**, the CO-rich solvent is flashed to break up the complex, the gaseous CO exiting from the flasher through line **21** and the solvent being taken off through the bottom line **29**.

The carbon monoxide, together with any entrained solvent, is passed into a condenser **22**, and subsequently through line **32** into the bottom of a gas scrubber **34**, operated at about 50 psia, where it is washed with propionitrile entering through line **48** before leaving the scrubber through line **36**. The CO removed through this line is of very high purity, and may be passed to storage units for subsequent use.

FIGURE 84: PROCESS FOR CARBON MONOXIDE RECOVERY BY CUPROUS CHLORIDE SCRUBBING

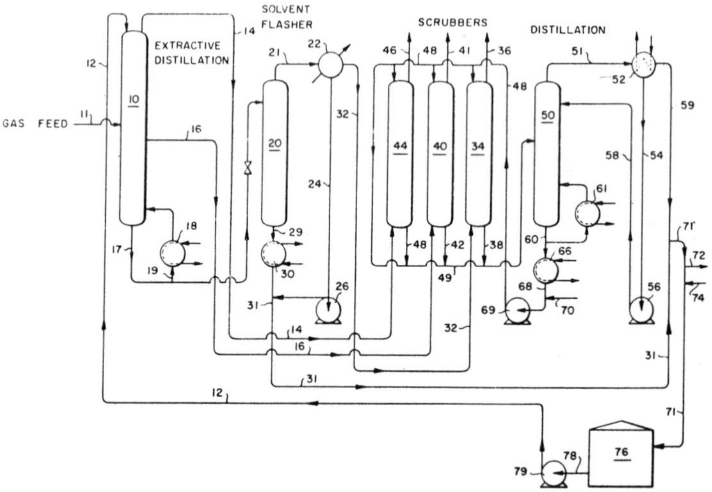

Source: G.C. Blytas and E.R. Bell; U.S. Patent 3,415,615; December 10, 1968

Since cuprous chloride also complexes with unsaturated hydrocarbons, but less readily than with carbon monoxide, if desired, an unsaturate-rich gas side stream containing propionitrile may be taken off from the extractive distillation column **10**, and the unsaturate recovered from it. Column **10** may be provided with a take-off line **16**, through which this unsaturate-propionitrile stream is removed. The unsaturate-propionitrile stream is passed through line **16** into the bottom of a gas scrubber **40**, where it is washed with liquid propionitrile, the unsaturate leaving the scrubber through line **41**, and the propionitrile wash liquid with condensed propionitrile being returned to line **49** via line **42** from the bottom of the column.

The gas scrubbers, **34**, **40** and **44**, may contain suitable inert packing, and are required to remove any propionitrile, which is relatively volatile, which has been entrained in the effluents from these columns. The fresh propionitrile is cycled to these scrubbers through line **48**, and withdrawn from the scrubbers through line **49**, through which it is fed into a solvent distillation column **50** for purification. The propionitrile is taken off the top of column **50** via line **51** and passed through the column condenser **12** into line **59**. Pump **56** and lines **54** and **58** return a portion of condensed propionitrile to the top of the column **50**. The bottoms from column **50** are taken off through line **60**, passed through reboiler **66** and line **68**, and are fed via pump **69** back into line **48** for recycle to the gas scrubbing units. A reboiler **61** may be provided if desired. Additional quantities of the propionitrile may be added as make-up through line **70**.

The solvent mixture from the solvent flasher **20** is taken off through bottoms line **29**, cooler **30**, and line **31**, and is admixed with propionitrile coming in at line **59** and introduced into line **71** discharging into storage means **76**. A gas purge **72** may be provided. Additional quantities of CuCl/propionitrile solvent may be added through line **74**. Solvent is recycled through surge line **78**, pump **79** and line **12**.

Recovery by Formate Formation and Decomposition

A process developed by *J.B. Lynn and A.H. Singleton; U.S. Patent 3,716,619; February 13, 1973; assigned to Bethlehem Steel Corporation* is one in which carbon monoxide is recovered from a gas by compressing the gas containing carbon monoxide and by reacting the carbon monoxide with a lower alkyl alcohol to form alkyl formate. The formate is separated from the gas, and the carbon monoxide is regenerated by decomposition of the formate. Figure 85 shows a suitable form of apparatus for the conduct of this process.

FIGURE 85: CARBON MONOXIDE RECOVERY BY FORMATE FORMATION AND DECOMPOSITION

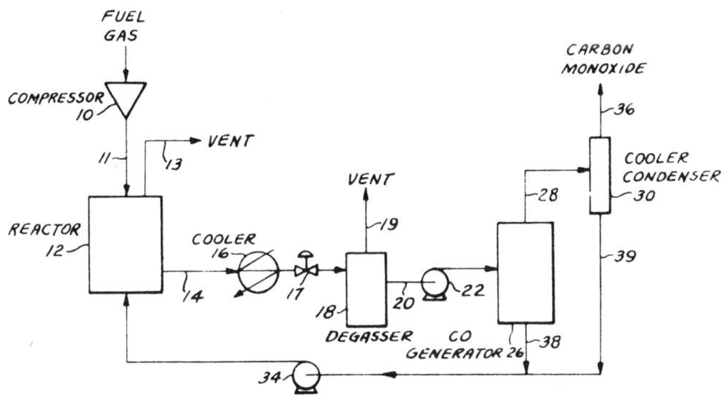

Source: J.B. Lynn and A.H. Singleton; U.S. Patent 3,716,619; February 13, 1973

As shown in the figure, a fuel gas comprising approximately in molar percent, 55% nitrogen, 25% carbon monoxide, 4.5% hydrogen and 0.5% methane, and from which carbon dioxide and water have been previously removed, enters compressor **10** where it is compressed to a pressure between about 2,500 and 5,000 psig (the partial pressure of carbon monoxide present is from about 625 to 1,250 psig). From the compressor the fuel gas enters reactor **12** via line **11**. The reactor contains, in addition to the entering fuel gas sodium methoxide ($NaOCH_3$) dissolved in methanol (CH_3OH) in the ratio of about 1 part by weight sodium methoxide per 100 parts methanol.

The temperature in the reactor is maintained at about 60° to 100°C to synthesize methyl formate by carbon monoxide esterification. During esterification, the fuel gas residue is continuously vented through line **13**. Methyl formate in a methanol solution leaves the reactor via line **14** and enters cooler **16** where the temperature is lowered to about 15° to 20°C and proceeds from there through pressure reducing metering valve **17** to degasser **18**.

In the degasser, the methyl formate-methanol solution pressure is lowered to cause degassing of dissolved contaminant gases. The fuel gas residue thus released is vented from the degasser by line **19**.

The methyl formate-methanol solution leaves the degasser by line **20** and proceeds through pump **22** to carbon monoxide generator **26**. In the carbon monoxide generator, the methyl formate-methanol solution is heated to from about 35° to 200°C to cause the methyl formate ester to decompose, regenerating carbon monoxide and methyl alcohol. The pressure in the carbon monoxide generator is preferably maintained the same as in the degasser. Gaseous products in the carbon monoxide generator, which include carbon monoxide, methyl formate vapor and methanol vapor, leave the carbon monoxide generator via line **28** to cooler/condenser **30**. Liquid methanol leaves the carbon monoxide generator via line **38** and is recycled by pump **34** to reactor **12**.

In the cooler/condenser, the gases entering through line **28** are cooled to purify, by condensation of the methanol and methyl formate, the carbon monoxide which emerges from the cooler/condenser at line **36**. Methanol and methyl formate condensate leave the cooler/condenser by line **39** and are recycled through pump **34** to reactor **12**.

In a specific example of this process blast furnace gas comprising in mol percent, 55% nitrogen, 40% carbon monoxide, 4.5% hydrogen and 0.5% methane, with water and carbon dioxide removed, was compressed to 2,600 psig and reacted with methanol and sodium methoxide in the ratio of 1 part by weight sodium methoxide per 100 parts methanol as described above. The solution product of the reaction analyzed, on a mol percent basis, as follows: methyl formate, 67%; methanol, 25%; sodium methoxide, 1.0%; carbon monoxide, 5.0% (dissolved); nitrogen, 2.0%; hydrogen, negligible; and methane, negligible.

The product solution from above was degassed by first lowering the temperature to 20°C, followed by lowering the pressure to about 300 psig. The product solution then analyzed, on a mol percent basis, as follows: methyl formate, 70.5%; methanol, 26.3%; sodium methoxide, 1.0%; carbon monoxide, 2.0% (dissolved); and nitrogen, about 0.1 to 0.2%.

Heating the product solution to 55°C and removing the gaseous product while maintaining solution pressure at 300 psig gave a regenerated gas stream which was then cooled to 15° to 18°C. On analysis the gaseous product was found to contain, in mol percent: carbon monoxide, 95%; methyl formate, 4.5%; and nitrogen, 0.5%. Further cooling of the gaseous product reduced the methyl formate content, producing a final product gas stream which analyzed 99.5% carbon monoxide.

CHLORINATED ISOCYANURATES

Recovery from Chlorinated Isocyanuric Acid Process Wastes

A process developed by *R.H. Carlson, R.N. Mesiah and H.R. Chancey; U.S. Patent 3,878,208; April 15, 1975; assigned to FMC Corporation* is one in which aqueous waste streams containing dissolved chlorinated isocyanurates and sodium chloride are contacted with hydrogen peroxide at a pH of about 0.5 to 12.0 whereby the chlorinated isocyanurates are converted to substantially insoluble cyanurate values which are separated by filtration or otherwise removed.

The cyanurate values consist mainly of cyanuric acid at the lower pH range and sodium cyanurates at the upper pH range. The resulting filtrate is much lower in organics than the original waste streams. The treatment constitutes an effective pollution control measure which can operate in conjunction with existing processes of manufacturing chlorinated isocyanurate chemicals.

CHLORINATED SOLVENTS

Recovery from Gaseous Wastes

Recovery from Textile Cleaning and Metal Degreasing: A process developed by *M.R. Zimmermann; U.S. Patent 3,486,305; December 30, 1969* involves the recovery of solvents from air, especially of per- and trichloroethylene, by the introduction of a filter containing activated charcoal into an air supply canal for the air containing the solvent for the purpose of adsorption of the solvent from the air onto the activated charcoal.

The process also involves subsequent switching over of the filter chamber for the purpose of desorption of the solvent from the activated charcoal by means of flushing steam by connection on one side to an individual steam generator provided for the desorption and on the other side to a condenser for the purpose of subsequent separation of flushing steam and solvent by condensation, the steam generator being directly connected, by means of its steam chamber, to the filter chamber.

A process developed by *H. Führing and J.H. Sieber; U.S. Patent 3,883,325; May 13, 1975; assigned to Böwe Böhler & Weber Maschinenfabric, Germany* is one in which an air stream from a treatment vessel, e.g., the drum of a dry cleaning machine or a device for the degreasing of metals using an organic solvent, is passed through a cooler to reduce condensable component level in the air stream before being introduced into an adsorber in which solvent residues are removed.

Between the cooler and the adsorber, the air stream is reheated so that the relative humidity of the gas entering the adsorber is well below 100%. Preferably, the air stream is circulated through the cooler and the heater for a period sufficient to reduce the level of condensable components in the air stream to the point that the adsorber will not be immediately saturated thereby, whereupon the adsorber is brought into play.

Recovery from Liquid Wastes

Recovery from Plastic Solutions in Photoresist Manufacture: A process developed by *A.M. Essex and R.B. Heiart; U.S. Patent 3,666,633; May 30, 1972; assigned to E.I. du Pont de Nemours and Company* provides for the recovery of polymer- and/or monomer-laden organic solvents which permits maximum recovery of the solvent without the build-up of the impurities on the distillation equipment. The process comprises passing steam directly through a batch of spent solvent, condensing the mixture of steam and pure solvent which is given off, and separating the water and solvent in a decanter. The cycle is then repeated by drawing off the polymer- and/or monomer-laden residue together with condensed steam and injecting another batch of spent solvent which in turn dissolves any solid residue left over from the first batch.

The apparatus for carrying out this process comprises a container for holding the spent solvent, a steam source, a condenser, a separation decanter, means for drawing off the polymer- and monomer-laden residue after each batch of solvent is processed, means for injecting additional spent solvent into the container, and electrical means for automating the entire apparatus.

The process and apparatus eliminates fouling of the working parts due to the build-up of residue in two ways. First, no solid heating surfaces are used, which minimizes the tendency of the polymer and monomer to solidify on those surfaces. A slight build-up occurs initially on the pipe conveying the steam into the container, but this is limited since the residue itself gradually becomes a heat insulator and no more build-up can occur. Second, the use of a batch process determines that any residue left after a given batch of solvent is purified will be dissolved by the next batch of spent solvent introduced into the container. This process allows recovery of 80 to 98% of the spent solvent. A suitable form of apparatus for the conduct of the process is shown in Figure 86.

FIGURE 86: APPARATUS FOR RECOVERY OF CHLORINATED ALKANE SOLVENT

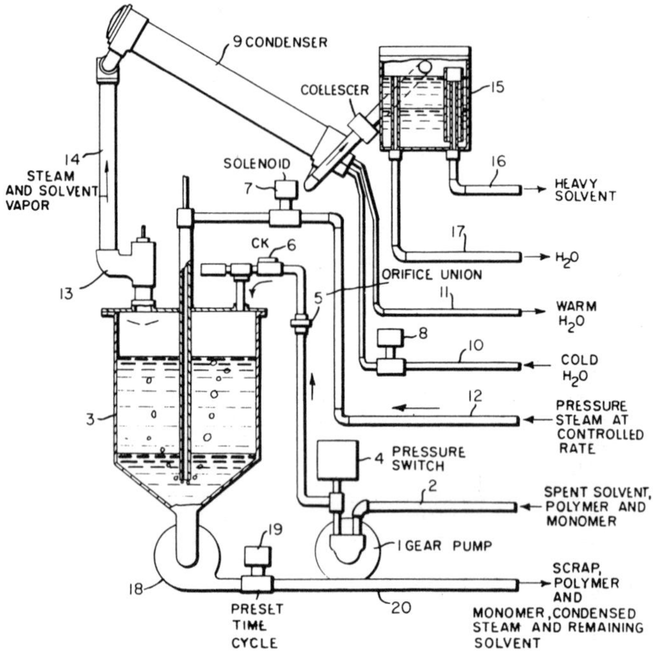

Source: A.M. Essex and R.B. Heiart; U.S. Patent 3,666,633; May 30, 1972

The process is begun with the activation of a positive displacement gear pump **1** which draws in the spent solvent with dissolved polymer and monomer from an outside source through pipe **2**. The pump forces the solvent into the container **3** for a preset time based on the amount of solvent which the vessel or container is designed to hold. For example, if the container is to be charged to 5 gallons, the pump could introduce the solvent at a rate of 2½ gallons per minute for 2 minutes. The entrance line **2** is equipped with a pressure switch **4** and an orifice union **5** which provides for automatic stopping of the system if no liquid pressure is detected for 5 seconds during the fill cycle.

The solvent passes through a check valve **6** which prevents backflow of the solvent when steam is entering the container. At the end of the filling time, the pump automatically stops and the fill cycle is over. At that instant, solenoid valves **7** and **8** open. Water flows through valve **8** from pipe **10** through the shell and tube type condenser **9** and out of pipe **11** in order to cool the condenser and liquefy the vapor that will be entering it. Valve **7** admits steam from pipe **12** under pressure at a controlled flow rate.

The steam is bubbled through the solvent and after a few minutes distillation begins. A thermo-switch **13** senses the rise in temperature as the distillation proceeds. The distillation temperature is less than 100°C and gradually climbs as the steam to solvent ratio increases during distillation. The nonvolatile impurities are left behind in the container **3**. The mixture of steam and solvent vapor travels up pipe **14** and into condenser **9**. There it liquefies. The liquefied mixture then travels to the decanter **15** where the heavier liquid is separated by gravity from the lighter liquid and is drained through and collected from pipe **16**. The water is drained through pipe **17**.

It can be determined experimentally at what temperature, detected by the temperature switch **13**, optimum recovery of the solvent is obtained. For example, using methyl chloroform as the solvent to be recovered, 90% of the solvent will have been distilled and recovered when the vapor temperature at the thermo-switch **13** reaches 70°C. At this point, the temperature switch closes valves **7** and **8**, thus ending the steam or distillation cycle, and activates a centrifugal pump **18** and opens solenoid valve **19**. Pump **18** sucks out the scrap polymer and monomer left behind in the container after distillation along with the condensed steam and the remaining amount of solvent. The waste leaves through pipe **20**.

The pump operates on a preset time cycle, then stops. Solenoid valve **19** closes. At this point, the gear pump is activated and another batch of spent solvent is introduced into the container and the process begins again as just described. The new batch of solvent dissolves any scrap polymer and monomer which the pump was not able to remove. Thus, all working parts remain clean throughout recovery and the apparatus may be used continuously without fear of gumming or fouling by residue build-up. When the parameters of operation have been settled upon, the apparatus works on its own normal cycle without manual intervention.

A specific example of how this process is used follows. A photoresist coating was laminated to a copper clad, epoxy-Fiberglas board. The laminate was then imagewise exposed, producing soluble and insoluble areas in the photoresist film coating. The board was then developed using methyl chloroform as a solvent. The unexposed, soluble areas of the photoresist film were dissolved in the methyl chloroform and washed from the board leaving the insoluble resist image on it.

The methyl chloroform containing the dissolved photoresist was then purified and recovered in the following manner. The gear pump **1** drew in the spent methyl chloroform and charged the container **3** at a rate of 2.5 gallons per minute. The desired amount of solvent in the container was 5 gallons, thus time delay relay **25** was set to time out in 2 minutes, ending the fill cycle. As described above, the steam cycle automatically begins.

Steam was introduced into the solvent in the container at the rate of about 30 pounds per hour. After approximately 23 minutes, the vapor temperature as detected by the thermo-switch **13** had risen from 67° to 70°C, the point at which optimum recovery of solvent was obtained. The switch automatically turned off the steam and ended the steam cycle. The pump out cycle began. Pump **18** drew out the residual polymer-monomer mixture and steam condensate. A time delay relay had been set for 35 seconds. When it timed out, the pump stopped, ending the pump out cycle.

The next batch of spent solvent was then introduced by the pump into the container, dissolving any scrap that the pump had not emptied and thus cleaning the apparatus as well. The process went on as before. The recovery and purification of the methyl chloroform was 88%, i.e., 88% of the spent solvent was purified.

Recovery from Degreasing Plant Sludge-Oil Mixtures: A process developed by *J.G. Miserlis and A. Petrou; U.S. Patent 3,803,005; April 9, 1974; assigned to Silresin Chemical Corporation* is a process involving the removal of trichloroethylene solvents from a degreasing plant sludge oil mixture. This process comprises employing an evaporation two-stage stripping operation to include a directly contacted heat interchange involving recirculation of a liquid oil solvent stream to avoid exceeding the critical decomposition temperature of the trichloroethylene. The process provides for the separation of solvent-free oil containing less than 50 parts per million solvent, the separation and recovery of the trichloroethylene solvent.

A process developed by *J.G. Miserlis and A. Petrou; U.S. Patent 3,893,893; July 8, 1975; assigned to Silresin Chemical Corporation* also involves recovery of trichloroethylene from oil waste by a steam distillation technique.

CYCLODODECANOL/ONE

Recovery from Oxidation Process Washwaters

A process developed by *H. Grasemann and H. Rose; U.S. Patent 3,793,373; February 19, 1974; assigned to Chemische Werke Huels AG, Germany* involves the recovery of unreacted and desired reaction products contained in alkaline washwater which was used to treat oxidation tailings in a process for the oxidation of aliphatic and cycloaliphatic hydrocarbons to alcohols and ketones. The alkaline washwater is cooled to a temperature below about 40°C, preferably between about 20° and 40°C, and the unreacted hydrocarbon, alcohols and ketones are thereby separated from the washwater and subsequently recovered therefrom.

ETHYL ACETATE

Recovery from Sulfite Liquor Evaporator Condensate

A process developed by K.W. Baierl, N.L. Chang, B.F. Lueck, A.J. Wiley and R.A. Holm has been described in EPA-660/2-73-030, Washington, D.C., U.S. Environmental Protection Agency (December 1973).

Pilot studies were conducted on two routes for recovery of volatile components in condensates from evaporation of spent liquors from acid sulfite pulping. The condensates were steam stripped prior to adsorption on activated carbon. The first route utilized fractional distillation and solvent extraction from the carbon, while the second route used low temperature thermal regeneration. Relatively pure SO_2, methanol, furfural, and ethyl acetate were recovered. Estimated process economy, based upon recovery of saleable volatiles, favorable mass and heat balances, and credits for BOD removal, indicates the first route may provide a favorable method for the removal of pollutants from condensates.

Low temperature regeneration studies failed due to mechanical design problems, but this second route continues to be considered as technically feasible. In an important auxiliary study, a substantial number of condensate samples from cooperating mills were analyzed. Differences in quality, quantity, and dilution of the condensates from different sources were apparent. Individual design of processing facilities for each mill is indicated, and especially so in those mills where backwashing of the evaporators contaminate the condensate with nonvolatile materials.

ETHYLENE DICHLORIDE

Recovery from Process Off-Gases

A process developed by *J.E. Panzarella; U.S. Patent 3,589,104; June 29, 1971; assigned to The Dow Chemical Co.* for the recovery of off-gases from chemical processes comprises (1) contacting the process off-gases with a cold surface upon which a film of water-miscible liquid is maintained; (2) separating the noncondensible gases from the condensate which forms upon contact of the off-gases with the cold wall and the water-miscible liquid; (3) separating the water-miscible liquid, water and condensate from one another; (4) returning the water-miscible liquid to the first step to form the film on the cold wall and (5) recovering the condensates free of the water-miscible liquid and water.

Figure 87 shows a suitable form of apparatus for the conduct of the process as applied to ethylene dichloride recovery. Off-gas and an ethylene glycol-H_2O mixture are led through a conduit **10** into the top of the cooling zone **11**. The liquid mixture is dispersed at the top of the cooling zone so that it covers the interior walls of the cooling zone. The liquid-gas mixture is then removed from the zone by way of a conduit **12**, and led into a gas-liquid separator **13**. The gases so separated are vented to the atmosphere via a conduit

14, and the liquids removed via a conduit **15** to a separator **16** which removes the condensed organics, which exit via a conduit **17**. It is to be noted that if a water-miscible liquid is used which is miscible with the condensate as well, one or more stills would be required in the place of the phase separator **16**. The glycol-H_2O mixture, now diluted over that introduced into zone **11**, is removed by a conduit **18** to a concentrator **19**, wherein sufficient H_2O is removed to raise the glycol concentration to the desired level. The excess H_2O is removed via a conduit **20**, and the concentrated glycol-H_2O is removed via a conduit **21**. It is to be understood that the process may be run in a countercurrent fashion with equal, and often increased efficiency.

FIGURE 87: APPARATUS FOR ETHYLENE DICHLORIDE RECOVERY FROM PROCESS OFF-GASES

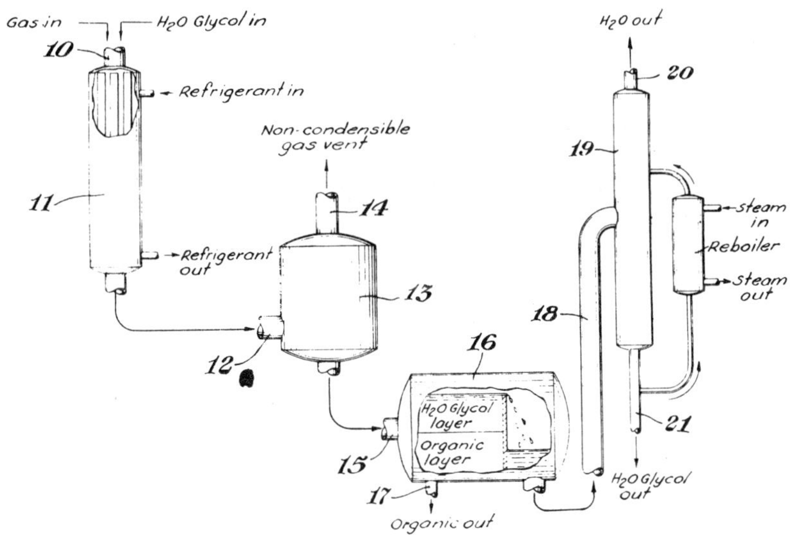

Source: J.E. Panzarella; U.S. Patent 3,589,104; June 29, 1971

ETHYLENE GLYCOL

Recovery from Polyester Plant Wastes

A process developed by *M.M. Mattia; U.S. Patent 3,625,886; December 7, 1971; assigned to Day & Zimmermann, Inc.* involves recovering organic compounds present in aqueous streams at concentrations between 50 and 5,000 parts per million. The process is continuous in that alternate adsorbers are interchangeably used. Economic steam desorption of high boiling compounds is permitted by use of a thermocompressor to recycle regenerating steam through the adsorbers. Steam condensing in a separator is kept to a minimum in order to obtain the highest possible organic content in the condensate.

Low concentrations of relatively high boiling point compounds in effluent streams constitute one of the major industrial water pollution problems. These compounds frequently prevent reuse of water because they contribute to product contamination and/or lower operating efficiencies. The compounds are not volatile enough to recover by distillation and are too low in concentration to permit economic recovery by evaporation. Typical examples of such compounds which may be recovered by this process include ethylene glycol from polyester plant effluent streams, phenol from catalytic cracking operations, phenol in effluent streams from phenolic resin plants, dimethylformamide in washwaters from polyurethane processes, etc. Figure 88 is a flow diagram showing the essential details of the process.

FIGURE 88: PROCESS FOR GLYCOL RECOVERY FROM POLYESTER PLANT WASTEWATER

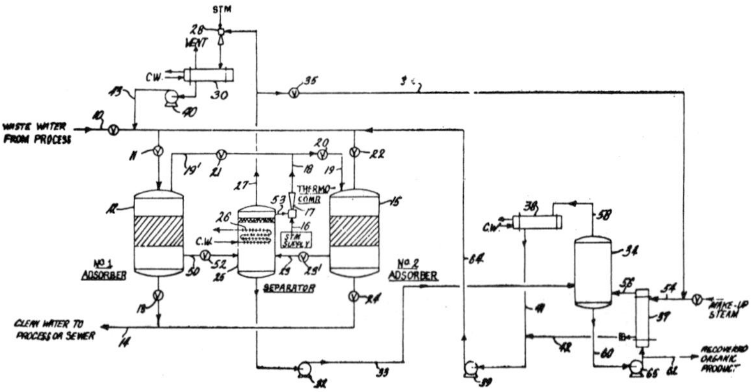

Source: M.M. Mattia; U.S. Patent 3,625,886; December 7, 1971

Wastewater from any one of a variety of processes, including the above mentioned processes, and containing any of the aforesaid organic compounds having a boiling point above 212°F, and a concentration of between 50 and 5,000 parts per million by weight, which create water pollution problems, enters the system through conduit **10** and is diverted to either side of the No. 1 adsorber **12** or the No. 2 adsorber **15** which are disposed in parallel. Valve **11** controls flow to adsorber **12** whereas valve **22** controls flow to adsorber **15**. The outlets from the adsorbers are connected to a conduit **14** which directs the clean water back to the process or to a sewer outlet. Discharge from adsorber **12** is controlled by valve **13**; discharge from adsorber **15** is controlled by valve **24**.

While adsorber **12** is on stream, adsorber **15** is being regenerated and vice versa. While adsorber **12** is on stream, valves **22** and **24** are closed. Valve **20** in conduit **19** is open, and valve **21** in conduit **19'** is closed. High pressure steam is admitted from a supply through conduit **16** to the thermocompressor **17**. Steam then flows through conduits **18** and **19** to the adsorber **15**.

Each of the adsorbers **12** and **15** is provided with a volume or supply of an adsorber material such as activated carbon, alumina, silica gel or other suitable adsorbent. During the regeneration of adsorber **15**, valve **23'** in conduit **23** is open. Valve **52** in conduit **50** is closed. Hence, stripping steam flows from the adsorber **15** to the condenser-separator **25**. Cooling water flows through coils **26** at a controlled rate to condense the minimum amount of steam. Uncondensed steam flows to the thermocompressor **17** via conduit **53**. Thereafter, the uncondensed steam is recirculated back to the adsorber **15**.

The steam from line **23** carries with it the organic compounds stripped from the adsorbent in adsorber **15** and into the condenser-separator **25** where it is condensed by the cooling coils **26**. This condensation results because the organic compound is higher boiling than the water and hence condenses preferentially. Organic compound and water collected at the bottom of the condenser-separator is transferred by pump **32** through conduit **33** to an evaporator **34** where it is further purified by removal of water. Pressure in the system can be controlled either by the flow of cooling water to the condensing coils, or by steam flow to the thermocompressor, or by the flow of excess steam through conduit **27**.

If regeneration occurs below atmospheric pressure, noncondensables must be removed from the system continuously. This is accomplished by the steam eductor **28**. Noncondensables plus a small amount of steam are drawn into the eductor through conduit **27**. The steam and any noncondensables in conduit **27** may flow to condenser **30** where the steam is condensed. Any noncondensed gases from condenser **30** are vented to atmosphere. The condensate from condenser **30** is reintroduced into conduit **10** via pump **40** and conduit **43**.

When desorption occurs at pressures above atmospheric pressure, any excess steam can be released from the system by way of conduit **36** containing valve **35** to conduit **54**. This conduit permits introduction of make-up steam into a reboiler **37** where it provides part of the heat required to further concentrate the organic compounds. Efficiency of the system is improved by utilizing the heat of the steam from conduit **36** to vaporize the condensate recirculated from the evaporator **34**, through conduit **60** and pump **65**, through reboiler **37** and conduit **56** back to the evaporator **34**.

Steam from the evaporator is withdrawn by way of conduit **58** and condensed in condenser **38**. The condensate from this condenser is withdrawn through conduit **41** connected to the inlet side of pump **39**. Condensate from the reboiler is withdrawn by way of conduit **42** by pump **39**. Conduit **42** contains a steam trap designated **T**. The condensate pumped by pump **39** is returned to the adsorber on stream by way of conduit **64**.

The level of liquid within the evaporator **34** is maintained at a predetermined point by a level control, not shown, which enables recovered organic product to be diverted to any storage vessel via conduit **62**. As previously mentioned, the system may be operated continuously. When it is desired to strip adsorber **12** and place adsorber **15** on stream, valves **11, 13, 20** and **23'** are closed. Also, valves **22, 24, 21** and **52** are opened. Each of the valves may be operated by solenoids so that switching of the adsorber from on stream to stripping may be accomplished by merely manipulating a single electrical switch.

If desired, a packed section or distillation trays can be provided below the cooling coils in the condenser-separator to produce a greater degree of fractionation in the condenser-separator. Frequently, the motive steam rate for thermocompression will exceed the desired steam condensing rate in the condenser-separator. In this case, excess steam will leave the condenser-separator under pressure control and flow to the reboiler where it will provide part of the heat to further process the recovered organic compounds. It will be apparent to those skilled in the art that in place of an evaporator, other structures may be utilized such as a distillation column. The separator **25** may be provided with a demister above the heating coils if desired.

By way of example, economic results may be obtained utilizing steam from the steam supply at 100 psig at approximately 400 pounds per hour. This results in a circulation of 1,000 pounds per hour of steam at a desorption pressure of 7 psia and a system pressure drop of 2 psi. For this operation, 400 pounds per hour of steam would also be condensed by the cooling coils so that the net flow of steam to the thermocompressor **17** via conduit **53** would be 600 pounds per hour. Steam temperature leaving the thermocompressor was 230°F. Lower pressure drops or higher motive steam pressure would naturally decrease steam requirements.

Broad parameters in respect to the steam supply may be used. For example, the steam supply pressure may vary from approximately 100 to 300 psi. The pressure of the steam as it enters the adsorbers **12** and **15** may vary from approximately 5 to 50 psia. The pressure drop in the system may vary from ½ to 3 psi.

Either vacuum or pressure regeneration can be employed. Each method has its specific advantage. Vacuum regeneration is used with heat sensitive materials where high temperature might either decompose or polymerize the adsorbed material. Desorption is also improved at lower pressures. High pressure regeneration, on the other hand, would improve heat transfer to the adsorbent, because of the greater density of the stripping steam. Also, at higher pressures, the excess steam exhausted from the condenser-separator can be used more effectively in subsequent purification operations.

A process developed by *J.W. Cox and M.C. Wilkes; U.S. Patent 3,878,055; April 15, 1975; assigned to PPG Industries, Inc.* is one in which ethylene glycol is recovered from spent glycol containing impurities such as dimethyl terephthalate, by subjecting the spent glycol to a sequential flash evaporation-distillation or distillation-flash evaporation-distillation procedure. Dimethyl terephthalate impurities are rendered nonvolatile by the addition of a small amount of alkali metal hydroxide prior to flash evaporation.

FORMALDEHYDE

Recovery from Chemical Process Wastes

A process developed by *J. Ackerman and P. Radici; U.S. Patent 3,843,643; October 22, 1974; assigned to Società Italiana Resine S.I.R. SpA, Italy* is one in which formaldehyde and phenol are recovered from waste liquors, with purification of the liquors, by addition of ammonia to the formaldehyde liquor to form hexamethylene tetramine, evaporation of water, addition of the concentrate to the phenol liquor, and recovery of the hexamethylene tetramine/phenol adduct formed.

FURFURAL

Recovery from Liquid Waste

Recovery from Waste Sulfite Pulping Liquor: The reader is also referred to the process described earlier in this volume under Ethyl Acetate, Recovery from Liquid Wastes and Recovery from Sulfite Liquor Evaporator Condensate.

A process developed by *M. Toivonen, J. Kettunen and E. Salunen; U.S. Patent 3,819,812; June 25, 1974; assigned to Savon Sellu Oy, Finland* is one in which ammonia, furfural, sulfur dioxide and volatile organic acids such as acetic and formic acids are recovered from ammonia based sulfite waste liquor by volatilizing them, drying the liquor to a powder, and separating them from the volatilized gases. Figure 89 is a block flow diagram showing the essential features of this process.

FIGURE 89: FURFURAL RECOVERY FROM WASTE PULPING LIQUOR

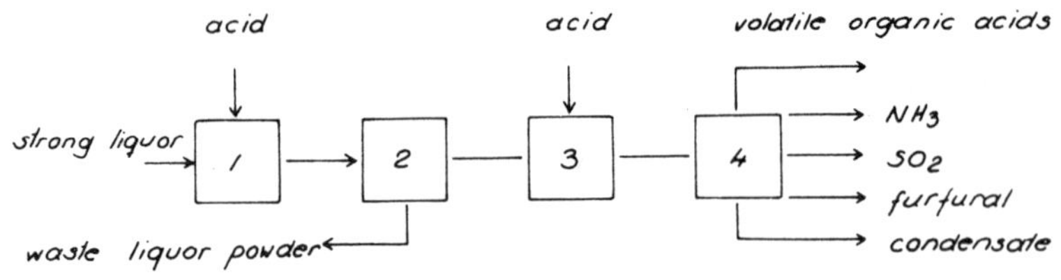

Source: M. Toivonen, J. Kettunen and E. Salunen; U.S. Patent 3,819,812; June 25, 1974

The waste liquor concentrate is conducted into an acidization and predistillation reactor **1**, into which mineral acid is also conducted. In addition to sulfuric acid, sulfur dioxide and phosphoric or nitric acid or different acid mixtures can be used as the mineral acid, the purpose of which is to release the simple fatty acids from their salts. The use of sulfur dioxide in the acidizaton is advantageous because it can be recovered for use as a cooking chemical later in the process.

By using phosphoric and/or nitric acid for the acidization, a waste liquor powder is obtained in the subsequent drying; this powder can be used advantageously for fodder or soil improving agent. In the ammonium sulfite process, in particular, acidization is not necessarily required because of the ability of the organic acids to dissociate ammonia salts thermally, or the acidization can be carried out advantageously with the said acids or their mixtures.

The acidized concentrate is conducted into the dryer **2**, for which roller, vacuum, furnace, spray, or other dryers, which are known as such and the temperature of which can be controlled, are used. The components which do not evaporate at the drying temperature of the waste liquor are recovered from the dryer in the form of waste liquor powder. The heat released in the cooling of this powder is used most advantageously for the heating of the burning air of the burner of the dryer.

The waste liquor powder is used for fodder, soil improving agent, lignin powder, or alternatively, for fuel. The waste gases obtained from the dryer, which besides water-vapor contain among other things, the acetic acid, the formic acid, the furfural, the sulfur dioxide and the ammonia, are condensed in the condenser **3** by a known method. The released heat is used most advantageously for the preheating of the drying gases.

The acetic acid, formic acid, furfural, sulfur dioxide, and ammonia are separated from each other in a separating device **4** which contains several unit operations of chemical technology, known as such. For example, the following method can be used. The furfural is separated from the liquor by a known process including distillation and evaporation operations. The acetic acid and the formic acid are separated from the liquor by leaching operations known as such (e.g., ethyl acetate or methyl ethyl ketone leaching) and separated from each other by azeotropic distillation known as such, and purified by known methods.

Either the sulfur dioxide is stripped from the liquor by methods known as such and returned advantageously to the preparation of the cooking liquid, or the liquor as such with its sulfur dioxide and ammonia is returned to the preparation of the cooking liquid. The said unit operations can be carried out in another order or be replaced by other known unit operations; condensing is not necessary in all cases.

Recovery from Solid Wastes

Production from Oat Hulls: The first commercial production of furfural as a nylon intermediate and as a solvent for petroleum refining was carried out by the processing of oat hulls as described by O.W. Cass in *Chemical Industries,* 60, 612 (1949). The furfural was converted to adiponitrile in nylon manufacture but petrochemical routes to adiponitrile have supplanted the process based on oat hulls.

HYDROCARBONS

Recovery from Gaseous Wastes

Recovery from Flare Systems: A process developed by *W.W. Ford and A.M.L. Kube; U.S. Patent 2,978,063; April 4, 1961; assigned to Phillips Petroleum Company* relates to a system for the recovery of valuable constituents in gases normally sent to the flare line. In plants for the separation and processing of hydrocarbons such as refineries, gasoline plants, petrochemical plants, and the like, large quantities of gases are sent to a flare as a result of upset conditions in portions of the plant or by leakage of high pressure gases through safety valves into the flare line system. It has been proposed to return such gases resulting from leakage to a low pressure point in the system and to provide a liquid trap in the flare line to prevent venting of gases to the flare line under normal operating conditions. The gases in a flare line system usually represent a heterogeneous mixture of gases as a result of their having originated from a variety of sources and it is not always desirable to introduce such mixture to a low pressure point in the system. Furthermore, a liquid trap has not always proven successful because a water trap has a tendency to freeze during unexpected cold weather and an oil trap has a tendency to accumulate condensable hydrocarbon materials within itself and to pass lugs of liquid to the flare as a result of such increase in volume through condensation of liquefied material.

In accordance with this process, provision is made for the absorption of condensable materials, from the gases being passed to the flare line, for further processing and reuse and for passing the noncondensable gases to the fuel line to be used for supplying heat to the plant under consideration. Figure 90 shows a suitable form of apparatus for the conduct of the process.

FIGURE 90: HYDROCARBON RECOVERY FROM PLANT FLARE SYSTEM

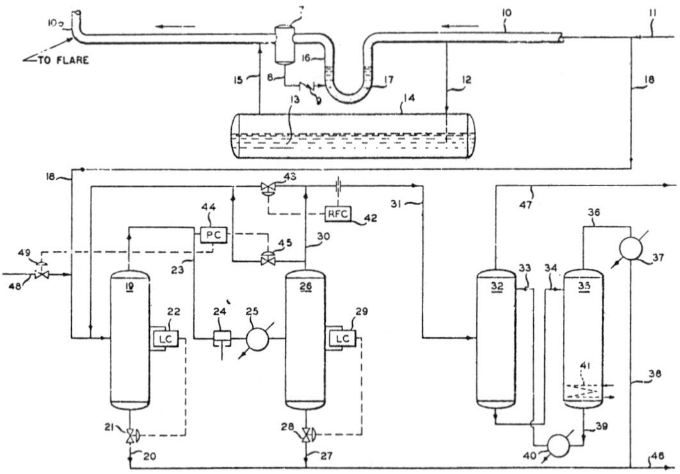

Source: W.W. Ford and A.M.L. Kube; U.S. Patent 2,978,063; April 4, 1961

Referring now to Figure 90, gases enter the flare line **10** from a gathering system (not shown), via conduit **11** and pass through an induction line **12** to beneath the surface of the liquid **13** in tank **14** so as to produce a pressure of 2 or 3 pounds in the flare line. Gases which escape through the liquid trap in tank **14** pass through line **15** to the flare via conduit **10a**. Liquid trap **16** is positioned in the flare line and is connected in parallel with the liquid trap of tank **14** and contains liquid **17**. The liquid head in trap **16** will normally be maintained at a level to produce a pressure 2 or 3 pounds higher than that produced by the liquid head in tank **14**.

A scrubber **7** can be placed in conduit **10** downstream from liquid trap **16** to recover liquid blown from trap **16** by a surge of high pressure in flare conduit **10**. The liquid is returned to trap **16** via conduit **8** containing check valve **9** if the scrubber is elevated sufficiently for gravity flow. The check valve can be replaced by a pump (not shown) if the scrubber is not elevated sufficiently for gravity flow.

A branch line **18** conducts gas from the flare line to a first liquid-gas separator **19** where any liquid which is present in the gas accumulates in vessel **19** and is removed by means of conduit **20** through motor valve **21**, which is actuated by liquid level control **22**. Gases removed from vessel **19** via conduit **23** are compressed by compressor **24**, cooled by cooler **25** and passed into a second liquid-gas separator **26**.

Liquid is removed from vessel **26** via conduit **27** through motor valve **28** which is controlled by liquid level control **29**. Gas from vessel **26** passes through conduits **30, 31** to absorber **32** where it is contacted with a liquid absorbent introduced via conduit **33**. The rich absorbent passes from absorber **32** through conduit **34** to stripper **35** from whence desorbed materials are removed via conduit **36**, cooled in cooler **37** and removed via conduit **38**. The lean solvent or absorbent is removed from stripper **35** via conduit **39**, is cooled in cooler **40** and returned to absorber **32** via conduit **33**. Heat is added to stripping vessel **35** by means of a heater **41**, which can be any conventional heating means such as steam, electricity and the like.

A recording flow controller **42** operatively connected to conduit **31** actuates a normally open motor valve **43** so as to limit the flow to absorber **32** to a predetermined maximum value. A pressure controller **44** operatively connected to conduit **23** actuates a normally closed motor valve **45** so as to open the valve when a pressure in the conduit falls below a predetermined value so as to recycle compressed gas back to the compressor inlet at times when the pressure in the flare line system is abnormally low. The condensed materials recovered from vessels **19, 26** and **35** are removed via a common conduit **46**. Noncondensed gases are passed to a fuel supply system via conduit **47**.

Positive control of compressor suction pressure can be obtained by supplying make-up fuel gas from an outside source via conduit **48**. If fuel gas is supplied via conduit **48**, pressure controller **44** will actuate normally closed motor valve **49** instead of, or in addition to, motor valve **45**. The water level in vessel **14** can be maintained by conventional means such as a water inlet valve operated by a liquid level control means. Heating means can be supplied to heat the water in the vessel during cold weather if such is desired.

The head of water in vessel **14** is selected so that a back pressure of from about 1 to 3 pounds per square inch gauge is maintained in the flare system and the liquid head maintained in liquid trap **16** is usually about 1 to 2 pounds higher than that of the liquid trap of vessel **14**. The gas in conduit **33** is compressed to some preselected value such as about 68 psig and the recording flow controller **42** is set at a preselected value, for example, about 3 millimeters standard cubic feet per day. The pressure controller **44** is set to open valve **45** at some predetermined minimum pressure in conduit **23**, for example, about ½ psig.

From the above it can be seen that all of the noncondensable gases which enter the flare line system are used for fuel under normal operation and that the condensable materials are recovered for further processing. When gases pass into the flare line system at an abnormally high rate, the excess over that which can be used as fuel is passed to the flare. Thus, the flare line is always capable of passing large quantities of gas at relatively low pressure and at the same time reverse flow of gas through the flare line is substantially completely prevented.

Recovery from Truck Loading Operations: A process developed by *G.P. McNamee; U.S. Patent 3,886,759; June 3, 1975* involves recovering hydrocarbon vapors from a hydrocarbon vapor-air mixture which is being expelled from a vessel during the filling of the vessel with hydrocarbons comprising the steps of: (1) controlling the hydrocarbon content of the vapor-air mixture by controlling the temperature thereof; (2) controlling the volume of the vapor-air mixture and passing the vapor-air mixture into a compressor; (3) compressing the vapor-air mixture to form a compressed mixture; (4) cooling the compressed mixture; and (5) passing the compressed mixture into a first absorber to form an absorbed mixture having a liquid hydrocarbon at a controlled temperature whereby the liquid hydrocarbon removes a substantial portion of the hydrocarbon vapor from the compressed mixture.

Vapor Recovery Systems for Service Stations: A system developed by *C.K. Viland; U.S. Patent 3,763,901; October 9, 1973* is a self-contained vapor recovery system for gasoline service stations and for similar applications. Displaced hydrocarbon gases are collected at the point of entry when a vehicle's fuel tank is being filled, or when the service station's main storage tanks are receiving a fresh loading of gasoline. These vapors are collected under controlled pressure conditions and passed through a refrigerated condensation or absorption zone, and the liquid is returned to the service station's storage tankage, preferably below the liquid level there.

The essentially hydrocarbon-free gas, now mainly air, is discharged into the atmosphere. The system not only helps conserve a valuable natural resource, the petroleum from which the gasoline is made, but also alleviates air pollution or smog formation, since unburned hydrocarbons in the earth's atmosphere react under the influence of sunlight with nitrogen oxides and carbon monoxide from any source of combustion, to form a typical smog blanket.

A process developed by *G.R. Onufer; U.S. Patent 3,881,894; May 6, 1975* provides a vapor emission control system for gasoline stations and the like. It recovers escaping gases or vapors at the connection between the filling hose nozzle and the end of the filler neck of an automobile gas tank by creating an annular inrushing atmospheric air curtain at the open end of the filler neck which sweeps escaping vapors into a vapor return line and delivers the same for disposal.

The disposal may involve either burning of the vapors, or the vapors may be returned to and taken up in gasoline for use in subsequent automobile refueling. The vapors may be taken up by the gasoline by first separating them from any atmospheric air with which they have been mixed and then passing them through microscopic passageways of the labyrinth immersed in liquid gasoline.

Recovery from Liquid Wastes

See under Oils in a section which follows.

NICKEL CARBONYL

Recovery from Metallizing Operations

Nickel carbonyl, which at room temperature is a liquid, is commonly employed in metallizing by thermal decomposition of the carbonyl in contact with the heated substrate to be metallized. Frequently such decomposition is incomplete, and undecomposed carbonyl, together with products of decomposition such as CO, and carrier gases such as CO_2 and N_2, form the exhaust gases of the operation. Carbonyl recovery is important as many of the carbonyls are expensive and even in the case of nickel carbonyl efficient recovery is an important factor in commercial aspects of the operation.

A process developed by *E.R. Breining and W.M. Bolton; U.S. Patent 2,985,509; May 23, 1961; assigned to Union Carbide Corporation* involves the recovery and purification of condensable metal bearing compounds from the exhaust gases of metallizing operations employing heat decomposable metal bearing constituents.

Heretofore the recovery of metal bearing compounds such as nickel carbonyl has been effected through the medium of relatively expensive equipment utilizing refrigeration techniques. It is a purpose of this process to provide a method of recovery which is efficient and economical and thus contributes to the over-all success of the metallizing operation, which does not require a costly initial installation or complex equipment, and which produces a recovered product free of contamination.

The process involves the steps of blowing the gases containing metal carbonyl into water while maintaining the temperature of the water sufficiently low to condense the metal carbonyl, permitting the heavier metal carbonyl to separate under the influence of gravity from the water, draining the liquid carbonyl from the water and contacting the same with a drying agent which absorbs water, mixing the drying agent thoroughly with the liquid carbonyl, and thereafter separating the liquid carbonyl from the drying agent without decomposition of the metal carbonyl.

OILS

Conversion of Oil Spills into Useful Asphalt Materials

A process developed by *J. Winkler; U.S. Patent 3,567,660; March 2, 1971* provides a method of conversion of oil spills into improved rubberized, fiber fortified asphaltic materials by coagulating oil spills with previously ground, spent automotive rubber tires which have preferably been previously premixed with powdery polystyrene or asphalt, the asphalt derived from naturally occurring asphaltite from petroleum or coal tars; harvesting by mechanical means the resulting nonliquid conglomerate reacting and concentrating this mixture by distillation to an asphaltic consistency.

Recovery of Oil from Marine Spills

The literature on recovery of oil from oil spills is voluminous and for additional detail the reader is referred to *Oil Spill Prevention and Removal Handbook* by M. Sittig, Noyes Data Corp., Park Ridge, New Jersey (1974).

In general, the handling of oil spills first involves the question of containment as reviewed in the above cited book and in *Oil Spill Containment Systems,* U.S. Environmental Protection Agency, Edison, New Jersey (January 1973).

Following containment, there is the question of the mechanics of oil removal and recovery. This is reviewed in the Sittig book cited above and in *Oil Skimming Devices,* U.S. Environmental Protection Agency, Edison, New Jersey (May 1970).

The various types of devices used, as described by Sittig, are: (1) weir devices, (2) floating suction devices, (3) solvent surface devices (drums and belts), (4) vortex devices, (5) screw devices, (6) jet devices, and (7) miscellaneous devices. Some few typical devices will be cited here in the interest of completeness of this volume.

One sorbent surface device which was developed by *M.H. Van Stavern, W.T. Jones, H.F. Cossey and W.J. Clark; U.S. Patent 3,612,277; October 12, 1971; assigned to Texaco Inc.* is a rotatable drum type of oil skimmer which is continually rotated to pick up a film of oil and water on the surface of the drum, having a supplemental or transfer drum located substantially above the oil slick and well out of contact therewith, so as to come into contact with film on the pickup drum and receive a portion of the film, product oil being recovered from both drums.

One vortex type device for recovering a substance floating as a sheet on the surface of a liquid mass was developed by *J.-C. Mourion and E.M.R. Dubois; U.S. Patent 3,635,342; January 18, 1972; assigned to Bertin and Cie, France.* In the method, the surface of a liquid on which a hydrocarbon spread as a thin sheet floating on the surface of the liquid is, in the vicinity of its surface, subjected to a local rotating movement to bring about the formation of a vortex type cavity which is open and has a vertical axis. The substance originating from the sheet accumulates in this vortex and the accumulated substance is extracted therefrom, the amount extracted being automatically replaced, as it is extracted, by further substance emanating from the sheet.

A sorbent belt device is that developed by *D.E. Sharpton; U.S. Patent 3,643,804; Feb. 22, 1972* in which an endless conveyor belt having an affinity for oil is mounted on the bow of a barge and projects downwardly and forwardly therefrom towards a water surface having oil floating thereon. Buoyancy means located at the lower end of the belt are adjustable such that the lower run of the belt can be selectively positioned slightly below the interface of the water and oil to assure maximum oil absorption with minimum water pickup.

A floating suction device is that developed by *M.F. Olsen; U.S. Patent 3,745,115; July 10, 1973* in which one or more floats are provided for immersion in an oil slick affected water area, the floats having a collecting compartment and a ballast compartment, and a limit valve for the ballast compartment, such that the floats will be partially submerged at the level of the collecting compartment so that the oil and water mixture may be collected.

Flexible tubes are also provided for the collecting compartment for transferring the collected oil and water mixture by suction to a separation tank. The separation tank has two ball float control valves, one of which permits the clean water to drain back into the environmental water area and the other of which permits the collected oil to be drained off for further use or refinement.

A combined containment and removal device for oil spills developed by *W.G. Ekdahl; U.S. Patent 3,747,760; July 24, 1973; assigned to Johns-Manville Corporation* is one wherein the water carrying the oil is flowing in a specific direction relative to a barrier and comprises containing the oil in the barrier to form a confined oil slick having a leading edge. Due to the flow of the water, a head wave is formed along the leading edge of the confined oil slick and oil removal apparatus is located at the head wave to remove oil from the head wave.

Another combined collection and removal device is that developed by *T. Muramatsu, K. Aramaki and Y. Kondo; U.S. Patent 3,771,662; November 13, 1973; assigned to Bridgestone Tire Company Limited, Japan.* It involves sweeping oil films on the water surface into a mobile U-shaped oil fence line. An oil collecting zone is formed in the U-shaped oil fence line, where all the oil films swept thereby are collected to form a comparatively thick oil film, so that the oil is efficiently removed from that zone.

An oil skimming vehicle developed by *R.P. Smith, Jr.; U.S. Patent 3,785,496; January 15, 1974; assigned to Craftmaster, Inc.* includes two independently pivotable booms connected together to form a V. A paddle wheel is connected to the two booms across the open end of the V for skimming the oil off the surface of the water and forcing the oil towards the apex of the V. A sump is located at the apex of the V for collecting the oil forced to the apex of the V by the paddle wheel. A flexible stabilizer bar is connected between the two booms for stabilizing the booms and limiting the maximum amount that the booms may pivot with respect to each other.

Recovery of Used Motor Oils

Little quantitative information has been available as to the ultimate disposal of the approximately 2.5 billion gallons per year of lubricating and industrial oils used in the U.S.; or of additional waste oils resulting from production, transportation, refining, and use of petroleum and its products. About 100 million gallons per year is known to be recycled by the re-refining industry, producing lubricating oils; and other oil is used as fuel, road oil, and a variety of other applications. Each of these applications, including re-refining, has substantial negative impacts on the environment.

Acid/clay treatment, the most commonly used re-refining process for waste lubrication oils, is not an attractive approach for expanding the re-refining industry because of its high cost and because of undesirable waste products produced (acid sludge, spent clay). More attractive processes are available, but further development work is required.

Use of vehicular waste oils for fuel, road oiling, and dust control, currently major uses, is unsound from the points of view of resource and environment conservation. During combustion, very fine lead particles are emitted to the atmosphere. Waste oils generally make poor road oils, resulting in rapid loss of oil to the surrounding land and waters, as described by N.J. Weinstein of Recon Systems, Inc. of Princeton, New Jersey in *Waste Oil Recycling and Disposal,* Report PB 236 148, Springfield, Virginia, Nat. Tech. Information Service (August 1974).

The recovery of waste oils has been discussed by V.S. Kimball in *Waste Oil Recovery and Disposal,* Noyes Data Corp, Park Ridge, New Jersey (1975).

As pointed out by J.C. Davis in *Chemical Engineering* magazine for July 22, 1974, pages 62, 64, 66 and 68, interest has redeveloped in the production of re-refined oil from waste lubricating oil in recent years. The re-refined oil looks increasingly attractive as a substitute for increasingly expensive petroleum products from Middle Eastern crudes. Thus, although the number of oil re-refiners decreased from about 150 to 40 in the 1964 to 1974 period, a reversal and growth in this field is expected.

A process developed by *W.R. Seelbach; U.S. Patent 3,879,264; April 22, 1957; assigned to Petroleum Reclaiming Corporation* utilizes an apparatus for refining used lubricating oil including separate readily serviceable oil heating or reactor and clean oil tanks and an electrical system for automatically controlling the heat purification of oil in the heating tank in association with refining solids, the transfer of heat purified oil and suspended refining solids to the clean oil tank through an associated filter to remove suspended refining solids from the oil and the shutting down of operation of the apparatus upon drying of the filtered solids.

A sump pump is employed to continuously withdraw oil and refining solids from adjacent the bottom of the heating tank and to direct same for flow in heat transfer surface, scrubbing relationship relative to the blades of an electrically powered immersion heater.

A process developed by *M.M. Johnson; U.S. Patent 3,879,282; April 22, 1975; assigned to Phillips Petroleum Company* involves reducing the ash content and the lead content of used motor oil by contacting the used oil with an aqueous solution of an ammonium phosphate. A used motor oil is treated in an integrated ecological method for removal of contained metal salts, tetraethyl lead, gasoline, ammonia, and hydrogen sulfide so that removal of each of these contaminates can be accomplished without contaminating the environment. Figure 91 is a flow diagram of the process.

FIGURE 91: RECLAIMING USED MOTOR OIL BY TREATMENT WITH AMMONIUM PHOSPHATE

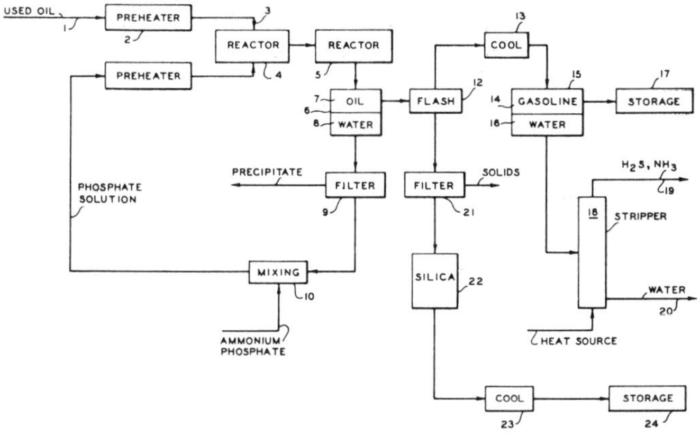

Source: M.M. Johnson; U.S. Patent 3,879,282; April 22, 1975

Referring now to Figure 91, used oil can be passed through line **1** to a preheater **2** where the temperature is raised from about 200° to 500°F and in one embodiment from about 325° to 425°F. The preheated oil is passed through line **3** into the first of a series of stirred reactors **4** where it is contacted with an aqueous solution of ammonium phosphate which has been preheated to a temperature in the same range. The pressure in the reactors and in the oil-water separator is sufficient to maintain the reaction mixture substantially completely in the liquid phase. In the reactor the mixture is agitated vigorously. The reaction mixture can then be passed to another stirred reactor **5** where the same reaction conditions and agitation are maintained.

The reaction mixture is passed to a phase separator **6** in which the oil layer **7** and the water layer **8** are allowed to separate. During this separation precipitated metal salts rapidly settle into the water layer.

The water layer with precipitate is passed through a filter **9** where the precipitate is removed and discarded or processed to recover the metal or metals contained therein. The filtered water is then admixed with ammonium phosphate in **10** to bring the concentration of ammonium phosphate into the range of about 5 to 15 weight percent and recycled to a preheater **11** where the temperature is raised within the reaction range and the material passed back to reactor **4**. By maintaining the phase separation, filtering of the water, and the addition of ammonium phosphate at temperatures somewhat reduced from the reaction temperature, the filtering operation and the dissolving of the ammonium phosphate are facilitated and a minimum of preheat is required in preheater **11**. Heating or cooling is adjusted as required to meet these objectives.

The oil layer **7** from phase separator **6** is passed into a flash chamber **12** where the pressure is reduced to atmospheric or slightly above and the gasoline and water present in the oil are flashed to a vapor. This vapor is then cooled in a cooler **13** to a temperature sufficiently low to condense the gasoline and water at the pressure in the vessels. Cooled liquid is passed to a phase separator **14** where a gasoline layer **15** and a water layer **16** are allowed to form. The gasoline layer can then be transferred to storage **17** and the water layer can be passed to a stripping operation **18** operated at conditions, such as temperature, pressure, flow rate, etc., well known in the art, sufficient to remove ammonia and hydrogen sulfide as overhead product **19**. These products can then be recovered or destroyed.

The stripped water **20** from the stripper **18** can then be recycled for use in the aqueous ammonium phosphate system or, at this point in the operation, is sufficiently free from harmful contaminates to be discarded into the environment. Any convenient heat source such as hot flue gases from the preheaters can be used for operating the stripping system.

The oil layer **7** which has been flashed is passed, while still close to reaction temperature, through a filter **21** to remove extraneous matter and then through a bed of silica gel **22** at a temperature above 350°F to remove the tetraethyl lead. Tetraethyl lead (TEL) in the oil resulting from contamination with gasoline containing it is not removed by the ammonium phosphate treatment.

For example, an autoclave test with gasoline containing TEL did not show a change in lead content on heating to 370°F for 1 hour with an aqueous solution of ammonium phosphate. However, it is known from U.S. Patent 2,745,793 that TEL is removed from hydrocarbons such as gasoline by passing the gasoline through a bed of silica gel at temperatures above 350°F. A similar effect is expected with the TEL present in used motor oil, giving a product containing very fine alkyl lead. The treated oil is then optionally cooled by heat exchange with a cooling medium **23** and passed to storage **24** or simply passed directly to storage **24**.

Recovery of Waste Industrial Oils

A process developed by *J.I. Wallover; U.S. Patent 3,527,696; September 8, 1970; assigned to Wallover Oil Company* is one in which waste industrial oil is filtered through a bed of activated earth to remove chemicals and extremely small foreign particles. The filtered oil is conducted away from the bed and to distilling apparatus if the oil contains distillates. Spent earth in the filter bed is removed periodically and passed through an incinerator to clean and reactivate it. Its place in the bed is taken by activated earth delivered from a substantially airtight reservoir, to which reactivated earth from the incinerator is returned. The smoke produced by the incinerator is burned and the resulting gases are scrubbed to remove solids from them in order to provide clean air which can be delivered to the atmosphere. The solids are collected for suitable disposal. Figure 92 is a flow diagram of the process.

FIGURE 92: APPARATUS FOR RECLAIMING WASTE INDUSTRIAL OILS

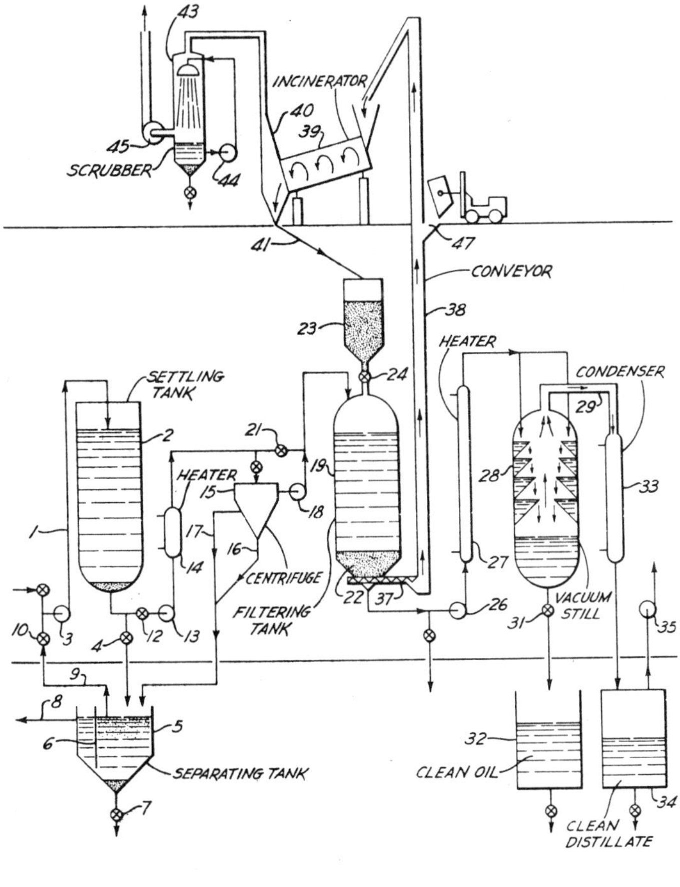

Source: J.I. Wallover; U.S. Patent 3,527,696; September 8, 1970

Referring to Figure 92, it is assumed that the waste industrial oil that is to be reconditioned or reclaimed contains distillates, chemicals and water, and solids in the form of foreign particles. The waste oil may be pumped from a storage tank through a line **1** to the top of a settling tank **2**. After the tank has been filled, the pump **3** is turned off and the oil in the tank is allowed to remain quiet for at least several hours. During this time, water and solids settle to the bottom of the tank. At the desired time a valve **4** is opened below the tank to draw off most of the settled solids and water and deliver them to a separating tank **5**. In this tank there is a vertical baffle **6** that divides the tank into a receiving chamber and a discharge chamber that are connected only through a passage beneath the baffle. Solids settle to the bottom of the tank and can be drawn off periodically through a valve **7** and then used for landfill or other purposes. Water will settle also and can be drawn off at **8** from the other side of the baffle and drained into a sewer, as it will be clean enough for that purpose. The fairly clean oil left in the separating tank can be conducted back through a line **9** and a valve **10** to the pump **3**, which will recirculate it to the settling tank **2**.

After solids and water have been drawn out of the settling tank through drain valve **4**, that valve is closed and a valve **12** is opened that leads to a pump **13** which pumps oil from the settling tank through a heater **14** that may be a heat exchanger heated by steam. The oil is heated to a temperature of about 150°F and is delivered to a centrifuge **15**, where most of the remaining solids and water are separated from the oil. The settling solids and some of the water leave the bottom of the centrifuge through a pipe **16**, and the rest of the water leaves through a lateral pipe **17**. These two pipes may join and they empty into separating tank **5**, where the solids and water again separate and are removed as explained above. Oil, now free of water and most solids, is removed from the upper part of the centrifuge by a pump **18** that delivers it to the top of a filtering tank **19**.

When the waste oil does not contain water and solids or when most of those two elements are removed in the settling tank, the centrifuge can be bypassed through a valve **21** connecting the heater **14** with the outlet of pump **18**. In either case, the oil, which still contains distillates and chemicals and perhaps fine solids, is delivered to the upper part of the filtering tank. In the bottom of the tank there is a bed **22** of granular filtering material that will remove chemicals and extremely small foreign particles from the oil. This material is an activated earth, such as fuller's earth, bauxite or alumina, for example. The filtering material for the bed is supplied from a substantially airtight reservoir **23** mounted on top of the filtering tank and connected with it through a normally closed valve **24**.

For best results, the activated earth should be of a size to pass through a 60 to 90 mesh screen. Finer material may plug the filter, while coarser material has a smaller total surface area and therefore requires more time for filtering.

The filtered oil is withdrawn from the bottom of the filtering tank by a pump **26** that pumps it through a heater **27** and into the top of a vacuum still **28**, where the distillates in the oil separate and pass out of an outlet pipe **29** at the top of the still. The clean, reclaimed oil in the bottom of the still is drawn off through a valve **31** that leads to a suitable storage vessel **32**. The vapors that leave the top of the still pass through pipe **29**, through a condenser **33**, and the resulting liquid distillate falls into a tank **34**, the top of which is connected with a vacuum pump **35** that forms the vacuum in the still.

The main feature of this process that makes the process economical and also avoids pollution problems is the provision for reconditioning or reactivating the spent activated earth in the filter bed so that it can be used over and over, thereby greatly reducing requirements for new filtering material and making it unnecessary to dispose of oil saturated material. Accordingly, suitable means, such as a screw conveyor **37** or the like, is provided for removing spent earth from the bottom of the filtering bed. This material is delivered to a conveyor **38** of any suitable form that will elevate it to a point above the filtering tank and discharge it into the upper end of an inclined rotary incinerator **39**.

In this incinerator carbonaceous material in the spent filtering material is burned out, thereby producing reconditioned or reactivated earth that is discharged from the lower end of the incinerator. This material and the products of combustion enter an afterburner **40**, in which the products of combustion are burned as completely as possible. The hot reactivated earth, now as good as new, leaves the bottom of the afterburner through a conduit **41** and is delivered to the top of the storage reservoir **23** on top of the filtering tank. Since the activated earth in the reservoir is quite warm or hot, there is no danger of condensation on it, which would deleteriously affect the material. The temperature of the incinerator should be as least as high as about 1000°F to burn the carbonaceous material out of the spent earth, but for fuller's earth it should not go above about 1100°F or it will destroy the filtering material. On the other hand, bauxite can be heated to 1600°F without harming it.

The smoke from the afterburner enters the top of a scrubbing tower **43**, where it is scrubbed with water pumped from the lower part of the tower by a pump **44** to remove solids and any chemicals from the smoke. The solids settle to the bottom of the tower and can be drawn off periodically and used for landfill or the like. The smoke, which is now clean air, is drawn out of the tower by an exhaust fan **45** that delivers it to the atmosphere.

Even though the activated earth is reused, some of it is lost in the operation and must be replaced occasionally. This can be done conveniently by providing the side of the vertical conveyor **38** with an opening **47**, through which new activated earth can be dumped into the conveyor when required. This material will be heated in the incinerator and will therefore be hot and dry when it enters the storage reservoir.

It will be understood that with the process and apparatus described above, waste oil reclamation becomes feasible. The cost of activated earth, which has made earlier attempts at reclamation uneconomical, is greatly reduced because the earth can be reactivated repeatedly and therefore used over and over again. This reuse has the additional great advantage that it no longer is necessary to dispose of spent earth, which was a great problem because of the water pollution factor. The solids that are removed during processing are clean and entirely suitable for landfill. Likewise, the removed water is clean enough to be acceptable in sanitary sewers, while the gases that are produced will not contaminate the atmosphere. Although a batch operation has been described, it can be made continuous by using several settling tanks and shifting to another as soon as one is emptied.

Use of Waste Auto Lubricating Oils as Fuel

A detailed consideration of the use of waste automotive lubricating oil as a fuel has been presented by S. Chansky, J. Carroll, B. Kincannon, J. Sahagian and N. Suprenant in EPA-600/5-74-032, Washington, D.C., U.S. Environmental Protection Agency (September 1974).

PHENOLS

Recovery from Gaseous Wastes

Recovery as Augmented Resin Solution: A process developed by *E.B. Baker; U.S. Patent 3,833,712; September 3, 1974; assigned to Borden, Inc.* is one whereby atmospheric pollution by volatile phenolic contaminants in industrial effluent gas streams is substantially reduced by absorbing the contaminants in an aqueous solution of a water-soluble alkaline phenolformaldehyde resin, condensing the mixture of absorbing resin and absorbed contaminants to form an augmented resin, optionally repeating the absorbing and recondensing steps finally to form a useful by-product resin suitable as a component of an adhesive.

Recovery from Liquid Wastes

Recovery from Industrial Wastes by Adsorption: A process for the recovery of phenol from aqueous effluents using ion exchange resins has been developed by Rohm & Haas Co. of Philadelphia, Pennsylvania.

Carbon adsorption, permanganate (or ozone) oxidation and biological oxidation are the other techniques available for phenol removal from aqueous effluents. None of these yield the phenol in recoverable form, however.

The polymeric adsorbents developed by Rohm & Haas are: (1) regenerable in place, (2) selective for phenols, (3) not subject to adsorbent losses upon regeneration, (4) not subject to a capacity loss in use, and (5) capable of yielding a recoverable product.

The phenol is removed completely from the ion exchange resin after use using methanol or acetone or with dilute sodium hydroxide. Thus, the phenol can be recovered as sodium phenolate or phenol in solution. If anhydrous phenol is desired, it can be recovered by simple distillation of the phenolic solution.

A process developed by *F.C. McCoy and R.C. Schlicht; U.S. Patent 3,812,031; May 21, 1974; assigned to Texaco Inc.* is based on the finding that effective separation of phenolic material from aqueous solution can be effected by adsorption upon polyurethane foam.

The adsorption takes place readily and the regeneration is simple and complete. The polyurethane foam adsorbent is commercially available in unlimited quantities at a relatively low cost, in fact, scrap foam is quite suitable for the proposed use. Since it also presents a very high surface contact area, it is relatively cheap to use, as well as readily adaptable to filling columns and other adsorption vessels which are designed to effect the necessary contact. Also, it is quite readily regenerated and may be recycled for successive adsorption and regeneration steps for an indefinite number of times. It is also relatively selective for the adsorption of phenols to the exclusion of other chemical materials.

The adsorbent material used, namely, the polyurethane foam, includes all of the so designated products, whether of the flexible type or the relatively rigid type. However, it has been found preferable to use the so called "reticulate" foam, by which term is meant to include such foams as have interstices so interconnected as to form more or less continuous passageways through the foam. This is to be distinguished from the nonreticulate foams in which the polyurethane has been expanded by a series of internal bubbles which are not necessarily interconnected and which, therefore, do not provide effective internal access by the liquids which are brought into contact therewith.

In order to further facilitate surface contact between the aqueous solution and adsorbent, the latter is subdivided or shredded, which may be readily effected by means of a Waring blender or similar device capable of shredding or chopping the foam. It is preferred, in the case of flexible foam, to conduct the shredding action in the presence of ice so that chunks of foam tend to be kept in more rigid condition by the cooling effect of the chopped ice and water in the blender. Thus, the particles of foam are preferably reduced to about 10 mesh in size. From the chemical standpoint, the polyurethane may take either the ester or ether form.

As to the adsorbent capacity of the polyurethane, this has been found, under the most favorable conditions, to equal or exceed the weight of the foam itself. This is in contrast to many other adsorbents which are able to adsorb only a small fraction of their weight of phenolic material. Considering the relatively higher costs of many of the other well known polymeric adsorbents, this results in a substantial economic advantage for polyurethane foam.

As previously intimated, regeneration of the adsorbent is greatly facilitated by the fact that it may be conducted with various low boiling solvents, as for example, acetone or methyl alcohol. Since acetone in particular is a solvent for many polymeric materials in conventional use, it is quite surprising that it can be used to regenerate polyurethane, which does not seem to be degraded by repeated regeneration, even after an indefinite number of adsorption and regeneration cycles. Therefore, this adsorbent is not only cheaper to use but is simpler to regenerate than the ion exchange resins which have been proposed for this purpose.

Recovery from Phenolic Resin Plant Wastes: A process developed by *J.H. Thayer and E.C.Y. Fan; U.S. Patent 3,736,292; May 29, 1973; assigned to General Electric Company* is one in which phenol content is markedly reduced in waste discharged from apparatus used for producing phenol-aldehyde resins. Aqueous solutions produced by the reaction between the phenol and aldehyde at different stages of the reaction and containing different proportions of phenol are mixed together at a temperature in the range of about 30° to 45°C to provide rapid separation into two solution layers, of which one contains low phenol concentration and may be discharged as waste, and the other contains high phenol concentration which is returned to the reaction vessel for further use in resin production. The low phenol waste solution thus discharged enables a reduction in pollution of streams into which products of the resin process are usually discharged.

A process developed by *S. Vargiu, S.S. Giovanni, G. Mazzoleni and S. Pezzoli; U.S. Patent 3,869,387; March 4, 1975; assigned to Società Italiana Resine S.I.R. SpA, Italy* is one in which phenol is extracted from wastewaters by reacting formaldehyde, urea and phenol containing wastewaters at a basic pH and high ratios of formaldehyde to urea, keeping the product at relatively acidic pH values for short reaction times, reacting the obtained product with additional urea and wastewater at slightly acidic pH values and low ratios of formaldehyde to urea and adjusting the obtained product to a basic pH. The condensates obtained are suitable for use as adhesives and binders for wood.

STARCH

Recovery from Potato Processing Wastes

A process developed by *C.J. Rogers and W.E. Coleman; U.S. Patent 3,829,363; August 13, 1974; assigned to the U.S. Administrator of the Environmental Protection Agency* is one in which fungal protein suitable as an additive for animal feeds and as a human food supplement is produced from starchy substrate materials, particularly potatoes and potato wastes, by homogenizing the starchy material and then subjecting it in a suitable mineral salt containing culture medium to the action of a strain of *Aspergillus niger*, and recovering the protein from the mold mycelia.

STYRENE

Recovery from Plastic Wastes

A process developed by *M.E. Banks, W.D. Lusk and R.S. Ottinger; U.S. Patent 3,829,558; August 13, 1974; assigned to the U.S. Secretary of the Department of Health, Education and Welfare* involves recovery of styrene monomer from the thermal decomposition of waste polystyrene plastic using an inert carrier gas as the source of heat. Figure 93 shows the scheme of operations for styrene recovery from a mixed plastic waste.

The mixed feed (polystyrene plus polyethylene plus PVC) is charged to a semicontinuous stainless steel reactor **60**. Nitrogen, used as an inert heat carrier, is preheated to 1230°C in a direct fired heater **62**, using oil/gas combination burners. Degradation of the waste plastics mixture is accomplished at a constant temperature in the reactor of 600°C.

After leaving the reactor, the gaseous decomposition products are cooled by heat exchange with incoming nitrogen being sent to the direct fired heater for preheating in heat exchanger **64**. The stream temperature is then lowered to 90°C with a water spray quench at **66** which removes the hydrogen chloride and water from the stream. The condensed hydrogen chloride-water stream is sent through a phase separator **68** for recovery of heavy hydrocarbons and then on to a hydrogen chloride distillation column **70** through preheater **72**.

The gaseous stream leaving the quench **66** is cooled to 50°C with a cold water heat exchanger **74**, sent to a phase separator **76** to remove organics other than styrene, and then refrigerated at 20°C by refrigerator **78**, which removes all but **0.6%** of the styrene monomer from the gas stream. After passing through phase separator **80**, the condensed styrene stream is further purified in a vacuum finishing column **82**, treated with an inhibitor, and sent to storage.

When using steam instead of nitrogen as the inert carrier gas, low pressure steam is sent through the direct fired heater and heated to 1230°C prior to being sent to the reactor. Basically, the same method steps are used for the recovery of the degradation products.

FIGURE 93: STYRENE MONOMER RECOVERY FROM MIXED PLASTIC WASTES

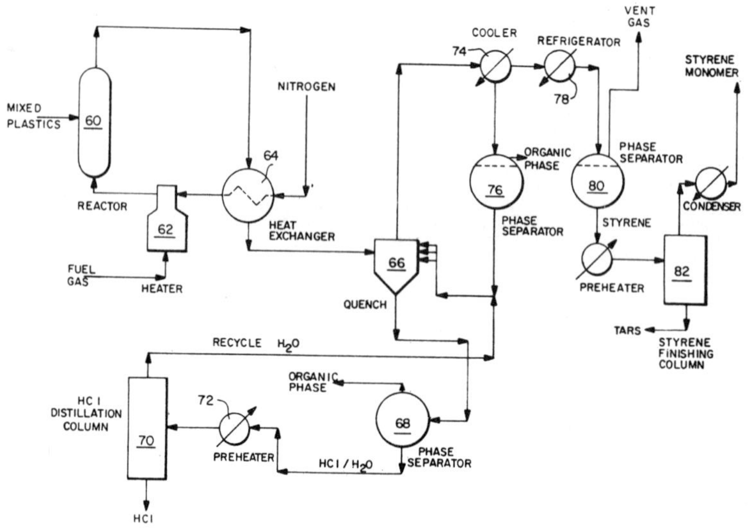

Source: M.E. Banks, W.D. Lusk and R.S. Ottinger; U.S. Patent 3,829,558; August 13, 1974

TERPENES

Recovery from Pulp Digester Blow Gases

A process developed by *C.F. Rosenblad; U.S. Patent 3,492,198; January 27, 1970* is one in which the relief gases and blow steam from the cooking of pulp in digesters on a batch basis are led into the blow tank where they are mixed. From the blow tank the mixture is led into a direct contact condenser where the main part is condensed by direct contact with cooling liquid, preferably from an accumulator containing condensate from a previously condensed mixture. The uncondensed remainder of the mixture continues on to an indirect condenser or gas cooler where the condensibles, including turpentine, are condensed out.

A process developed by *C.S. Gaillard; U.S. Patent 3,753,851; August 21, 1973; assigned to Westvaco Corporation* is a process for treating hot, odoriferous blow gases from a pulp digester whereby a significant portion of the available heat is recovered, turpentine is recovered, a portion of the odoriferous gases are removed and recovered and the remaining gases are burned resulting in further heat recovery.

The digester blow gases containing water vapor, turpentine vapor and gaseous odoriferous sulfur compounds are passed through a series of direct condensers to a gas storage means before being conducted to an indirect condenser. The uncondensed gases leaving the indirect condenser are scrubbed while the condensate mixture leaving the indirect condenser is conducted to a turpentine separating means.

A process developed by *R.E. Marks; U.S. Patent 3,816,239; June 11, 1974; assigned to Envirotech Corporation* involves recovering terpenes from a wet, gaseous stream containing terpene vapor and water vapor by condensing water from the stream by indirect heat exchange to produce an aqueous liquid terpene-lean condensate phase while leaving a first terpene-rich vapor phase.

The aqueous terpene-lean condensate is disengaged from the vapor phase, subcooled and used as the indirect heat exchange medium at boiling for condensing additional terpene-lean condensate from the wet, gaseous stream. During this indirect heat exchange, a portion of the terpene-lean condensate is flash vaporized; and these distillate vapors are combined with the first terpene-rich vapor phase for recovery of terpenes therefrom by condensation at a temperature below the dew point temperature of the terpene component.

TRIBUTYL PHOSPHATE

Recovery from Nuclear Fuel Reprocessing

A process developed by *W.W. Schulz; U.S. Patent 3,708,508; January 2, 1973; assigned to the U.S. Atomic Energy Commission* is one in which tributyl phosphate solutions used in reprocessing nuclear fuel are purified and removed by passing them through a bed of particles of a macroreticular anion or cation exchange resin whereby the chemical and radiolytic degradation products and fission products present in the used solution are adsorbed on the resin. The tributyl phosphate used as extractant may be mixed with a diluent, e.g., C_{10} to C_{14} n-paraffins. Utilization of a macroreticular resin bed may eliminate the need to wash the tributyl phosphate extractant with large volumes of dilute Na_2CO_3 solution and thus avoids generation and disposal of a large volume of radioactive waste.

VANILLIN

Recovery from Sulfite Waste Liquor

A process developed by *C.-T. Yang; U.S. Patent 3,790,637; February 5, 1974; assigned to American Cyanamid Company* is one in which vanillin is made by oxygenating sulfite waste liquor at a substantially neutral pH (i.e., between pH 5.5 and 8) in the presence of a water-insoluble solvent for vanillin to effect recovery of the vanillin.

UREA

Recovery from Urea Plant Waste Streams

A process developed by *D.F. Bress and C.A. Petrarca; U.S. Patent 3,798,021; March 19, 1974; assigned to Foster Wheeler Corp.* which is also referred to in this volume under "Ammonium Nitrate, Recovery from Gaseous Wastes" provides for urea recovery. Waste heat from the process is passed in indirect heat exchange relation with a stream of air and water entrained fertilizer solids. In this manner, it is possible to remove substantially pure water and air together with the fertilizer containing solution which can be used in the manufacture of the fertilizer.

WAXES

Recovery from Oily Process Sludges

A process developed by *E.L. Cole and H.V. Hess; U.S. Patent 3,711,400; January 16, 1973; assigned to Texaco Inc.* is one in which oily sludges are mixed with light hydrocarbons differing substantially in specific gravity therefrom to separate the sludges into an oil-hydrocarbon phase and a water-waxy solids phase. The oil-hydrocarbon phase is heated to a temperature above the critical temperature of the light hydrocarbons to remove the light hydrocarbons which are then recycled for mixing with additional sludge and the oil is recovered for further use. The water-waxy solids phase is treated to obtain water of reduced chemical oxygen demand which can be discharged in receiving bodies of water without polluting the same.

The waxy solids-water slurry which remains, in the second stage of the process, is dried and the resulting dry waxy solids are treated with a light aromatic solvent, such as benzene, to yield a solution of wax and the aromatic solvent and wax-free solids. After the aromatic solvent wax solution is stripped to remove the aromatic solvent, a valuable wax product remains. In a final step, the wax-free solids are heated to remove the last traces of the aromatic solvent therefrom yielding dry, hydrocarbon-free solids which are suitable for a wide variety of uses, such as, for example, for landfill.

PAPER AND WOOD MILL WASTES

BARK

It is the practice among most manufacturers of wood pulp to try to remove as much bark as possible from their pulpwood. This barking of the wood results in large quantities of waste material which, generally, have been of little commercial value. However, to dispose of this barking waste while at the same time avoiding pollution of various kinds, many pulp mills and especially those using wet barking methods, have had to burn this waste at some expense. Mills using dry barking methods can, of course, burn their waste somewhat more economically.

The components of barking waste can be classified and identified as fibrous and nonfibrous. The fibrous components consist of: pieces or particles of wood, which are dislodged from the woody section of the log during the barking operation; the gross fibrous fraction of the bark, often referred to as the inner bark or phloem; and finally that part of the outer bark or phellem which while somewhat corky in nature, still possesses a fibrous structure. The nonfibrous components consist mainly of the scaly outermost parts of the outer bark or bark cork.

The proportions and nature of the above components in the original barking waste and in the final products above will vary with such factors as:

(a) Method of barking, e.g., in the tumbling-friction method using rotating cylindrical drums (wet or dry) the proportion of wood can be quite high; whereas in high pressure water-jet methods the proportion of wood may be low.

(b) The species of wood, e.g., in balsam fir, spruce, poplar (and to a lesser extent in some of the pines) the inner bark is long and stringy, distinctly fibrous in nature and fairly abundant, whereas in Douglas fir the fibrous fraction consists of relatively short, needle-like fibers.

(c) The method by which the pulpwood is transported to the mills, e.g., river-driven wood can lose much of its whole or outer bark in transit, whereas truck or rail transported wood has most of its bark intact.

(d) The time of year during which the barking or wood cleaning is being done.

Preparation of Soil Conditioner from Bark

A process developed by *R.C. Sproull and C.D. Cone, Jr.; U.S. Patent 2,881,066; April 7, 1959; assigned to Southern Lumber Company* yields a nitrogenated and phosphorylated fertilizer, mulch and soil conditioner which requires as a raw material naturally occurring plant residues heretofore regarded as waste, such as barks, sawdust and other wood-like materials. The product is produced by treating comminuted wood-like material with an aqueous solution of urea and phosphoric acid. After the material is thus treated, it is dried at a temperature below the temperature at which carbonization and loss of nitrogen takes place.

Figure 94 is a flow diagram of the process. In carrying out the process, an organic wood-like material, such as bark or sawdust or a mixture thereof, is ground or pulverized in a suitable grinder **10** to reduce the same to small particle size. To provide fertilizer mixtures having different qualities, the wood-like material may be ground to different particles sizes. Storage bins **11** and **12** are provided for receiving the coarse and fine comminuted material, respectively. The comminuted wood-like material thus prepared is transferred to a batch scale **13** and is then conveyed to a rotary digester **14**.

FIGURE 94: PROCESS FOR PRODUCING SOIL CONDITIONER FROM BARK

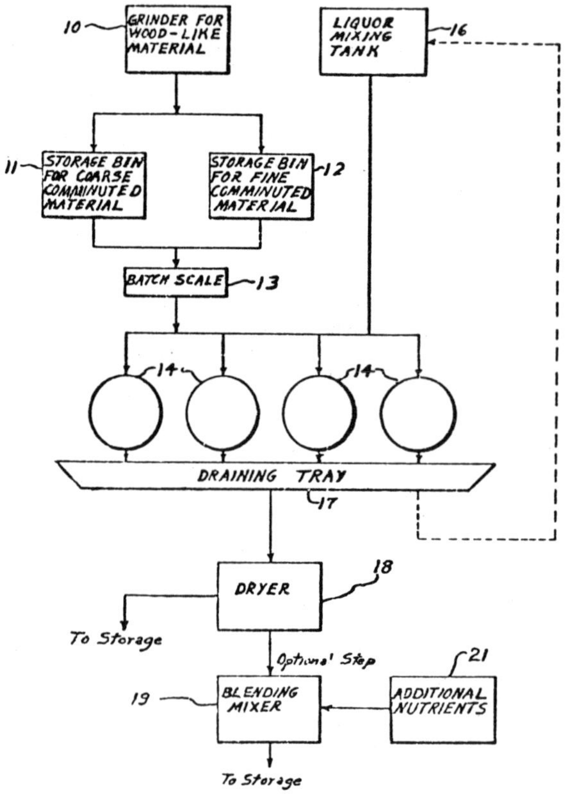

Source: R.C. Sproull and C.D. Cone, Jr.; U.S. Patent 2,881,066; April 7, 1959

A treating solution comprising the desired amounts of urea, phosphoric acid and water is supplied to the rotary digester **14** from a suitable liquor mixing tank **16**. In actual practice, a solution comprising, by weight, approximately 6.5% urea, 45.7% phosphoric acid (85% pure H_3PO_4) and 47.8% water is satisfactory in every respect and produces a product containing approximately 4% nitrogen and 15% phosphorus. Technical trade mixtures may also be employed. While the reaction containers are shown as being in the form of rotary digesters, it will be apparent that other reaction tanks may be employed in combination with suitable means for stirring the comminuted material into the solution.

Preferably, a plurality of rotary digesters **14** are employed whereby the full capacity of the grinder **10** and liquor mixing tank **16** may be utilized. Preferably, the weight ratio of the dried comminuted material to the treating solution should be approximately 0.083 to 0.15.

With the digester **14** charged with the wood-like material and the treating solution, the contents thereof are heated to a temperature of from 80° to 100°C, whereby chemical reaction takes place between the comminuted material and the treating solution. In actual practice, a reaction period of approximately one hour produces a satisfactory product. The material thus treated in digesters **14** is preferably washed and then drained of excess liquor by means of a suitable draining tray **17**. Next the treated material is transferred to a dryer **18** where it is dried up to a maximum temperature of approximately 110°C. Up to this temperature, no carbonization or loss of nitrogen takes place. It is very important that true carbonization be eliminated due to the fact that where true carbonization occurs, the product will not decay.

To insure complete utilization of the treating solution, the weak solution is recycled from the rotary digester **14** to the liquor mixing tank **16**. After drying the product in dryer **18** it is transferred to storage or for use. If desired, the dried product may be blended with other ingredients or plant nutrients in a blending mixer **19**, the additional nutrients being supplied from a supply tank **21**. The nutrients can be added by a dry blending operation, or if desired, the nitrogenated or phosphorylated material can be soaked in solutions of proper concentrations containing additional plant nutrients. The nutrients added by the latter method are thus absorbed into the interstices of the wood-like material and also are deposited thereon in a finely powdered state.

These additional water soluble ingredients are thereby rendered less subject to leaching and removal from the proximity of the plant by rainfall or watering. After blending in the mixer **19**, the material is transferred to storage or for use as a fertilizer, mulch and soil conditioner.

A process developed by *R.E. Burton; U.S. Patent 2,995,434; August 8, 1961; assigned to Union Lumber Company* relates to the treatment of sewage effluent and industrial waste by trickling the same through a bed of chopped bark, and more particularly relates to the use of chopped or shredded redwood bark for such purpose.

The process has the advantage of solving one of the problems in redwood logging and mill operations, namely, disposal of the bark which is stripped from the logs. The process has the further advantage that the resultant product is useful as a soil conditioner and fertilizer. Thus, the bark provides a low cost lightweight medium for trickling filters and is converted into a valuable end product.

A process developed by *W.J. Burgon; U.S. Patent 3,239,329; March 8, 1966; assigned to Canadian Forest Products Limited, Canada* is a process for producing a phosphate product having increased amounts of available phosphate thereon comprising, grinding together phosphate rock and moist bark to form a coating of rock particles upon the bark particles, the bark having a moisture content of 30 to 70% and being in the range of 20 to 800% of the weight of the phosphate rock on a dry basis. The next step is acidulating by uniformly distributing throughout the mixture of bark and rock 20 to 80% by weight of the rock of a solution of sulfuric acid and recycle liquor containing phosphate. The resulting mixture of bark, rock and acid is dried at a temperature of from 100° to 1250°F for from 3 to 30 minutes while agitating to cause the bark organic acids to undergo surface condensation and assist in binding the calcium in the phosphate rock to produce a product having a mineral acid, organic acid calcium, and water soluble phosphate coating thereon.

A further step involves extracting the product with an aqueous liquid selected from the group consisting of water and water containing phosphoric acid, to remove a portion of the phosphate therefrom, separating the resulting extract liquor containing phosphate and recirculating at least a portion of the extracted liquor to the acidulation step to concentrate the phosphate. Figure 95 is a flow diagram of the process.

FIGURE 95: ALTERNATIVE PROCESS FOR PRODUCING SOIL CONDITIONER FROM BARK

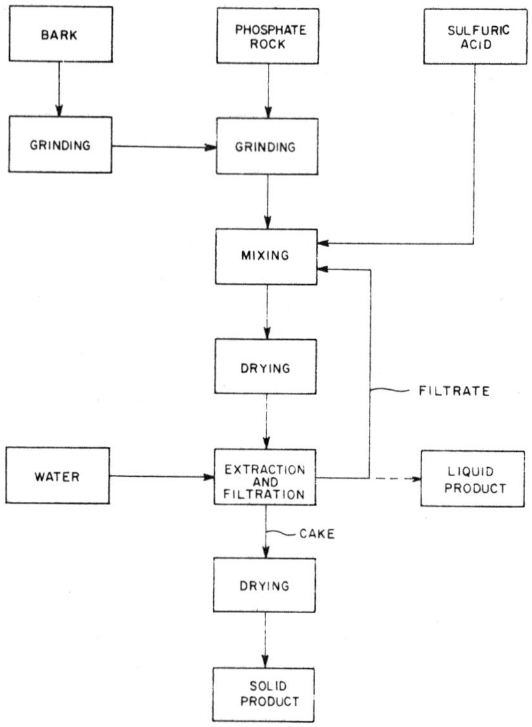

Source: W.J. Burgon; U.S. Patent 3,239,329; March 8, 1966

Production of Fiberboard

A process developed by *T.G. Brandts, D.E. Helleur, R.M. Hopkins and R.G. Boivin; U.S. Patent 3,224,925; December 21, 1965; assigned to Consolidated Paper (Bahamas) Limited, Bahamas* is one in which barking waste is separated into a fibrous wood fraction, a fibrous bark fraction and a nonfibrous bark fraction. The fibrous fractions are then blended together in the desired proportions and formed into a pressed, dried sheet or board. Depending on the fines content of the fibrous fractions it may be necessary to remove some fines to produce a satisfactory product.

CELLULOSE FIBERS

Recovery from Liquid Waste

It is conventional in papermaking operations to utilize large quantities of water as a carrying medium to facilitate the handling of fiber or pulp as well as to carry the fiber in a desired manner into the paper sheet being formed on a papermaking machine. Further substantial quantities of water are used both continuously and intermittently to wash unwanted fiber from various units or components of the mill equipment. In many mills a number of techniques are employed to permit the reuse of these large quantities of water so as to reduce plant water requirements and to reduce the volume of waste or effluent water from the mill.

Such techniques are particularly utilized with respect to the water drained from the newly formed sheet in the paper machine, known as white water. Commonly a system, known as a saveall system, is employed to separate the fiber which has carried through the sheet from the bulk of the white water liquid so as to permit reuse of the water for washing operations and to recover this fiber for reuse. However, these known water reuse techniques have been generally confined to the steady state and continuous operations of the mill. In contrast, at times of abnormal operations and when changing production runs, as well as to a lesser extent during normal operations, considerable quantities of water and fiber escape through floor drains and chest overflows.

Occasionally this loss is accidental as by spillage, but frequently the loss is intentional through the dumping of chests and other equipment during grade change and by washing operations. As a result the mill sewer carries away substantial quantities of water and fiber from the papermaking operation. This fiber-containing mill effuent represents a substantial loss of valuable material and presents a waste disposal problem.

Paper mills lose between 1 to 3% of their production to the waste effluent stream. Much of this is reusable fiber. These losses result from grade changes, clean-ups, trim, excess white water, and spills. These losses can be categorized as two types. The first type of loss is a continuous one in which a portion of the white water from the paper machine is sewered.

Normally every paper machine has a saveall device which will take a major portion of the white water and clarify it, returning the fiber to the papermaking process and the water to the machine for shower usage. However, no paper machine has a completely closed system and therefore a varying percentage of the total white water is untreated by the saveall and therefore ends up in the mill effluent stream.

While this flow will contain some good fiber, a major portion of these fibers will be less than 0.1 mm in length and therefore may not be desirable for reuse. Short fibers or fines give a slow draining characteristic to the paper machine furnish, in which case the speed and capacity of the paper machine are negatively affected.

The second source, and the one which is normally the major source of fiber loss from the papermaking process, occurs during a grade change. Because of the large, expensive nature of papermaking equipment, it is paramount that a grade change be completed in as short a time as possible to minimize machine downtime. To expedite grade changes the contents of various stock chests and white water chests are emptied to the sewer with no attempt for fiber recovery. All of this fiber constitutes good fiber and therefore is reusable.

Recovery by Use of Coagulants: A process developed by *E.K. Brax; U.S. Patent 3,873,418; March 25, 1975; assigned to Savo Oy, Finland* is one in which fibers and wood material which are present in white water from wood processing industries are recovered and used as additional raw material, and resulting clarified water is recycled for use as fabrication water. The pH of water is controlled during the recovery process and the material to be recovered is caused to agglomerate by the addition of agents whose presence in the final product is beneficial. A polyelectrolyte or the like can be added to enhance flock formation.

Recovery by Use of Flotation Agents: A process developed by *H. Sommer, H. Bestian and D. Bergmann; U.S. Patent 3,399,110; August 27, 1968; assigned to Farbwerke Hoechst AG, Germany* is one in which fibrous material is recovered from wastewaters in the paper, cardboard and cellulose industry, by adding to the wastewater a nitrogenous product obtained by condensation of urea and an 1,2-alkylene imine or oligomer thereof having a substitutable hydrogen atom at the nitrogen atom, the nitrogenous product being added in an amount between 0.1 and 4% by weight of the solid and suspended particles.

Recovery by Use of Screens: A process developed by *A. Boniface; U.S. Patent 3,833,468; September 3, 1974; assigned to Dorr-Oliver Inc.* provides a method wherein the effluent stream from a paper mill is treated on a sieve bend screen such as described in U.S. Patent 2,916,142. In such an application of the screen, the effluent stream is directed at a substantial velocity along the plane of a screening surface having a plurality of slots there through which extended substantially transversely to the direction of the feed flow.

In this manner, solid particles which are smaller than the size of reusable fiber along with most of the carrier liquid are deflected by dynamic forces through the slots. The longer reusable fiber in the range of plus 150 mesh or in the order of plus 0.1 mm fiber, and some of the carrier liquid are retained at substantial velocity on the screen surface and travel to the discharge end thereof. By this screening technique the solids loading of the mill effluent is reduced by 25 to 75% thus decreasing the requirements for waste treatment.

Several different techniques may be employed for utilizing the oversize or recovered fiber from the sieve bend screen. First, the recovered fiber may be directed to the broke system where it mixes with the pulp regenerated from broke. Subsequently, the recovered fiber along with the broke fiber may be introduced to the machine chest of the paper machine where it mixes with the other furnish.

Alternatively, as the recovered oversize from the sieve bend screen has a solids consistency of about 3 to 6%, this recovered fiber stream may be directed to the machine chest since the solids consistency of the stock therein is in this same range. Further, the recovered fiber stream from the sieve bend screen may be dewatered so as to obtain a valuable dry product which can be disposed of in the commercial secondary fiber market.

Recovery from Solid Waste

Cellulose is by far the most widespread and readily available of all solid organic materials. It comprises almost one-third of the weight of all trees, vines, grasses, and straws. Unlike other resources such as oil and minerals, cellulose is constantly replenishing itself by photosynthesis and growth. Vast quantities of this material accumulate as waste products from activities such as food processing, lumbering, papermaking, and cereal grain harvesting. Additionally, municipal and industrial wastes of paper, rags, boxes, wood, excelsior, grass, and leaves raise the available amount of cellulose-bearing material to astronomical proportions.

The ready availability of such a quantity of cellulose has been the impetus for voluminous research into novel and diverse methods of utilizing this material. Wallboard, door cores, and mulch have been made from cereal grain straws and sugarcane bagasse; chemical-grade cellulose from cotton linters, wood, and bast fibers; animal feed pellets from pea vines and other fibrous vegetable roughages; animal bedding from oat, rice, and wheat straw; furfural and brake shoes from sugarcane bagasse; various chemicals from corn cobs; and limitless other examples. Still, most waste cellulose is dumped, buried, burned, or used as a fuel supplement; and its chemical energy or physical utility is thereby wasted or only fractionally retained.

Conversion to Oil: The conversion of cellulose wastes to oil has been described by

H.R. Appell, Y.C. Fu, S. Friedman, P.M. Yavorsky and I. Wender in Report of Investigations No. 7560, Washington, D.C., U.S. Bureau of Mines (1971).

The Bureau of Mines reported experimentally converting cellulose, the chief constituent of organic solid waste, to a low-sulfur oil. All types of cellulosic wastes, including urban refuse, agricultural wastes, sewage sludge, wood, lignin, and bovine manure, have been converted to oil by reaction with carbon monoxide and water at temperatures of 350° to 400°C and pressures near 4,000 psig, and in the presence of various catalysts and solvents. Cellulose conversions of 90% and better (corresponding to oil yields of 40 to 50%) have been obtained.

A continuous reactor for use at maximum conditions up to 500°C and 5,000 psig has been operated successfully. Using sucrose as a feedstock, operation in this system has permitted a simplified and preliminary chemical study of the conversion process. Oil yields of over 30% have been obtained with this unit.

Hydrolysis to Yield Recoverable Sugars: Wood hydrolysis to yield sugars which can in turn be fermented to give alcohol dates to the early 1900s.

In World War I, as pointed out by R.N. Shreve in *Chemical Process Industries,* New York, McGraw-Hill Book Co. (1967), two American plants produced alcohol from sawdust, using sulfuric acid as the hydrolyzing agent. Because of low yields the plants were uneconomic and were closed after the war.

Then in the 1930s interest continued in foreign countries and some 30 foreign plants were in successful operation in 1941, according to Shreve. In World War II, under the stimulus of material shortages, additional attention was given to process development in this area in the United States.

A process, known as the Madison wood sugar process, was developed for hydrolyzing mixtures of wood waste with 0.5 to 0.6% sulfuric acid at 150° to 180°C by allowing the dilute acid to flow continuously through the charge of wood. Compared to the German Scholler process, hydrolysis was accomplished in less time because the sugars produced were removed more rapidly. Decomposition was less because the sugars were in contact with the acid for a shorter period of time and consequently, yields of sugar and alcohol were higher. Fewer by-products inhibitory to fermentation were produced, so that fermentations were more rapid. The sugar produced in 2.8 hours from a ton of dry, bark-free, Douglas fir wood waste yielded 64.5 gallons of 95% alcohol as compared to 3.2 hours for 58 gallons by the rapid cycle method developed earlier, (American modified process) and 13 to 20 hours for 55 gallons by the Scholler process as carried out in Germany.

A batch process pilot plant was designed by the Tennessee Valley Authority on the basis of original work at the Forest Products Laboratory at Madison, Wisc. The pilot plant was designed to hydrolyze wood chips to sugar and then concentrate the sugar solution to 50% molasses for use in animal fodder. The flow sheet for this process is shown in Figure 96 as taken from the report, *Conversion of Organic Solid Wastes into Yeast* by F.H. Meller, P.H.S. Publication No. 1909, Bureau of Solid Waste Management, U.S. Public Health Service (1969).

Subsequently, a continuous process was developed as described by Meller and the flow sheet for the continuous process plant is given in Figure 97.

Recovery as Single-Cell Protein: A process was then developed to produce yeasts from the recoverable sugars from wood hydrolysis, as further described by Meller. In addition to the sugars, other nutrients required include a nitrogen source, oxygen, small amounts of phosphate, and other minerals in even smaller amounts. Growth is carried out at temperatures in the range of 20° to 35°C in reactors provided with a means of agitation. This may be a stirrer of various types, or may simply be the introduction of large amounts of air into the fermentor through perforated pipes at the bottom of the vessel.

FIGURE 96: WOOD HYDROLYSIS BATCH PROCESS—FLOW SHEET

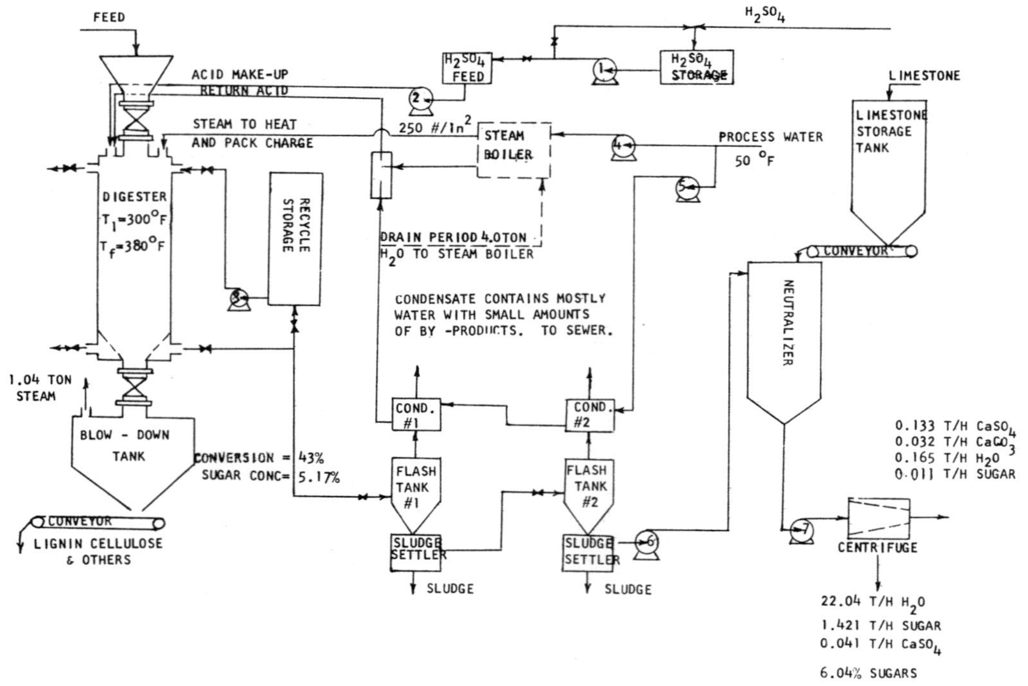

Source: PHS Publication No. 1909

FIGURE 97: CONTINUOUS WOOD HYDROLYSIS PROCESS–FLOW SHEET

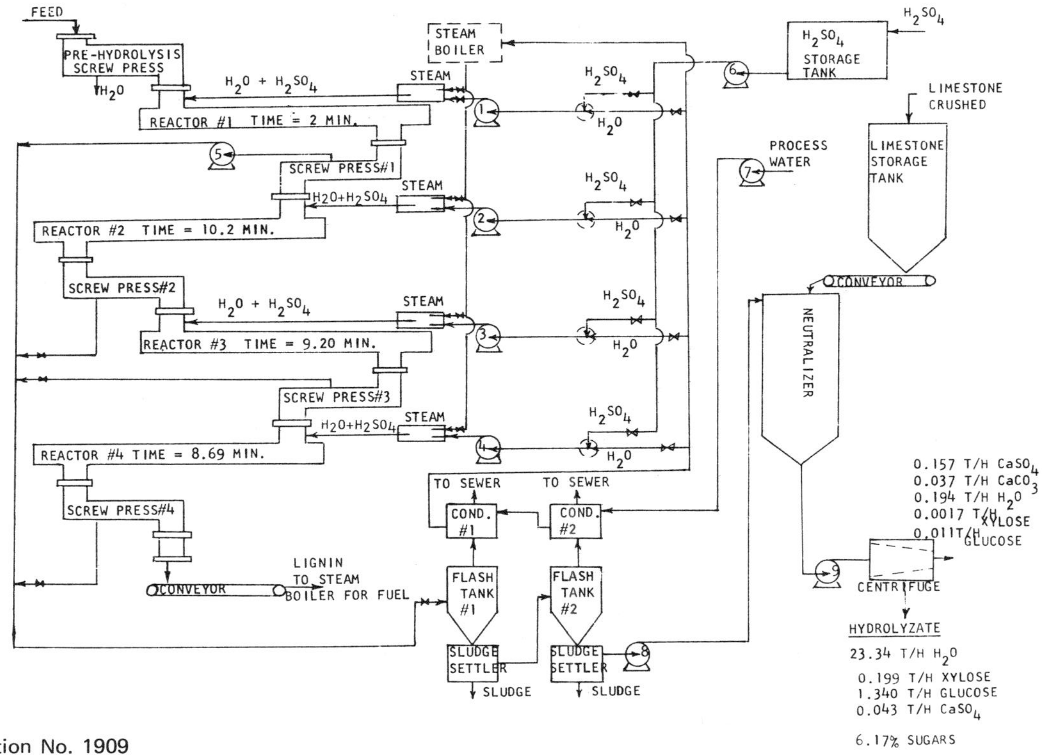

Source: PHS Publication No. 1909

When the yeast cell concentration is adequate, the broth and suspended cells may be harvested by passage through a centrifuge which since the cells have a density slightly greater than that of the broth and a size of approximately five microns, readily removes them. If the fermentor is operated continuously, a constant feed of fresh medium is pumped into the vessel, while a stream of broth and cells goes to the centrifuge. After centrifugation, the cells may be washed and dried, or may be dried directly.

Figure 98 is a flow diagram of the fermentation plant design which is used for the production of yeast from wood sugars.

FIGURE 98: FLOW SHEET–FERMENTATION PLANT FOR SINGLE CELL PROTEIN PRODUCTION FROM CELLULOSE

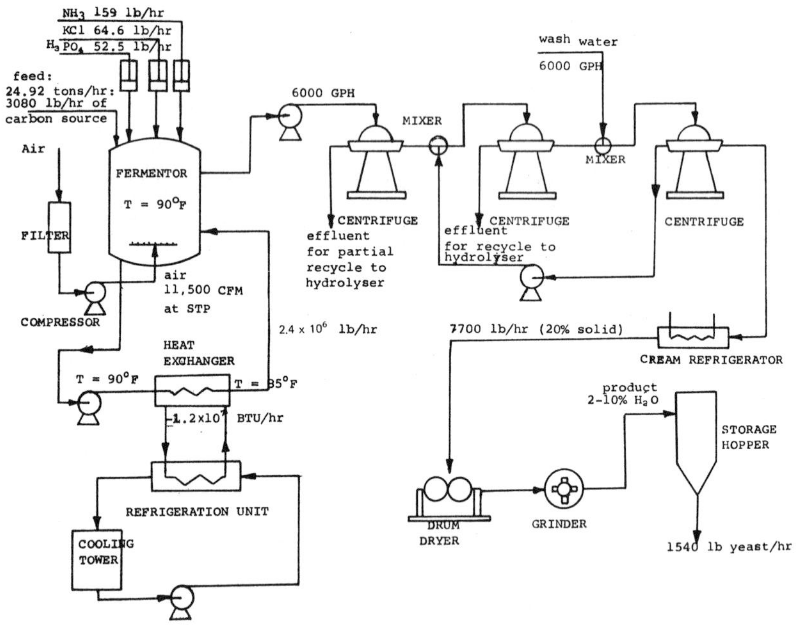

Source: PHS Publication No. 1909

Another approach has been described by C.D. Callihan and C.E. Dunlap of the Department of Chemical Engineering at Louisiana State University and is described in Report SW-24c, Washington, D.C., U.S. Environmental Protection Agency (1971).

A pilot plant was designed, constructed, and operated that produced microbial single-cell protein from waste sugarcane bagasse. Bagasse was ground and given a mild alkaline-oxidation treatment before fermentation. Bagasse was slurried at as much as 10 g/l dry weight in water with a simple nutrient salts mixture to form the fermenter feed stream. The process was operated in both batch and continuous-flow patterns.

Cellulomonas, sp. bacteria were used in a pure culture for most runs, but a mixed culture run of Cellulomonas and the symbiotic organism *Alcaligenes faecalis* showed much higher production capabilities. Maximum cell density obtained with pure Cellulomonas was 1.7 g dry weight per liter and 6.24 g per liter for the mixed culture. Culture mass-doubling times during log-phase growth were usually from 3.2 to 3.7 hr.

The maximum experimental volumetric production efficiency (VPE) of a continuous run using pure Cellulomonas was about 0.10 g of dry cell mass per liter of fermenter capacity per hour. The mixed culture run had a theoretical VPE of 0.512. Single-cell protein was produced as a light brown to yellow-brown powder and had a crude protein content of 50 to 55%. Figure 99 shows the pilot plant design developed for the conduct of this process.

FIGURE 99: PILOT-PLANT FLOW SHEET FOR SINGLE-CELL PROTEIN PRODUCTION FROM BAGASSE

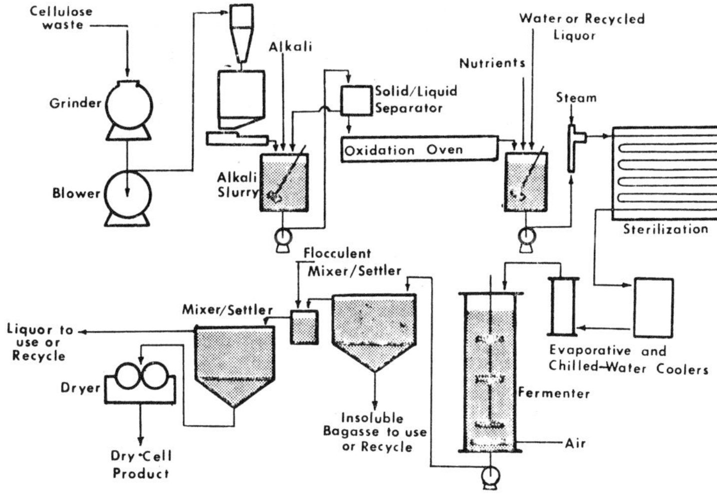

Source: EPA Report SW-24c

Recovery of Fibers from Municipal Waste: A process developed by *P.G. Marsh; U.S. Patent 3,736,223; May 29, 1973; assigned to The Black Clawson Company* is a process for recovering reusable papermaking fiber from municipal refuse. It includes mixing the refuse with liquid in a treatment vessel and reducing the relatively frangible portions of the refuse by subjecting the mixture to hydraulic and mechanical shear while separately removing relatively infrangible portions from the vessel.

The resulting slurry of frangible refuse and water is centrifugally cleaned to remove inorganics such as glass, and the centrifugally cleaned slurry is then screened to separate reusable papermaking fibers from the fine fibrous material and other organics, in excess of 25% of the paper fibers being rejected with the screened materials. The latter, which will usually be rich in synthetic resins, may then be treated under heat and pressure to produce hardboard, or merely dewatered and burned.

The fibers separated from the other portions of the slurry may be classified by means of a fiber selector, subjected to further centrifugal cleaning and dewatered, and the resulting dewatered fiber subjected to additional treatment such as digestion, deinking, bleaching, floatation and washing to upgrade the quality of the fiber for subsequent reuse.

Additionally, the slurry may be treated by refining equipment operating at low consistency and tackle clearances and then subjected to froth flotation and/or further centrifugal cleaning. The inorganics removed by centrifugal cleaning, as well as the relatively infrangible portions of the refuse, may be further classified for reuse, particularly the metal and glass fractions, or they may be disposed of by landfill methods.

A process developed by *R.G. McIntire, D.M. Cold and A.M. Moler; U.S. Patent 3,876,157; April 8, 1975* is one in which fibrous material having utility in the production of numerous and differing products is produced from refuse by a continuous process after separation and removal of undesirable constituents such as dirt, glass, ferrous and nonferrous metals from the refuse. Individual remaining particles are then reduced in size and suspended in a water slurry. The slurry is then subjected to centrifugal acceleration to remove any glass, metallic, or other relatively dense particles that may remain therein.

The resulting slurry is sterilized by continuous cooking to thermally destroy all major pathogens and/or parasitic organisms. The solids content is then increased and additives introduced into and mixed with the solids following which the mixture of solids and additives is rolled to produce granules of a desired size and the resulting product is dried.

The product produced may be useable to produce paper pulp, fibers for board production, fibers for roofing felt, fertilizer base, fibers for hydraseeded land reclamation, and a cattle feed supplement.

PAPER

Paper is one of the major manufactured materials consumed in the U.S. and the largest single component, 35 to 45% by weight, of collected municipal waste. In 1969, 58.5 million tons of paper were consumed. By 1980, this is projected to increase to 85.0 million tons (Figure 100). Paper, paperboard, and construction paper and board are three major categories accounting for 51.5, 40.8, and 7.7%, respectively, of the 1969 paper consumption.

FIGURE 100: PAPER TRENDS: CONSUMPTION, DISPOSAL, AND RECYCLING

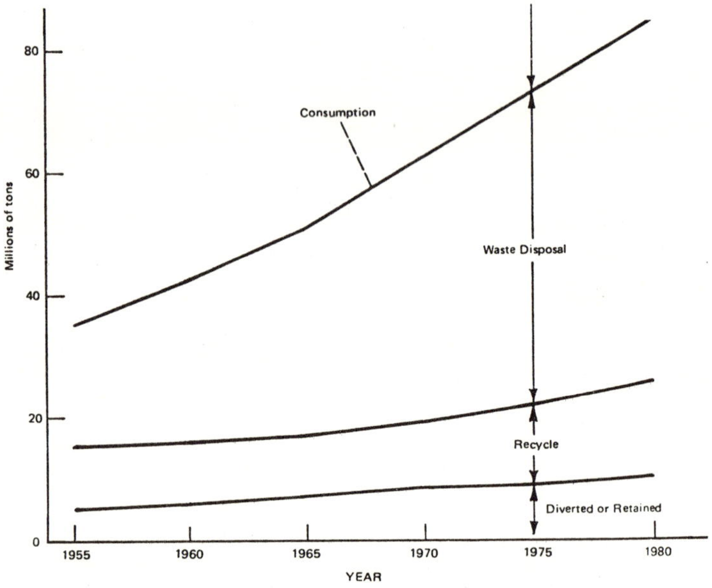

Source: EPA Report SW-118

Only 17.8% (10.4 million tons) of the paper consumed in 1969 was recovered for recycling compared with 23.1% in 1960 and 27.4% in 1950. Most of the remainder was discarded as waste (put in landfills or dumps, incinerated, or littered), and a portion was diverted, obscured, or retained in other products. Trends in disposal and recycling (Figure 100) show that the proportion of paper recycled after consumption has been steadily decreasing. The downward trend in recovery ratio, coupled with an increase in consumption, has resulted in an accelerated rate of wastepaper disposal. Between 1956 and 1967, wastepaper disposal increased nearly 60%, from 22 million tons per year to 35 million tons per year.

Paper recycling and the use of chemicals for the removal from wastepaper of ink, plastic coatings, wax, adhesives, etc. has been reviewed by M. McDonald in *Paper Recycling and the Use of Chemicals,* Park Ridge, N.J., Noyes Data Corp (1971).

Sources of Waste

Wastepaper can be classified into four major grades: mixed, news, corrugated, and high grades, accounting for 27.4, 19.8, 32.6, and 20.2%, respectively, of wastepaper recovered. This wastepaper comes from residential, commercial, and conversion sources, which account for 16.6, 43.6, and 39.8%, respectively, of the paper recovery. Table 21 shows the relationships between the waste grades and sources. The recovery pattern of paper wastes follows directly from the characteristics of each wastepaper source.

Wastepaper generated in conversion operations, where paper and board are made into consumer products, is almost totally recovered. It is easily accessible and generally uncontaminated, and almost half of such waste consists of desirable high grades. This waste is often baled onsite by the converter and never enters the waste stream.

At the opposite extreme, paper waste from residential sources is widely dispersed and highly contaminated with adhesives, coatings, and other materials in the waste stream that are costly and difficult for papermills to remove. Thus, almost none of the mixed paper in residential waste is recovered.

TABLE 21: WASTEPAPER RECOVERY BY GRADE AND SOURCE*
(thousands of tons)

Grade	Residential	Commercial	Converting	Total
Mixed	70	1,860	850	2,780
News	1,610	50	345	2,005
Corrugated	--	2,300	998	3,298
High grades	--	200	1,841	2,041
Total	1,680	4,410	4,034	10,124
Percent of all wastepaper recovery	16.6	43.6	39.8	100.0

*Source: Darnay, A., and W. E. Franklin, Salvage markets for materials in solid wastes. Washington, U.S. Government Printing Office, 1972. p.45-23.

Note: Net exports add another 176,000 tons derived from converting operations.

The only paper recovered in significant quantities from residential waste is old news. Recovered newspapers are separated from other waste by homeowners and are usually collected by charitable organizations. Some municipalities have begun experimenting with collecting newspapers along with the regular refuse collection by placing them in special racks on the collection vehicles. This offers promise for increasing the recovery of old newspapers for residences.

Commercial waste consists largely of business papers, mail, and packaging materials, especially corrugated boxes, and is usually concentrated at commercial/retail centers. It is obviously more accessible and desirable than mixed papers from residential sources but generally less so than conversion wastes. Corrugated boxes comprises about 52% of the commercial wastepaper recovered. They are usually baled or at least kept separate from other waste by the store or office. Significant quantities of mixed papers are also recovered, because they often occur at commercial establishments in high concentrations with few contaminants.

Additional quantities of waste are potentially recoverable from residential and commercial sources. Based on Midwest Research Institute (MRI) estimates, in 1967 there were 35.2 million tons of paper discarded as waste but not recovered: 6.3 million tons were newspapers; 8.6 million tons were corrugated; and 20.3 million tons were all other types. Of course, not all of this waste is potentially recoverable. A portion of the waste is discarded in rural or remote locations and will never be practicably recoverable. A portion is lost in litter or is burned, and a portion would be unusable for technical reasons. The MRI study estimated that the recoverable volume is most likely 10.2 million tons, or 29% of the presently unrecovered paper waste (Table 22). Recovery of this additional amount would have meant an increase in recycling of over 100% in 1967.

Approximately half of the additional recoverable tonnage is made up of newspapers and corrugated board. The recycling of these wastes can be facilitated by creating a demand for materials so that they will be collected before being discarded. Prior separation and separate collection of these wastes holds the possibility of a relatively quick and efficient means of increasing the recycling of substantial quantities of wastes.

The remainder of the tonnage that is potentially recoverable is mixed paper, which would require further processing before recycling. A promising technology for recovery of paper from mixed residential waste has been developed by the Black-Clawson Co. In this process, 400 lb of paper fiber is recovered from each ton of mixed waste input. Ferrous metals and glass are also recovered during processing.

TABLE 22: ADDITIONAL WASTEPAPER RECOVERY POTENTIAL FROM SOLID WASTE (million of tons)*

	Unrecovered and discarded as waste	Most likely recoverable	Recoverable as a % of presently unrecovered paper waste
Newspapers	6.3	2.2	35.0
Corrugated	8.6	3.0	35.0
All other	20.3	5.0	24.6
Total	35.2	10.2	29.0

*Source: Darnay, A., and W. E. Franklin. Salvage markets for materials in solid wastes. Washington, U.S. Government Printing Office, 1972. p.45-24.

Markets

From a waste utilization point of view, the paper industry is made up of an integrated segment that primarily uses woodpulp and an independent segment that primarily uses wastepaper (called paperstock by the industry). Most recycling takes place in the independent sector. Major products made from paperstock, and these are major products of the independent segment, include combination board (e.g., cereal, detergent, and shoe boxes), deinked newspapers, and construction paper.

Combination board production has grown at a substantially slower rate than that of its direct competitor, solid wood pulpboard, which is made almost entirely from virgin pulp. From 1959 to 1969, total paperboard production increased by 65%; solid wood pulpboard, by 112%; and combination board, by only 5%. Herein lies the major reason for the decrease in the wastepaper recyling ratio.

There has been only one major new market for wastepaper in recent years, the deinking of old newspapers to make newsprint. Newspaper deinking is a very promising market for old news, and increased newspaper recycling will be influenced strongly by this market.

Issues and Problems

There are many interrelated factors that have contributed to the decline in the percentage of paper recycled; however, the two primary direct causes are the lack of new markets and the declining share in the combination board market.

It is technically feasible to substitute paperstock for woodpulp in many paper products (Table 23); however, this is not practiced extensively because of economic factors and the high reliance of the dominant integrated industry on virgin pulp. Key items that discourage use of wastepaper are discussed in the following paragraphs.

TABLE 23: TECHNICAL LIMITS FOR RECYCLED MATERIAL FROM PAPER AND PAPERBOARD*

Material	Recycle limits (% paperstock)
Paperboard:	
Unbleached kraft	10–25
Semichemical pulp	100
Bleached kraft	5–15
Combination board	90–100
Paper:	
Newsprint	100
Office, communications	10–80
Publishing, printing, converting	10–80

*Source: Midwest Research Institute. Economic studies in support of policy formation on resource recovery. Unpublished report to the Council on Environmental Quality, 1972.

(1) Logistics: Paper must be collected from diverse sources, transported to a processor, and then transported to a consuming mill. Combination board mills are usually within reasonable distances of wastepaper sources, but the integrated mills are generally located in the South or West, near forests, but far from cities, where waste is generated. Thus, the high costs of collection and transportation work to the detriment of paper cycling.

(2) Contaminants: Contaminants in wastepaper have affected recycling economics unfavorably, and they have also influenced industry orientation. Separation of wastepaper by grade and removal of contaminants are labor intensive and thus costly.

(3) Prices: Waste paper prices have a history of wide fluctuation due to the relative rigidity of supplies and the marginal costs of acquiring new supplies in periods of upswinging demand.

(4) Technology: Improvements in woodpulping technology have enabled the paper industry to tap abundant virgin raw materials at increasingly lower costs.

(5) Integration: Most papermills own their own forests, and most paper equipment installed since 1945 has been woodpulp based and located close to these virgin raw materials. The mills are designed as continuous operations starting with wood, going into pulp, and ending with the finished product. Through this integration, papermills have also been able to exercise control over the supply and price of their raw materials.

(6) Tax Treatments: The cost of virgin woodpulp can be kept down significantly by two tax treatments: a cost depletion allowance (credit against income taxes, based on timber owner's invested capital in a forest and percentage of reserves sold) and a capital gains allowance (profit from sales of timber treated as a capital gain if the timber has been owned for more than 6 months).

Economics

Most of the above problems have a negative effect on the economics of wastepaper use. If one examines the economics of using wastepaper in the manufacture of certain paper and board products, it is obvious that increasing the amount of paperstock in these products increases the cost of manufacturing them.

There are significant changes in environmental impact when wastepaper is substituted for virgin woodpulp in the production of paper products. Table 24 summarizes the environmental impacts resulting from the manufacture of 1,000 tons of pulp from recycled fiber rather than from virgin woodpulp. The use of recycled fiber requires 61% less process water and consumes 70% less energy.

Although deinking and bleaching may be required to upgrade secondary fibers for high quality finished products, recycling still produces environmental benefits in almost every category. Table 25 which compares virgin pulp with recycled, deinked pulp, indicates that 15% less water and 60% less energy are required and that 60% less air pollutants are generated. However, waterborne wastes increase significantly. The increase in solid wastes generated in processing is more than offset by the recovery of paper from municipal solid waste.

The comparative economics of using supplemental wastepaper in existing mills for manufacturing certain paper products are shown in Table 26. These examples are by no means exhaustive of the many paper industry products, but these cases, which represent three products with different economic characteristics, support what would seem to be obvious from the industry's current orientation. The cost "penalty" for increasing the use of paper stock is $2.50 per ton for corrugating medium, $3.75 per ton for linerboard (corrugating medium and linerboard are the materials used to make corrugated boxes), and $20 to $30 per ton for printing/writing paper. The latter cost differential results from the substantial upgrading of wastepaper that would be required to produce a product of the high quality required in the printing/writing grade of paper.

TABLE 24: ENVIRONMENTAL IMPACT COMPARISON FOR 1,000 TONS OF LOW GRADE PAPER*

Environmental effect	Unbleached kraft pulp (virgin)	Repulped wastepaper 100%	Change from recycling (%)+
Virgin material use (oven-dry fiber)	1,000 tons	0	−100
Process water used	24 million gal	10 million gal	−61
Energy consumption	17×10^9 Btu	5×10^9 Btu	−70
Air pollution effluents (transportation, manufacturing, and harvesting)‡	42 tons	11 tons	−73
Waterborne waste discharged, BOD‡	15 tons	9 tons	−44
Waterborne wastes discharged, suspended solids‡	8 tons	6 tons	−25
Process solid waste generated	68 tons	42 tons	−39
Net postconsumer waste generated	850 tons§	−250 tons¶	−129

*Source: Midwest Research Institute. Economic studies in support of policy formation on resource recovery. Unpublished report to the Council on Environmental Quality, 1972.

+Negative number represents a decrease in that category resulting from recycling.

‡(Biological oxygen demand.) Based primarily on surveys conducted in 1968-70.

§This assumes a 15 percent loss of fiber in papermaking and converting operations.

¶This assumes that 1,100 tons of wastepaper are needed to produce 1,000 tons of pulp. Therefore, 850 - 1,100 = −250 represents the net reduction of postconsumer waste.

TABLE 25: ENVIRONMENTAL IMPACTS RESULTING FROM THE MANUFACTURE OF 1,000 TONS OF BLEACHED VIRGIN KRAFT PULP AND EQUIVALENT MANUFACTURE FROM DEINKED AND BLEACHED WASTEPAPER*

Environmental effect	Virgin fiber pulp	Deinked pulp	Change from recycling (%)†
Virgin material use (oven-dry fiber)	1,100 tons	0	−100
Process water used	47 million gal	40 million gal	−15
Energy consumption	23×10^9 Btu	9×10^9 Btu	−60
Air pollution effluents (transportation, manufacturing, and harvesting)‡	49 tons	20 tons	−60
Waterborne waste discharged, BOD‡	23 tons	20 tons	−13
Waterborne wastes discharged, suspended solids	24 tons	77 tons	222
Process solid waste generated	112 tons	224 tons	100
Net postconsumer waste disposal	850 tons§	−550 tons¶	−165

*Source: Midwest Research Institute. Economic studies in support of policy formation on resource recovery. Unpublished report to the Council on Environmental Quality, 1972.

†Negative number represents a decrease in that category resulting from recycling.

‡Based on surveys conducted in 1968-70.

§This assumes a 15 percent loss of fiber in papermaking and converting operations.

¶This assumes that 1,400 tons of wastepaper are needed to produce 1,000 tons of pulp. Therefore, 850 - 1,400 = −550 represents the net reduction of postconsumer solid waste.

TABLE 26: COMPARATIVE ECONOMICS OF PAPER MANUFACTURED FROM RECYCLED AND VIRGIN MATERIALS*

Product	Linerboard	Corrugating medium	Printing/ writing paper	Newsprint
Baseline case (recycled fiber content) (%)	0	15	0	0
Baseline average operating cost ($/ton)	78.50	79.50	80–120	125
Supplemental fiber use (recycled fiber content) (%)	25	40	100	100
Operating cost with increased use of recycled fiber ($/ton)	82.25	82.00	100-150	98
Net cost of increased recycled fiber usage ($/ton)	3.75	2.50	20-30	27

*Source: Midwest Research Institute. Economic studies in support of policy formation on resource recovery. Unpublished report to the Council on Environmental Quality, 1972.

The cost of newsprint, manufacture, however, is lowered by using 100% recycled fiber (deinked newsprint). This has been the only major new market for wastepaper in recent years.

The economics of constructing new mills based on either virgin or secondary fibers also show why the industry has preferred to build plants utilizing virgin fiber. An analysis of folding boxboard (combination board made from secondary fiber versus solid wood pulpboard made from virgin pulp) found that the return on investment for the virgin-based plant was 8.1% whereas that for a plant based on wastepaper (combination board) was only 4.5%. Under such circumstances, construction of new combination board mills is highly unlikely.

Sources and Uses of Recycled Paper

Paper recycling involves the collection of discarded paper and its reuse as a fiber source by the paper industry. Roughly 40% of the paper recycled today is referred to as "converting" waste and is generated in industrial operations where paper and paperboard are fabricated into products. The remaining 60% of the paper recycled comes from discarded postconsumer waste. Corrugated boxes, mixed office and high-grade papers, and newspapers account for roughly 35%, 35%, and 30%, respectively, of recovered postconsumer waste. A large amount of wastepaper is not recycled and contributes to the nation's solid waste problem.

Acquisition of Wastepaper: Postconsumer wastepaper is recovered from solid waste in two ways: salvage industry collections of old corrugated boxes and office papers from industrial and commercial establishments and municipal or private collection of old newspapers from residences. Collection of this source segregated waste provides a clean, usable material for the paper industry. Commercial and industrial establishments, which would otherwise have to pay for wastepaper removal and disposal, generally are willing to separate paper from

TABLE 27: WASTEPAPER UTILIZATION IN PAPER AND PAPERBOARD MANUFACTURE, 1970

Type of Paper	Total Paper Production	Total Wastepaper Consumption	Type of Waste Consumed: Mixed Paper	Newspaper	Corrugated Paper	Pulp Substitutes and High-Grade Deinked Paper
Total for all grades and molded pulp (10^3 tons)	53,329	12,021	2,639	2,235	4,080	3,067
Total paper (10^3 tons):	23,409	2,228	33	455	108	1,632
Newsprint	3,345	371	-	371	-	-
Printing, writing, and related	11,023	736	-	–	-	736
Tissue	3,595	971	7	76	69	819
Other	5,446	150	26	8	39	77
Total paperboard (10^3 tons):	25,465	8,330	1.766	1,473	3,779	1,312
Unbleached kraft and solid bleached	15,036	285	48	8	162	67
Semichemical	3,460	754	42	28	622	62
Combination	6,969	7,291	1,676	1,437	2,995	1,183
Construction paper and board, molded pulp, and other (10^3 tons)	4,455	1,463	840	307	193	123
Distribution (percent)	-	100.0	22.0	18.6	33.9	25.5

Source: Paper, paperboard, and woodpulp capacity, 1970-1973. Washington, American Paper Institute, 1971.

other materials in exchange for free collection by the secondary materials industry. Separation of newspapers by homeowners is becoming more widely accepted, and there are more than 90 cities with active programs. For newspaper collection, cooperative action by city governments, residents, and paper dealers is required. Cities generally have been able to achieve separate collection without adding manpower or equipment to collection services by using existing men and trucks more efficiently.

Another potential method of paper recovery is by mechanical processing of mixed municipal waste in large recovery plants. Very little paper recovery presently takes place using such systems. Both wet and dry separation techniques are under development. The Hydrasposal/Fibreclaim system, a wet process designed by the Black-Clawson Company, is being demonstrated in Franklin, Ohio, by EPA. An experimental air separation system is undergoing testing by the Forest Products Laboratory.

Recovered paper is consumed by various segments of the paper industry for use in manufacturing paper and paperboard products. Table 27 contains a breakdown of wastepaper consumption by various segments of the paper industry.

In general, wastepaper competes with virgin fiber at the final product stage, products made from wastepaper compete with similar products made from virgin fiber. Some substitution of wastepapers occurs in mills that use primarily virgin wood pulp and vice versa. In general, however, mills built to use virgin fiber consume virgin fiber for the bulk of their outputs, and secondary mills consume wastepaper.

Status and Trends of Paper Recycling

Historical data show that domestic wastepaper recycling as a percent of consumption has declined steadily for a number of years, from 27.4% in 1950 to 17.7% in 1972. A recent study conducted for the American Paper Institute predicted that in the absence of Federal policy intervention or significant changes in consumer demand for products made from wastepaper, industry fiber input from wastepaper would drop to even lower levels than that of 1972.

However, it appears that because of a combination of economic factors, there was an increase in wastepaper demand in 1973. These factors include high demand for paper products and associated full capacity production by the paper industry; weather conditions preventing normal logging operations; a rapid increase in foreign demand for wastepaper exports from the United States, related in turn to devaluation of the U.S. dollar and foreign fiber shortages; and increasing consumer pressure for recycling, reflected in part by the wastepaper purchase specifications of the General Services Administration.

A major unknown at this point is whether the current high level of demand for wastepaper will continue. If paper production, which has increased at an annual rate of about 6% for the past year, returns to its historical 3 to 4% growth, pressures for increased use of wastepaper may decrease. Foreign wastepaper demand is likely to continue at relatively high levels into the future but may not grow as rapidly as in the past year.

Exports account for around 6% of wastepaper demand, and high levels of export can have important impacts on regional wastepaper supplies, especially on the East and West Coasts. Purchase specifications such as those of the General Services Administration, which require inclusion of postconsumer wastepaper in products, could have significant long-term impacts if they are widely adopted by cities, states, and industry. General Services Administration purchases of paper products (mostly packaging) account for less than 2% of total U.S. paper consumption and quantitatively are not significant in themselves. However, they could have a significant indirect impact on wastepaper consumption because General Services Administration specifications are widely used by others.

In the long term, demand for wastepaper will depend to a large extent on the price and availability of virgin wood pulp. Timber growth currently exceeds removal on a national scale; in 1970, net growth was 18.6 billion cubic feet against removals of 14.0 billion cubic feet. However, softwood timber removals, preferred for most uses of timber as well as wood pulp, were only 1 billion cubic feet less than net annual growth (9.62 billion and 10.67 billion cubic feet, respectively). Estimates by the U.S. Forest Service and others indicate that in the future the gap between timber growth and removals is likely to narrow, increasing the pressure on timber supplies and price.

New trends in utilization of timber will also influence pulpwood availability. The use of logs with smaller diameters to make saw timber for lumber and plywood production decreases the availability of wood for pulping (small diameter logs were formerly used exclusively for pulpwood). There are also indications that labor shortages and land protection regulations may increase the cost of timber harvesting and cause virgin fiber prices to rise.

Not all of these developments will translate directly into increased wastepaper demand. Utilization of residues from logging and lumber manufacturing is expected to increase. In 1970, almost one-third of the 8.9 billion cubic feet of residues generated were utilized in making pulp and related products. Residue use is expected to reach 45% of available tonnage by 1985.

The net impact of these trends is difficult to predict. However, even if there is no shortage of pulpwood supply in the near future, the cost of pulp is likely to increase. It is likely that new trends on the whole will make wastepaper a more attractive raw material to the paper industry, and the long-term decline in recycling relative to consumption might be arrested. Some industry observers feel that a gradual increase in wastepaper recycling is possible but none of the indications points to an increase in recycling to anywhere near the maximum potential level discussed in the following sections.

Paper Recycling Potential

In 1970, wastepaper recycling of old newspaper corrugated paper, and office paper was 7.4 million tons. (See Table 28). An estimated 32 million tons of these materials were discarded into the solid waste stream in that year. As indicated in Table 28 an additional 5 to 12 million tons could have been collected in SMSA's (Standard Metropolitan Statistical Area) in 1970 and would have increased the recycling of these materials by 65 to 160%.

An estimated additional 15 million tons of other paper, primarily packaging materials, were also discarded into the waste stream in 1970. These materials are difficult to collect for recycling into other paper products and are generally discarded in combination with other materials in the mixed municipal waste stream. As was indicated in previous sections, conversion to energy is currently the most economical way of recovering value from discarded paper that cannot be separately collected for recycling.

TABLE 28: WASTEPAPER AVAILABILITY (RECOVERABLE GRADES), 1970

	Paper (10^6 tons)						
					Recoverable		
Type of paper	Consumed	Discarded to waste stream	Recovered	Generated in SMSA's	Maximum (75 percent of that generated in SMSA's)	Minimum (50 percent of that generated in SMSA's)	Additional increment recoverable
Newspaper	9.8	9.7	2.2 (22.4 percent)	7.4	5.5	3.7	1.5-3.3
Corrugated	13.3	13.2	2.6 (20.0 percent)	10.0	7.5	5.0	2.4-4.9
Mixed and high grade (primary office papers)	11.1	9.1	2.6 (23.6 percent)	8.3	6.2	4.2	1.6-3.6
Total	34.2	32.0	7.4	25.7	19.2	12.9	5.5-11.8

Source: Franklin, W.E. Paper recycling; the art of the possible. Washington, American Paper Institute, 1973.

Barriers to Increased Paper Recycling

Increased wastepaper recycling is inhibited by a number of factors including consumer preference, paper manufacturing economics, technology, and raw material availability. Supply of wastepaper may be an important constraint in the short term, but as was indicated previously, there are sufficient quantities of wastepaper available in the waste stream to increase consumption significantly. The most important barrier in the long run is the uncertain demand for wastepaper, which is related to the economics of paper and paperboard production.

Wastepaper Demand: Some of the more important reasons why wastepaper demand has consistently fallen short of available paper in waste are as follows:

(1) Customers have shown a preference for products made from virgin fibers, usually because of differences in appearance and quality; this has resulted in loss of a market share by products made from wastepaper and has necessitated sale of these products at a discount. Today the market image of wastepaper is changing, and public preference for "ecology paper" is a factor in the current supply shortage, especially for wastepaper grades acceptable in writing and printing stocks.

(2) The paper industry is vertically integrated into timber production and owns and manages forest resources to provide long-term stability and certainty of raw material supply. In comparison, a high degree of risk is associated with wastepaper supply. A company can seldom ensure ownership of wastepaper resources as it can of forest stands.

(3) Large, concentrated sources of timber have encouraged large production complexes with economies of scale located near forests, making woodpulp-based mills favored over smaller, less economical wastepaper-based mills.

(4) Federal tax policy (in particular, the capital gains treatment of timber) has encouraged the development of long-range timber supplies and the technology to use these supplies. There is no similar policy to encourage development of wastepaper supplies.

Most of these factors add up to economics that favor use of virgin resources. Estimates have been made of the comparative economies of wood pulp and wastepaper use, based on average cost and price conditions in 1971. These estimates, presented in Table 29 are useful as general indicators of the comparative economics when conditions exist that are similar to those assumed in the analysis. The condition of the paper industry deviates from these estimates: a high level of demand exists for both wood pulp and wastepaper, and the high market prices for paper products alter this relationship. Although detailed cost comparisons under present market conditions are not available, it is believed that the cost penalty of using wastepaper rather than wood pulp has decreased.

TABLE 29: COST BARRIERS TO INCREASED WASTEPAPER RECYCLING*

Type of paper	Mill status	Wastepaper composition (percent)		Cost differential in favor of virgin material† (dollars/ton of product)		
		"Virgin" case	Increased recycling case	Production (approximate)	Product value‡	Total
Linerboard	Old	0	25	2.50-6.75	0	2.50-6.75
	New	0	25	2.50	0	2.50
Corrugating medium	Old	0	100	(2.00)-1.00	4.50	2.50-5.50
	New	15	35	(1.00)	5.50	4.50
Folding boxboard	Old	0	100	3.00-22.00	9.00	12.00-31.00
Printing and writing	Old	0	10	(8.00)-15.00	0	(8.00)-15.00
Newsprint	Old	0	10	(20.00)-(30.00)	10.00	(10.00)-(20.00)

*Source: Franklin, W. E. Paper recycling; the art of the possible. Washington, American Paper Institute, 1973.

†Parentheses indicate negative numbers.

‡Approximate amount by which the market value of the product of virgin material exceeds that of the product of wastepaper.

Wastepaper Supply: An established supply system for wastepaper exists, the wastepaper segment of the secondary material industry, which supplies the major portion of the wastepaper now recycled. The salvage industry collects old corrugated and office papers directly from commercial establishments and buys newspapers from volunteer groups, schools, and cities that collect them.

Recently, the secondary material industry has experienced difficulties in obtaining wastepaper supplies to meet increased demand, and wastepaper prices have risen rapidly. Although it is believed that this is only a short-term situation, it appears that there are significant institutional barriers that impede the rapid increase of supplies. In addition, there is a lack of experience on the part of many municipalities in designing and implementing newspaper collection systems. (This inexperience sometimes includes the function of drafting appropriate ordinances and contracts with buyers.) Many municipal officials are wary of instituting ordinances requiring residents to separate newspapers from other waste and fear that demand will dry up, leaving them with stacks of bundled newspapers and a disappointed, irate public.

Possible increases in collection costs also deter initiation of newspaper collection programs. However, several case studies conducted by EPA indicate that by increasing the efficiency of existing municipal collection systems (e.g., use of idle equipment and full employment of underutilized workmen), a city can generally collect newspaper with little or no additional expenditure. The ultimate economies depend on the type of collection system utilized, the community's disposal cost, and the price received for the wastepaper.

Conclusions and Recommendations

Paper recovery using separate collection techniques is economically and technically feasible. Recovery of paper by mechanical separation has been demonstrated and is economically attractive when alternative disposal costs are high.

The United States is in a situation of fiber shortage; the paper industry is operating at capacity and straining; wastepaper and pulp prices are up; foreign demand is up; and industry speaks of supply shortages.

Given this situation, and the considerable uncertainties associated with its probable duration, only a limited number of conclusions can be drawn with confidence. It appears clear that for the immediate future, activities that increase the supply of wastepaper from post-consumer sources are desirable. The conclusions of EPA's studies are that wastepaper supplies can be obtained to meet significant increases in wastepaper demand, but such supply increases will call for innovative action on the part of secondary material dealers, communities, and paper users alike.

Federal actions in this area, in the form of technical assistance services and planning assistance, appear to be appropriate at this time. EPA has developed extensive knowledge and data in this area as a consequence of detailed contract investigations, direct technical assistance work with communities, and interaction with the principal wastepaper consuming corporations. In the coming years, EPA will undertake a technical assistance program to aid both communities and industry in developing and marketing wastepaper and supplies.

Proprietary Processes

Paper from Corrugated Wastepaper: A process developed by *T.L. Murphy, Jr.; U.S. Patent 3,440,134; April 22, 1969; assigned to M. T. Reed Construction Co.* provides a method of utilizing waste corrugated cardboard and paper in the production of acceptable stock for use in conventional paper making. The method comprises the steps of comminuting waste corrugated cardboard and paper, forming a slurry of the waste, adjusting the solids content of the waste in slurry form to approximately 20%, digesting the comminuted waste slurry with caustic soda in an amount consisting of 10% by weight of the waste, heating the slurry with direct steam to approximately 160°C for a period of approximately

3 hours, then pulping the digested slurry at a temperature of approximately 35°C for between 3 to 5 minutes, increasing the solids content of the resulting pulp to approximately 30%, and bleaching the pulp as required to achieve a desired brightness thereof.

Figure 101 shows an acceptable form of apparatus for use in the process. The following is an example of operation of the process, described with reference to that figure.

One hundred pounds of corrugated waste paper, after having been suitably hand sorted; for instance, is charged into a conventional paper shredder where it is chopped into approximately three inch squares. From there the shredded material is conveyed to a conventional macerator into which 500 pounds of water is added while the comminution is being undertaken. The comminuted material in slurry form is transferred through a conventional screw conveyor operating in a trough as shown diagrammatically at **10**, there being plurality of spray heads **11** disposed above same for the introduction of a 50% solution of sodium hydroxide so that the caustic is mixed as the slurry is propelled towards a digestor **12**.

Sufficient caustic is sprayed over the slurry so that the total quantity of caustic results in a level of approximately 10% by weight of the waste material on a dry basis, namely, approximately 10 pounds of sodium hydroxide to the 100 pounds of waste starting material. The digestor **12** is of a suitable size to accommodate the mixture, inasmuch as the operation of the process is carried out in a batch manner in this specific example but need not be provided suitable equipment is utilized for continuous operation. Steam is introduced through a suitable conduit **13** which opens directly into near the central portion of the digestor **12**.

FIGURE 101: APPARATUS FOR REPULPING CORRUGATED WASTE PAPER

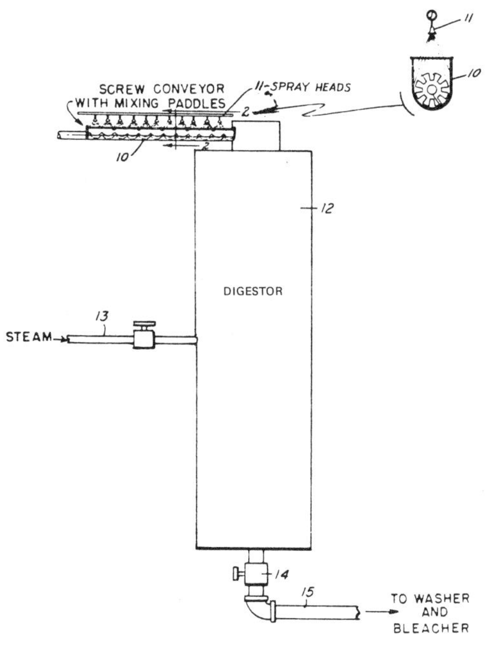

Source: T.L. Murphy, Jr.; U.S. Patent 3,440,134; April 22, 1969

In order to provide for suitable mixing, conventional agitation means is provided in the digestor; or the digestor may be of the tilting variety. The steam is introduced under high pressure so that the contents in the digestor are increased to 160°C and are held at this level for approximately three hours. Approximately 8.75% caustic was consumed during this three-hour digestion step. At the end of the digestion period a blow valve **14** is opened so that the contents in the digestor are discharged through conduit **15** for further treatment.

This further treatment takes the form of conventional pulping with the introduction of additional water so that the resultant slurry has a solids content of 3 to 4% achieved for the most part by the condensation of the steam during the three-hour digestion step.

Due to the vigorousness of the caustic digestion period it is found that the pulping time which is carried out at 35°C is as little as five minutes in which to achieve defibration of the solids material in the slurry. Thereafter, the pulp is centrifuged to remove water to thicken the slurry to 30% solids content. The yield of the resulting pulp is approximately 86% based upon the waste materials introduced. The potassium permanganate number is 14.2/25 ml.

Based upon the potassium permanganate number the pulp is bleached in a conventional three-stage method of chlorine concentration extraction and hypochlorite to achieve a brightness of approximately 80 points. A paper made by conventional methods from the pulp from the process is an acceptable material for many purposes, so that the corrugated wastepaper and cardboard may be utilized for other purposes than reproduction as corrugated material.

Paper from Printed Waste by Deinking: A process developed by *R.H. Illingworth; U.S. Patent 3,354,026; November 21, 1967; assigned to Garden State Paper Co., Inc.* is based on the discovery that waste printed paper may be deinked by utilizing, as the active deinking agent, a product formed by sulfating the fatty acyl derivative of 4-amino-2-butanol. The product thus formed has the general formula

$$R\text{—}CONH\text{—}CH_2\text{—}CH_2\text{—}\underset{\displaystyle CH_3}{\underset{|}{CH}}\text{—}OSO_3Na$$

where R is a member of a long chain alkyl group having 9 to 19 carbon atoms. The detergent may be used alone or in combination with such materials as alkali phosphates, polyphosphates, hydroxides, sulfonates, and the like, to enhance the detersive action.

In carrying out the deinking process, water is introduced to a reactor or hydropulper which is equipped with means for converting bulk cellulosic waste to pulp. The deinking agent described hereinabove is mixed with the water to form a detergent solution. The quantity of active deinking agent employed is between about 0.1 and 3.0%, based upon the dry weight to the cellulosic stock to be deinked.

Temperatures of between about 100° to 165°F are entirely satisfactory. If desired, the water may be softened prior to making up the deinking solution. The deinking solution should be alkaline. While any alkali or alkaline earth hydroxide or salt may be used, for this purpose sodium hydroxide, potassium hydroxide and soda ash are preferred. Sufficient alkali should be added to the deinking solution to raise the pH of the solution into the range of between about 9 and 10.5.

The printed paper, scrap or junk is next introduced to the solution. It may be desirable to shred the stock before subjecting it to treatment. The paper stock to be deinked is preferably added to the solution in its naturally dry condition; that is, without first subjecting it to water or moisture treatment.

The amount of paper charged, by weight of the aqueous treating solution, should generally be below about 10%. It has been found that good results are obtained when the deinking solution contains less than about 6.0% by weight of the scrap paper and this amount is preferred. The scrap in the treating solution is retained in the reactor until substantial defiberization takes place. Following this treatment, the defibered material is dropped to a chest or other suitable reservoir, after which it is diluted with water which is maintained at normal room temperature, until a solid content of between about 0.5 and 2.0% is obtained.

The resulting pulp is then separated from the solution, as by settling and decantation, filtration and similar techniques well known in the art. The recovered pulp is then formed into a web in a well-known manner. The recovered pulp may now be blended with fresh virgin sulfate or sulfite pulp if desired.

Another process developed by *R.H. Illingworth; U.S. Patent 3,377,234; April 9, 1968; assigned to Garden State Paper Co., Inc.* is a process for deinking waste printed paper which comprises pulping printed cellulosic material in an aqueous solution comprising, as the active deinking agent, an admixture of 20 to 30 weight parts of sodium lauryl sulfate and 1 to 4 weight parts of sodium dodecyl benzene sulfonate, the amount of printed cellulosic material being below about 10% by weight of the aqueous solution, and the amount of the active deinking agent being between about 0.3 and 3%, based upon the weight of the cellulosic material, continuing the pulping until the ink separates from the fibers, and recovering the resulting pulp from the aqueous solution.

Still another process developed by *R.H. Illingworth; U.S. Patent 3,392,083; July 9, 1968; assigned to Garden State Paper Company, Inc.* involves deinking waste printed paper by pulping the waste paper with a polyol such as a glycol and a nonionic surface active agent containing a polyoxyalkylene chain of at least two alkenoxy groups and derived from alkyl phenolic compounds in which the total number of alkyl carbon atoms is between 4 and 24.

Figure 102 is a block flow diagram of this process. As indicated in the drawing, water from an appropriate source is charged to a suitable reactor or pulper. The reactor or pulper used in the process is equipped with a stirrer or agitator of any appropriate shape which will agitate and defiber the cellulosic material. If desired, baffle plates may be attached to the interior of the reactor to assist the agitation action.

FIGURE 102: PROCESS FOR DEINKING WASTEPAPER WITH NONIONIC DETERGENT AND A POLYOL

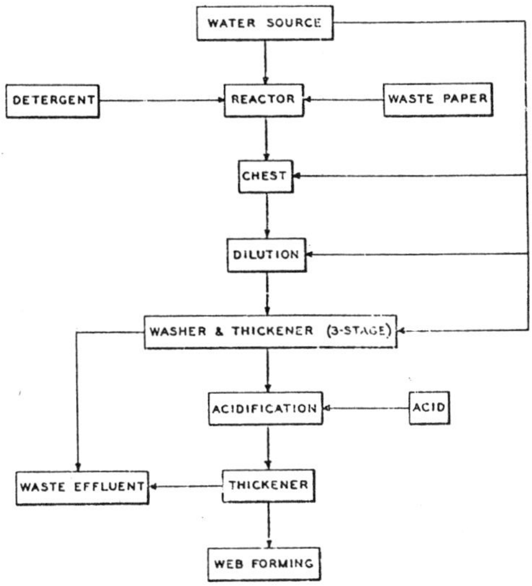

Source: R.H. Illingworth; U.S. Patent 3,392,083; July 9, 1968

After the water has been charged to the reactor, the combined deinking agents described are added and agitation continued until the agents are dispersed and/or dissolved. Wastepaper, junk, or other printed cellulosic material are then added to the reactor.

After a suitable period of time in the reactor, the mixture is dropped to a storage chest which is preferably equipped with a suitable agitator. If desired, water may be charged to the chest to reduce the solid content of the mixture therein. The mixture from the chest is then diluted to the solid content indicated hereinabove, and washed and thickened in a well known manner as, for example, by a Lancaster three-stage washer and thickener equipped with a 45 mesh wire screen. The pulp may be thickened to about a 5% solid consistency, or between about a 3 to 8% solid consistency in this manner.

Cocurrent or countercurrent washing, alone or in combination, may be used. The resulting pulp is then acidified to the pH desired by addition thereto of a dilute solution of a suitable acid, as for example, alum, sulfuric acid, SO_2, and so forth. The resulting pulp may be finally thickened and formed into a web.

A process developed by *T.S. Mestetsky and B.G. Webster; U.S. Patent 3,846,227; Nov. 5, 1974; assigned to GAF Corporation* is one in which waste paper is effectively deinked by contact with an aqueous treating composition containing an N-alkyllactam solvent at a temperature of from about 110° to about 210°F. The N-alkyllactam solvent is not consumed in the treating process and can be effectively separated from the waste material removed from the paper being treated for reuse in the contacting of additional quantities of waste paper.

The use of N-cyclohexyl pyrrolidone, N-t-butyl pyrrolidone and N-t-octyl pyrrolidone is particularly advantageous in that simplified regeneration techniques of appreciable technical and economic significance can be employed therewith. The aqueous treating composition will generally contain from about 0.5 to about 200 parts by weight of water, from about 0.05 to about 1 part by weight of a water-soluble, nonvolatile, inorganic ionic compound, e.g., NaOH, and from about 1 to about 99 parts by weight of the N-alkyllactam solvent. The deinking is carried out at a temperature of from about 110° to about 210°F. Difficult-to-deink grade material, such as waste paper containing thermoplastic and elastomeric coatings, laminates and adhesives, can be effectively and conveniently deinked by the N-alkyllactam solvent composition.

Figure 103 is a flow diagram of this process. In the drawing, waste paper to be deinked is passed to conventional pulper **10** through line **12** for contact therein with an aqueous treating composition containing an N-alkyllactam solvent at a composition cooking temperature of from about 110° to about 210°F. Treated pulp and cooking liquor are passed from pulper **10** through line **14** to press **16** wherein a major portion of the aqueous treating composition is squeezed from the pulp for recycle through line **18** to pulper **10** for contact with additional quantities of waste paper to be treated. The pressed pulp leaving press **16** through line **20** passes to washing zone **22** for countercurrent contact therein with wash water to remove N-alkyllactam solvent, together with waste material removed from the paper being treated, from the treated pulp, which is withdrawn from washing zone **22** through line **24**.

The wash water solution removed from washing zone **22** through line **26** thus contains lactam solvent advantageously to be recovered and reused in the deinking process. The wash water can be concentrated, as in vacuum evaporator **28**, to produce a hot, concentrated wash solution that is passed through line **30** to quiescent zone **32** in which the highly advantageous phase separation, achievable when employing the preferred solvents specified, is accomplished. If necessary or desired, the hot wash water concentrate from evaporator **28** can be further heated to reach the cloud point temperature under the particular circumstances pertaining to a given concentration of the particular preferred N-alkyllactam solvent employed and a given amount of particular waste materials associated therewith. Water removed from evaporator **28** in the concentration of the wash solution can be condensed and recycled through line **34** to washing zone **22** for use in the washing of additional quantities of pulp.

FIGURE 103: PROCESS FOR DEINKING WASTE PAPER WITH AN AQUEOUS N-ALKYLLACTAM

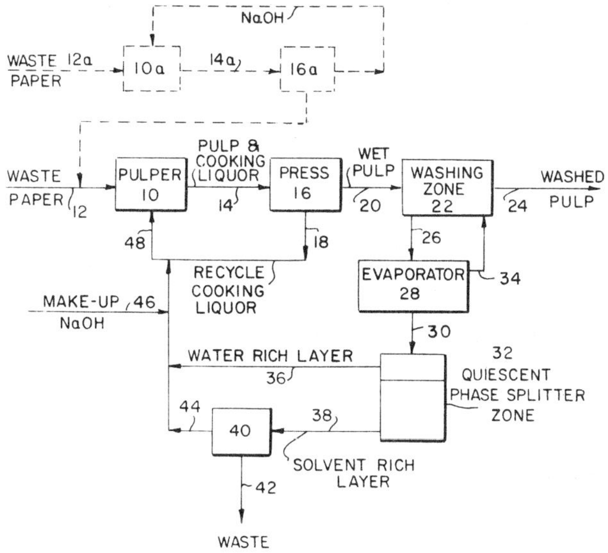

Source: T.S. Mestetsky and B.G. Webster; U.S. Patent 3,846,227; November 5, 1974

The upper, water-rich layer in zone **32**, having a small amount of solvent and some accompanying waste material present herein, can be withdrawn through line **36** for recycle to pulper **10**. The bottom, solvent-rich layer in zone **32** is withdrawn through line **38** for solvent purification treatment by the removal of waste materials therefrom in solvent purification zone **40**. Separated waste materials are withdrawn from the zone **40** through line **42** for discharge. Such wastes constitute the accumulation of inks, coatings and the like removed from the treated paper, including the synthetic, thermoplastic and elastomeric substances used in the inks, coatings, laminates and adhesives associated with difficult-to-deink grade material treated.

A purified solvent/water stream is withdrawn from the zone **40** through line **44** for recycle to pulper **10**. As shown in the figure, therefore, the purified solvent stream in line **44**, the water-rich stream in line **36**, and makeup caustic or other inorganic admixture agent in line **46** can all be fed, at appropriate rates, to pulper **10**, together with recycle cooking liquor in line **18**, as is illustrated diagrammatically in the drawing by common recycle line **48** to pulper **10**.

As is shown also in the figure, the waste paper to be deinked can also be passed through line **12a** to pulping zone **10a** for pulping with an aqueous alkaline solution, e.g., a caustic solution, generally at room temperature, prior to deinking with the aqueous treating composition at the more elevated temperatures herein disclosed.

The thus pretreated pulp, in this embodiment, is advantageously passed through line **14a** to press **16a** for removal of a portion of the aqueous pretreatment solution prior to passage, as through line **14b** to contacting zone **10** for cooking with the aqueous treating composition containing N-alkyllactam solvent. The alkaline pretreatment solution removed from the pulp in the press **16a** can be recycled to pulping zone **10a** for contact with additional quantities of waste paper. It is within the scope of the process to employ any one of several convenient techniques for the removal of waste materials from the solvent-rich stream passed to solvent purification zone **40**. For example, it is found that filtration of the solvent-rich stream results in the removal of sufficient waste material to provide a purified solvent-water recycle stream suitable for recycle and reuse in the treatment of additional quantities of waste paper.

The provision of a carbon adsorption bed in the zone **40** can likewise be employed in the purification of the solvent-rich stream withdrawn from quiescent zone **32**. Solvent purification zone **40** can also comprise an evaporation zone from which a purified solvent-water stream can be withdrawn and condensed for the recycle and reuse. The zone **40** may also, if desired, comprise a distillation zone although the simpler, less expensive methods of purification hereinabove indicated are generally preferred from an economic viewpoint. When less preferred lactam solvents, are employed in the process, the concentrated wash water solution will generally be distilled to produce a purified solvent stream for recycle to pulper **10**.

Paper from Waxed Paper Waste: A process developed by *A. von Koeppen and R.T. Carrera; U.S. Patent 3,822,178; July 2, 1974; assigned to Wright Chemical Corporation* involves repulping waxed stock such as waxed paper or paperboard for use in Fourdrinier, or cylinder papermaking processes or wet-lap or screw-press dewatering processes. Waxed stock material, either alone or with nonwaxed stock is charged to a pulper and combined with water and certain water soluble nonionic emulsifiers with mechanical agitation and at a temperature sufficiently high to melt the wax in the waxed stock. A dispersed wax-fiber slurry is thereby formed which can be diluted for further processing in a Fourdrinier or cylinder machine or, the thus formed dispersed wax-fiber slurry may be processed directly in a wet-lap machine.

Mills which manufacture waxed paper necessarily produce substantial quantities of waste waxed paper from trimmings, waxed paper which does not meet test standards or other specifications, and waxed paper which has been damaged or contaminated. Until this time, these mills have not been able satisfactorily to recycle waxed stock because of process and equipment difficulties. As a result, these mills have been confronted with serious disposal problems for this waste waxed paper and similar stock which have generally necessitated the burning thereof, or paying waste dealers to haul it away.

This necessarily involves waste of wax as well as waste of otherwise usable quality paper fiber. Since waste waxed paper which is discarded each month, in substance, is replaced with virgin pulp or high quality pulp substitute, such waxed paper which is wasted in these mills, in addition to presenting disposal problems, represents a waste of an equivalent quantity of relatively expensive virgin pulp, or other high quality expensive pulp substitute.

In available commercial papermaking and board making operations there are at least three kinds of operations which are of major significance; namely, Fourdrinier, cylinder and wet-lap processes. These processes, though similar in principle, entail considerably different operation parameters.

Generally, in each of these processes, a cellulosic fiber slurry having a solids consistency of from approximately 4 to 6% by weight is charged to a pulper such as Hydropulper, Hollander, or the like. In both the Fourdrinier and cylinder processes the resulting slurry is then diluted substantially, usually to provide a solids consistency of approximately 0.5% and the cellulosic fibers are then deposited onto a papermaking machine. In the wet-lap operation cellulosic fibers are only partially processed to an intermediate holding stage in which so-called wet-lap is formed. The wet-lap, having from about 60 to 75% water, for example, may be maintained at that water level for substantial periods of time.

In this process waxed stock either by itself or together with nonwaxed stock is charged to a beater and combined with water and certain water dispersible nonionic emulsifiers with mechanical agitation and at a temperature sufficiently high to melt the wax constituent in the waxed stock, for example, from approximately 150° to 190°F. An emulsified wax-fiber slurry is formed having a solids consistency of from approximately 4 to 6% by weight.

In the Fourdrinier and cylinder processes, this slurry is then further diluted to produce a resultant slurry having a solids consistency of approximately 0.5 to 1% by weight and charged to the particular sheet-forming machine being employed wherein the fibers are formed into a mat-like product and the initial liquid removal performed. Thereafter, suitable drying operations are employed to produce a paper or board product having the desired moisture content.

In the application of this method to a wet-lap process, the emulsified wax-fiber slurry from the beater is usually transferred directly to a wet-lap machine wherein a wet-lap product is produced which, after pressing, characteristically has a moisture content of from 60 to 75% by weight.

While the practice of the process encompasses the use of 100% stock having a wax content up to 30% by weight, it should be noted that normally in mill installations, preferred practice will involve charging the pulper with an admixture of both waxed stock and nonwaxed or virgin stock. In these instances, the stock being fed to the pulper will consist of from 10 to 50% of waxed stock having a wax content of from approximately 15 to 30% by weight, with the remainder of the stock being fed to the pulper being composed of non-waxed or virgin pulp.

Products made in accordance with this process are characterized by the complete absence of glossy specks even though they were formed from 100% high wax content stock. Typically, these products will have a retained wax content of from 2 to 4% by weight, although substantially lower retained wax contents in these products are achievable. Figure 104 is a block flow diagram of this process.

FIGURE 104: PROCESS FOR REPULPING WAXED PAPER USING A WATER SOLUBLE NONIONIC EMULSIFIER

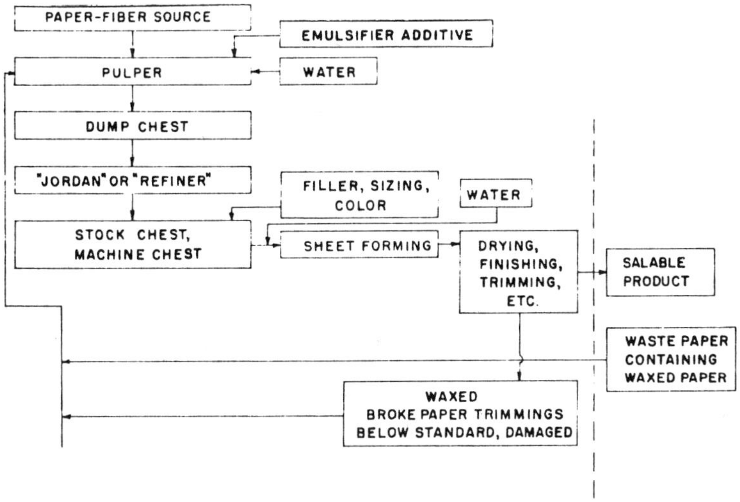

Source: A. von Koeppen and R.T. Carrera; U.S. Patent 3,822,178; July 2, 1974

Paper from Wet Strength Paper Waste: A process developed by *G.V. Ireland, Jr.; U.S. Patent 3,619,347; November 9, 1971; assigned to Salvox Mfg. Company* is one in which wastepaper treated typically with urea resins or the like to impart high wet strength is recovered by treating it in a standard paper pulper in an aqueous medium with a mixture of sulfuric and phosphoric acid and, advantageously, an emulsifying agent.

Paper and Fiberboard from Mixed Waste: A process developed by *S.H.W. Brooks; U.S. Patent 3,741,863; June 26, 1973; assigned to The Rust Engineering Company* is a process for making medium density fiberboard and/or paper from municipal and industrial waste products.

The waste material is first pulverized or shredded into small pieces and then dried to remove excess moisture and to sterilize the material. The dried waste material is passed through one or more separators to remove the glass, metals, and other noncellulosic materials therefrom, leaving the cellulosic materials for further treatment. After further heating the cellulosic materials in the presence of a nonflammable medium, such as steam, to raise its temperature and further sterilize the material, the cellulosic pieces are abraded under a pressure and temperature sufficient to break down the hydrogen bonds and to soften any lignin and other resins present, in what is referred to as a a dry-refining process. This frees the fibers and fiber bundles from one another, and provides relatively long, thin, soft fiber structures.

The fibers and fiber bundles are then intimately mixed with a resin and formed into a mat, after which the mat is compressed under heat and pressure to form a fiberboard. Alternatively, to make paper, the fibers are dispersed in water, formed into a sheet, compressed, and then dried.

A process developed by *F.D. Iannazzi, J.W. Rafferty, D.B. Sparrow and R. Strauss; U.S. Patent 3,833,460; September 3, 1974; assigned to Arthur D. Little, Inc.* involves separating mixtures of waste paper into two components using the differences in strength and/or water-receptivity characteristics of the two components which are typically the liner board and corrugated medium of corrugated paper or a mixture of magazines and newsprint.

In separation, the material of the paper mixture is cut into pieces ranging in size from about one to six inches in either dimension, treated with a debonding agent and then subjected to differential size reduction wherein one of the mixture components is reduced to small pieces, e.g., one quarter inch in either dimension, while the other remains in pieces of essentially the same size. Separation of the two distinctly different sized components is then effected.

A process developed by *P.G. Marsh and D.E. Chupka; U.S. Patent 3,849,245; Nov. 19, 1974; assigned to Black Clawson Fibreclaim, Inc.* is a process for recovering reusable papermaking fibers from municipal refuse containing substantial quantities of garbage and other organic waste. A slurry consisting essentially of cellulose fibers contaminated by grease is subjected to a centrifugal flotation treatment for separating the cellulose fibers from the other solid materials in a conical vessel having a tangential inlet port and outlet ports at its apex and base, the supply flow and pressure to the vessel being maintained sufficiently high and the discharge flows through the outlet ports being controlled to effect discharge of substantial volumetric flows through both outlet ports under pressure drop conditions equal to or higher than those used in conventional centrifugal cleaning operations in a vortex separator of comparable size.

The discharge flow through the apex port contains a large majority of the fibers of good paper making quality, while the discharge flow through the base outlet contains the majority of the reject material, the most significant proportion of which is solvent-soluble grease and/or wax capable of separate recovery therefrom, the remainder being largely plastic and paper fiber fines.

A process developed by *T. Baehr and H. Thumm; U.S. Patent 3,854,667; Dec. 17, 1974;*

assigned to J.M. Voith GmbH, Germany utilizes a receptacle which can accommodate a suspension of waste paper and contaminants having low specific gravity. A rotor is provided on the peripheral wall of the receptacle so as to shred the waste paper in the suspension in order to obtain a pulp. In the region of the rotor the wall is provided with an outlet for the pulp, and in a portion of the wall opposite the rotor there is provided an additional outlet for discharging of matter having low specific gravity. A sleeve is provided on the rotor substantially concentric with the axis of rotation thereof, for directing and centering a vortex core of suspension which is being agitated by the rotor.

A process developed by *C.R. Gleason and G.A. Thomas; U.S. Patent 3,865,684; Feb. 11, 1975; assigned to World-Wide Paper Reclamation, Inc.* involves the reclamation of purified and reusable pulps from various types of waste papers by first shredding the paper into fine pieces, boiling the shredded paper in a chemical solution containing an alkali metal hydroxide, an alkali metal carbonate, an alkali metal bicarbonate and sodium borate, water washing the chemicals from the cooked paper and then bleaching the pulp with a chemical bleach at elevated temperature.

A process developed by *D.E. Chupka; U.S. Patent 3,873,410; March 25, 1975; assigned to The Black Clawson Company* is one in which waste paper-containing materials, e.g., commercial "waste paper" or municipal waste, are treated for recovery of reusable paper therefrom by slushing in a pulper followed by screening of the entire output extracted from the pulper, with or without an intermediate centrifugal cleaning operation, in a screen having fine perforations sized to accept only substantially fully defibered paper.

The reject flow from this screen is maintained at an adequately high rate, e.g., 20 to 80% to assure freedom from blocking of the screen perforations. A deflaker is connected to receive this reject flow, and the output of the deflaker is separated by a tailing screen into accepts which are recycled to the primary screen and rejects which are eliminated from the system.

Major advantages of this method and system include the use of only one piece of apparatus for each of the essential operations of pulping, screening and deflaking, the elimination of low specific gravity reject from the system at a single station, and distribution of power requirements in such manner that the ratio of pulping power to deflaking power is 1.5 or less.

Figure 105 is a flow diagram of the process. The system shown has at its slushing station a pulper comprising a treatment vessel or tub **10** wherein a rotor **11** rotates above an extraction plate **12** provided with perforations of appropriate size, which should be relatively large, namely of the order of 1 to 2 inches in diameter. A conveyor **13** carries the waste paper material to the tub **10**, and it is to be understood that this charge may be selected from its full range of waste paper containing materials from commercial waste paper to municipal solid waste. A junk remover **15** is positioned adjacent the tub and communicates therewith through a conduit **16**, and water is continuously supplied to the tub by a line **18** through the junk remover **15** and its connecting conduit **16** into the tub.

Preferred examples of pulpers suitable for use in this station include those shown in Felton et al U.S. Patent 3,339,851 and Blakley et al U.S. Patent 3,595,488, and details of the junk remover are shown in Baxter U.S. Patent 3,549,092. The essential characteristics of the pulper are that it be capable of subjecting mixed waste paper materials and water to forces of mechanical and hydraulic shear and mechanical impact which will quickly and effectively reduce the solid material in the tub to fragments capable of passage through the perforations in extraction plate **12**.

The pulper **10** operates continuously, with additional waste paper and water being added at rates which will maintain the solids content of the mixture in the desired range, 2 to 8% being practical and 3 to 5% being preferred for practical reasons. A slurry of water and fragmented material is continuously withdrawn from below the extraction plate **12** by a pump **20**, and if this slurry contains a significant amount of high specific gravity

FIGURE 105: APPARATUS FOR REPULPING OF WASTE PAPER

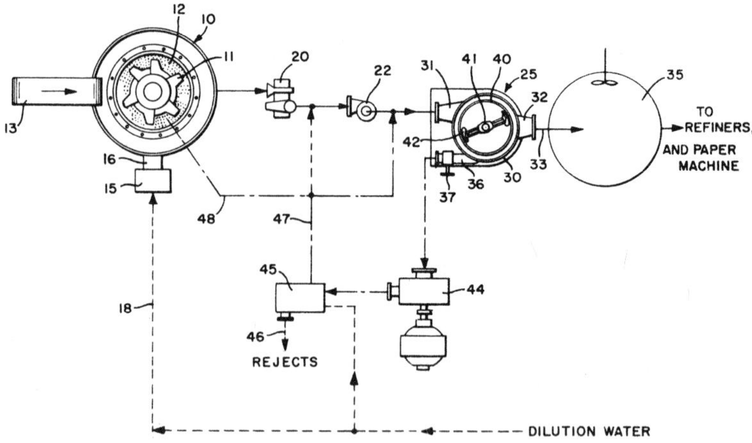

Source: D.E. Chupka; U.S. Patent 3,873,410; March 25, 1975

contaminants such as metal and glass, which can be expected if the charge is municipal waste, it is preferably supplied by pump **20** to a centrifugal cleaner **22** for removing the high specific gravity contaminants to simplify the subsequent screening as well as to prevent possible damage to the screen.

The cleaned slurry from the cleaner 22 will contain a considerable proportion of relatively large undefibered pieces of paper as well as pieces of plastic and the like of comparable size, but when the initial slushing operation is carried out in a pulper of the characteristics noted above, the result will also be that a major proportion of the paper will be substantially completely defibered.

The process therefore provides that rather than a series of coarse and fine screening operations and dilute consistencies, this slurry be subjected immediately, and at its maintained high consistency, to a screening operation which will result in acceptance of at least the majority of the substantially completely defibered paper and the rejection of the undefibered paper fragments, plastic and other pieces of reject material.

The screen **25** is shown as comprising a main housing **30** having a tangential inlet port **31** at the upper end thereof, an outlet port **32** for accepted stock from which a line **33** leads to a storage chest **35**, and a discharge port **36** for reject stock, the port **36** being provided with a control valve **37** for regulating the relative proportions of the flow through the accepts and reject ports. Within the housing **30** is a screen cylinder **40**, and a rotor **41** operates within the cylinder **40** and carries vanes **42** spaced inwardly from the surface of the screen cylinder by a substantial distance in the range of $\frac{3}{16}$ to $\frac{1}{2}$ inch. The perforations in the screen cylinder **40** may comprise slots running axially of the cylinder or cylindrical holes.

The perforations in the screen cylinder **40** should be sized to accept only substantially completely defibered paper and particles of similar sizes, and preferred results have been obtained with a screen constructed as described wherein these perforations are slots of the order of 0.022 inch in width or perforations of the order of 0.062 inch in diameter, i.e., in the range of approximately 0.050 to 0.079 inch. In the operation of such a screen, the rotor **41** is driven at relatively high speed providing vane speeds of the order of 3,000 to 10,000 feet per minute, the higher speeds being best for handling incoming stock of high consistency in the range of 3 to 5%.

Under these conditions, the rotating vanes form a tubular layer of stock in the space between them and the screen surface, and create a substantially tangentially oriented hydraulic shear field in that tubular layer wherein the shear rates are sufficient to cause elongated particles to be aligned generally tangentially of the screen member and thereby to flow past rather than through the perforations. Undefibered pieces of paper will therefore be retained inside the screen member for passage through the outlet port **36**, and the valve **37** is set to maintain a high reject rate, namely in the range of about 20 to 80% depending upon how dirty the stock is, so that there will be minimum tendency for undefibered paper particles to block the screen perforations.

The reject flow from screen **25** will therefore contain the undefibered pieces of paper and the plastic and other reject material too small for acceptance by screen **25**, and this reject flow is conducted to a deflaker **44** of any satisfactory construction such, for example, as a disk refiner. In general, it appears that if the consistency of the stock supplied to the screen **25** is in the range of 3 to 5%, the reject flow will be of slightly high consistency, namely in the range of 4 to 7%, which is quite satisfactory for defibering purposes.

The flow from the deflaker **44** is preferably supplied to a tailing screen **45** of any suitable type for removing undefibered material, which will include at least the major part of the plastic, and also a major portion of the dirt and other reject material. For example, the screen may be a vibrating flat bed screen of the Jonsson type with perforations larger than screen **25**, e.g., ⅛ inch, and the reject therefrom is discharged at **46**. The accepts flow from screen **45** does not require further pulping and can be returned at **47** to the supply line to the screen **25**, either upstream or downstream of the cleaner **22**, but if the screen **45** requires a large volume of dilution water for proper operation, as in the case of a Jonsson screen, at least a substantial portion of the accepts flow should be recycled to the pulper **10**, as indicated at **48**, to maintain the proper water balance in the system. The screened stock from chest **35** is ready for refining or other treatment in preparation for reuse.

A process developed by *R.A. Daane and S.T. Han; U.S. Patent 3,891,497; June 24, 1975; assigned to Beloit Corp.* provides a nonaqueous defiberizing process for waste paper in the presence of steam and a water-immiscible solvent such as tetrachlorethylene. The product is a dry recovered fiber.

Recovery from Polyolefin-Coated Paper Waste: A process developed by *R.B. Goss; U.S. Patent 3,595,741; July 27, 1971; assigned to Riverside Paper Corporation* involves a solvent extraction method for reclaiming stock from waste materials containing natural or synthetic plastic or resin material by extracting the plastic or resin material with a suitable solvent. The miscella is separated from the stock and the solvent is recovered by distillation during which a hydrocarbon carrier is added to the miscella to form a favorable liquid waste residue which is readily recoverable from the distillation apparatus.

Coated and/or impregnated paper, paperboard, and boxboard is used extensively in the manufacture of various packaging materials and containers, such as milk cartons, frozen food and juice containers, and the like. The paper pulp used in the manufacture of coated and impregnated containers for the most part is a relatively high quality pulp, and it would be desirable to reclaim, for repulping the paper stock present in the coated waste paper generated in such manufacturing.

Pulp formed totally or partially from such reclaimed paper stock can be used in the manufacture of high quality paper. However, in order to use the reclaimed paper stock in the manufacture of high quality paper, all coating and impregnating materials must be substantially completely removed from the paper. As little as a small fraction of a percent of residual coating material in the reclaimed paper can be detrimental in the subsequent papermaking process, or may render the resultant paper product unuseable or aesthetically unacceptable.

A major source of waste paper is broke and rejects obtained in the manufacture of coated milk cartons and frozen food and juice cartons, although it is readily apparent that other waste paper products can be utilized. In addition to the coating and/or impregnating material, the waste paper may also include ink applied either to the paper or to the coating.

Previously, wax was used extensively as a coating material for paper products. In recent years, however, improved plastic coatings and coating methods have been developed and plastic coatings, principally polyethylene and polypropylene has substantially replaced wax as a coating material for paper products.

This process may be practiced in either a batch or a continuous manner, both of which are described in the above-cited patent. Figure 106 is a flow diagram of the continuous process scheme.

The fibrous waste material is delivered from a storage bin **41** through one or more pressure locks **43, 45** into the lowermost end of an inclined digester **47**. The waste material is propelled upwardly in the inclined digester **47** by means of a suitable screw conveyor, chain conveyor or the like, not illustrated, and the digester may contain suitable baffles and/or other means for causing mixing and agitation of the waste material during extraction.

Solvent is continuously withdrawn from a solvent storage tank **49**, preheated in a suitable heat exchanger **51**, and introduced into the inclined digester adjacent the uppermost end thereof. It will be seen that the solvent and waste material travel countercurrently within the inclined digester **47**, the waste material being carried upwardly by means of the conveying apparatus, and the solvent percolating downwardly by gravity. Suitable means, not shown, may be utilized for pressurizing the digester in order that the extraction may be carried out at an elevated temperature.

FIGURE 106: SOLVENT EXTRACTION PROCESS FOR RECLAIMING STOCK FROM PLASTIC-COATED PAPER

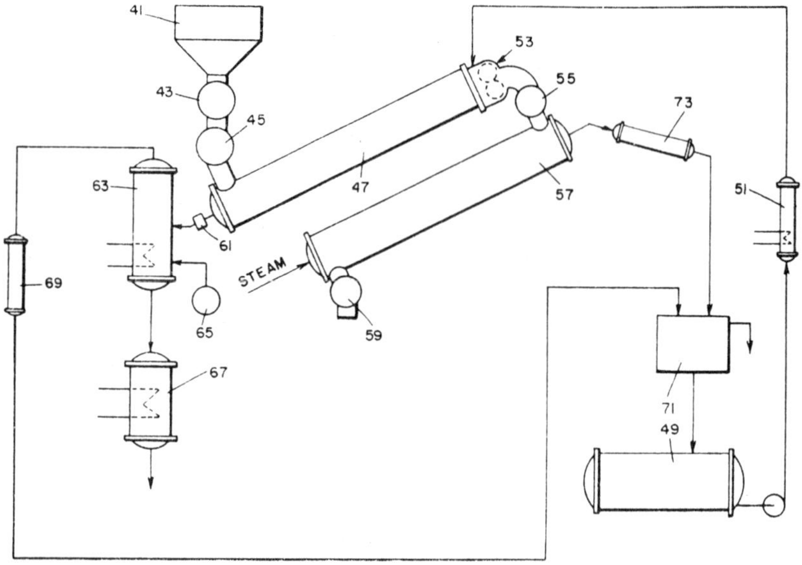

Source: R.B. Goss; U.S. Patent 3,595,741; July 27, 1971

The size and the shape of the inclined digester **47**, and the volume and flow rate of solvent and waste paper passing therethrough may be conveniently adjusted in order to insure that substantially all of the plastic will be extracted in the digester, and that the stock will be substantially free of plastic when it reaches the upper end of the digester.

Residual solvent remaining in the stock may be partially removed by means of compression flights in the screw conveyor or other compressing means such as rollers indicated diagrammatically by numeral **53**, within the upper end of the digester **47**, the solvent squeezed from the stock being returned to the digester. To remove the remaining residual solvent, the stock is removed from the digester **47** through a pressure lock **55** and delivered into an inclined stripping column **57**. The stock is caused to travel downwardly in the still **57** and is exposed to a countercurrent flow of steam for stripping of the remaining solvent from the stock. The solvent free stock is removed from the lower end of the still through pressure lock **59**.

The miscella is withdrawn from the lower end of the inclined digester **47** through a suitable pressure valve **61**, and is introduced into a solvent recovery vacuum still **63**. A hydrocarbon carrier is continuously introduced to the solvent recovery still **63** from a storage tank **65**. The waste residue suspension of hydrocarbon carrier and plastic is withdrawn from the bottom of the still **63** and delivered to a finishing still **67** for removal of entrained solvent. The solvent free waste residue is withdrawn from the finishing still **67** and is disposed of, as by burning.

The vaporized solvent from the stills **63** and **67** is condensed in a condenser **69** and the condensate is delivered to a condensate separator **71** where the water is separated from the solvent by decantation. Similarly, the steam and residual solvent vapors are withdrawn from the upper end of the stripping column **57** and are condensed in a condenser **73** and delivered to the condensate separator **71** for reuse in the process. Purified solvent is returned from the condensate separator **71** to the solvent storage tank **49**.

In an example of the operation of the continuous process for reclaimed paper stock from polyethylene coated waste paper, 5,000 pounds per hour of polyethylene coated broke and rejected cartons, containing 7.6% by weight polyethylene, were continuously introduced into an inclined screw conveyor digester having a length of 16 feet and a diameter of 2 feet. 32,000 pounds per hour of trichloroethylene, preheated to a temperature of 210°F were introduced into the upper end of the inclined digester and the digester maintained at a pressure of 12 psig by means of compressed air, and at a temperature of 210°F by a steam jacket. The flow of trichloroethylene downwardly through the inclined digester was controlled so that the digester was substantially flooded with trichloroethylene.

The average residence of the waste paper stock and trichloroethylene in the digester was 30 minutes. The extracted paper stock was compressed in the uppermost portion of the inclined digester to partially remove solvent, and was then passed through a pressure lock into the upper end of an inclined stripping column having a length of 16 feet and a diameter of 2 feet. The paper stock was conveyed downwardly within the stripping column by means of a screw conveyor and a countercurrent flow of 1,500 pounds per hour of steam was introduced into the lower end of the stripping column for removal of residual solvent from the paper stock. The average residence time of the paper stock within the stripping column was 30 minutes.

The paper stock, freed of residual solvent, was removed from the lowermost end of the stripping column and repulped for use in manufacturing high quality tablet paper. The miscella withdrawn from the lowermost end of the inclined digester was continuously delivered into a solvent recovery distillation column for vaporization of trichloroethylene. The distillation column was operated at a vacuum of 8 inches of mercury and was heated, by means of steam coils, to a temperature of 185°F. 620 pounds per hour of No. 2 fuel oil were continuously introduced adjacent the bottom of the distillation column and a fluid suspension of fuel oil and polyethylene was continuously withdrawn from the bottom of the distillation column at a rate of 1,250 pounds per hour and introduced into a finishing still. The finishing still was operated at a vacuum of 10 inches of mercury and was heated to a temperature of 180°F by means of steam coils. A substantially solvent free stream of fuel oil and polyethylene was removed from the finishing still at a rate of 1,000 pounds per hour.

The vaporized trichloroethylene withdrawn from the distillation column and the finishing still was condensed in a water cooled condenser and delivered to a condensate separator. Similarly, the mixture of steam and vaporized trichloroethylene withdrawn from the upper end of the stripping column was condensed in a water cooled condenser and delivered into the condensate separator. In the condensate separator the water and trichloroethylene were separated by decantation, the trichloroethylene being returned to the storage tank, and the water being wasted.

Separation from Plastic Waste for Recycling: A process developed by *J.F. Laundrie; U.S. Patent 3,814,240; June 4, 1974; assigned to U.S. Secretary of Agriculture* involves transporting, in suspension, a mixture of wastepaper and thermoplastic film in a hot gas stream causing the thermoplastic film to contract upon itself and form particles, thereby decreasing its specific surface to such an extent as to allow removal of the contracted plastic particles from the mixture by standard mechanical methods, such as air classification or screening. The process effectively removes all thermoplastic film from the wastepaper thus allowing for more efficient paper utilization through higher recycling rates.

PULPING LIQUOR

The reader is also referred to the section on Sodium Sulfite under Inorganic Materials for references on chemical recovery in sulfite pulping.

Production of Fuel Gas from Kraft Liquor

A process developed by *D.L. Brink and J.F. Thomas; U.S. Patent 3,718,446; February 27, 1973; assigned to The Regents of the University of California* is one involving the high temperature distillation and pyrolysis of an organic material such as kraft black liquor, formed during conventional chemical wood pulping, to break down the material to noncombustible solids and to a stable gaseous clean burning fuel.

The process involves introduction of malodorous-containing gaseous emissions from the pulping operation as a source of oxygen for attaining the temperature necessary for the pyrolysis but insufficient to effect stoichiometric (complete) combustion of the organic material. These kraft mill malodorous emission gases are otherwise occasionally burned with inherent incomplete combustion, leaving significant amounts of obnoxious gases to pollute the atmosphere.

Recovery of Sodium-Base Liquor by Carbonation

A process developed by *C.G. Anderson; U.S. Patent 3,826,710; July 30, 1974; assigned to Owens-Illinois, Inc.* provides an integrated closed carbonation system for recovering sodium base sulfite pulping liquor which is of interest chemically, thermally and ecologically. The system includes in combination (1) smelt solutions containing from 33 to 67 mol percent sodium sulfide with the balance primarily sodium carbonate, (2) self-sufficiency in carbon dioxide for displacement of the sulfide, (3) temperatures during carbonation of from 140° to 180°F, (4) an intermediate concentration of CO_2 for carbonation, (5) systems for cycling various gas streams to conserve and reuse carbon dioxide while still purging the system of nitrogen and in particular the recycling of purge gases from a precarbonation zone to a main carbonation zone, (6) production of gases containing H_2S at a concentration greater than 6% by volume suitable for direct combustion to SO_2, and (7) purge gases essentially free of either H_2S or SO_2.

Figure 107 shows the essential elements of this process. There are several known methods based on the combustion of a spent sodium base sulfite liquor in a reductive recovery furnace to produce an inorganic smelt containing sodium sulfide and sodium carbonate. In the more common of these, this smelt is treated with water and carbon dioxide to remove the sulfur as hydrogen sulfide and form a carbonate. Thereafter, the H_2S is oxidized to SO_2 which is combined with the sodium carbonate to reconstitute the sodium sulfite for reuse in a pulping liquor. The Stora process as it is known today, is characterized by

(1) countercurrent treatment of a clarified smelt solution containing sodium carbonate and sodium sulfide at elevated temperatures in a plate column with essentially pure carbon dioxide to drive off a mixture of carbon dioxide and hydrogen sulfide and to form a solution of sodium bicarbonate and sodium carbonate,

(2) the reaction of this solution with a solution of sodium bisulfite in the so-called decarbonation reaction to form sodium sulfite and to liberate carbon dioxide, which is used in the first step,

(3) reaction of the mixed carbon dioxide and hydrogen sulfide gases from the first step with essentially pure sulfur dioxide in a Claus catalytic reactor to form molten sulfur and a gaseous product consisting primarily of carbon dioxide with a small quantity of sulfur dioxide, which may be returned for use in the first or carbonation step,

(4) combustion of the sulfur in a sulfur burner to form sulfur dioxide,

(5) absorption and desorption of sulfur dioxide from water to prepare concentrated sulfur dioxide for use in step (3), and

(6) absorption of sulfur dioxide in sodium sulfite from step (2) to form sodium bisulfite for use in step (2), and/or for use in pulping.

This process has evolved over a period of years, and many alternatives have been suggested. U.S. Patent 2,909,407 employs the main steps of the Stora process but also includes the reaction of hydrogen sulfide with sulfur dioxide to form sulfur in the presence of water and the use of heat for coagulation and separation of the sulfur.

The Mead recovery process is characterized by an integration of the recovery furnace and the carbonation system in a continuous adiabatic operation, with utilization of the flue gases from the combustion of the spent liquor for the carbonation of the smelt solution, the combustion of the released hydrogen sulfide in the recovery furnace and the scrubbing of the sulfur dioxide from the combined flue gases with the carbonated solution.

The critical carbonation step is carried out in a countercurrent packed tower at elevated temperatures, using a concentrated, clarified smelt solution. The process has been further developed by the inclusion of a precarbonation step, using a portion of the hydrogen sulfide containing off-gases for precarbonation, so as to decrease the volume and increase the concentration of hydrogen sulfide in the gases returned to the recovery furnace. This improved version is described in U.S. Patents 2,788,273 and 2,849,292 with a modification thereof in U.S. Patent 3,026,240 wherein precarbonation is performed directly with SO_2 free flue gas in parallel with the main carbonation operation instead of in series with it.

The Weyerhaeuser process set forth in U.S. Patent 3,005,686 describes a sodium sulfite recovery process applicable to the liquors from kraft, neutral sulfite, or acid sulfite pulping. In this process, a smelt solution containing sodium carbonate and sodium sulfide is treated in a first carbon dioxide and hydrogen sulfide absorption stage, followed by a carbonation and hydrogen sulfide stripping stage in successive countercurrent packed towers.

The gas used consists principally of nitrogen and carbon dioxide, which is recirculated from a catalytic conversion stage. The hydrogen sulfide, at a concentration of less than 6% in the product gases, is converted to sulfur or sulfur dioxide by catalytic oxidation in the presence of air. The carbonated liquors are treated in a decarbonation stage with sodium

FIGURE 107: CARBONATION SYSTEM FOR RECOVERY OF SODIUM BASE PULPING LIQUOR

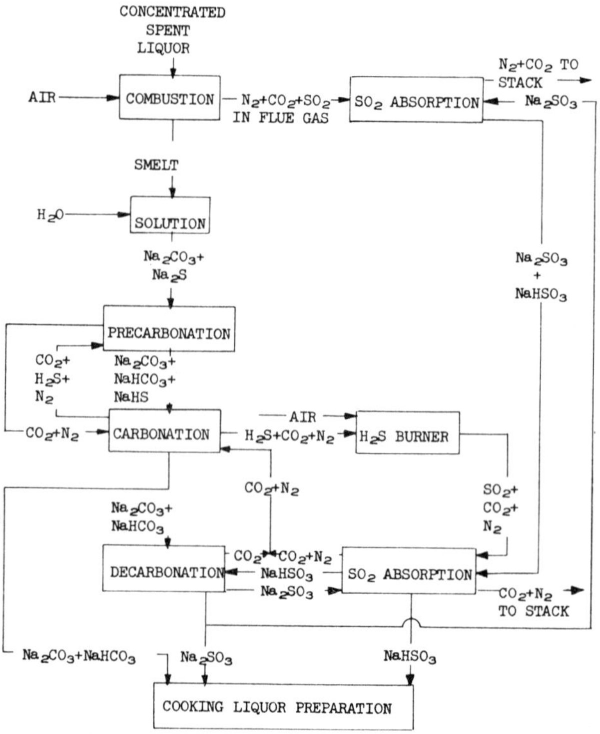

Source: C.G. Anderson; U.S. Patent 3,826,710; July 30, 1974

bisulfite to liberate carbon dioxide, which is also used in the carbonation step. The sodium bisulfite is produced in a sulfiting tower by reaction of the product liquor from the decarbonation reaction with the sulfur dioxide from a sulfur burner.

Still other methods of carbonation in sodium base pulping liquor recovery processes are described in *Paper Trade Journal,* vol. 154, issue 25, pages 39-49 (1970). In both the Mead and Weyerhaeuser processes described above in other prior art, gases formed from a precarbonation or preabsorption step are vented directly to the atmosphere or if further treatment is desired to an additional absorption tower, a furnace or the like.

In essence, the process involves an improvement on carbonation recovery systems for converting green liquor composed principally of Na_2S and Na_2CO_3 to $NaHCO_3$ and Na_2CO_3, which may be used in the subsequent preparation of Na_2SO_3 the latter of which is adapted for reuse in a pulping liquor. Thus, in a system in which the green liquor is first contacted with a gaseous mixture containing hydrogen sulfide and carbon dioxide in a preabsorption tower, converting the liquor to a mixture of NaHS, $NaHCO_3$ and Na_2CO_3, which is then passed through a reaction tower countercurrent to a gas composed principally of carbon dioxide and nitrogen, in which reaction tower the liquor is stripped of H_2S, the liquor at the same time being absorptive of CO_2 whereby the liquor converts to $NaHCO_3$ and Na_2CO_3, most of the hydrogen sulfide in the gases to the precarbonation tower is selectively absorbed in that tower, leaving a gaseous mixture containing nitrogen and carbon dioxide, which is then recycled for use in the reaction tower. By recycling these gases to the reaction tower it is possible to avoid releasing to the atmosphere (or further treating them to remove) any hydrogen sulfide which they may contain. They also serve as an additional source of both nitrogen and carbon dioxide for the carbonation and hydrogen sulfide stripping actions in the reaction tower.

Thus it is possible to lead off from the carbonation system a single gas stream emanating from the reaction tower, which stream has a relatively high concentration of hydrogen sulfide suitable for further processing for the recovery of sulfur or sulfur dioxide. The absorption of hydrogen sulfide by the green liquor in the preabsorption tower materially increases the vapor pressure of hydrogen sulfide of the product liquor not only in absolute terms but also relative to the vapor pressure of carbon dioxide over that liquor.

This effect is even more marked when the hydrogen sulfide in the gases to the preabsorption tower is preferentially absorbed with a minimum absorption of the carbon dioxide in these gases. Such preferential absorption is favored by limiting the contact time between the liquor and the gas in the preabsorption tower, as by use of sprays or a drip grid packing. In this manner, the hydrogen sulfide vapor pressure of the liquor in the reaction tower may be increased, permitting a higher concentration of hydrogen sulfide in the gas emanating from that tower.

Particular features of novelty of this process are (1) the recycling of gases leaving a precarbonation zone to a main carbonation zone; while (2) operating the precarbonation zone so as to favor selective absorption of hydrogen sulfide in that zone. This eliminates the potential discharge of hydrogen sulfide to the atmosphere from the precarbonation zone; while maintaining the advantage of a precarbonation system in producing a higher concentration of hydrogen sulfide in the gases from the main carbonation zone.

SAWDUST

Sawdust may be utilized for its heating value. In Maine, for example, there are sawdust-fired residential heating units in use in some cases. Some examples of the production of utilitarian products from sawdust are given in the paragraphs which follow.

Production of Building Blocks

A process developed by *C.L. Kranz and T.A. Groner; U.S. Patent 2,899,325; August 11, 1959* involves the fabrication of building blocks from sawdust as one major component. The basic materials of the composition are sawdust, sand and cement which provides the majority of bulk and strength to the material. Many of the outstanding characteristics of the material are furnished by an admix of finely powdered burned gypsum and powdered pine rosin which are mixed with the above basic materials.

A typical composition would be comprised of, by volume: 5 parts sawdust; 1 part sand; 1 part cement; 0.85 to 1.25 parts of finely powdered burnt gypsum and 0.15 to 0.25 part of powdered pine rosin.

The above materials are thoroughly mixed together and sufficient water is added to produce a stiff workable plastic mass. The plastic mass may be poured in a concrete block mold, vibrated, and compressed to a uniform density. The blocks are then steam cured or permitted to dry and after seasoning are ready for construction purposes. As will be apparent, the above material requires no special equipment as regular cement mixers and block forming machines may be used to form the blocks. The above quantities have proven very satisfactory and provide a light weight building block which is used in the same manner as a concrete block. The admix of burned powdered gypsum and powdered pine rosin acts as a filler, permitting the block to be sawed, dried, nailed and screwed into, yet provides strength and high insulating qualities. The admix also functions as a preservative helping to maintain the blocks termiteproof, fireproof and resistance to freezing, thawing and other temperature and humidity changes. Also, the block texture is such that paints and plasters readily adhere and the blocks have acoustical absorption characteristics serving to soundproof structures constructed thereof.

As the materials of the blocks are inexpensive and available, the cost of the improved composition compares favorably with that of concrete blocks. However, with the improved weight, working, insulating, finish and other characteristics, a more economical and universal building block is obtained.

Production of Fertilizers

A process developed by *R.L. Formaini and D.P. Schutz; U.S. Patent 3,146,087; August 25, 1964; assigned to Allied Chemical Corporation* involves preparing fertilizer from finely divided cellulosic materials by oxidation and ammoniation.

Numerous processes have been developed for making useful products from waste cellulosic material such as sawdust. Processes in which the cellulosic material is oxidized such as is that described in U.S. Patent 2,232,990 lead to the production of oxy cellulose which is suitable for the production of transparent films or sheets. Other process such as the one described in British Patent 347,641 can be used to convert sawdust into fertilizer. This latter is accomplished by the simultaneous oxidation and ammoniation of the cellulosic material.

While this latter process has considerable merit in that it can lead to fertilizer products containing 10 to 20% nitrogen, the process is slow (lengthy reaction time), hazardous from an explosion viewpoint and leads to products which have been found to be unsatisfactory in view of their uncertain activity characteristics in use.

The process is carried out by oxidizing finely divided cellulosic material with nitrogen dioxide at temperatures of about 20° to 120°C until the carboxylic acid content is brought to at least about 5% by weight and subsequently treating the oxidized product with ammonia at pressures of at least 700 psig and temperatures between about 175° and about 265°C.

Production of Soil Conditioners

A process developed by *E. Farber; U.S. Patent 2,735,756; February 21, 1956; assigned to Timber Engineering Company* involves enhancing the effectiveness of materials produced by heating lignocellulosic compositions with acids to a temperature above the boiling point of water but below the charring point of the compositions, as agents for improving characteristics of soils as seed germination and plant growing media.

The process involves the steps of heating a particulate lignocellulosic material and a mineral acid at a temperature in excess of the boiling point of water but below the charring temperature of the material to produce a substantially dry soil improving agent. Then contacting the agent, at a temperature in excess of the boiling point of water, with an aqueous medium to generate steam in situ in the agent and enhance its effectiveness for improving soils as seed germination and plant growing media.

WOOD WASTE

Production of Charcoal Briquettes

The entrainment drying and carbonization of wood waste has been described by C.C. Boley and W.S. Landers in Report of Investigations No. 7282, Washington, D.C., U.S. Bureau of Mines (August, 1969).

The Bureau of Mines dried and carbonized two wood wastes, one softwood (redwood) and one hardwood (oak), using continuous, vertical entrainment techniques previously developed for use with coal. It was demonstrated that these techniques are readily applicable to wood waste and that appreciable changes in the yields of carbonization products can be obtained by varying operating conditions.

A limited amount of experimental work was done on briquetting the charcoals. Firm briquettes were formed with oak charcoal, although greater quantities of starch binder than used commercially for kiln-produced charcoal were required. Redwood charcoal did not briquette successfully under the test conditions employed.

Production of Clean Wood Chips for Further Processing

A process developed by *N.S. Lea and J.S. Brawn; U.S. Patent 3,367,495; February 6, 1968; assigned to Scott Paper Company* involves the recovery of relatively bark-free wood chips from chipper fines and/or sawdust. In the preparation of paper from wood, parts of logs with much, but not all, of the bark removed are reduced to chips of varying size in apparatus known as a chipper. Among the chips are particles of wood, dirt, bark, knots and compression wood that are very small, e.g., small enough to pass through a ¼" screen. If good quality paper is desired, these should be separated by any suitable method or means from the larger, normal-sized wood chips. Following such separation, the larger chips are subjected to any of various treatments, such as chemical or semichemical pulping to convert them to wood pulp. By various subsequent treatments which are well-known to those skilled in the art, paper is manufactured from the pulp.

The very small particles separated from the larger chips are known as "chipper fines." Conventionally, they are separated from the larger chips by screening and are then burned to recover whatever heating value they may possess. If the good wood could be recovered from the chipper fines substantially free of the other chipper fines components, especially bark, substantial economic benefits would result, for the potential value of the paper that could be produced from the good wood exceeds the value of the chipper fines as fuel.

Various obstacles have discouraged the recovery and utilization of the good wood contained in the chipper fines. First of all, the separation of the small particles of good wood from the bark and other impurities in the chipper fines can be difficult and costly. Secondly, attempts at producing wood-pulp from the chipper fines without removal of impurities have resulted in undesirably low grades of paper. Thirdly, it has been generally believed that if normal-sized chips are cooked along with very small chips, such as might be recovered from the chipper fines, the yield increase would not be sufficient to justify the cost of recovering the good wood from the chipper fines.

Exemplary of various factors contributing to the problem of separating good wood from the back and other impurities in chipper fines is the reported observation that both bark and wood chips tend to float on the surface of water. This frustrates attempts to use simple flotation or "float-sink" separation processes in which a mixture of two solid materials having respective specific gravities greater than and less than water are delivered to the surface of a body of water and are then separately recovered therefrom after the heavier material has sunk beneath the surface.

Of course, it has been observed that if chips of certain kinds of wood and bark are delivered to the surface of water and allowed to stand long enough, the wood will become waterlogged and sink before the bark does. However, the period of time required for the wood to absorb the necessary amount of water by normal capillary action unduly protracts the treatment and renders rapid separation impossible. Recently, it has been proposed that special float-sink separation procedures involving vacuum and heat treatments and specialized flotation media be used to hasten the separation of mixtures of bark and wood chips.

As is well known, however, the employment of heat, vacuum and specialized media introduce additional complexities and cost factors which it would be desirable to avoid, if possible. Consequently, there is still a need for additional methods and apparatus of a simple, economical rapid and efficient nature for separating good wood fragments from fragments of bark and other impurities.

According to the process, mixtures of fine particles of waste wood and impurities, e.g., particles small enough to pass through a screen with openings of approximately ¼", are submerged in an aqueous bath; a generally downwardly directed force is applied to the mixture in the bath for forcibly conducting the mixture downwards in the bath; the mixture is then released from the downwardly directed force. Upon completion of the above operations, the good wood has a tendency to rise in the bath while the impurities sink or at least remain an appreciable distance below the surface.

It has been found that the performance of the above operations will cause a predominant proportion of the good wood in the mixture to rise to the surface while a predominant proportion of the impurity or impurities fails to so rise. Consequently, the process facilitates the recovery from the surface of the bath of good wood freed of a substantial proportion of the impurity or impurities in the original mixture.

A process developed by *L.-H. Tee; U.S. Patent 3,878,994; April 22, 1975; assigned to Urban Wood & Fiber Products, Inc.* involves treating waste wood to produce usable end products. Waste wood received in a dump pit is broken up by a drop weight, and may, if desired, be introduced by a grapple into a flotation tank for separation from intermixed debris that is denser than water. The floating waste wood is then picked up by the grapple and placed on a conveyor for introduction into a hog grinder wherein the waste wood is ground into wood chips.

Magnetic material intermixed with the wood chips is separated therefrom by a magnetic pulley conveyor. The chips are then run through a screen separator for separation by size into oversize chips, acceptable chips and undersize chips. The acceptable chips may be stored to await transport to users or may be further processed. Additional processing includes moisturizing the chips in a chip compactor and washing them in a water bath to remove foreign material. Undersize chips may further be separated by size to reclaim usable wood material.

Production of Humus Carrier for Plant Nutrients

A process developed by *D.R. Boomer, E.T. Clapperton and R.C. Phillips; U.S. Patent 2,901,339; August 25, 1959; assigned to Wooduse, Inc.* is a continuous process for the preparation of agricultural nutrient carriers and soil conditioning agents from ligneous and cellulosic wastes.

It has heretofore been proposed to use lignocellulose waste materials, such as sawdust, bagasse and the like, as agricultural nutrient carriers and as soil conditioning agents similar to humus. However, it is known that these materials as such are not suitable for use in soils for the reason that the cellulosic content thereof supports the growth of bacteria and fungi with resultant depletion of the nitrogen and other fertilizer materials present in the soil.

It is also known that the nitrogen-depleting tendencies of lignocellulose materials may be partially or completely overcome by heating the materials in the presence of a mineral acid to convert part or substantially all of the cellulosic fraction to an unreactive or inert form which will not support any substantial bacterial or fungicidal action. However, some difficulties have been encountered in carrying out such processes on a commercial scale, due to problems of controlling the operations to produce a satisfactory product.

Moreover, it has heretofore been necessary to utilize raw materials having substantially uniform particle sizes in order to produce a uniform product. Additionally, the processes heretofore proposed are not wholly satisfactory from an economic standpoint in that they require substantial capital investments for equipment and relatively high costs of operation.

One of the principal objects of this process is, therefore, to provide a process for the treatment of lignocellulose materials, which is not subject to the above and other disadvantages of the processes heretofore proposed.

In this process the lignocellulose material in comminuted form is mixed with an aqueous solution of a mineral acid such as sulfuric acid, phosphoric acid, hydrochloric acid or mixtures thereof, and the mixture conveyed, in fluid form, through a reaction zone wherein the desired conversion takes place. In this process, the mixture is conveyed through the reaction chamber by means of a stream of hot gases, the gases also functioning as the heat transfer medium to bring the material to the desired reaction temperature.

These gases, mixed with gaseous and volatilized reaction products, are separated from the solid, comminuted conversion product and preferably scrubbed with water to remove valuable constituents thereof such as acetone, furfural, etc., which may be recovered as such or utilized as fuel in the process. The solid product is quenched, preferably with water, to stop the reaction, and then preferably treated with ammonia and/or other desirable food supplements. This process is particularly advantageous in that the apparatus required is inexpensive to install and operate, the thermal efficiency is high, and it is capable of producing a uniform product, i.e., a particulate product wherein each particle is uniformly treated, even though the particle sizes of the raw material vary over a wide range.

A process developed by *C.S. Renwick, Jr.; U.S. Patent 3,034,882; May 15, 1962* is a process for treating wood chips with organic fertilizers of nitrogen, phosphorus, potassium or other desired plant nutrients or chemicals to render them useful as a fertilizer storage medium.

The method of producing a humus carrier for plant nutrients involves the steps of (1) forming to appropriate size a chip of wood having a myriad of voids therein, (2) removing a substantial amount of moisture from the body, (3) treating the chip with a compound of chemical plant nutrients in solution including phosphoric acid to saturate the chip and fill the voids therein, (4) treating the saturated chip with a silicate of soda solution to effect a reaction between the phosphoric acid and the silica of soda to produce on the chip a gelatinous orthosilic acid precipitate, and (5) subsequently dehydrating the orthosilic precipitate to effect the deposition on the chip of a fine grained silica coating sealing the plant nutrients within the chip.

Production of Soil Conditioners

A process developed by *C.W. Gay, Jr. and D.D. Dwyer; U.S. Patent 3,561,943; Feb. 9, 1971* is one in which finely divided wood waste material such as ground pine bark or sawdust or the like, separately or in admixture, is mixed with a microbial source such as ground natural manure, sewage sludge or industrial wastes. To the mixture there is also added water and the total mixture is then heated, with agitation, in the presence of a molecular oxygen containing gas, in amounts sufficient to cause aerobic decomposition.

In a preferred embodiment of the process the wood waste material is transformed into a valuable soil conditioner in a continuous fashion by delivering the mixture of wood waste, microbial source material and water to one end of a hollow, tilted open-ended, elongated cylinder, revolving the cylinder about its longitudinal axis. At one end of the cylinder an aeration medium is introduced comprising a molecular oxygen containing gas, venting the aeration medium adjacent the other open-end of the cylinder, removing excess moisture and withdrawing the treated mixture from the end of the cylinder opposite the mixture feed end thereof.

If desired, the excess moisture removed from the treated mixture and/or processed wood waste material in the proportion of about 5 to 15%, preferably 10 volume percent can be recycled for admixture with the wood waste material as a convenient supply of microbial source material and water.

Production of Soda Pulp from Knotter Rejects

A process developed by *N.S. Lea and D.E. Kotzerke; U.S. Patent 3,393,121; July 16, 1968; assigned to Scott Paper Company* involves treating acid sulfite stock and acid sulfite knotter rejects for the separation of a selected wood fraction from the stock and knotter rejects followed by soda cooking of the fraction to obtain pulp fiber therefrom.

In the preparation of paper by the well-known acid sulfite pulping process, a certain amount of the cooked material discharged from the digester is undesirable and must be removed from the pulp stock before it can be made into commercially acceptable paper products. The amount of such material will vary widely and includes knots, uncooked and partially cooked wood chips, pieces of bark, dirt and bark carried by the wood, and fiber bundles or shives. Their separation is normally accomplished on the basis of particle size, although there is increasing use of the centrifugal principle which separates particles on the basis of density.

Usually, in normal acid sulfite pulping processes, the stock from the blowpits or washers is diverted to a screen, an example of which would be a Jonsson Knotter, with ¼" diameter drain openings for separation of the large particles of undesirable materials from the pulp fibers. These undesirable materials are generally termed "acid sulfite knotter rejects."

In accomplishing the separation described above, there is some loss of good fibers along with the knotter rejects. To minimize the loss of good fiber it is conventional practice to further subject the knotter rejects to another Jonsson Knotter screen (¼" diameter openings) where additional fiber is gleaned from the rejects before they are sent to the boiler house for burning to recover their fuel content. Included in the rejects which are sent to the boiler house are knots, pieces of bark, uncooked and partially cooked good wood chips, shives and fiber bundles. If the fiber bundles and good wood fraction could be separated from knotter rejects and thus be rendered substantially free of the most undesirable constituents such as large knots and large pieces of bark, substantial economic benefits would result, for the value of the paper that could be produced from the recovered fiber bundles and good wood fraction exceeds the value of the rejects as fuel.

According to the process, to obtain the fraction acceptable for soda cooking, a special screen of a relatively large mesh is designed to remove the most undesirable constituents (knots in particular) is used, followed by screening of the thus obtained accepted fraction by a screen of a substantially smaller mesh to separate fibers from the fraction to be subjected to soda cooking.

A preferred embodiment of such a process involves the steps of: (a) screening acid sulfite knotter rejects to remove the most undesirable constituents (large knots, pieces of bark, etc.); (b) screening the accepted fraction which passes through the screen of step (a) to remove pulp fibers; and (c) soda cooking the fraction retained on the screen of step (b) to obtain pulp.

The foregoing procedure results in a pulp which is of low bleachability and high quality. These results, surprisingly, are attained with a much lower caustic consumption than is used on fresh wood chips by normal alkaline processes. In fact, it is possible to operate such a process in an economically beneficial manner without recovery of the cooking chemicals.

PLASTICS AND FIBERS

COTTON FIBERS

Recovery from Cotton-Synthetic Fiber Blends

A process developed by *A.T. Mays; U.S. Patent 3,801,273; April 2, 1974; assigned to Johnson & Johnson* provides a means of recovering waste cellulose fibers from mixtures of waste cellulosic fibers, waste polyester and/or acrylic fibers, and synthetic, crosslinked resin materials which comprise: heating a mixture of waste cellulosic fibers, waste polyester and/or acrylic fibers, and synthetic, crosslinked resin materials within the range of from 212° to 275°F for a period of from ¾ to 5 hours in an aqueous treating solution containing an alkali metal hydroxide and one or more added normally liquid chemical agents such as ketones, alcohols, lactones, and sulfoxides, which initiate the decomposition or solubilization of the waste polyester and/or acrylic fibers and the synthetic, crosslinked resin materials; adding a neutral or alkaline oxidizing agent to the mixture of waste fibers and synthetic, crosslinked resin materials; heating the mixture of waste fibers and synthetic, crosslinked resin materials in the presence of the neutral or alkaline oxidizing agent to complete the decomposition or solubilization of the waste polyester and/or acrylic fibers and the synthetic, crosslinked resin materials; and recovering the waste cellulosic fibers.

A process developed by *A.H. Drelich; U.S. Patent 3,843,321; October 22, 1974; assigned to Johnson & Johnson* involves very similar techniques.

PLASTICS

Plastics are becoming an increasingly important material in our society, and their growth continues at an impressive rate. From 1960 to 1970, plastics consumption increased at an average annual rate of 11.8% and totaled 8.5 million tons in 1969. By 1980, consumption is expected to reach 19 million tons. Very little plastic scrap is recycled other than that reused within the manufacturing plant in which it is generated. This, however, is a fairly significant quantity. Plastics fabricators, for example, consumed an amount of internal scrap equal to about 1.5 million tons in 1970. There is essentially no recovery of plastic waste from obsolete products.

The plastics reprocessor is the recycling channel for all industrial plastics recycled outside originating plants. About 500,000 tons of waste plastics were handled by reprocessors in 1970. Of the plastics recycled through reprocessors, about 55% came from resin producers;

30%, from fabricators; and 15%, from converters. There are two types of plastics, thermoplastics and thermosetting plastics. The thermosets, 20% of plastics consumption, cannot be softened and reshaped through heating and, thus, are not recyclable. In addition, most of the plastics used as coatings and adhesives are impossible to recycle. Thus, about 75% of the plastics consumed are potentially recyclable.

As a general rule, scrap plastic has to be used in an end application having wider specification requirements than the product yielding the scrap. The primary markets for scrap plastic include such items as hose, weather stripping, toys, cheap housewares, pipe, and similar applications. These are areas where (1) plastic properties and performance are not paramount; (2) relatively noncritical processes are used, such as compression molding or heavy extrusion; (3) the cost of plastic resin is a high proportion of total product cost.

Issues and Problems in Recycling

Technology: There is a fundamental difference between the nature of plastics recycling and that of metals, paper, glass and other materials. Metals production, for example, begins with an impure ore that is progressively concentrated, smelted, refined, and freed from impurities. Plastics production, on the other hand, begins with a high-purity virgin polymer to which various additives, colorants, and reinforcements are added. Thus, in the metals industries, there is a background of technology designed to purify and upgrade ores and concentrates. Such technology can also be applied to the upgrading of scrap. In the plastics industry, where the basic raw material is progressively contaminated in production, little technology has been developed that can be applied to purify waste plastics.

Compatibility: The principal difficulty in recycling plastics is that different polymers (polyethylene, polyvinyl chloride, etc.) are not compatible with each other and must be separated, a very difficult and costly task.

Economics: The continually decreasing cost of basic plastic materials has made scrap plastic less competitive with its main competitor, off-grade virgin resin. For example, since 1961, the price of low-density polyethylene has decreased from 24 to 13 cents per pound. Scrap plastic, which is restricted by rising labor and distribution costs, did not drop as rapidly; the price of the scrap is now only about 1 cent per pound under the off-grade resin price, versus about 3 cents in 1961.

Logistics: This problem, which is common to recycling of all materials, is important to plastics recycling. The extremely low density of plastics makes transportation very costly.

Separation: Separation of plastics from other waste is extremely difficult, making recovery of plastics from municipal waste almost impossible unless the plastics can be diverted from the waste stream and kept separate.

The general topic of recycling and reuse of plastics has been treated at length in a study by Arthur D. Little, Inc., Report PB 214 045, Springfield, Va., National Technical Information Service (1972). The reader is also referred to *Pollution Control in the Plastics and Rubber Industry,* by M. Sittig, Park Ridge, N.J., Noyes Data Corp. (1975).

Recovery of Metal Ingredients from Plastic Products

A process developed by *F. Ensslin, G. Frenzel and E. Löbermann; U.S. Patent 3,776,147; December 4, 1973; assigned to Preussag AG-Metall, Germany* is designed to eliminate plastics waste and recover the metals and metal compounds in the waste. The plastics waste, whether in a solid, pasty or liquid form, is converted into a combustible fluid-like form and is directed through a burner into a combustion chamber. The combustion of the waste results in combustion gas and fly ash which is cooled and directed through a separator to obtain the metals and metal compounds from the fly ash in the form of oxides.

In order to convert the waste into a combustible form, the waste when solid or pasty is melted and heated to a temperature favorable for combustion. In some cases, depending on the nature of the waste and the concentration of fillers, thinners and the like, it can be advantageous to add to the waste before combustion substances for promoting the combustion, particularly oils or other easily combustible organic substances.

The liquefied waste in liquid or molten form is preferably burnt by spraying it with the help of a gas containing oxygen, preferably preheated air. By burning in preheated primary air with the addition of secondary air it becomes possible to control precisely both the combustion temperature and the rate of combustion.

The resulting mixture of combustion gases and fly ash is preferably subjected to a preliminary cooling in a heat exchanger, final cooling being effected in a final cooler. The fly ash is separated from the cooled mixture by known methods, preferably by precipitation in a dust filter. The separated fly ash contains, in concentrated form, the metals or metal compounds initially present in the waste, the metals occurring in the fly ash in the form of oxides, chlorides or sulfates. These metal compounds can be recycled in the known way by adding them to the raw materials used for manufacturing the plastics.

The method is suitable, in principle, for destroying all combustible plastics, for example, polyolefins, polyesters, polyamides, or polyurethanes, their manufacturing by-product wastes and wastes which have been made combustible by adding combustion promoters, as well as many other materials. The process allows the catalysts used in the manufacture of these plastics to be recovered, as well as additives, particularly stabilizers, the recovered product taking the form of oxides or other inorganic compounds of, for example, titanium, antimony, tin, cobalt, germanium or the like.

Recovery of Mixed Plastic Scrap

A process developed by *H.J. Brown; U.S. Patent 3,852,046; December 3, 1974* is one in which waste plastics comprising a mixture of two or more thermoplastic materials are treated by grinding the products to form a particulate thermoplastic mixture, washing the particulate mixture to remove nonplastic materials, placing the thermoplastic mixture in a mold or shaped vessel, and heating the mixture to a temperature above the softening point of the lower-melting thermoplastics and below the temperature of the softening point of the highest-melting thermoplastic of the mixture to form a plastic product.

In another embodiment, particulate thermoplastic is used as a filler for asphalt compositions. In a further embodiment, waste thermoplastics are mixed with combustible cellulosic materials and formed into logs and the like which thermoplastic mixture has been treated to remove halogen-containing thermoplastics such as polyvinyl chloride.

A process developed by *S. Ooba and S. Hirayama; U.S. Patent 3,857,799; December 31, 1974; assigned to Fuji Photo Film Co., Ltd., Japan* involves regenerating waste plastics by mixing the waste plastics with an ether-type polyester and at least one member selected from the group consisting of a homopolymer of vinyl acetate, an ethylene-vinyl acetate copolymer or a tacky polyolefin; and if necessary a foaming agent, and then molding the mixture into a desired shape.

Recovery and Process Reuse of Manufacturing Scrap

A process developed by *J.D. Robertson; U.S. Patent 3,797,702; March 19, 1974* involves a scrap recovery and feed system where scrap plastic is recovered from a plastic manufacturing process, such as in the manufacture of sheet, film, formed or molded objects, and the scrap plastic is chopped up and fed back to the processing machine along with the base material, such as a virgin plastic material and various additives. The scrap is fed with an auger from an upper hopper toward the outlet opening of a lower hopper containing the base material. The materials are mixed just prior to entry in the extruder or other processing machine.

Separation of Individual Plastics from Mixtures

The recycling of plastics from urban and industrial refuse has been reviewed by J.L. Holman, J.B. Stephenson and M.J. Adam in Report of Investigation No. 7955, Washington, D.C., U.S. Bureau of Mines (1974). Figure 108 shows a sink-float scheme which was devised to separate waste plastic concentrates into various fractions. The integration of the sink-float process into an overall proposed separation scheme is shown in Figure 109.

FIGURE 108: FOUR-STAGE SINK-FLOAT SCHEME FOR SEPARATING WASTE PLASTICS MIXTURE

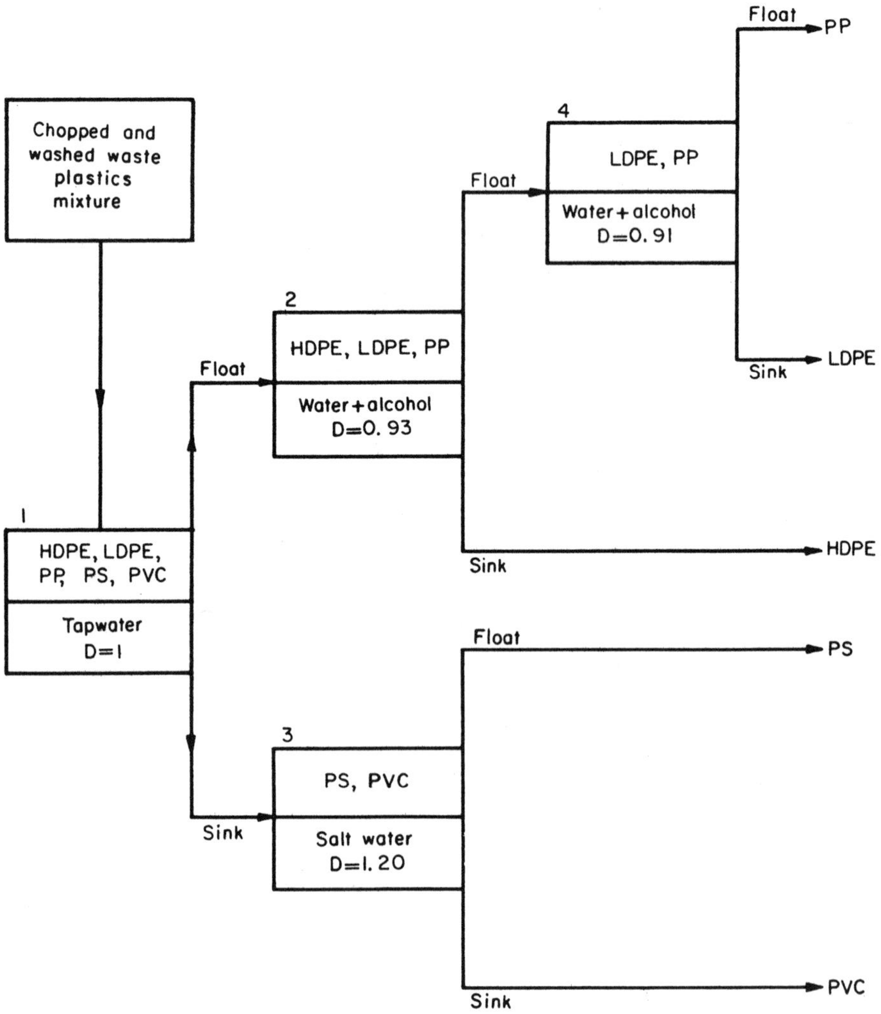

Source: Bureau of Mines R.I. 7955

FIGURE 109: PROPOSED FLOW DIAGRAM OF PROCESSING SYSTEM FOR RECLAIMING WASTE PLASTICS

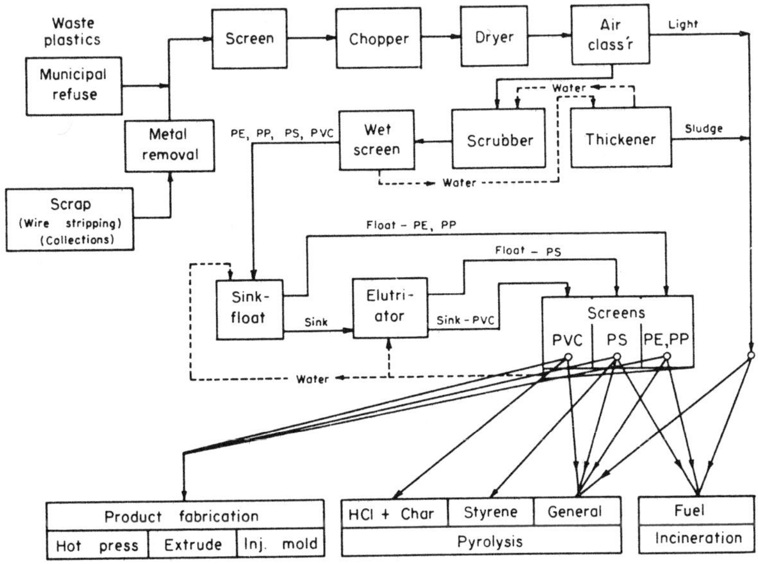

Source: Bureau of Mines R.I. 7955

POLYAMIDES

Recovery from Process Waste Fibers

A process developed by *G.P. Monet; U.S. Patent 3,069,465; December 18, 1962; assigned to E.I. du Pont de Nemours & Co.* involves recovering from nylon the salt-forming components from which it is made. The normal operation of a nylon plant results in the unavoidable collection of waste fiber and polymer at various points in the process. This material is generally not suitable for remelting and extrusion into high-grade fibers. It is therefore desirable that the salt-forming components of nylon be recovered from this material for the production of new polymer.

In the prior art, adipic acid and hexamethylenediamine have been recovered from nylon polymer by repeated hydrolysis of the polymer in sulfuric acid, recovering the adipic acid by crystallization after each hydrolysis step and the diamine by distillation after neutralizing the acid. In that process, the hydrolysis and crystallization of the adipic acid have to be carried out in a stepwise fashion, requiring frequent filtrations and a great deal of time for optimum recovery of the adipic acid. Such a process, then, is not efficient for handling the greatly increased volume of waste which has resulted from the continually increasing production of nylon fiber.

The process, as shown in the block flow diagram in Figure 110 involves continuously hydrolyzing nylon waste with an aqueous mineral acid of from 30 to 70% concentration, continuously feeding the resulting hydrolysate to a crystallization zone, continuously crystallizing the acid component and removing the crystals from the hydrolysate, continuously adding calcium hydroxide to the resulting mother liquor to substantially neutralize

FIGURE 110: PROCESS FOR RECOVERY OF ADIPIC ACID AND HEXAMETHYLENEDIAMINE FROM NYLON WASTE

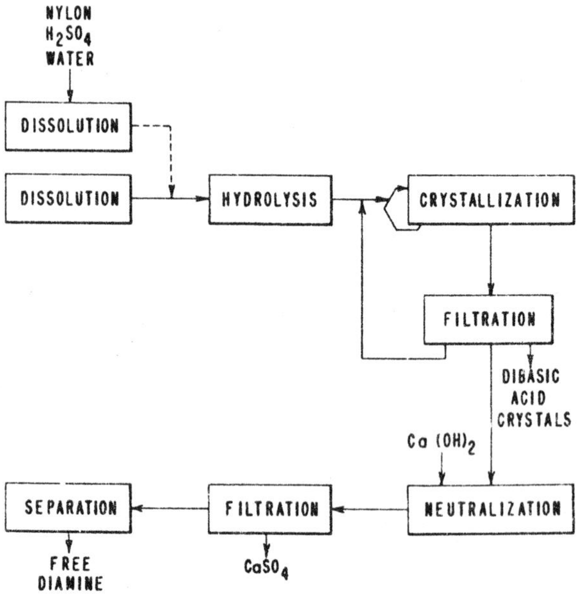

Source: G.P. Monet; U.S. Patent 3,069,465; December 18, 1962

it and liberate at least about 95%, but less than 100%, of the diamine component, and separating this component from the mother liquor. In order that continuous recovery may be accomplished, adipic acid crystals having an average diameter above 40 to 50 microns must be provided. To provide such crystals, the crystallization is effected by continuously introducing the hot hydrolysate containing between 10 and 20% adipic acid into a crystallization vessel, maintaining a substantial volume of liquid in the vessel, agitating the contents of the vessel while maintaining an average temperature of 20° to 30°C and removing slurry containing the adipic acid crystals from the tank at a rate sufficient to maintain the desired volume in the tank.

Usually, the hydrolysate from the hydrolysis step contains more than 20% adipic acid. Optimum crystal size is achieved when the hydrolysate is diluted to 10 to 20% adipic acid concentration, preferably with filtrate obtained in the removal of the adipic acid crystals. The 20° to 30°C temperature range is also critical in the production of adipic acid crystals of optimum size. This temperature range is preferably accomplished by vacuum cooling and where such cooling is utilized the hydrolysate is introduced well below the surface of the liquid to avoid too rapid evaporation of the water.

The slurry from the crystallization vessel is filtered to remove the adipic acid crystals, the filtrate containing the diamine sulfate being continuously neutralized by the addition of calcium hydroxide. The insoluble calcium sulfate formed in the neutralization reaction is removed by filtration and the diamine recovered and purified by distillation in the conventional manner. Preferably, the amount of calcium hydroxide added is such that 98 to 100% of the diamine sulfate is reacted to form free diamine, any excess of calcium hydroxide being avoided to prevent filtration difficulties.

Recovery from Scrap Tire Cord Material

A process developed by *J.C. Elgin; U.S. Patent 3,468,974; September 23, 1969; assigned to U.S. Rubber Reclaiming Co., Inc.* is one in which a molding composition which contains from 64 to 91% of linear polyamide material and from 3 to 36% of vulcanized rubbery hydrocarbon elastomeric material is provided by extrusion pelletizing tire cord material from which a major portion of elastomeric material has been removed.

POLYESTERS

Recovery from Spinning and Processing Wastes

See also under Synthetic Fibers, page 410.

A process developed by *J.T. MacDowell; U.S. Patent 3,222,299; December 7, 1965; assigned to E.I. du Pont de Nemours and Company* involves the continuous production of linear terephthalate polyesters utilizing reprocessed waste polymer.

In the manufacture of terephthalate polyester articles such as fibers and films, it is inevitable that a certain amount of waste polymer is produced as a by-product of the process. In the manufacture of fibers, for instance, an appreciable quantity of fibrous waste results from the spinning of filaments having physical properties outside of the standardized property limits advertised for sale. Representative fiber properties are denier, break elongation, tenacity, uniformity, and the like. Waste fibrous product is also produced in the packaging process or when some piece of equipment fails to operate in the proper manner.

In attempting to reclaim this polyester waste, it has been found that the simple process of remelting the polymer and extruding it a second time does not give a satisfactory product. On the other hand, procedures requiring the complete breakdown of the polymer to the original starting materials followed by purification of these materials and repolymerization, results in an expensive process making it uneconomical to utilize the waste polymer.

An alternative procedure for utilizing waste polymer which has enjoyed a measure of success involves the heating of the waste polymer with glycol to convert the polymer into monomer, i.e., the diglycol terephthalate, and then injecting the monomer, without further purification, in relatively small amounts into fresh monomer being fed to a continuous polymerization system. The mixture of the reclaimed monomer and fresh monomer are polymerized in the usual fashion, and extruded into satisfactory fibers, films, or other shaped structures.

Using this system, however, it has been found that, for reasons not understood, the amount of reclaimed monomer injected into the polymerization system must be kept below a level of about 20% of the total monomer if the polymer produced is to have satisfactory properties. When amounts of recovered monomer higher than about 20% are injected, it is found that the rate of polymerization of the mixture is reduced to such an extent that the output of the equipment must be cut back to a prohibitively low level. Further, it has been found that the addition of increased amounts of catalyst does not alleviate this condition.

It might be assumed that the rate of polymerization could be increased by further reducing the pressure on the polymerization vessel and raising the temperature. However, it has been found that a significant increase in temperature above that normally used results in an increase in polymer color, thus making the product unacceptable for sale. With respect to pressure, it is generally true that polymerization vessels for polyesters are designed to operate at a particular pressure, e.g., about 1 to 3 mm of mercury, and any requirement for an appreciable reduction in this pressure means that new equipment of a more expensive design must be fabricated to achieve such a result.

It has been found that in a polymerization process in which a diglycol ester of terephthalic acid is fed to a reaction vessel and there polymerized under conditions of reduced pressure and elevated temperature, and in which at least a part of the diglycol ester is derived from a reclaimed terephthalate polyester, the operability and efficiency of the process is greatly enhanced by the presence in the polymerization mixture of a metal salt of a sulfonated aromatic ester-forming compound.

Therefore, the process comprises mixing scrap linear terephthalate polyester with a glycol and heating the mixture until the polymer is converted for the most part to the diglycol terephthalate ester, mixing this diglycol terephthalate ester from reclaimed polymer with an appreciable quantity of diglycol terephthalate ester prepared from purified starting materials, and thereafter polymerizing the mixture under conditions of reduced pressure and elevated temperature to produce a fiber or film-forming linear terephthalate polyester, characterized by the presence in the polymerization mixture of a minor amount of a metal salt of a sulfonated aromatic ester-forming compound. A flow diagram of the process is given in Figure 111.

FIGURE 111: PROCESS FOR RECLAIMING POLYESTER FIBER WASTE

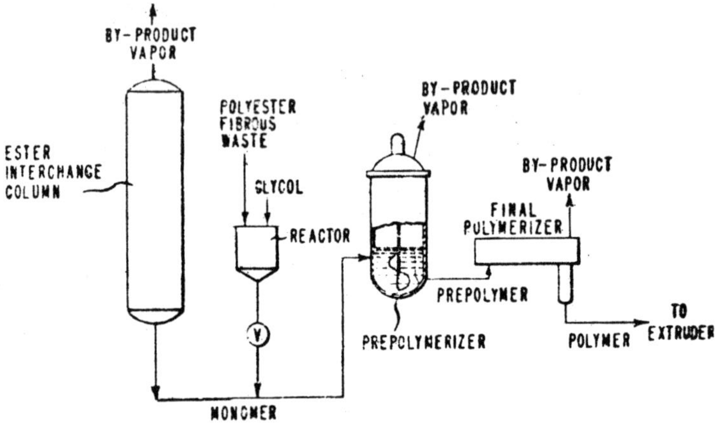

Source: J.T. MacDowell; U.S. Patent 3,222,299; December 7, 1965

A process developed by *K.T. Barkey, E.B. Lefferts and D.C. May; U.S. Patent 3,488,298; January 6, 1970; assigned to Eastman Kodak Company* for recovering polyester scrap for reuse involves initially degrading the polyester with a lower alkyl alcohol and subsequently recovering glycol, dicarboxylic diester, and alcohol from the resulting reaction mixture by distillation. It has been discovered that substantially higher yields of desired components result when a small amount of a phosphorus-containing material that is capable of reacting with the ester exchange catalyst(s) in the scrap is blended into the reaction mixture at the end of the alcoholysis step of the processes.

A process developed by *D.E. Morton; U.S. Patent 3,703,488; November 21, 1972; assigned to The Goodyear Tire & Rubber Company* involves depolymerizing polyester resin such as high-molecular-weight scrap by charging the resin with a small amount of a material selected from the group consisting of materials from which the polyester resin is derived into a worm extruder and extruding the mixture at a temperature of from 260° to 320°C. The high-molecular-weight resin is reduced in viscosity to a degree of polymerization in the range of from 10 to 40 and may be reused in the preparation of new resin.

A process developed by *H.S. Ostrowski; U.S. Patent 3,884,850; May 20, 1975; assigned to Fiber Industries, Inc.* is a process for continuously depolymerizing high-molecular-weight linear polyester scrap the polyester being the polycondensation product of a dihydric alcohol and a dicarboxylic acid comprising dissolving the polyester scrap at an elevated temperature in a solvent which is liquid at the dissolving temperature, the solvent being selected from the group consisting of monomeric ester, biester, low-molecular-weight esters and oligomers of the high-molecular-weight polyester to form a depolymerization mixture.

The process then involves continuously adding with the scrap polyester, dihydric alcohol corresponding to the dihydric alcohol of the polyester scrap in a stoichiometric excess sufficient to produce the desired degree of depolymerization to at least an oligomer, effecting the depolymerization under atmospheric pressure at a temperature above the boiling point of the dihydric alcohol but below the melting point of the polyester and continuously recovering lower-molecular-weight depolymerizate.

Recovery from Waste Photographic Films

A process developed by *E.R. Hittel and F.W. Rennie; U.S. Patent 3,652,466; March 28, 1972; assigned to E.I. du Pont de Nemours and Company* is a process for the recovery of polyester base material from films, including photographic films, having at least one macromolecular organic polymer coating comprising (1) treating pieces or flakes of such films with a caustic alkali solution to form a slurry of flakes, (2) feeding the slurry into a classification column and allowing flakes to move downwardly to a rising column of aqueous liquid to separate the polyester material from the coating material, (3) removing flakes settling to the bottom of the column, and (4) recovering the polymeric coating material from the top of the column. The flakes are then dried and used as a source of polyester material.

FIGURE 112: PROCESS FOR POLYESTER RECOVERY FROM SCRAP PHOTOGRAPHIC FILM

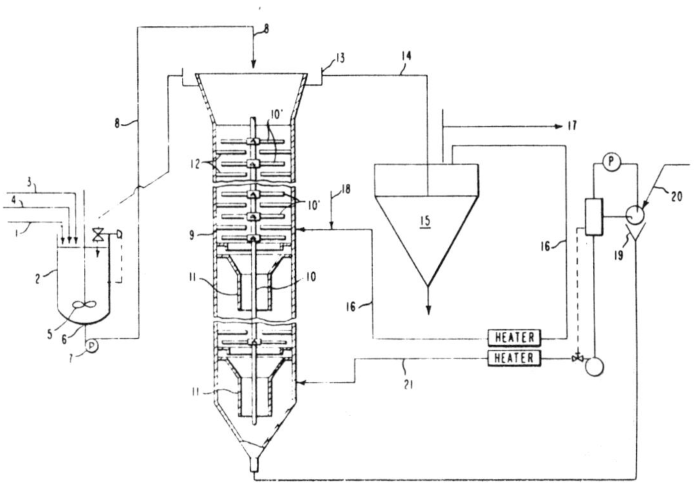

Source: E.R. Hittel and F.W. Rennie; U.S. Patent 3,652,466; March 28, 1972

Figure 112 shows a suitable form of apparatus for the conduct of the process. Referring to the drawing, small pieces or chips of scrap polyester film base in the form of squares having sides ¼" to ¾" are fed continuously through conduit **1** to a treatment vessel **2** that is provided with conduit means **3** for feeding an aqueous solution containing a surfactant into the vessel and conduit means **4** for feeding hot caustic liquid, such as sodium or potassium hydroxide, into the vessel. The vessel is provided with an agitator **5** and has an outlet **6** at the bottom for withdrawing a slurry of the caustic material and the pieces and any removed coating material. The pieces are maintained in the tank for a period sufficient to dissolve the coatings, e.g., gelatin, and to release the silver compounds from any photographic emulsion coating. The slurry is forced by pump **7** and conduit **8** to the top of classification column **9**.

The classification column comprises a cylindrical casing, which in a practical size, can be 3 to 4 ft in diameter and about 27 ft in length. It is provided with a centrally disposed agitator **10** mounted in suitable bearings, not shown. Conduit **8** having a diameter about one-fourth or less that of the column allows for nonturbulent flow of the slurry into the column. Funnel-shaped members **11** having a lower diameter about one-third that of the column are provided in the column. They aid in controlling the flow of liquid and the countercurrent movement of the pieces through the column.

Metal agitator blades **10'**, about 1½" x ½", attached to the shaft and extending to within a few inches of the wall of the column, are spaced along the shaft column. The agitator blades are set about 1' apart. Between each two sets of agitator blades there are disposed metal baffle vanes **12**, about 2" x ⅜" in size. They are attached to and extend from the wall of the column to within a few inches of the shaft. Overflow liquid containing the elutriated coatings that were removed from the polyester base leaves the top of the column through the overflow launder **13** and through conduit **14** into settling tank **15**.

The agitator and baffle elements prevent bridging of the pieces in the column, keeping the pieces from packing and preventing flow; and they help to loosen the coatings from the polyester base. The blades are adapted to rotate very slowly and do not force the movement of the flakes through the column, allowing the polyester base pieces to settle slowly to the bottom of the column, freed from its coatings, by the differences in the settling rates of the coatings and base. The column serves several purposes:

(1) The major portion of any silver halide material entering the column is separated from the polyester pieces or flakes which settle in the column.

(2) The flakes are contacted with a countercurrent flow of liquid which serves to elutriate the polymer skins, silver solids, and whatever other coating materials may be present.

(3) Slight agitation in the column enhances the separation of coatings from the polyester flakes.

(4) As the flakes pass from one section to another, they are washed by the upward flow of water in the column.

The slurry of coated polyester flakes entering the column moves counter to a current of water which is recycled from the bottom of the column after the polyester flakes have been filtered off. The flakes descend in the column where the upward moving liquid separates the polyester chips from the coatings. The flakes settle in the column and the coating components, having a slower settling rate, overflow from the column, passing to the settling tank where the major portion of the silver compounds are removed by sedimentation. Part of the liquor is reheated and returned to the column through conduit **16** as elutriation liquid in the upper portion of the column. The remainder of the liquor passes to waste treatment facilities **17** to recover silver, neutralize caustic, and remove gelatin. An alternate feed of caustic liquor is fed into the column through conduit **18** to replace the caustic removed as waste.

Flakes settling in the bottom of the column pass to a filter **19** for removal of water. After the flakes have been filtered from the solution, make-up water is added to the main flow of water through conduit **20** to replace the water removed as waste, and the combined make-up and filtrate water are reheated to about 100°C at atmospheric pressure and returned to the bottom of the column through conduit **21**. This returned liquor washes the flakes as the water passes upward through the column. The upward flow rate can be 0.5 to 1.5 ft/min. The filtered flakes after being washed free of caustic with water may be dried and stored or melted for reuse. The following is a specific example of the operation of the process.

Example: About 1,500 lb of polyethylene terephthalate film having the composition 74.00% polyester base, 1.03% resin, 11.75% silver halide, and 13.22% gelatin which had been prepared as disclosed in U.S. Patent 2,465,319 and coated as described in U.S. Patent 2,627,088 with a copolymer prepared from 85 parts vinylidene chloride, 15 parts methyl acrylate, and 2 parts itaconic acid, was reduced to small flakes about ¼" square. The flakes were treated in the apparatus shown in the drawing.

Thus, the flakes were slurried in a tank with a 3% (weight/weight) solution of sodium hydroxide and a 0.5% aqueous solution of sodium dodecyl sulfate. The slurry was fed into a classification wash column where the flakes moved countercurrent to water heated to 100°C, the water flowing upward at about 1'/min, separating the emulsion coating and subbing layer from the film base and washing the polyester flakes. After a 45-minute treatment in the wash column, the polyester material was removed and filtered from the aqueous solution and washed with water. The flakes were dried at room temperature.

The polyethylene terephthalate recovered was in a form pure enough to be reused immediately in preparing film base by adding portions of it to molten polyethylene terephthalate prepared in the usual manner for extrusion into a film. Alternatively, the flakes themselves can be melted alone and extruded into film or fibers.

A process developed by *J.T.K. Woo, D.E. Glowe and J.S. Thornton; U.S. Patent 3,873,314; March 25, 1975; assigned to Horizons Incorporated* is a process for the recovery of valuable constituents of photographic film consisting of a polyester film base coated on at least one surface with a subbing layer of vinylidene terpolymer material to which a gelatin silver halide coating has been applied as an overcoating.

This process consists of: preheating small pieces of the film to a temperature sufficient to remove water from the gelatin layer; contacting small pieces of the previously heated film with a polar aprotic solvent for dissolving the terpolymer subbing layer; maintaining the pieces of film in contact with the polar aprotic solvent at about room temperature until dissolution of the terpolymer subbing layer is completed and the gelatin silver halide overcoating has been detached from the film base; thereafter stirring the resulting mixture violently to break up the detached gelatin silver halide coating; and separately recovering the (1) polyester film base, (2) detached gelatin coating and the silver therein, and (3) the terpolymer subbing material.

POLYOLEFINS

Conversion of Scrap to Useful Liquid Hydrocarbon Products

A process developed by *H.G. Schutze and J.A. Dinwiddie; U.S. Patent 3,736,271; May 29, 1973; assigned to Esso Research and Engineering Company* is one in which waste polyolefins are converted to useful products by converting the polymer into a liquid state either by heating the polymer to a temperature above its melting point or dissolving the polymer in a suitable solvent and then initiating a free radical decomposition reaction by the addition of a suitable initiator. Inorganic materials in the polymer are removed while the polymer is in the liquid state by centrifugation, settling, filtration or solvent washing.

Figure 113 illustrates the application of the process to a polypropylene slurry. The polypropylene slurry is in admixture with a solvent. This slurry is introduced by line **1** into a distillation column **2** at a temperature of about 390°F and the slurry contains approximately

FIGURE 113: PROCESS FOR RECOVERY OF USEFUL HYDROCARBON LIQUIDS FROM POLYOLEFIN WASTES

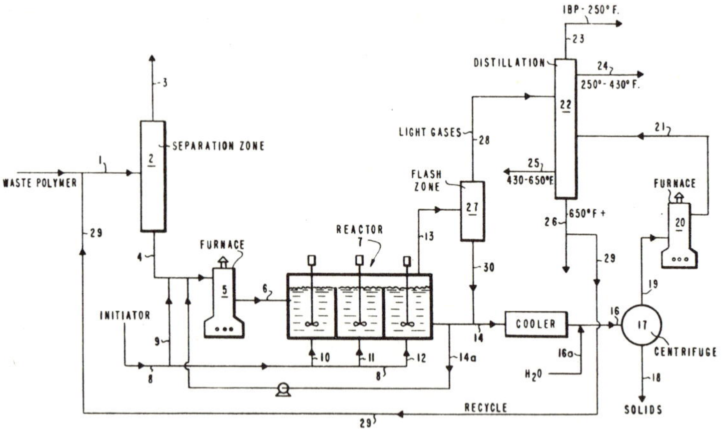

Source: H.G Schutze and J.A. Dinwiddie; U.S. Patent 3,736,271; May 29, 1973

2/1 polypropylene to solvent. If the solvent is present in sufficient amount to have value when recovered or if the solvent leads to undesirable end products, the solvent is removed overhead by line **3** by distilling at temperatures of 450° to 500°F. The polymer, either essentially free of solvent or with small amounts of solvent if the solvent does not lead to undesirable end products, is removed by line **4** and introduced into a furnace **5**. A recycle stream may be required to aid in the removal of the polymer from the bottom of the distillation column **2** and reduce the viscosity of the polymer sufficiently so that the polymer may be pumped to the furnace **5**.

The polymer is heated in the furnace **5** up to temepratures with the range of 550° to 900°F or higher, preferably 600° to 700°F, and then introduced by line **6** to a three-stage reactor **7**. The reactor **7** is illustrated having three stirring sections to approach a plug flow reactor, although other configurations may be used. Initiator is introduced by line **8** to the reactor **7**. It may be preferred to add initiator by line **9** before the polymer is introduced to furnace **5**. If added through line **9**, initiator may be added in an amount between 0.5 and 5.0%, preferably 1 to 3%, by weight based on the polymer. Initiator, preferably a sulfide or disulfide and more specifically toluene disulfide, may be added to each section of the reactor **7** by lines **10, 11** and **12** respectively. Initiator may be added to each section in an amount between 0.5 and 5.0% by weight.

The conditions in the reactor **7** may be maintained at a temperature between 550° and 900°F, preferably the temperatures may be maintained between 600° and 700°F. The total time for the polymer to be maintained in the reactor is from 1 to 4 hours depending on the nature of the useful products desired. During this time, decomposition of the polymer occurs and in some instances with some light gases being formed which may be removed from the reactor by line **13**. The polymer after it has been decomposed in the reactor is removed as a decomposition solution by line **14**. Since the reaction is endothermic, some of the decomposition solution is recycled from the reactor through line **14a** and returned to the furnace **5** in order to maintain the desired temperature in the reactor.

The polymer, after it has been decomposed in reactor **7**, is removed as a decomposition solution by line **14** and passed to a cooler **15** where the reaction solution is cooled. In the case of polypropylene, cooling may be at a temperature of about 150° to 250°F and any pressure buildup reduced to a range of 50 to 100 psig. After the decomposition solution is cooled, it is passed by line **16** to a means for separating the inorganic material, which is preferably a solid bowl centrifuge **17** where the inorganic solids are separated and removed by line **18**. The polymer in the liquid state is removed by line **19** which may contain therein an electrical desalter (not shown) to remove any inorganic solids not removed by the centrifuge **17**, where it may be heated in a furnace **20** and passed by line **21** to a distillation column **22**. In the distillation column **22**, various fractions may be obtained such as, for the purpose of illustration, the fraction up to 250°F, a fraction of 250° to 430°F boiling point, another fraction of 430°F, and if desired a fraction above 650°F. These separate fractions are removed by line **23, 24, 25** and **26** respectively.

The gases which are removed from the reactor **7** by line **13** may be introduced into a flash vessel **27** where the light gases are removed by line **28** and introduced into the distillation column **22**. The fraction boiling above 650°F may be used as a recycle stream and solvent for the process and accordingly recycled by line **29**. Any liquid which is obtained in the flash vessel **27** may be removed by line **30** to be introduced back into line **14** to aid in the pumping of the decomposition solution or to aid in reducing the viscosity of the decomposition solution.

Thus various distillate fractions are obtained as products of the process. Some of these fractions may be utilized by adding to feed materials the usual petroleum refining processes. For example, a 200° to 450°F fraction may be added to cat naphtha to be processed in the usual manner. More specifically the products from the decomposition process are illustrated in two possible alternatives, (1) in the absence of aromatic solvents and (2) in the presence of aromatic solvents.

In the absence of aromatic solvents: (a) The fractions boiling below 650+°F may be hydrogenated and if necessary treated to remove sulfur, to produce a highly branched paraffin. Such a product is an unusual and valuable solvent. (b) The 650+°F fraction may be distilled into the usual lube fractions and hydrogenated, if necessary, to produce high-quality lube oils with no extraction or dewaxing. (c) Using an aliphatic solvent boiling below about 600°F in the decomposition process, the polymer may be decomposed so that it is just barely soluble at room temperature. After separation of the inorganic solids, then the solvent is distilled to produce a product similar to the polybutenes which are useful as adhesives. (d) Alternatively, if a high-melting petrolatum (170° to 195°F) is used as the solvent and the decomposed polymer and solvent are not separated, a product is obtained after fractionation of the light ends which is similar to the polyolefin and high-melting microwax blends used as adhesives and paper-coating products.

If an aromatic solvent is present in the decomposition process, fractions of predominately aromatics are obtained which are high-quality lube materials. Furthermore, the aromatic fractions may be used as specialty oils such as power and communication electrical oils and as viscosity modifiers.

Recovery and Process Reuse of Manufacturing Scrap

A technique developed by *P.J. Boeke; U.S. Patent 3,551,242; December 29, 1970; assigned to Phillips Petroleum Company* is one whereby recovery of waste plastic materials in industrial manufacturing operations is enhanced by a process which comprises applying a coating material such as polyvinyl chloride to a substrate such as polyethylene in a predetermined geometrical pattern. Subsequent manufacturing operations affect only the coated portion of the substrate. The noncoated substrate outside the processing zone is readily recovered in condition for reuse.

POLYSTYRENE

Recovery of Foamed Polystyrene Scrap

A process developed by *C.P. Ronden; U.S. Patent 3,535,408; October 20, 1970; assigned to Cupples Container Company* is one in which scrap foamed vinyl aromatic polymeric material, and particularly scrap foamed polystyrene, is recovered by a procedure including forming the scrap into dense granules, impregnating the granules with a normally liquid low-boiling hydrocarbon blowing agent, blending the blowing agent-containing granules with fresh foamable polymeric particles, and subjecting the resulting mixture to a processing step, such as sheet extrusion, involving heat plastification of the polymeric material.

FIGURE 114: RECOVERY AND UTILIZATION OF SCRAP IN PRODUCTION OF FOAMED VINYL AROMATIC POLYMERIC PRODUCTS

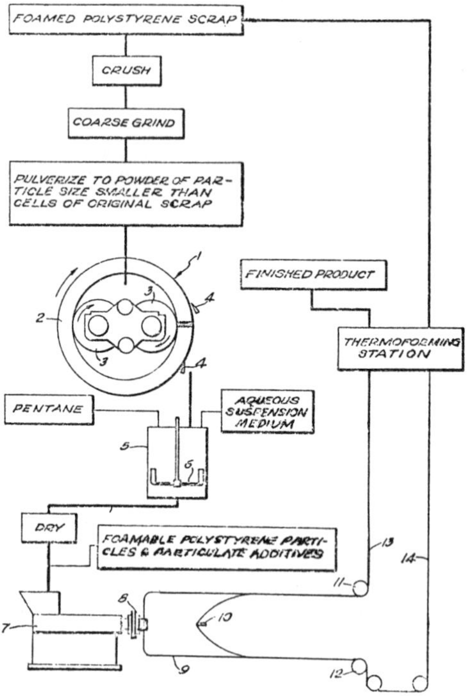

Source: C.P. Ronden; U.S. Patent 3,535,408; October 20, 1970

Figure 114 is a schematic flow diagram of the process. The blend of recovered scrap and fresh foamable polymeric material is delivered continuously to the feed hopper of an extruder **7** equipped with a die **8** having a peripheral orifice, the extruded product being in the form of a tube **9** which is slit continuously at diametrically opposed points by cutter **10**. The two halves of the tube are then led over rolls **11**, **12**, respectively, so providing two flat running webs of foamed sheet, indicated respectively at **13** and **14**. The running webs are led to a thermoforming station equipped with continuously operating thermoformers for forming finished articles from the sheet and separating the articles by a trimming operation. The scrap resulting from the trimming operation is delivered to the crushing apparatus, along with such other scrap as reject products, and the process commences again as hereinbefore described. The process is most advantageous when practiced in connection with manufacturing procedures which employ foamable beads, such as are produced by suspension polymerization, or foamable granules, such as are obtained by extruding filaments of foamable polymeric material and severing the filaments into granules, both the beads and the granules falling within the term foamable polymeric particles as used herein. The foamable particles of fresh polymeric material can be in the range of 0.008" to 0.125".

A process developed by *R.P. McKenzie and C.P. Ronden; U.S. Patent 3,883,624; May 13, 1975; assigned to Grandview Industries, Ltd., Canada* is one in which foamed thermoplastic polymeric scrap, typically scrap from thermoforming polystyrene products from extruded foamed sheet, is converted to essentially virgin resin form and recycled. The scrap is ground to fragmentary form and deposited as a continuous bed of fragmentary material on a metal conveyor belt operated to advance the bed successively through a heating zone, where the foamed scrap fragments are fused and volatiles are eliminated, and a cooling zone, where the fused material is rapidly cooled to the form of a thin reticular web.

The polymeric material has a coefficient of thermal expansion at least several times that of the metal of the belt, and rapid cooling causes the polymeric web to shrink to a greater extent than does the metal. As a result, the polymeric web snaps free from the belt as the belt bends around a roller at the end of the conveyor, and the free web is fed directly into a grinder for conversion to particulate form, the particulate material being recycled to the extruder.

POLYURETHANES

According to *Chemical Engineering,* June 9, 1975, p. 46, about 10% of the rigid polyurethane product produced ends up as scrap during foam fabrication.

Recovery as Liquid Products by Chemical Means

One approach as described in the above-cited article involves deesterification by treatment with a liquid containing 90 to 95% of a glycol and 5 to 10% of a dialkanolamine at about 200°C for about 24 hours. The viscous liquid product can be mixed with new polyol and reacted with polymeric isocyanate to make new rigid foam.

A process developed by *W.R. McElroy; U.S. Patent 3,117,940; January 14, 1964; assigned to Mobay Chemical Company* involves dissolving polyurethanes and more particularly dissolving polyurethanes in active hydrogen-containing compounds to prepare improved solutions containing scrap or waste polyurethane plastics which are suitable for the preparation of new and useful materials.

It has been proposed heretofore in U.S. Patent 2,937,151 to dissolve a cellular polyurethane plastic prepared from a linear polyester and an organic polyisocyanate in a liquid active hydrogen-containing polymeric material by heating a mixture of the cellular polyurethane and the liquid material which has free hydroxyl groups. As stated in that patent, temperatures above about 250°C result in discoloration of the solution of polyurethane plastics. If temperatures above 250°C are avoided in the process, it is necessary to heat the polyurethane for such long periods of time that the process is impractical commercially. Moreover, the resulting solutions of polyurethanes in organic compounds which contain terminal hydroxyl groups are often cloudy and the polyurethane is only partially solubilized.

In the process, the polyurethane is heated in the primary amine in the presence of a tertiary amine catalyst such as, for example, triethylenediamine, N-ethylmorpholine, N-methylmorpholine, N-cocomorpholine, 1-methyl-4-dimethylaminoethylpiperazine and the like. The resulting liquid resins are useful as plasticizers, for example, in the preparation of moldings, coatings for wood and metal, in the preparation of cellular polyurethane plastics and in adhesive compositions.

Pyrolysis to Recover Polymer Raw Materials

A process developed by *N. Matsudaira, S. Muto, Y. Kubota, T. Yoshimoto and S. Sato; U.S. Patent 3,404,103; October 1, 1968; assigned to Bridgestone Tire Company Ltd., Japan* is one in which polyurethane made from a polyether and an isocyanate is thermally decomposed in the presence of (a) an amine and (b) an alkali metal oxide or hydroxide or an alkaline earth metal oxide or hydroxide. The products obtained are (1) the polyether and (2) an amine corresponding to the isocyanate used to prepare the polyurethane where the —NCO groups of the isocyanate are substituted with amino groups.

Use as a Filler in Plastics and Rubbers

A physical approach involves using recycled polyurethane as a reinforcing filler. In this mode of operation, the polyurethane to be recycled is ground to a powder and blended into one of various polymers such as thermoplastic polyurethane, or nitrile rubber or ethylene-propylenediene terpolymer.

POLYVINYL CHLORIDE

Polymer and Plasticizer Recovery by Solvent Extraction

A process developed by *W.B. Sussman, C.D. Belcher and G.E. Brown, Jr.; U.S. Patent 3,624,009; November 30, 1971; assigned to Fiber Process, Inc.* is one in which scrap fabrics containing synthetic resin are first washed with solvent under controlled heat, pressure and agitation to produce as a first product clean, dried, resin and solvent-free fibers. The polymer solution extracted from the washing stage is treated in one or more stages to produce as a second product granular, clean, dried resin compound. Clean solvent is recovered for recirculation through the system.

A very desirable commercial product can be produced by combining a fiber fabric (natural or synthetic, singly or in combination) with a synthetic resin. Such fabric finds widespread use in the automobile industry for the production of upholstery and interiors. The fiber of such a fabric is typically rayon or cotton, and the resin is polyvinyl chloride (PVC).

This material is supplied in rolls or bolts which are then cut for use by the manufacturer. The scrap problem, particularly in the automobile industry, is of enormous proportions. First of all, many thousands of tons of this scrap material are generated. Because of the chemical composition of the material, it is impossible to burn the scrap without causing harmful air pollution. Therefore, the bulk of the scrap is used today as landfill, which is the only practical means of disposal yet discovered.

Rapidly developing technology has produced effective solvents for such synthetic resins. Stripping processes utilizing these solvents have been proposed, but never successfully developed for a variety of reasons. In the first place, the cost of the resin and processing in addition to the cost of scrap isn't justified by the reclamation of cleaned fibers alone.

The process overcomes these problems, however. The operation will be described with reference to the flow sheet in Figure 115. Suitable scrap raw material from the hopper **10** is delivered to the hogger **12** which serves to reduce the raw material to pieces of small, relatively uniform size. Preferably, the hogger should be designed to reduce the scrap material to a size not larger than approximately 8 square inches, with the bulk (on the order of 75%) of material in a size range 1 square inch to 4 square inches.

The sized material from the hogger is delivered on the conveyor **14** past the dryer **16** and into the feed hopper **18**. While the starting material for the process is dry in the common sense of the word, the dryer is utilized to drive off ambient or residual moisture from the scrap material. The elimination of residual moisture at this step is important in minimizing azeotrope formation of the solvent as will be explained hereinafter.

FIGURE 115: FLOW DIAGRAM FOR PVC RECOVERY FROM COATED FABRIC

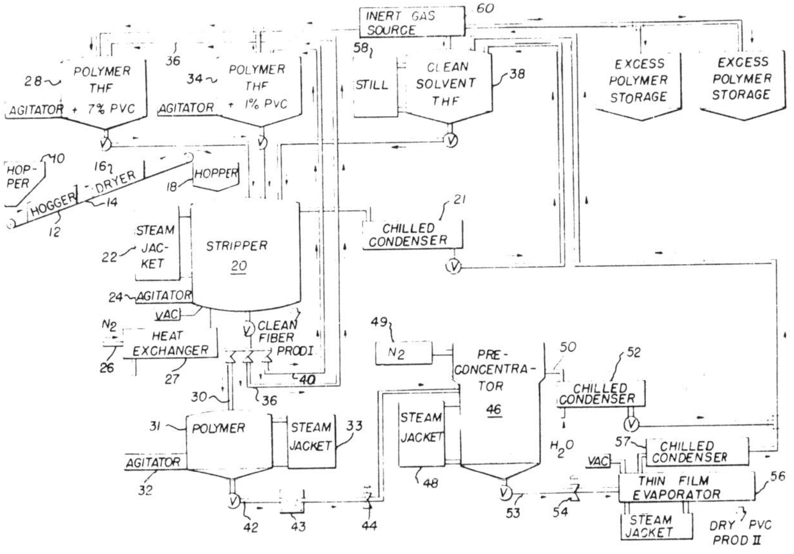

Source: W.B. Sussman, C.D. Belcher and G.E. Brown, Jr.; U.S. Patent 3,624,009; Nov. 30, 1971

Material from the feed hopper **18** is delivered in suitable batches to the stripper shown schematically at **20**. The stripper is provided with a steam jacket **22** or other suitable heat source, and an agitator **24** which can provide constant or periodic agitation of the contents of the stripper. Once the raw material is suitably delivered to the stripper, the unit is sealed and charged with inert gas from a source indicated schematically at **26**. Additional heat may be supplied to the stripper by the heat exchanger **27** in the inert gas supply. Temperatures during the stripping operation are maintained below 200°F in order to prevent degradation of the natural fibers.

The first washing step is accomplished by supplying a polymer solution from the vessel **28**. The polymer solution from this vessel is a mixture of tetrahydrofuran (a known solvent for PVC, hereinafter referred to as THF) containing about 7% PVC. A predetermined quantity of this washing solution will be charged into the stripper **20** and the first washing cycle carried out. At the conclusion of the first cycle, the liquid withdrawn from the stripper via the conduit **30** (containing 10% PVC) is taken to the vessel **31** for further processing as described hereinafter.

The second wash cycle is carried out utilizing a polymer solution from the vessel **34**, which includes a mixture of THF and about 1% PVC. This washing cycle, in terms of time, temperature and agitation will be substantially the same as the first described washing cycle. At the conclusion of this second wash, the liquid will be withdrawn via the conduit **36**. This material, containing about 7% PVC is then delivered to the vessel **28** for use in the first washing cycle in the next batch of material.

The third washing cycle utilizes clean solvent (commercially pure THF) from the vessel **38**. This third wash cycle will be the same as the first and second wash cycles in terms of temperature and agitation. At the conclusion of the third washing cycle, the liquid will be withdrawn via the conduit **40**. This withdrawn liquid contains about 1% PVC, and is delivered to the vessel **34** for use in the second wash cycle of a subsequent batch of material. After the third wash, additional clean solvent is used to flush the residual PVC from the fibers. Flushing is continued until the solvent withdrawn via the conduit **40** is clear as determined visually by a sight glass or the like in the conduit.

It will also be understood that in order to prevent contamination of the withdrawn solvent with small fiber particles, conventional filters will be employed in each of the conduits **36** and **40**. Any solvent which is vaporized during the washing cycles is passed through the chilled condenser **21** and returned to the clean solvent storage vessel. After extraction of the liquid following the third wash cycle, the clean fabric fibers are dried completely by continued application of heat and heated inert gas under a vacuum. When dry, the fabric fibers may be removed from the bottom of the stripper **20** in a fully dry condition, substantially free from both PVC and trace THF. This product (designated product I) finds a ready market in, for example, the production of high-rag-content papers.

It will be recalled that the first liquid extraction from the stripper, a polymer mixture solution containing approximately 10% PVC, was conveyed to the vessel **31**. It should be noted that this vessel includes an agitator **32** and steam jacket **33** to prevent settling of PVC compound. This material may continuously be fed through the conduit **42**, and the filter **43** to the flow regulator **44**, and then to the preconcentrator **46**. This preconcentrator which does not form a part of the process per se, is essentially a vertical tube rising film evaporator. It is provided with a steam jacket or other suitable supply of heat **48**, and a supply of inert gas **49** which will be utilized for purging prior to start up. Volatilized THF is taken off the preconcentrator at **50** and led to the chilled condenser **52** where the solvent vapor condenses to clean liquid solvent. This clean liquid solvent is, of course, returned to the storage vessel **38** for further use in the process.

After this preconcentrating step, the liquid taken off the bottom of the preconcentrator **46** contains on the order of 20% PVC. This liquid mixture is then fed by gravity through the conduit **53** and flow regulator **54** into the thin film evaporator **56**. This thin film evaporator utilizes means for distributing a thin film of the polymer solution on the inner surface of a heated drum or cylinder. The unit is under a slight vacuum and the solvent is continuously flashed off as a vapor which is condensed back into clean liquid solvent. The chilled condenser can be designed as an integral part of the thin film evaporator, or could constitute an entirely separate unit as shown at **57**. The clean liquid solvent is then transported back to the vessel **38** for further use in the process. At the same time, dry PVC is continuously extracted in solid form from the thin film evaporator **52** which represents product II of the process.

Actual tests have established that the dried PVC product just described contains less than ½% residual THF. There are no residual fibers of a length in excess of 0.1" in the reclaimed PVC. The starting chemical properties of the PVC in the scrap material are unchanged in that plasticizers, stabilizers and fillers are substantially intact. Detailed tests have established that the physical properties of the PVC are in fact improved because of the thorough homogenization caused by the reduction of the PVC to solution during the process. As such, the reclaimed product II (PVC) is equal or superior to the original PVC, and as such is a highly valuable product.

It will be observed from the foregoing description that the solvent utilized in the process travels in a fully closed system. Specifically, the solvent passes from the clean solvent storage vessel **38** through the stripper **20** to the vessel **34**, again through the stripper **20** to the vessel **28**, and the third time through the stripper **20** to the vessel **32**. By this time, the clean solvent has picked up on the order of 10% PVC in solution and it proceeds through the preconcentrator **46** and thin film evapotator **56**. The application of heat during the washing cycles in the stripper and in the preconcentrator will flash off certain quantities of solvent in vapor form which are reclaimed in the chilled condensers **21** and **52** respectively. The remaining solvent is condensed at the thin film evaporator **56**, and all clean solvent is returned to the vessel **38** for reuse. It will be observed that an azeotrope still **58** is combined with the vessel **38** to continually remove trace moisture from the solvent.

Total system loss of solvent during the process is less than ½%. It is important that the entire process be carried out with a water-free atmosphere. The system should therefore be purged with inert gas prior to start up. While the drawing refers to N_2, it will be understood that any available inert gas may be utilized. As previously described, inert gas may be supplied to the stripper **20** as at **26** and to the preconcentrator **46** as at **49**. Ordinarily inert gas from a supply **60** will be supplied to other parts of the system such as to the vessel **28, 34** and **38**.

A process developed by *E.A. Hafner; U.S. Patent 3,836,486; September 17, 1974; assigned to Hafner Industries, Inc.* is one in which reusable vinyl chloride polymer and copolymer and other components such as plasticizers are reclaimed from scrap plastics by contacting the scrap plastics with a solvent for the vinyl chloride polymer, contacting the resulting liquid phase with a substantially nonaqueous nonsolvent for the polymeric portion, and then separating the resulting insoluble phase containing the vinyl chloride polymer.

Following the removal of this polymer portion, the filtrate is fractionally distilled to recover the solvents, nonsolvents, plasticizers and other ingredients. Typical solvents are methyl ethyl ketone, methyl isobutyl ketone, cyclohexanone, dimethylformamide, tetrahydrofuran and ethylene dichloride. Typical nonsolvents are methanol, ethanol, isopropanol and n-butanol. The recovered vinyl chloride polymer compositions may contain pigments and fillers, and exhibit improved performance on recycling in the fabrication of plastic articles.

FIGURE 116: FLOW DIAGRAM FOR PVC RECOVERY FROM SCRAP PLASTIC

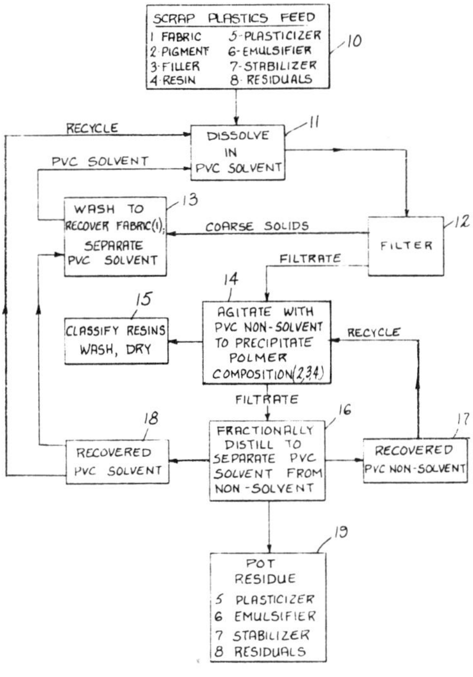

Source: E.A. Hafner; U.S. Patent 3,836,486; September 17, 1974

Figure 116 is a block flow diagram of the process. The vinyl chloride polymer scrap plastics feed **10** may typically contain the eight listed ingredients where the resin component is a vinyl chloride polymer. The first essential stage of the process is the solvent addition step **11** which is followed by a separation step represented by a filtration **12**. The coarse solids (solvent insolubles) resulting from the filtration may then be further treated in a recovery stage **13**, as by washing to remove fabric and associated insolubles, and to separate the solvent. The filtrate from stage **12** may be concentrated to remove solvent, and then centrifuged, settled, or further fine-filtered to remove components such as pigments and fillers.

In a preferred embodiment, the pigments and fillers are left in the filtrate and are then coprecipitated at stage **14**, while the filtrate is hot and with suitable agitation, with a nonsolvent, coincident with precipitation of the vinyl chloride polymer composition. By suitable choice of nonsolvent and quantities thereof it is also possible at this point to classify the polymer composition into resins of various molecular weights or, when mixed plastics are in the feed, to separate the resins by type of polymer, followed by washing and drying of the resins as desired, as at stage **15**.

The filtrate from the nonsolvent addition step **14** is then fractionally distilled at stage **16** to separate and recover the solvent and nonsolvent for removal to holding tanks **17** and **18**. The solvent then may be recycled to washing stage **13** and solvent addition stage **11** and the nonsolvent may be recycled to the nonsolvent addition stage **14**.

SYNTHETIC FIBERS

Recovery from Bonded Nonwoven Fabric Waste

A process developed by *A.H. Drelich, D.K. Smith and U.P. Strauss; U.S. Patent 3,873,411; March 25, 1975; assigned to Johnson & Johnson* is a process for recovering reusable fibers from mixtures of waste fibers and/or fabrics and synthetic, crosslinked polyacrylate resins. The steps involved in the process are:

(1) heating a mixture of waste fibers and/or fabrics and synthetic, crosslinked polyacrylate resins in an aqueous saponifying solution containing an alkali metal hydroxide and a water-soluble salt containing a monovalent cation, which solution saponifies the synthetic, crosslinked polyacrylate resins and initiates their decomposition or solubilization without undesirably swelling the same;

(2) treating the mixture of waste fibers and/or fabrics and synthetic, crosslinked, saponified, polyacrylate resins in an aqueous washing solution containing a water-soluble salt containing a monovalent cation to remove alkali therefrom; and

(3) washing and recovering the reusable fibers from the mixture of waste fibers and/or fabrics and decomposed or solubilized synthetic, crosslinked polyacrylate resins.

Recovery from Rayon Blends with Polyesters or Acrylics

See Recovery from Cotton-Synthetic Fiber Blends under Cotton Fibers, page 391.

Recovery from Spinning and Processing Wastes

In the spinning and processing of man-made fibers, such as fibers of thermoplastic materials, there is usually a large amount of waste created, which is due to a number of reasons. Such waste may be due to roll wraps; from starting up a processing line; or as a result of malfunctions in some of the processing equipment. The fiber waste is usually in continuous lengths, and may either be lubricated or nonlubricated, depending at which stage of the processing the waste occurred.

Attempts have been made to segregate the lubricated waste fiber from the nonlubricated, and then baling each type for subsequent disposition. Sometimes these attempts have not been successful, especially if the waste occurs over an area of a processing line including a lubrication booth with the result that part of a waste length would be nonlubricated, and the other part would be lubricated. In the latter situation it is not economical to attempt to cut one portion from the other as it is more vital that the operators on a processing line get the operating line back into operation and make certain that the fibers passing along the processing line are moving as they should.

One manner of handling the waste bales of fibers previously has been to burn them since no successful and commercially practical manner of recovering the waste material was known. The fiber and any lubricants in the fiber have created quite a pollution problem, and thus presented undesirable conditions as well as making prohibitive the cost of the facilities and personnel for handling the burning.

Another manner employed was to store the bales until some commercially effective method was developed for recovering the waste. In the meantime, however, since each bale weighs in the vicinity of about 500 lb, storage space became a serious and costly problem, as well as being costly to move the waste bales from the processing lines to the storage area. Still another manner involved using the waste for landfill.

Various approaches have been investigated for recovering the waste fiber in an attempt to reuse it, but these have usually been disadvantageous either due to being too expensive and thus not commercially practical, or the degradation of the material was such that the resulting product was of inferior quality and color.

One approach with respect to polyester waste such as polyethylene terephthalate (PET) waste, has been to treat fiber waste with superheated steam at a temperature below the melting point of the polyester as disclosed in U.S. Patent 3,098,046. The fiber waste so treated becomes brittle and then can be reduced to a powder form. The powdered PET is next suspended in a liquid medium, such as methanol or glycol, and the resulting dimethyl terephthalate is then used again for the production of a polycondensation product.

Another approach as disclosed in Defensive Publication T870,010, published January 13, 1970 in the Official Gazette of the U.S. Patent Office (870 O.G. 393), is to pass the waste PET tow lengths through a tow dryer and to raise the temperature from a normal 140° to 215°C. The result is that the waste tow becomes highly crystallized and brittle. This material is then passed through a hammermill and reduced to particles. The resulting product, however, has a low bulk density of 15 to 20 lb/ft^3, whereas virgin polymer powder has a bulk density of about 50 lb/ft^3. One reason for the low bulk density is that there is still some fibrous-like material in the product; also the particles are not entirely reduced to powder.

It is very expensive and time-consuming to continue processing the material in the hammermill until it is totally reduced to powder. It is difficult to blend the resulting material of lower bulk density in with virgin polymer powder. The extruder into which the powder is fed for melt extrusion into a fiber is designed to take the higher bulk density virgin polymer powder, but the recovered lower bulk density material cannot be fed into the extruder fast enough to fill up the extruder screw flights so as to maintain a constant feed.

Additionally, a certain amount of undesirable color is introduced, and the recovered material still has a certain amount of lubricant remaining so that it is not possible to spin a first-quality fiberfill which can be used for such things as matresses, pillows and air filters. When attempt was made to recover waste polyester material of poly(1,4-cyclohexylene-dimethylene terephthalate) (PCHDMT), on passing of the waste material through the tow dryer at the required elevated temperature, such as 265°C, the material was caused to become undesirably brown in color, and thus was not suitable for commercial use. In attempts to reduce the waste materials by use of chemicals, the cost of the chemicals was too much and the process was found to be too costly to operate.

A process developed by *N. Rosenstein and A.J. Rosenstein; U.S. Patent 3,429,018; Feb. 25, 1969; assigned to Spunize Company of America, Inc.* is one in which a number of supply packages of waste synthetic filamentary material containing knots, excessive twist, excessive finish etc. are creeled and the ends of such packages are fed in the form of a web of spaced ends between counter-rotating rollers. The rollers are pressure-loaded sufficiently to substantially destroy the knots and open the individual filaments. After passing the rollers the material is stuffer-box-crimped while still in web form and the crimps set. The resultant material is cut into staple lengths to produce a spinnable staple having three-dimensionally crimped fibers.

A process developed by *R. Hester and H.P. Pierson; U.S. Patent 3,703,347; November 21, 1972; assigned to Eastman Kodak Company* involves recovering waste thermoplastic fiber and particularly a large otherwise unmanageable, entangled mass of continuous length filamentary material, and converting it into reuseable feedstock. The continuous length filamentary material is converted into random length staple fiber, the staple fiber is then mechanically compacted and melted into a melt flow, and the melt flow is finally extruded as the reuseable feedstock.

A very similar process is described by these same inventors in *U.S. Patent 3,795,468; March 5, 1974; assigned to Eastman Kodak Company.*

TEXTILES

The U.S. textile industry consumed approximately 5 million tons of textile fiber in 1970, an increase since 1960 of 61.5%. Far more significant for textile recycling was the change in the type of fiber consumed, with a major shift occurring from use of natural to man-made fibers. In 1960, natural fibers constituted 69% of fiber consumption; man-made fibers, 31%. In 1970, the figures were 39% for natural fibers and 61% for man-made. By 1980, the ratio of natural to man-made fibers is expected to be 25 to 75.

In 1970, an estimated 0.8 million tons of waste textiles were processed by waste textile dealers and sold (recycled) to various markets. In addition, an undetermined amount of used clothing that potentially would enter the waste stream was collected by social welfare agencies and redistributed.

Textiles represent only a small portion of municipal solid waste. In 1968, textiles in collected municipal solid waste totaled 1.2 million tons, or 0.6% of the total. Most of the textile consumption that does not appear in the municipal waste stream is either collected by social welfare agencies, disposed of, sent to secondary textile dealers by industry, or is being accumulated in households.

Figure 117 represents the major sources and markets for textile waste. The mill waste is the home scrap of the textile industry; the manufacturing waste, the prompt portion, and consumer discards, the obsolete.

In contrast to most of the other materials discussed in this volume, the home scrap (mill waste) is not reused within the generating plant, but instead, passes through the secondary textile dealers. Mill waste accounted for about one-third of the material handled by waste textile dealers in 1970.

Waste from fabrication (prompt) is a considerably less important source of recycled textile waste than is the case with many other materials. It has been estimated that waste recovered from fabrication is only 60% of that generated. Fabrication waste accounted for an estimated 20% of the waste handled by waste textile dealers in 1970. Obsolete waste accounted for the remaining 45% of recycled textile waste. This waste is provided mostly by social welfare agencies and institutions such as Goodwill Industries from items deemed unsuitable for reuse as clothing.

FIGURE 117: WASTE TEXTILE UTILIZATION FLOWS

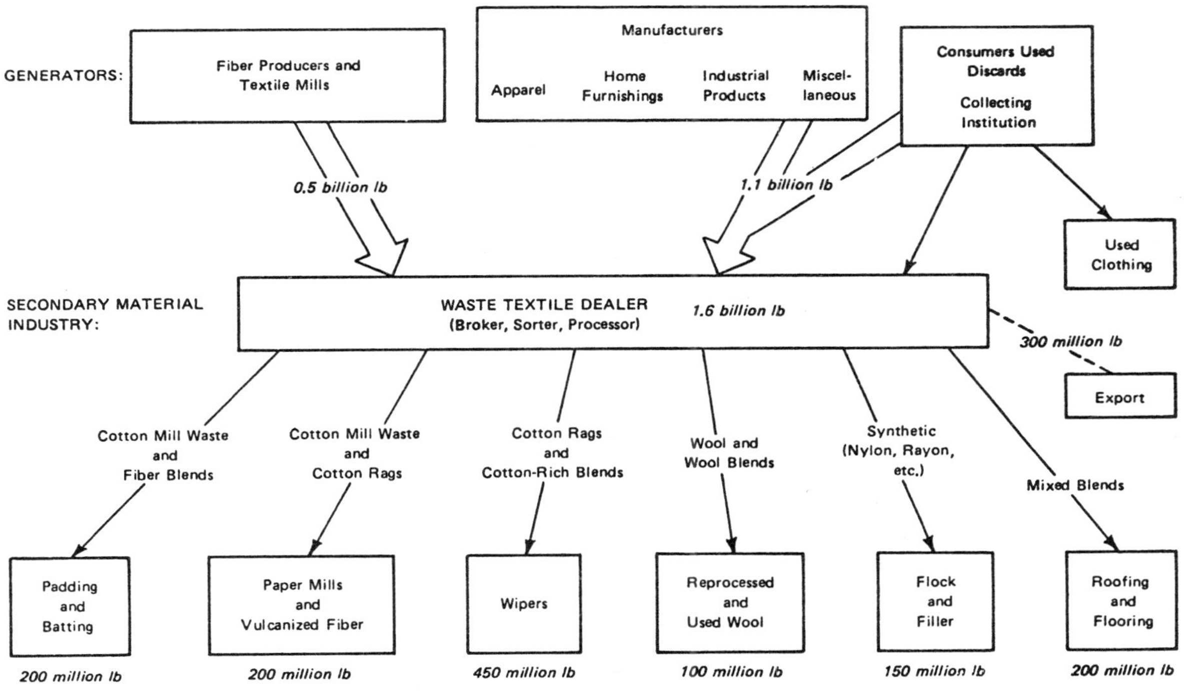

Source: PB 212 731

RUBBER

The reader is also referred to *Reclaiming Rubber and Other Polymers* by J.A. Szilard, Park Ridge, N.J., Noyes Data Corp. (1973) for more information on this topic.

THE STATE OF SCRAP RUBBER UTILIZATION

In 1971, 266 million tires for passenger cars, trucks, and motorcycles were shipped by domestic tire manufacturers or imported for consumption in the United States. In that year, tires taken out of service totaled 240 to 250 million.

Of the tires taken out of service, approximately 46 million were retreaded, 7 million were consumed by the rubber reclaiming industry, and roughly 2 million were consumed by tire splitters. The remaining 185 to 200 million tires, weighing approximately 2.4 million tons, were disposed of by retailers, retreaders, or consumers or were left on discarded vehicles. The most common disposal method is land disposal, but many tires become litter or are left to accumulate at various locations.

Tires are one of the most difficult consumer wastes to dispose of properly. In sanitary landfills, whole tires cannot be effectively compacted and tend to work their way up to the landfill surface. They also consume more landfill space per unit weight than other items and are not biodegradable. Tires on the surface landfills or dumps or littered in urban or rural areas provide nesting places for rodents, flies and mosquitoes.

Shredding, splitting, or otherwise reducing the size of tires overcomes the problem of landfilling to a large degree. Shredding is the most feasible method of physically altering the tires but is presently employed only in a few locations. Cost is the primary deterrent, and neither public nor private entities are likely to pay the price of processing in the absence of disposal regulations.

A small number of tires can be combined with large quantities of refuse in incineration as long as they do not comprise more than 5% of the charge. Larger percents of tires result in damage to furnace walls and also require flue gas control. Tires represent only 1.5% of the municipal waste stream but are often delivered to a disposal site in large batches rather than in an even day-to-day flow. Shredding of the tires would allow the percent of tires in the incinerator charge to be increased above 5% if the shredded rubber is thoroughly mixed with other waste materials. There are three means by which rubber tires are presently recycled: retreading, conversion by rubber reclaimers into new products, and physical conversion by tire splitters into new products.

Retreading: As percents of new tire shipments, retreaded passenger tires dropped from 25% in 1963 to 17% in 1968. Truck tire retreads dropped from 32 to 28% in the same period. Tire performance and consumer preference are two critical factors inhibiting the expanded usage of retreads. Poorly constructed retreads do not perform as well as new tires, especially at high speeds. This is primarily because the bond between the carcass and tread can fail, causing the tread to come loose from the tire. This bond could be improved if the precise chemistry and carcass dimensions of every retreaded tire were known. However, tires of various manufactures have slightly different chemistry or size, making the development of a perfect bonding agent and exact tread fitting difficult.

Consumers have a preference for new tires because retreads are felt to be inferior. Manufacturers sell an increasing variety of low-priced new tires, which in essence compete with retreads. A factor that would ultimately restrict retreading is the technical suitability of old carcasses for retreading. Only about 35% of the discarded passenger tires are suitable for retreading. This is largely because consumers allow tires to become too worn before replacing them and because driving on underinflated tires weakens the tire carcass. Better tread bonding, leading to consistent tire performance, and consumer education and cooperation would be required to increase tire retreading.

Rubber Reclaiming and Tire Splitting: Rubber reclaiming and tire splitting are limited by the market potential for the products they produce. Reclaimed rubber is not technically suitable for making new tires. The total production of other lower-grade rubber products produced by reclaimers and tire splitters (e.g., doormats, hoses, and belts) is extremely small relative to tire discards. In 1969 the entire output of these industries was only about 290,000 tons.

THE RECLAIMED RUBBER INDUSTRY

The present status and environmental impacts of the reclaimed rubber industry have been reviewed in EPA 440/1-74/030, Washington, D.C., U.S. Environmental Protection Agency (August 1974). As noted there, the quantity of scrap rubber being reclaimed and reused and the number of rubber reclaiming plants operating in the United States have both steadily declined over the past decade. This decline has occurred despite increased production and use of rubber products. As a percent of new rubber produced, reclaimed rubber fell from 19% in 1958 to 10% in 1968.

Some of this reduction is probably due to development of new rubbers not compatible with reclaimed rubber, but, undoubtedly, the major decreases were caused by cost, quality, and environmental reasons. Competitive materials, such as rugs and colored plastics, have reduced usage in automotive mats and mechanical parts from 105 million pounds in 1960 to approximately 55 million pounds. Similar reductions are noted for other mechanical goods, hose, shoe heels and soles, and hard rubber products.

Reclaimed rubber is the product resulting from the treatment of ground scrap tires, tubes, and miscellaneous waste rubber articles with heat and chemical agents whereby a substantial devulcanization or regeneration of the rubber compound to its original plastic state is effected, thus permitting the product to be reprocessed, compounded, and revulcanized. The term "devulcanization" which is frequently associated with reclaiming is a misnomer. Actually all the commercial reclaiming processes employed are based on depolymerization of the rubbers. This depolymerization can occur either by promoting thermal scission or breaking of the polymer chain or by oxidation at points other than at cross-linking sites. Some scission of the existing cross-links may also occur.

There are two fundamental factors which determine the type of reclaim. The first, and the most important, is the type of scrap from which the reclaim is made. The second is the process by which the scrap is reclaimed. By far the most important source of raw material is tire scrap. The supply is plentiful and well distributed so that it is relatively easy to collect. The quality of rubber in tires is high, giving an unusually high percentage of rubber

FIGURE 118: FLOW DIAGRAM OF TYPICAL MECHANICAL, PAN (HEATER), AND WET DIGESTER RECLAIM PROCESSES

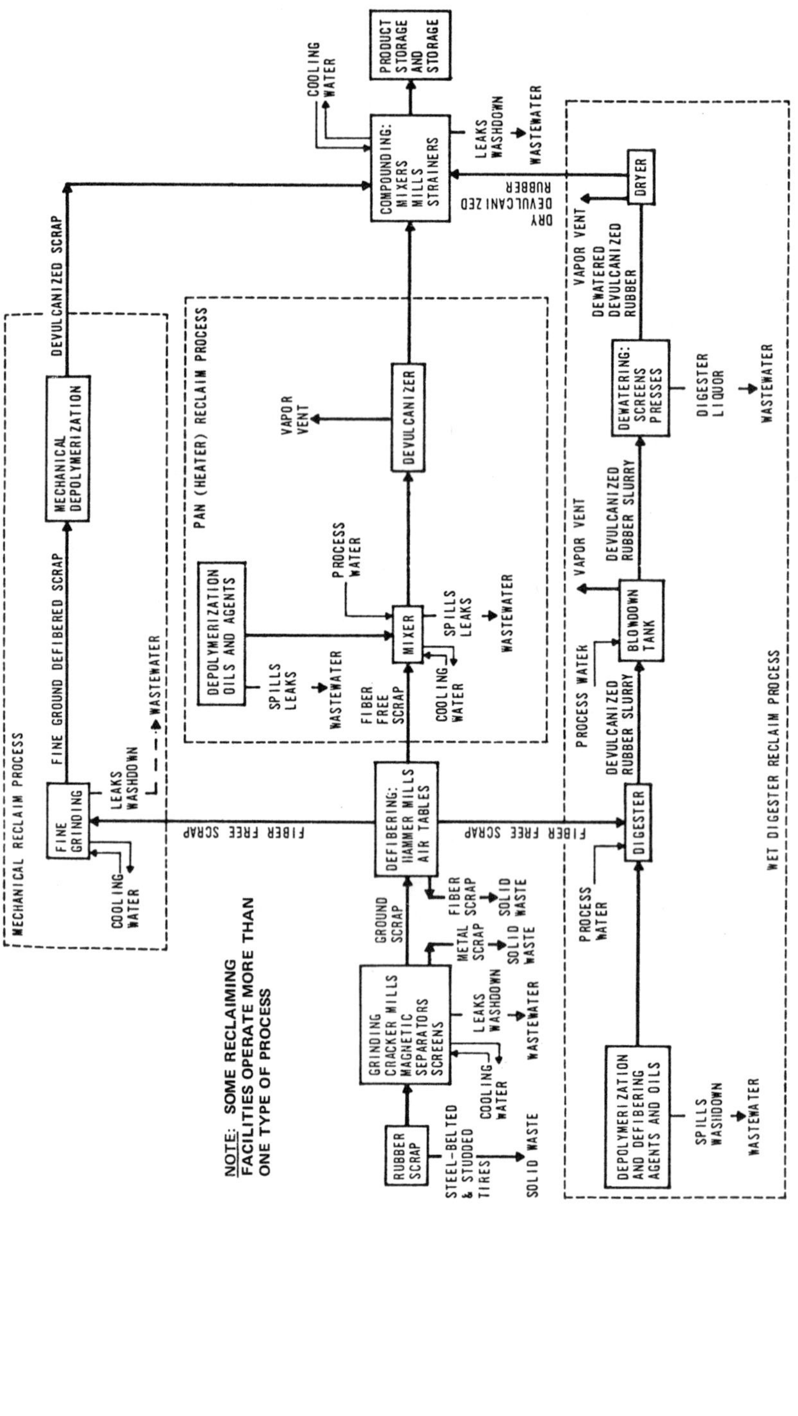

Source: EPA Report 440/1-74/030

hydrocarbon at low cost. The whole tire creates problems due to the tire-cord fiber contained in the carcass portion. This fiber has to be removed either by mechanical means or by chemical methods such as those used in the digester process.

Three basic techniques are used at existing plants to produce reclaimed rubber: the digester process, the pan process, or the mechanical process. A generalized material flow diagram for the three processes is shown in Figure 118. Broadly, the reclaiming process can be divided into three major parts, two of which are mostly mechanical and the other predominantly chemical. The rubber scrap is first separated and ground, then given heat treatment for depolymerization, and finally processed by intensive friction milling. All three processes employ similar rubber-scrap separation and size-reduction methods. They differ in the depolymerization and the final processing steps.

Rubber Separation and Size Reduction

The rubber scrap is first sorted, and then reduced by mechanical chopping or cracking on a very heavy cracker mill to a suitable size for the particular depolymerization step being used. A cracker mill consists of two horizontal heavy steel rolls revolving at different speeds. The roll surfaces are corrugated, giving a scissor-cutting action. In the case of tires, the wire beads are broken in the cracker mill and removed with magnetic separators along with any other ferrous material which may have been picked up during their use. The cracked ground stock is conveyed to a vibrating screen of a given mesh size. The oversized material is returned to the crackers for further grinding. The stock which passes through the screen is conveyed to storage bins to await use in the depolymerization process.

Most reclaiming plants require fiber-free scrap in the depolymerization process. A series of screens, air separators, and sizing equipment are used to remove fiber from ground rubber scrap. First the rubber scrap is passed through hammer mills and beaters which removes fiber from the rubber. This screened mass then goes to an air-flotation table where the final separation of fiber and rubber takes place. After passing through magnetic separators, the rubber crumb which results goes to storage bins. Rubber scrap separation and size reduction is followed by the appropriate depolymerization process.

Depolymerization

Digester Process: The digester process consists of placing the ground scrap, water, and reclaiming agents into a steam-jacketed agitator-equipped autoclave (digester). The batch is then cooked for 5 to 24 hours at 370° to 405°F. During this time the rubber becomes partially depolymerized, and attains the consistency of soft granules. Reclaiming agents that are used include petroleum and coal tar-base oils and resins, as well as various chemical softeners, such as phenol alkyl sulfides and disulfides, thiols (mercaptans), and amino compounds. Reclaiming agents or oils are used to speed the depolymerization, to impart desirable processing properties. Natural rubber can be reclaimed in the absence of any reclaiming agents. SBR types, however, require the assistance of these reclaiming agents or catalysts to produce a reclaim. The reclaiming agents generally function by catalyzing the oxidative breakdown of the polymer chain and oxidatively disrupting sulfur cross-links.

Rubber scrap which has not been defibered mechanically requires chemical degradation of the fibers in the digestion process. Defibering agents, such as caustic soda or the chlorides of zinc and calcium, and plasticizing oils are added to the digester to complete the charge. The fiber from tires and other rubber scrap products is hydrolyzed and goes into solution. The presence of synthetic rubber such as SBR in the tire scrap necessitates the use of the metallic chlorides in place of the caustic soda since the caustic solution produces a heat-hardening effect with SBR instead of softening.

At the end of the digestion period the contents of the digester are blown down under internal pressure into a blowdown tank. Water is added to the soupy mass to facilitate the subsequent washing operations. After thorough agitation, the mixture is discharged onto continuous vibrating screens where a series of spray nozzle showers wash the rubber free from the digester liquor. The washed rubber, which is in the form of a slurry, is then passed through a dewatering press which forces out much of the occluded water. The dewatered rubber is dried in a hot-oven or tray dryer prior to further processing.

In cases where chemical defibering is carried out in the digestion process, the rubber has to be washed free from the decomposed fiber as well as the digestion liquor. This washing procedure generates wastewater which can be alkaline or laden with metal chlorides and fine sludge-like particles of hydrolyzed fiber and rubber. Chemical defibering and the subsequent washing process create an effluent problem. At present, mechanical fiber separation is used to reduce the wastewater problem inherent in the digester process.

Pan Process: The finely cracked scrap, which is usually free from fiber, is reduced to a finer particle size by grinding on smooth steel rolls. The finely ground scrap is blended in an open mixer with the correct amount of reclaiming oils and is then placed in open pans which are stacked on a carriage and placed in a large horizontal heater. The heater is a single-shell pressure vessel into which live steam is passed. Depolymerization is carried out at about 365°F for 2 to 18 hours. After this treatment the heater is vented, the pans discharged, and the cakes of rubber sent on for further processing. Since the condensate from this operation is highly contaminated with gums, resins, etc., it cannot be returned to the boilers and, therefore, must be treated as a waste. This waste is similar in composition to the digester alkali and chloride wastes.

The pan process is relatively inexpensive because the equipment is simple, and the washing and drying steps are eliminated. The supply of cheap fiber-free scrap rubber can be a limiting factor in the operation of the pan reclaiming process.

Mechanical Process: The mechanical reclaming process, unlike the other two preceding processes, is continuous. The fine ground, fabric-free rubber scrap is fed continuously into a high-temperature, high-shear machine. The machine is a horizontal cylinder containing a screw for forcing and working the material along the chamber wall at 350° to 400°F in the presence of reclaiming agents and catalysts for depolymerization. The rate of depolymerization is controlled by the speed of screw while the compression and temperature is maintained constant. The discharged reclaimed rubber needs no drying and is ready for further processing.

Final Processing

The final stage of the reclaiming operation, namely milling, first involves the mixing and blending of the material from any of the various depolymerization processes in a Banbury internal mixer with small amounts of reinforcing materials such as clay, carbon black, and softeners. This aids in smoothing dried stock and obtaining uniformity. The reclaim is then given a preliminary refining on a short two-roll mill having a high-friction ratio between the roll surfaces. The sheet thickness after the first refiner pass is about 0.01 inch. The reclaim is then strained to remove foreign matter before going to the final thickness of 0.002 to 0.005 inch. The strainer is an extruder which contains a wire mesh screen held between two strong perforated steel plates in the head of the machine. The strainer not only removes foreign matter, but also plasticizes and blends the reclaim.

Each reclaimer may complete his reclaiming operations in either of two ways—by sending his product to the customer in the form of slabs stacked on pallets or bales. Slabbed reclaim is made on a mill and the discharged sheet is wrapped on a rotating drum of a specified diameter, until the proper thickness is obtained. The wrapped layers or sheet are then cut off the drums, forming a solid slab of a certain length, width, and weight. The slabs are then dusted with talc to prevent sticking to each other, tested, and shipped to the customer. Baled reclaim is also made on a mill, except the thin milled sheet is conveyed to a baler, where the rubber is compacted to form a bale. The bale is then encased in a bag, stacked on a pallet, tested, and sent to the customer. The effluent wastewaters occurring during the various reclaiming processes are identified in Table 30.

TABLE 30: PROCESS-ASSOCIATED WASTEWATER SOURCES FROM RUBBER RECLAIMING

Plant Unit or Area	Source	Nature and Origin of Waste Water Contaminants
Wet Digester Reclaim Process		
Grinding	Washdown, spills, leaks	Bearing oil from machine drives. Oil from seals on milling equipment.
Depolymerization, Defibering and Oil Storage	Spills, runoff	Oil, solids, caustic and organics.
Blowdown Tank	Air pollution equipment	Oils and organics.
Dewatering	Digester liquor	Oil, solids, caustic and organics from spent depolymerization and defibering agents and excess oil. High concentrations of fibrous material removed from tires.
Dryers	Air pollution equipment	Oils and organics.
Compounding	Washdown, spills, leaks, air pollution equipment	Bearing oil from machine drives. Oils from seals on milling equipment. Solids from soapstone dip tanks. Solids from air pollution equipment.
Pan (Heater) Reclaim Process		
Grinding	Washdown, spills, leaks	Bearing oil from machine drives. Oil from seals on milling equipment.
Depolymerization Agent and Oil Storage	Spills, runoff	Oil, solids, caustic and organics.
Devulcanizer	Air pollution equipment	Oil and organics.
Compounding	Washdown, spills, leaks, air pollution equipment	Bearing oil from machine drives. Oils from seals on milling equipment. Solids from soapstone dip tanks. Solids from air pollution equipment.

Source: EPA Repot 440/1-74/030

PROPRIETARY PROCESSES

Reclaiming Heat and Mineral Values from Scrap Rubber

A process developed by *G. Crane, E.L. Kay and J.R. Laman; U.S. Patent 3,890,141; June 17, 1975 assigned to The Firestone Tire and Rubber Company* involves treating solid waste vulcanized rubber by the steps of heating the solid waste vulcanized rubber in contact with from 10 to 1,000% of a hydrocarbon liquid at 400° to 800°F for 0.1 to 50 hours to convert the solid waste vulcanized rubber to a fluid, transferring the fluid thus produced to combustion apparatus, burning the fluid, producing heat energy and ash, and collecting the ash and recovering zinc and titanium therefrom.

Reclaiming Tires by Conversion of Oil Spills to Asphalt

The reader is referred to the section on Oils, Conversion of Oil Spills into Useful Asphaltic Materials (page 333).

Reclaiming Rubber by Mechanical-Chemical Treatment

A process developed by *H.A. Merges; U.S. Patent 3,460,769; August 12, 1969* is one in which vulcanized rubber scrap is reclaimed by heating and intimately mixing it with additives after the scrap has been pulverized, and this heated mixture is subjected to centrifugal forces while being exposed in a matter of seconds to friction, milling, shearing and crushing forces in a pulsating manner.

Reclaiming Rubber by Mechanical-Thermal Treatment

A process developed by *S. Ficker; U.S. Patent 3,873,474; March 25, 1975* is one in which polymeric scrap material, such as rubber from used vehicular tires, is reclaimed by comminuting the scrap and then advancing it under substantially atmospheric pressure through a heating chamber with the aid of two or more heated feed screws having interleaved threads, one screw rotating at half the speed of its neighbor and having a double thread with twice the pitch of the single thread of the latter. Residence time in the heating chamber may range between 3 and 30 minutes, at operating temperatures of 140° to 300°C. By relatively accelerating or retarding one of two adjoining screws, their interleaved threads are moved into a nearly contacting relationship in which they scrape adhering incrustations from each other.

Recovering Resins from Rubber by Thermal and Chemical Treatment

A process developed by *L.B. Wakefield, D.T. Roberts, Jr. and E.L. Kay; U.S. Patent 3,880,807; April 29, 1975; assigned to The Firestone Tire and Rubber Company* yields a solid rubber tackifier or stabilizer resin which is the reaction product of 5 to 50 parts by weight of furfuraldehyde and 100 parts by weight of tire pyrolysis oil obtained by distillation of vulcanized tire scrap at a temperature in the range of 1000° to 2500°F which oil has the following analysis (volume percent).

	Percent
Alkyl Benzenes	15-25
Phenols	0.1-5
3-Ring Aromatics	0.1-5
Biphenyl and Acenaphthene	5-25
Alkyl Naphthalenes	5-35
Indenes	1-20
Styrene	1-15
Alkylstyrenes and Indans	1-25

Recovering Rubber by Hydroconversion

A process developed by *S.B. Alpert; U.S. Patent 3,704,108; November 28, 1972; assigned to Hydrocarbon Research, Inc.* is one in which waste natural and synthetic rubbers are converted in the presence of a catalyst and in a hydrogen environment by passing a particulate rubber slurry and hydrogen upwardly through a reactor maintained under temperatures in the 700° to 850°F range and pressures in the 500 to 3,000 psig range. The particulate rubber is converted in the reactor to hydrocarbon liquids while unconvertible solids originally compounded with the rubber leave unchanged in the product. The hydrocarbon stream can have a wide range of properties depending on the objectives of the user of the process.

Recovering Rubber in the Form of Charcoal

A process developed by *Y. Sanga; U.S. Patent 3,875,077; April 1, 1975* is a process for producing activated charcoal comprising the steps of cutting waste tires into small pieces subjecting these cut pieces of tires to high-temperature heating and burning in an atmospheric shut off from air to perfectly carbonize them, and then activating the charcoal thus produced.

SOME FUTURE TRENDS

MUNICIPAL WASTES

Unquestionably the major future trend in resource recovery will be toward huge integrated plants for processing municipal wastes, recovering glass, nonferrous metals, ferrous metals and utilizing much of the organic content of the waste as fuel. Disposal on a grand scale will undoubtedly help to iron out the irregularities in local recycling center problems and will alleviate the shortage of adequate landfill space facing many cities.

The major barriers to aluminum and glass recovery are related to the economical extraction of these materials from mixed municipal wastes. Once extracted, there is sufficient demand to facilitate significant increases in recycling of these materials. Technology is under development for aluminum and glass recovery but has not proven to be economically feasible to date. Beverage container control legislation could have significant impact on the need or prospects for aluminum and glass recovery in the future. EPA will continue to monitor developments in this area and will provide technical assistance and information.

The most logical approach to plastics recovery is as an energy source because currently it is not technically feasible to separate plastics from mixed wastes in an economical manner. However, the polyvinyl chloride fraction could pose significant environmental problems, especially if this fraction grows significantly in the future. EPA will continue to monitor trends in this area and make recommendations as necessary to insure the prospects for environmentally sound energy recovery.

Plastics presently constitute approximately 3.8% of municipal solid waste, but plastic consumption is growing rapidly. Essentially no recovery of plastics as a material from mixed waste now takes place. Of all the materials in mixed waste, plastics are probably the most difficult to extract. Recycling of certain plastic wastes from fabrication plants is practiced. In industrial plants it is possible to keep different types of plastics (e.g., polyethylene, polystyrene, and polyvinyl chloride) separated at the source. Once plastics have entered use, and especially after they have been discarded into the waste stream, they are extremely difficult to separate.

Supply of waste plastics is simply not possible with existing technology because plastics cannot be separated from paper and other materials in mixed waste with similar physical characteristics. Experiments have been conducted with new separation techniques but are a long way from full-scale application. Significant value may be recovered from plastics in energy recovery systems. Plastics have the highest British thermal unit content of any of the materials in mixed waste and thus make a valuable contribution to the heat value of the waste. The heat content of plastics is about 11,000 British thermal units per pound, approximately the equivalent of coal.

The only potential difficulty with recovery of energy from plastics is the presence of polyvinyl chloride, a small proportion of total plastic production (13%) and hence an even smaller proportion of solid waste (0.42%). However, when burned, it emits hydrogen chloride, a toxic gas. When this gas is combined with moisture, as in the wet scrubbing sections of incinerator emission controls, hydrochloric acid is formed, which will corrode the metal parts of the control equipment and other metal incinerator parts, resulting in increased maintenance costs and potentially increased air pollution emissions. In addition, the water discharged from these scrubbing systems has a low pH and would present a threat to natural waters if not properly treated.

RUBBER

There is a variety of new potential means of recycling rubber tires; the most promising include chemical decomposition, incineration for steam production, road building, and use as offshore reefs.

Chemical Decomposition: Destructive distillation, carbonization, and hydrogenization are thermal conversion processes that may be used to recover the chemical constituents of tires. The first two processes are both forms of pyrolysis—thermal decomposition in a low oxygen atmosphere; hydrogenization is a process of chemical synthesis involving addition of hydrogen. The primary product of carbonization (high-temperature destructive distillation) is carbon black, a major tire raw material. Lower temperature destructive distillation yields a mixture of oils, gases, and carbon residue with significant fuel value. Hydrogenization yields hydrocarbon products that can be used in tire manufacturing.

Pilot-plant-scale carbon black processes utilizing scrap rubber have been tested by some of the major rubber companies. It was reported that the processes are feasible but economically unattractive at present, with costs of about three to four times that of carbon black production by conventional means from petroleum. Destructive distillation of tires has been examined by the Bureau of Mines in conjunction with a major rubber company. The products consist primarily of heavy oils, light oils, gases, and char. It has been reported that the value of these products from 100 pounds of tires is approximately $1.50 (i.e., $30 per ton of tires processed). It has been further suggested that even at this value the process is uneconomical if shipping, handling, storage, and preparation costs are included.

Hydrogenization of the products of destructive distillation has been examined by the Hydrocarbon Research Institute at the pilot plant scale, and preliminary projections were made that such a plant could operate at a profit if tires could be obtained for $5 per ton.

Incineration for Steam Generation: Incineration of tires for steam generation in special furnaces has been examined by some of the major tire manufacturers and appears to be a promising disposal and recovery method. One such plant has been built. When in full operation, the plant will handle about 1 million tires per year. The incinerator is reported to be completely odorless and pollution-free. The major drawback appears to be the high capital cost of the equipment.

Road Building: Road building and repair offers a large potential outlet for old shredded tires. Rubber that has been ground into small particles can be added to asphalt as a binder. The rubber is reported to enhance the resilience of the road in cool weather and reduce the flowing characteristics in hot weather. Rubber can reduce the tendency of asphalt to bleed to the road surface where it presents a skidding hazard, and the rubber in the mix allows more asphalt to be added for better aging and reduced raveling tendencies. However, estimated costs of adding rubber to asphalt mix (usually in concentration of 3 to 8%) range from $1.50 per ton to $2.50 per ton of mix. The justification for this cost will depend on whether the cost can be offset by lower maintenance costs or longer road life. Tests of rubberized roads to quantify benefits are presently under way in several locations. The State of New York is already using reclaimed rubber with hot asphalt to seal cracks and joints, and Arizona has laid rubberized asphalt on streets and airport aprons.

Reef Building: Tires have been used to construct artificial reefs for fish spawning on the East and Gulf coasts in response to increasing interest in sports-fishing. It is reported that water action on the West coast is too strong for construction of artificial reefs.

TEXTILES

There are not sufficient historical data available to show trends in textile recycling. However, it is known that secondary textile consumption in many traditional markets has been declining and that other markets, such as the important wiping cloth market, have been growing at a slower rate than total textile consumption. Thus, it is almost certain that the rate of textile recovery (waste recovered versus textile consumption) has been declining.

The increasing trend toward the use of cotton-polyester and wool-polyester blends probably represents the major problem of textile recycling of the 1970s. These blends are not only generally unusable as wastes in themselves, but they also tend to become mixed with other, usable waste textiles and thereby reduce the economic value of the total waste supply. This has caused problems, particularly in the three major markets for cotton waste: rag paper, vulcanized fiber, and wiping cloths. In the case of the first two markets, contamination of cotton is limited to a maximum of 1 to 2%. Thus, increase in blends means greater control by the textile processors, resulting in increased cost. It also greatly reduced the usable yield from used textiles.

Fiber blends have essentially the same effect on the wiping cloth business. Wipers are less sensitive to small percentages of polyester fiber, but fiber blends with over 50% polyester do not have satisfactory absorption characteristics. (Garments with polyester/cotton blends of 50/50 and 65/35 are extremely common.) The present percent of such blends in mixed rag bundles is unknown, but the increased replacement of manmade fibers by synthetics is testimony that they are likely to increase, thereby reducing usable yields.

Another major problem of textile recycling is that used textiles are losing ground in many traditional markets. Wool markets are one of the most serious problems, due mainly to the Wool Labeling Act (the effect has been psychological, on consumers who perceive that virgin wool is cleaner or purer) and increased competition from secondary wool from foreign sources. Also, virgin-based materials are replacing used textiles in some markets. The incentive for using secondary textiles as paddings, filler, etc., has traditionally been their low cost. Now, development of virgin-based products such as urethane foams at competitive prices has resulted in decreasing used textile markets.

METAL AND MINERAL WASTES

A projection of quantities and modes of handling of some industrial metal and mineral wastes has been made by E. Aleshin, S.A. Bortz and T.B. Wade of IIT Research Institute, Chicago, Illinois in IITRI Report No. G-6027 to the U.S. Bureau of Mines (October 18, 1973). Some of their conclusions are quoted in the paragraphs which follow.

Aluminum

The data in Table 31 show increases in secondary aluminum production, along with production of waste. Present slag output (1972) is about 135,000 tpy which will increase almost fivefold by the year 2000. Less than 50% of aluminum available for recycling is actually used. Thus, if all available metal is recycled, slag production will exceed 1,200,000 tpy by the year 2000, greatly increasing the problem of disposal within the industry. If all available scrap aluminum were recycled, and the remaining aluminum extracted from the slag (including free metal and that contained in alumina (considering that slag contains up to

35% metal), demand on primary producers would be reduced by more than 1,500,000 tons, or mearly 40% of total primary aluminum production. This would not only conserve domestic supplies but would reduce import of bauxite. In addition, since the energy required to recover aluminum from scrap is about 5% that required for primary production, an added benefit is conservation of energy. It takes about 6,000 Btu/ton to convert ore to aluminum, and about 300 Btu/ton to convert scrap to usable aluminum.

TABLE 31: ESTIMATION OF WASTE SLAG PRODUCTION IN THE SECONDARY NONFERROUS METALS INDUSTRY, 10^6 TPY

Material	1967	1980	1990	2000
Aluminum scrap	0.956	1.903	3.451	5.215
Slag	0.117	0.233	0.422	0.638
Copper scrap	1.389	2.336	3.784	5.648
Slag	0.154	0.259	0.420	0.627
Lead scrap	0.485	1.332	2.980	4.743
Slag	0.066	0.182	0.407	0.647
Zinc scrap	0.082	1.029	2.577	4.341
Slag	0.005	0.062	0.155	0.260

Source: IITRI Report No. G-6027

At present much scrap is not recycled because the aluminum is generally mixed with other metals and difficult to separate. Research should thus be conducted to develop methods for efficient and economical separation of these materials.

Copper

Table 31 shows that total slag developed in the production of secondary copper will exceed 600,000 tpy at the turn of the century. Although slag produced is not great compared to other mineral wastes, this amount does add to the overall picture. In addition, if 100% of the available secondary copper is recycled, then the problem will be more acute at over 1,000,000 tpy. Use of all secondary copper could reduce the demand for primary copper by about one-half, thus conserving this resource. Production of primary copper, which creates large amounts of copper slag, was not considered in this study.

Reduction of slag may be implemented by extracting the more abundant materials contained. For example, if iron oxide could be extracted efficiently, slag would be reduced by about 25%. In addition, the lime contained therein might then be used in agricultural applications. Extraction of iron may also result in recovering some other metals, thus adding to the economy of the process. Extraction of minor metals alone may be marginal. However, studies should be initiated to determine amounts available and methods of extraction.

Lead

Table 31 also shows the projections for lead waste. Total slag from the production of secondary lead will exceed 600,000 tons in the year 2000. It is known that only about 40% of the total lead theoretically available for recycling is actually reprocessed. Thus, if all the available scrap lead were to be recycled, total slag production would be more than 1,500,000 tons, a sizeable quantity which would present more than a local problem. Slag from production of primary lead would add considerably to this total. Research should thus be initiated to determine more economical ways to extract lead from unused scrap. Furthermore, studies similar to those suggested for copper would be beneficial, regarding extraction of oxides and minor metals.

Recovery of all available lead scrap would considerably reduce demand on primary producers, and would reduce future dependence on imported metal. Dissipative uses, such as pigments and chemicals, and increased battery use, will still require replenishment with primary metal—amounting to about 30% of total lead used.

Zinc

After 1967, projected estimates indicate an increase in secondary zinc production amounting to about 260,000 tons of slag in the year 2000. Production of primary zinc, from smelting operations using carbonaceous reducers, would add to this total. If all available zinc is recycled, the demand on primary producers would be reduced more than 50%, thus increasing reserves. However, slag production would increase to about 500,000 tons. This would present problems identified with other nonferrous slags regarding extraction and use. Thus, as previously indicated, research should be initiated to extract metals and find further uses for the remaining residue, especially those which are water-soluble.

Iron Foundry Wastes

In 1969, total iron foundry emissions were about 1,755,000 tons with castings amounting to about 17,000,000 tons. This represents more than 400,000 tons of iron lost. The projections shown in Table 32 indicate that iron foundry emissions will increase about 3.6 times the present rate, an equivalent of about 1,500,000 tons of iron. In addition, vast quantities of waste sand, possibly amounting to over 60,000,000 tons, as well as quantities of slag, will be produced in the year 2000. Thus, a high priority should be placed on developing means to recover and use iron from melting dusts. In addition, efficient means to gather these materials in a central location must be devised. Furthermore, uses or better disposal methods for waste sand must be developed.

TABLE 32: ESTIMATION OF WASTE PARTICULATE EMISSIONS IN THE IRON FOUNDRY INDUSTRY, 10^6 TPY

Material	1967	1980	1990	2000
Particulates	1.647	2.594	4.142	5.906
Sand	17.0	26.8	42.7	61.0

Source: IITRI Report No. G-6027

Presently, most finely divided iron foundry effluents are pelletized (if at all treated) and disposed in landfill areas. Molding sand is generally used in landfills; however small amounts are allocated for cement manufacture. In addition, it has been found that arc furnace dusts may have application as undercoat pigment and as aggregate for road fill and ballast. Other uses indicated are soil conditioning (cupola effluent), sand reclaimer dust as soil stabilizer and foundation leak preventive. As indicated previously a large research effort should consider recovery of iron from melting dusts. The study should also include means to gather dust from more remote locations and to determine synergistic effects among various wastes combined with those from foundries. Lastly, an ever increasing problem of waste sand disposal should be given high priority.

It should also be pointed out that fine particles often have unique properties which are beneficial in applications such as thickening agents, pigments, suspension aids, cementatious materials, and fillers for polymeric materials. Fine particles are also beneficial in fabrication of small extruded products. Since much of the waste from foundry operations is in the form of fine particles, research on their uses in existing commercial processes or new industries would be advisable.

Conclusions

The data indicate that considerably more research must be initiated in determining added values for secondary metal slags. This includes determining whether metals or oxides can be extracted, and whether other uses can be put to the slags, such as agricultural products and aggregate, or extraction of chemical values from, for instance, chloride slags. Since all scrap is not utilized, the factors in nonutilization should be studied. These factors include: separation and extraction costs, market potential, transportation and tariffs, socio-economic factors, regulations, and environmental considerations. An important consideration for using more scrap is conservation of power. Recovery of aluminum from scrap can result in a 95% energy saving over converting ore to metal. Recycling other nonferrous metal scrap should show similar effects. To be considered also is the interrelationship of secondary scrap and slags in the total recycling picture including synergistic effects with other wastes.

POLLUTION CONTROL 1975 IN THE ASBESTOS, CEMENT, GLASS AND ALLIED MINERAL INDUSTRIES

by Marshall Sittig

Pollution Technology Review No. 19

The mineral or silicate industries are engaged in the processing or manufacture of asbestos, bricks, cement, clay, glass, fritted glass and ceramics, fiber glass, mineral wool and stone, sand and gravel, etc. They generate serious air and water pollution aggravated by disposal problems of heavy solid wastes which fortunately are biologically and chemically inert for the most part.

Here dust is the major problem and dust reclamation at the source is an economic necessity. Submicron particles are barely influenced by gravity. Once they enter the atmosphere, they stay there for a long time.

The asbestos industry, in particular, is faced with the twin threats of asbestosis and lung cancer as a result of airborne asbestos dust. Now, in addition, waterborne asbestos from processing of the mineral is being assessed as a potential hazard in drinking water.

There are, of course, the usual industrial pollution problems as well, such as stack gas emissions from glass-producing furnaces, phenolic emissions from fiber glass manufacture, and many more.

Reports of accomplishments and proposed solutions to these problems are afforded by this book which is based on various government-sponsored surveys with practical examples detailed from late U.S. patents. A partial and condensed table of contents is given here. Chapter headings are enumerated and followed by examples of important subtitles.

ISBN 0-8155-0578-7 **333 pages**

RECYCLING AND RECLAIMING OF MUNICIPAL SOLID WASTES 1975

by F. R. Jackson

Pollution Technology Review No. 17

More and more cities, large and small, are exploring new technological systems for recycling solid municipal wastes.

Naturally the greatest interest centers upon those recycling systems that permit profitable reclamation of saleable items and materials from the town's refuse.

Articles made of metal can be cleaned, washed, compressed, and sold to the scrap metals industry. Glass can be comminuted and even separated as to color.

Paper and cardboard is a promising target for recycling, because paper fiber is the largest single component of municipal wastes and potentially very valuable.

Water-insoluble organic matter, such as plastics and wood, can be reclaimed and processed for conversion into energy in a variety of ways.

As these new technologies are developed, they will add to the alternatives available to a community, when it decides to recycle some of its wastes as part of its total solid waste management program. Numerous alternate ways and means are needed, because local conditions vary widely, and no single line of approach is capable of meeting every community's needs.

This Pollution Technology Review surveys information available up to the middle of 1974, and is based primarily upon studies conducted by industrial and engineering firms under the auspices of the EPA and other government agencies. Forward-thinking municipal administrators will find this book most useful, because it collates comprehensively the results of governmental and industrial research.

A partial and condensed table of contents follows here.

ISBN 0-8155-0560-4

342 pages

POLLUTION CONTROL
IN THE ORGANIC CHEMICAL INDUSTRY
1974

by Marshall Sittig

Pollution Technology Review No. 9

Wastes from plants manufacturing identical compounds may be quite dissimilar, because of the difference in processes and raw materials. Since the organic chemical industry is now largely petrochemical, preference has been given to waste treatment from such operations.

Detailed treatability evaluation of each waste stream has become a prerequisite for any profitable product.

Physical methods for waste and by-product removal include gravity separation, air flotation, filtration, evaporation and adsorption techniques.

Chemical methods include precipitation, polyelectrolyte treatment, oxidation and other chemical conditioning such as neutralization.

Biological treatment methods comprise activated sludge and its modifications, trickling filters, aerated lagoons, and waste stabilization ponds.

This book intends to assist the chemical engineer in treating chemical waste products and effluents in conjunction with prudent raw material choices and thus bring about process economics. It presents condensed vital data from government and other sources of information that are scattered and difficult to pull together.

A partial and condensed table of contents follows here.

Dinitrotoluene from Nitrotoluene

Disulfoton from Ethanol

Dyes and Pigments from Aromatics

Epichlorohydrin from Allyl Chloride

Ethylbenzene from Benzene

Ethyl Chloride by
Hydrochlorination of Ethylene
Chlorination of Ethane
Hydrochlorination of Ethanol

Ethylene or Propylene from Paraffins

Ethylene Dichloride from Ethylene

Ethylene Glycol from Ethylene Oxide

Ethylene Oxide from Ethylene

Formaldehyde from Methanol

Long Chain Alcohols from
Ethylene Oligomers

Malathion® from Methanol

Methanol from Natural Gas

Methylamines from Methanol + Ammonia

Methyl Bromide from Methanol

Methyl Methacrylate from
Acetone Cyanohydrin

Nitrobenzene from Benzene

Nitrochlorobenzene from Chlorobenzene

Nitroparaffins from Paraffins

Oxo Products from Olefins

Parathion from p-Nitrophenol

Perchloroethylene from Propane

Phenol from Chlorobenzene

Phenol from Cumene

Phorate from Ethanol

Phosgene from CO + Cl_2

Phthalic Anhydride from
Naphthalene or Xylene

Propylene Glycol from Propylene Oxide

Propylene Oxide from Propylene

Styrene from Ethylbenzene

SNG from Crude Oil

2,4,5-T from Trichlorophenol

Terephthalic Acid from Xylene

Tetraethyl Lead from Ethyl Chloride

Trichloroethane from Vinyl Chloride

Trichloroethylene by
Chlorination, then by
Dehydrochlorination of Acetylene
or by Oxyhydrochlorination
of Dichloroethane

Trifluralin from Perfluoromethylchlorobenzene

Vinyl Acetate from
Ethylene and Acetic Acid

Vinyl Chloride from Acetylene

Vinyl Chloride from Ethylene Dichloride

AIR POLLUTION CONTROL
Halogen Acids
Halogenated Hydrocarbons
Halogens
Hydrocarbons
Particulates
Sulfur Compounds
Coking of Coal

WATER POLLUTION CONTROL
Primary
Oil Separation
Equalization
Neutralization
Sedimentation
Flotation
Flocculation
Nutrient Addition
Secondary
Activated Sludge
Extended Aeration
Trickling Filters
Aerated Lagoons
Waste Stabilization Ponds
Chemical Oxidation
Nitrification—Denitrification
Tertiary
Chemical Precipitation
Gas Stripping
Microstraining
Carbon Adsorption
Electrodialysis
Ion Exchange
Evaporation
Reverse Osmosis
Chlorination
Rapid Sand Filtration
Sludge Handling
Aerobic Digestion
Anaerobic Digestion
Wet Oxidation
Thickening
Lagooning
Sand Drying Beds
Vacuum Filtration
Filtration
Land Disposal
Incineration
Sea Disposal
Ultimate Disposal
Thermal Oxidation
Deep Well Disposal

OVERALL WATER POLLUTION CONTROL MODELS
BPCTCA: Best Practicable Control Technology Currently Available
BATEA: Best Available Technology Economically Available
BADCT: Best Available Demonstrated Control Technology

THE ECONOMICS OF POLLUTION CONTROL

FUTURE TRENDS

ISBN 0-8155-0536-1

305 pages